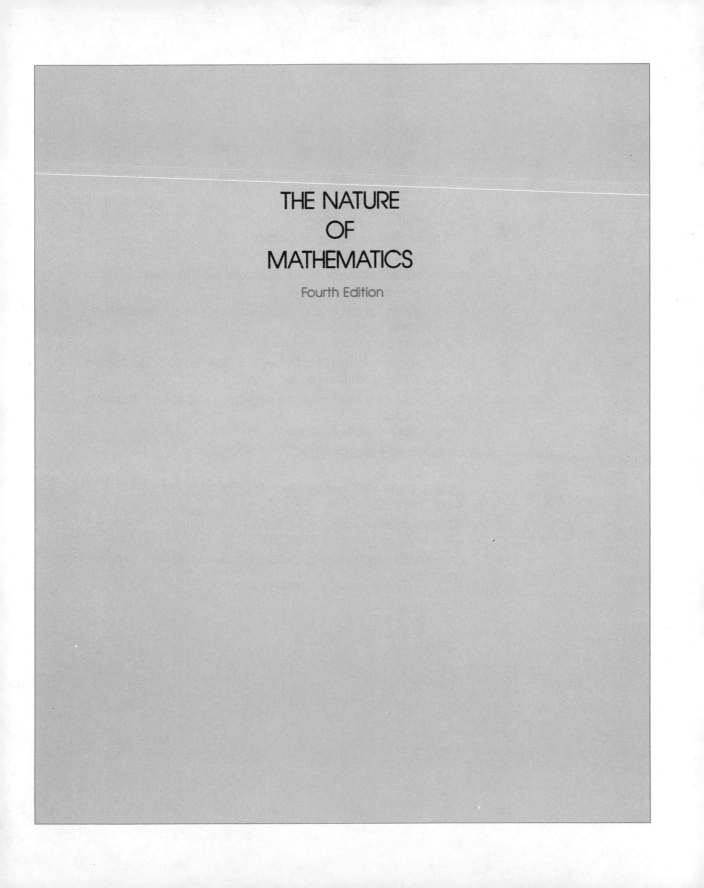

THE NATURE
OF
MATHEMATICS
Fourth Edition

Other Brooks/Cole Titles by the Same Author

Basic Mathematics
Karl J. Smith

Mathematics: Its Power and Utility
Karl J. Smith

Beginning Algebra for College Students, Third Edition
Karl J. Smith and Patrick J. Boyle

Study Guide for Beginning Algebra for College Students, Third Edition
Karl J. Smith and Patrick J. Boyle

Intermediate Algebra for College Students, Second Edition
Karl J. Smith and Patrick J. Boyle

Study Guide for Intermediate Algebra for College Students, Second Edition
Karl J. Smith and Patrick J. Boyle

Essentials of Trigonometry
Karl J. Smith

Trigonometry for College Students, Third Edition
Karl J. Smith

College Algebra, Second Edition
Karl J. Smith and Patrick J. Boyle

Precalculus Mathematics: A Functional Approach, Second Edition
Karl J. Smith

THE NATURE
OF
MATHEMATICS

Fourth Edition

KARL J. SMITH
Santa Rosa Junior College

Brooks/Cole Publishing Company
Monterey, California

Brooks/Cole Publishing Company
A Division of Wadsworth, Inc.

Printed in the United States of America

10   9   8   7   6   5

Library of Congress Cataloging in Publication Data

Smith, Karl J.
  The nature of mathematics.

  Rev. ed. of: The nature of modern mathematics.   3rd ed.
© 1980.
  Includes index.
  1. Mathematics—1961-   . I. Smith Karl J.
The nature of modern mathematics. II. Title.
QA39.2.S6   1984   512'.1   83-14455

ISBN 0-534-02806-3

Sponsoring Editor: Craig Barth
Production Service: Phyllis Niklas
Production Coordinator: Joan Marsh
Manuscript Editor: Carol Dondrea
Permissions Editor: Andrea Smith
Interior Design: Janet Bollow
Cover Design: Jamie Sue Brooks
Cover Photo: © Pete Turner, The Image Bank West
Interior Illustration: Reese Thornton
Photo Editor: Reese Thornton
Typesetting: Allservice Phototypesetting Company of Arizona
Cover Printing: Lehigh Lithographers
Printing and Binding: R. R. Donnelley & Sons Company

To my wife, Linda
and
our children,
Melissa and Shannon

Preface

This book was written to create a positive attitude toward mathematics—a realization that mathematics is not an endless procession of dull manipulations, theorems, proofs, and irrelevant topics. Rather than simply presenting the technical details needed to proceed to the next course, the book attempts to give insight into what mathematics is, what it accomplishes, and how it is pursued as a human enterprise.

I frequently encounter people who tell me about their unpleasant experiences with mathematics. I have a true sympathy for these people, and I recall one of my elementary school teachers who assigned additional arithmetic problems as punishment. This can only create negative attitudes toward mathematics, which is indeed unfortunate. If elementary school teachers and parents have a positive attitude toward mathematics, their children will see some of the beauty of the subject. I want students to come away from this course with the feeling that mathematics can be pleasant, useful, and practical—and enjoyed for its own sake.

Now in its fourth edition, this text continues to be addressed to the liberal arts student, but with important changes for the liberal arts student in the 1980s. It now includes material on significant interest rates in saving and borrowing, on metric measurements in geometry, on problem solving in a variety of situations, and on mathematical modeling. The student is asked not only to "do math problems," but also to "experience math." In fact, the slogan "Mathematics Is Not a Spectator Sport" is not just an advertising slogan, but an invitation which suggests that the only way to succeed in mathematics is to become involved with it. This text can also be adapted for teacher preparation courses as well as finite mathematics courses, as shown in the table of course outlines on the next page.

As you can see from the course outlines, this text is designed so that the instructor has many options. Chapters are written to be almost independent of one another, and can be covered in any order appropriate to a particular audience. One of the most frequently occurring suggestions that I received from reviewers and users of the third edition was to rearrange the chapters. Nearly everyone had his or her "own arrangement." This simply points out one of the strongest features of this book—that the order in which the material is presented depends on the preference of the instructor and the makeup of the students taking the course. The material can be easily adapted to almost any order or arrangement. A few sections mention sets, but the ideas used are, for the most part, general elementary school knowledge. Those that require the material of Sections 2.1 and 2.2 are labeled in the text. Other sections require exponents from Section 4.3 (Sections 5.1, 6.3, and 6.5), integers from Section 4.5 (Sections 6.2, 7.5, and Chapter 11), or specific ideas that are indicated by a footnote at the beginning of

"A problem must involve the students; they must search for the answer. Perhaps they will not reach the goal, but the search itself may prove more important than the goal."

F. Jacobson

Course Outlines	*With Computers*	*Without Computers*
One-Semester Courses		
Liberal Arts	1, 3, 4, 5, plus one additional chapter from 2, 6, or 7	1, 2, 4, 5, 6
Teacher Training	1, 2.1, 2.2, 3, 4, 6, 7	1, 2.1, 2.2, 4, 5, 6, 7
Finite Mathematics	2, 3, 8.1–8.4, 9.2–9.4, 10, 11	2, 8, 9, 10, 11
Combination Course	1, 2.1, 2.2, 3, 8, 9, 10	1, 2.1, 2.2, 4, 8, 9, 10
Two-Quarter Courses		
Liberal Arts	1, 3, 4, 5, 8, 9	1, 2, 4, 5, 8, 9
Teacher Training	1, 2, 3, 4, 6, 7	1, 3, 4, 5, 6, 7
Combination Course	1, 2.1, 2.2, 3.2–3.5, 4, 5, 6, 7	1, 2.1, 2.2, 4, 5, 6, 7, 8.1, 9.2, 9.3
Two-Semester Courses		
Liberal Arts	1st: 1, 3, 4, 5, 6 2nd: 2, 7, 8, 9, 10	1st: 1, 2, 4, 5, 6 2nd: 7, 8, 9, 10, 11
Combination	1st: 1, 4, 5, 6, 7 2nd: 2, 3, 8, 9, 10	1st: 1, 4, 5, 6, 7 2nd: 2, 8, 9, 10, 11

the section. Except for these, the instructor can feel free to pick and choose or jump around to suit the needs of the class. Each chapter concludes with an optional section that can be used as time and interest demand.

Historical notes and history charts are provided to give the student some insight into the humanness of mathematics. Rather than being strictly biographical reports, these notes focus on the people. Nearly every major mathematician will have some part of his or her life to tell on the pages of this book. I hope that a glimpse of the people of mathematics will provide a glimpse of the nature of mathematics.

A wide range of problems is offered, since students who enroll in a survey course come from a wide range of backgrounds. There are quite a number of problems of the routine or drill type, as well as a large selection of thought-provoking problems. Much interesting material appears in the problems, and the reader can learn quite a bit just by reading them. The answers to the odd-numbered problems are given in the back of the book, and complete solutions to all the problems, as well as sample tests, are available in the Instructor's Manual. The "Mind Bogglers" and "Problems for Individual Study" sections should be utilized as interest demands. They are not to be forced on the student but should be investigated for their own sake. Hopefully, these sections will stimulate class discussion and still other avenues of investigation.

I thank the many people who wrote me offering suggestions for revising the book: Joseph M. Cavanaugh, East Stroudsburg State College; Gerald Church, Southwest Oklahoma State University; Richard Freitag, Edinboro College; Gerald E. Gannon, California State University, Fullerton; Robert L. Hoburg, Western Connecticut State College; M. Kay Hudspeth, Pennsylvania State University; Vernon H. Jantz, Seward County Community College; Nancy J. Johnson, Broward Community College; John LeDuc, Eastern Illinois University; Adolf Mader, University of Hawaii; Cherry F. May, Santa Fe Community College; Allen D. Miller, California Polytechnic State University; Mickey G. Settle, Pensacola Junior College; and Donald G. Spencer, Northeast Louisiana University.

I also appreciate the suggestions and comments of the reviewers of the first three editions: Jerald T. Ball, George Berzsenyi, Jan Boal, Chris C. Braunschweiger, T. A. Bronikowski, Charles M. Bundrick, T. W. Buquoi, Eugene Callahan, Michael W. Carroll, James R. Choike, Thomas C. Craven, Ralph De Marr, Charles Downey, Samuel L. Dunn, William J. Eccles, Ernest Fandreyer, Gregory N. Fiore, Ralph Gellar, John J. Hanevy, Caroline Hollingsworth, James J. Jackson, Charles E. Johnson, Martha C. Jordan, Judy D. Kennedy, Linda H. Kodama, Daniel Koral, Helen Kriegsman, C. Deborah Laughton, William Leahey, William A. Leonard, George McNulty, John Mullen, John Palumbo, Gary Peterson, James V. Rauff, O. Sassian, James R. Smart, Arnold Villone, Clifford H. Wagner, Barbara Williams, and Stephen S. Willoughby.

I would especially like to thank my editor, Robert J. Wisner of New Mexico State, for his countless suggestions and ideas, and Craig Barth, Joan Marsh, and Jamie Brooks of Brooks/Cole, as well as Jack Thornton for his sterling leadership and inspiration.

The production of this book was a true team effort, and I appreciate Phyllis Niklas for her care and involvement, Reese Thornton for his illustrations and art research, Andrea Smith for all her work with permissions, Roger Breen for preparing a test manual as well as a Florida manual on implementing the CLAST requirements using this textbook, and Hal Andersen for working all of the problems, offering suggestions, and acting as my mentor on this edition.

Finally, my thanks go to my wife, Linda, whose suggestions and help were invaluable. I will never be able to repay her and my family for those hours lost while I was working on the manuscript and for those evenings she spent "proofing." Without her, this book would exist only in my dreams.

Karl J. Smith

Contents

*Optional sections.

4 The Nature of Numbers   164

5 The Nature of Daily Arithmetic   242

6 The Nature of Algebra   284

*Optional sections.

*Optional sections.

11 The Nature of Mathematical Modeling   542

*Optional sections.

THE NATURE
OF
MATHEMATICS

Fourth Edition

Date	Cultural History
3000 BC	3110 BC: First dynasty of the Ancient Kingdom of Egypt 2800 BC: The Great Pyramid 2580 BC: Cheops Pyramid

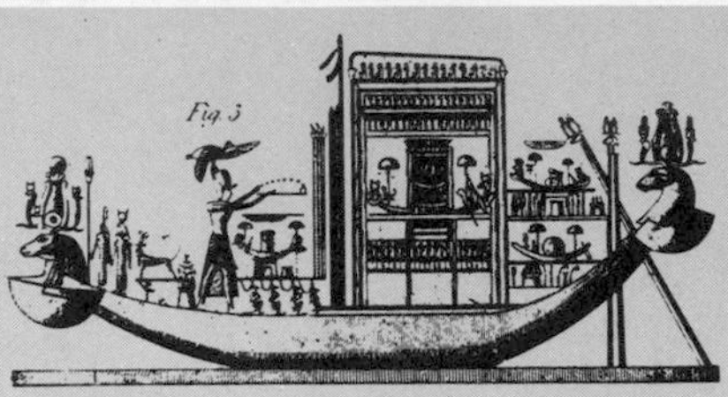

Sacred ship of Egypt

2500 BC	2500 BC: Isis and Osiris cult in Egypt

Amun, Isis and Osiris

2000 BC	1900 BC: Epic of Gilgamesh 1700 BC: Stonehenge

1500 BC	1495 BC: Obelisk of Thothmes at Karnak  1300 BC: Approximate beginning of the iron age 1250 BC: Moses leads exodus from Egypt 1200 BC: Trojan war

Thothmes' obelisk

1000 BC	1000 BC: Phoenicians invent modern alphabet 850 BC: Homer's *Iliad and Odyssey* 753 BC: Rome founded 538 BC: Persians capture Babylon

Odysseus and the sirens

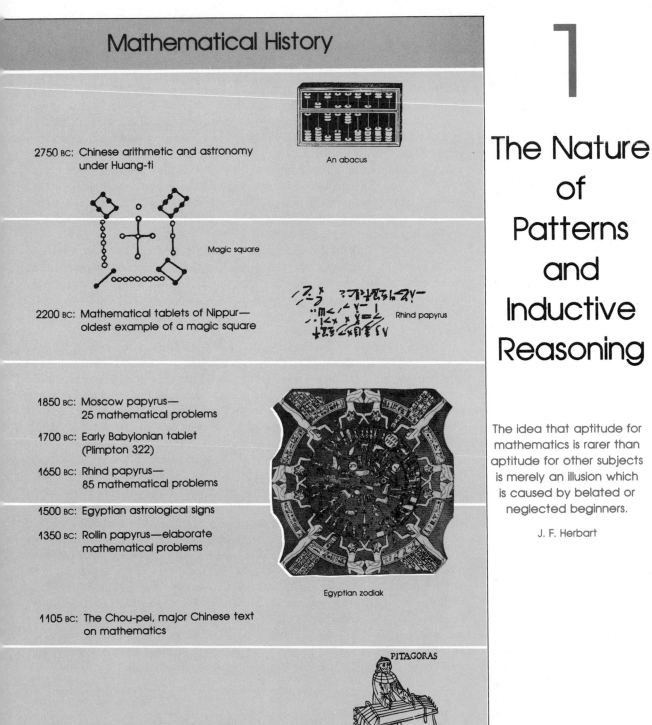

Mathematical History

An abacus

2750 BC: Chinese arithmetic and astronomy under Huang-ti

Magic square

2200 BC: Mathematical tablets of Nippur— oldest example of a magic square

Rhind papyrus

1850 BC: Moscow papyrus— 25 mathematical problems

1700 BC: Early Babylonian tablet (Plimpton 322)

1650 BC: Rhind papyrus— 85 mathematical problems

1500 BC: Egyptian astrological signs

1350 BC: Rollin papyrus—elaborate mathematical problems

Egyptian zodiak

1105 BC: The Chou-pei, major Chinese text on mathematics

585 BC: Thales, founder of Greek geometry

540 BC: The teachings of Pythagoras

PITAGORAS

Pythagoras

1

The Nature of Patterns and Inductive Reasoning

The idea that aptitude for mathematics is rarer than aptitude for other subjects is merely an illusion which is caused by belated or neglected beginners.

J. F. Herbart

1.1 Problem Solving

There is a common belief that mathematics is a difficult, dull subject that is to be pursued only in a clear-cut, logical fashion. This belief is perpetuated by the way mathematics is presented in most textbooks; often mathematics is reduced to a series of definitions, methods to solve various types of problems, and theorems. These theorems are justified by means of proofs and deductive reasoning. I do not mean to minimize the importance of proof in mathematics, for it is the very thing that gives mathematics its strength. But the power of imagination is every bit as important as the power of deductive reasoning. As the mathematician Augustus De Morgan once said, "The moving power of mathematical invention is not reasoning but imagination."

It is often the artificial organization of mathematics that bewilders the beginning student. Textbooks rarely show the long history in the development of a concept or any of the blind alleys that were taken. The fact is that the mathematician seeks out relationships in simple cases, looks for patterns, and only then tries to generalize. It is usually much later that the generalization is proved and finds its way into some textbook.

It is this *process* of problem solving that is lacking in most mathematics textbooks. As a mathematics teacher I often hear the comment, "I can do the mathematics, but I can't solve word problems." There *is* a great fear and avoidance of "real-life" problems on the part of most students, but we can't turn our back on these problems because they do not fit into the same mold as the "examples in the book." Few practical problems from everyday life come in the same form as those you study in school.

In this book we will consider some of the great ideas of mathematics, and will then look at how these ideas can be used in an everyday setting to build your problem-solving abilities. The most important prerequisite for this course is an openness to try out new ideas, a willingness to experience the suggested activities rather than to sit on the sideline as a spectator. You will find it difficult if you wait for the book or the teacher to give you answers—instead, be willing to guess, experiment, manipulate, and try out the problem without fear of being wrong.

The model for problem solving that we will use was first published in 1945 by the great, charismatic mathematician George Polya. His book *How to Solve It* (Princeton University Press, 1971) has become a classic. In Polya's book you will find his problem-solving model as well as a treasure trove of strategy, know-how, rules of thumb, good advice, anecdotes, history, and problems at all levels of mathematics. The problem-solving model is as follows:

First:	You have to *understand* the problem.
Second:	Find the connection between the data and the unknown. You should eventually obtain a *plan* of the solution.
Third:	*Carry out* your plan.
Fourth:	*Examine* the solution obtained.

These steps can be amplified by some individual strategies that might be used at appropriate moments:

1. If you cannot solve the proposed problem, look around for an appropriate related problem.
2. Work backward.
3. Work forward.
4. Narrow the condition.
5. Widen the condition.
6. Seek a counterexample.
7. Guess and test.
8. Divide and conquer.
9. Change the conceptual mode.

Let's apply this procedure for problem solving to the map shown in Figure 1.1.

Historical Note

George Polya (1888–) was born in Hungary and attended the universities of Budapest, Vienna, Göttingen, and Paris. He is presently a Professor Emeritus of Mathematics at Stanford University. Polya's research and winning personality have earned him a place of honor not only among mathematicians, but among students and teachers as well. His discoveries span an impressive range of mathematics: real and complex analysis, probability, combinatorics, number theory, and geometry. Polya's *How to Solve It* has been translated into 15 languages. His books have a clarity and elegance seldom seen in mathematics, making them a joy to read. For example, here is his explanation of why he is a mathematician: "It is a little shortened but not quite wrong to say: I thought I am not good enough for physics and I am too good for philosophy. Mathematics is in between."

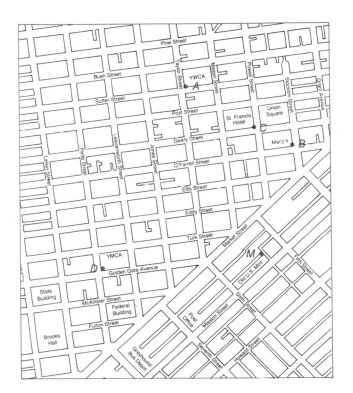

FIGURE 1.1   Portion of a map of San Francisco

Linda lives at the YWCA (point *A*) and works at Macy's (point *B*). She usually walks to work. How many different routes can Linda take if she doesn't backtrack—that is, if she always travels toward her destination? Where would you begin with this problem?

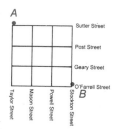

FIGURE 1.2 Simplified portion of Figure 1.1

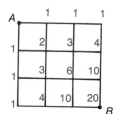

FIGURE 1.3 Map with solution

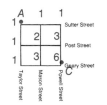

FIGURE 1.4

Step 1. *Understand the problem.* Can you restate it in your own words? Can you trace out one or two possible paths?

Step 2. *Devise a plan.* Simplify the question asked. Consider the simplified drawing shown in Figure 1.2.

Step 3. *Carry out the plan.* Count the number of ways it is possible to arrive at each point.

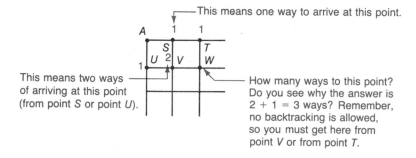

Now, you can fill in all the possibilities on Figure 1.2 as shown in Figure 1.3.

Step 4. *Look back.* Does the answer 20 different routes make sense? Do you think you could fill in them all?

Example 1: How many different ways could Linda get from the YWCA (point *A*) to the St. Francis Hotel (point *C*)?

Solution: Draw a simplified version of Figure 1.1, as in the figure in the margin. There are 6 different paths.

Example 2: How many different ways could Linda get from the YWCA (point *A*) to the Old U.S. Mint (point *M*)?

Solution: If the streets are irregular or if there are obstructions, you can still count blocks in the same fashion, as shown in Figure 1.4.

There are 52 paths from point *A* to point *M* (no backtracking).

Let's formulate a general solution. Consider a map with starting point *A*:

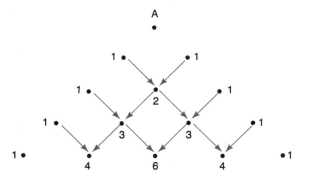

Do you see the pattern for building this figure? Each new row is found by adding the two previous numbers, as shown by the arrows. This pattern is known as *Pascal's triangle*. Notice in Figure 1.5 that the rows and diagonals are numbered for easy reference.

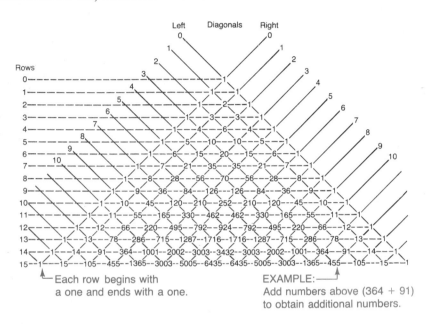

FIGURE 1.5 Pascal's triangle

How does this apply to Linda's trip from the YWCA to Macy's? It is 3 blocks down and 3 blocks over. Look at Figure 1.5 and count out these blocks as shown in Figure 1.6.

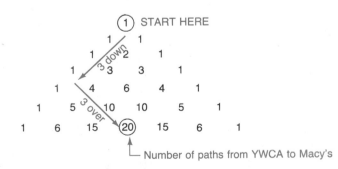

FIGURE 1.6

Example 3: How many different ways could Linda get from the YWCA (point A in Figure 1.1) to the YMCA (point D)?

Solution: Look at Figure 1.1; from point A to point D it is 7 blocks down and

3 blocks left. Use Figure 1.5 to see that there are 120 paths:

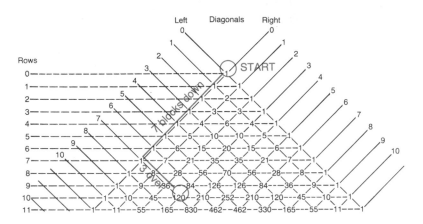

If there are alleys or irregular paths, you will need to use the map as shown in Example 2 rather than Pascal's triangle as shown in Example 3.

Problem Set 1.1

1. List the steps in Polya's problem-solving model.

2. Given Pascal's triangle as shown in Figure 1.5; write down the next row.

3. Study Pascal's triangle as shown in Figure 1.5. Name several patterns you notice about the triangle.

4. Describe the location of the numbers 1, 2, 3, 4, 5, . . . in Pascal's triangle.

5. Describe the location of the numbers 1, 4, 10, 20, 35, . . . in Pascal's triangle.

6. a. What is the sum of the numbers in row 1 of Pascal's triangle?
   b. What is the sum of the numbers in row 2 of Pascal's triangle?
   c. What is the sum of the numbers in row 3 of Pascal's triangle?
   d. What is the sum of the numbers in row 4 of Pascal's triangle?

7. How many 3¢ stamps are there in a dozen?

8. If you take 7 cards from a deck of 52 cards, how many cards do you have?

9. Oak Park cemetery in Oak Park, New Jersey, will not bury anyone living west of the Mississippi. Why?

10. Two U.S. coins total 30¢, yet one of these coins is not a nickel. What are the coins?

11. How many outs are there in a baseball game that lasts the full 9 innings?

12. If posts are spaced 10 feet apart, how many posts are needed for 100 feet of straight line fence?

Use the following map to determine the number of different paths from point A to the point indicated in Problems 13–19:

13. *E*   **14.** *F*   **15.** *G*   **16.** *H*

17. *I*   **18.** *J*   **19.** *K*

20. Two volumes of Newman's *The World of Mathematics* stand side by side in order on a shelf. A bookworm starts at page i of Volume I and bores its way in a straight line to the last page of Volume II. Each cover is 2 mm thick, and the first volume is $\frac{17}{19}$ as thick as the second volume. The first volume is 38 mm thick without its cover. How far does the bookworm travel?

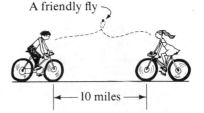

21. A boy cyclist and a girl cyclist are 10 miles apart and pedaling toward each other. The boy's rate is 6 miles per hour, and the girl's rate is 4 miles per hour. There is also a friendly fly zooming continuously back and forth from one bike to the other. If the fly's rate is 20 miles per hour, how far does the fly fly by the time the cyclists reach each other?

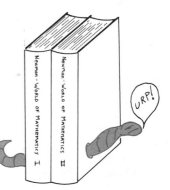

A friendly fly

← 10 miles →

22. A farmer has to get a fox, a goose, and a bag of corn across a river in a boat that is large enough only for him and one of these three items. If he leaves the fox alone with the goose, the fox will eat the goose. If he leaves the goose alone with the corn, the goose will eat the corn. How does he get all the items across the river?

Mind Bogglers

Each section of the book has one or more problems designated as Mind Bogglers. These problems are loosely tied to the material of the section, but may require additional insight, information, or effort to solve. I hope you will make it a habit of reading these problems and attempting to work those that interest you, even though they may not be part of your regular class assignment.

Problem 23 is dedicated to Bill Leonard of Cal State Fullerton because 23 is his favorite number. (His book, *No Upper Limit,* Creative Teaching Associates, 1977, has 23 chapters. It is a delightful book, and I recommend it to you.)

23. A very magical mathematics teacher had a student select a two-digit number between 50 and 100 and write it on the board out of view of the instructor. Next, the student was asked to add 76 to the number, producing a three-digit sum. If the digit in the hundreds place is added to the remaining two-digit number and this result is subtracted from the original number, the answer is 23, which was predicted by the instructor. How did he know the answer would be 23?

24. A magician divides a deck of cards into two equal piles of 26 cards each and places the piles face down on a table. He then picks up one of the piles, counts down from the top to the seventh card, and shows it to the audience without looking at it himself. These seven cards are replaced face down in the same order on top of the pile. He then picks up the other pile and deals the top three cards up in a row in front of him. He forms three columns of cards by making them add up to 10 (face cards count 10, and others count at face value). That is, if the card in the first column is a 4, he counts out six more cards; if the card is a queen, no additional cards are needed. The remainder of this pile is placed on top of the first pile. Next the magician adds the values of the three face-up cards, and this number of cards down in the deck is the card that was originally shown to the audience. Explain why this trick works.

MAGIC AND MATH

Many magic tricks, especially those performed with cards, are really disguised mathematical problems. You might enjoy trying the tricks described in Problems 23–25 even if you can't explain why they work. Speaking of cards, here is an interesting number fact concerning cards. Consider the number of letters in each of the names of the cards:

Ace	3
Two	3
Three	5
Four	4
Five	4
Six	3
Seven	5
Eight	5
Nine	4
Ten	3
Jack	4
Queen	5
King	4

The total is 52, the same as the number of cards in the deck!

25. A magician begins a trick by giving someone a sealed envelope. She gives a deck of cards to another person and asks that individual to riffle-shuffle the cards twice. Now she obtains a six-digit number by going through the deck, cards face up, and removing the first six spades that have digit values. The digits turn out to be, in order,

$$1, \quad 4, \quad 2, \quad 8, \quad 5, \quad 7$$

She arranges the cards in a row on a table and rolls a die to obtain a random digit from 1 to 6. The audience multiplies this digit by

$$142,857$$

while the magician retrieves and cuts open the sealed envelope. It contains the correctly predicted product. Explain how this trick was done.

Problems for Individual Study

At the end of many sections you will find problems requiring some library research. I hope that as you progress through the course you will find at least one topic that interests you so much that you will want to do additional reading on that topic. For most of the topics listed in this section, I've provided one or two references to get you started.

26. *Recreational mathematics.* What are some mathematical puzzles, tricks, magic stunts? What is the mathematical explanation for number tricks? What games are based on mathematical problems? What puzzles have mathematical explanations?

How is mathematics related to chess, card games, athletics?

Exhibit suggestions: puzzles, tricks, games of mathematical nature; mathematical analysis of games and tricks.

References: William Schaaf, *A Bibliography of Recreational Mathematics* (Washington, D.C.: National Council of Teachers of Mathematics, 1970).

See also the *Journal of Recreational Mathematics.*

27. *Pascal's triangle.* Do some more research on Pascal's triangle, and see how many properties you can discover. You might begin by answering these questions.

a.   What is a binomial expansion?

b.   How is the binomial expansion related to Pascal's triangle?

c.   What relationship do the patterns in the figures below have to Pascal's triangle?

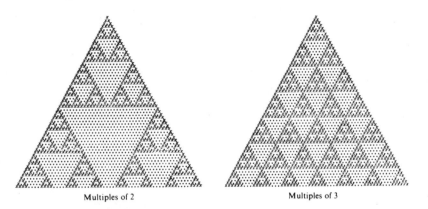

Multiples of 2   Multiples of 3

References: Peggy A. House, "More Mathemagic from a Triangle," *Mathematics Teacher,* March 1980, pp. 191–195.

Margaret J. Kenney, *The Incredible Pascal Triangle* (Chestnut Hill, Mass.: Boston College Mathematics Institute, 1981).

Karl J. Smith, "Pascal's Triangle," *Two Year College Mathematics Journal* 4 (Winter 1973).

1.2 Magic Squares

For centuries magic squares have intrigued both amateur and professional mathematicians. A **magic square** is an arrangement of numbers in the shape of a square, with the sum of each vertical column, each horizontal row, and each diagonal all equal. The number of rows and columns are the same, and this number is called the **order** of the magic square. Some magic squares are shown in Figure 1.7. A **standard magic square** is made up of the consecutive counting numbers starting with 1.

Historical Note

The first known example of a magic square comes from China. Legend tells us that around the year 2000 B.C. the Emperor Yu of the Shang dynasty received the following magic square etched on the back of a tortoise's shell:

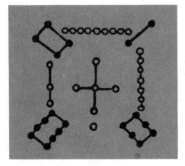

This incident supposedly took place along the Lo River, so this magic square has come to be known as the Lo-shu magic square. The even numbers are black (female numbers) and the odd numbers are white (male numbers). You are asked to translate this third-order magic square (three rows and three columns) into modern symbols in Problem 1.

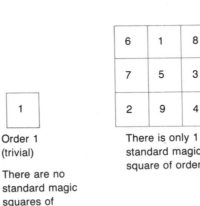

1

Order 1
(trivial)

There are no standard magic squares of order 2.

There is only 1 standard magic square of order 3.

There are 880 standard magic squares of order 4.

FIGURE 1.7   Standard magic squares. If one magic square can be obtained from another by rotating or by reflecting, then it is not counted as a different square.

If you know one magic square, you can construct other magic squares by adding the same number to each entry (for example, if you add 2 to each entry in the order-3 standard magic square, the top row will be   8   3   10). You can also construct other magic squares using a *rotation* as shown in Example 1.

Example 1:   Rotate the 3-by-3 magic square shown in Figure 1.7 in order to fill in the blanks.

		2
8		6

Solution: You can cut out a piece of paper with the magic square shown in Figure 1.7 and actually rotate it to help you fill in the blanks.

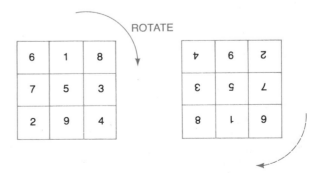

Answer:

4	9	2
3	5	7
8	1	6

A second way of obtaining another magic square is to use a *reflection*, as shown in Figure 1.8.

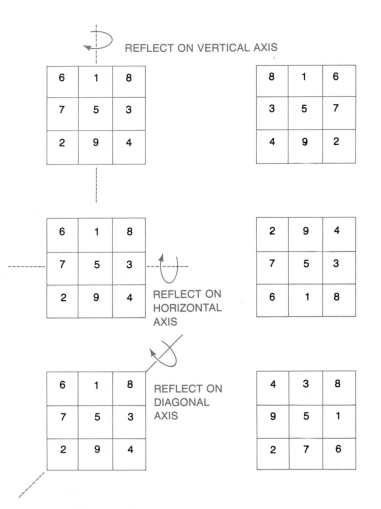

FIGURE 1.8   Reflections of a magic square

Melancolia engraving by Dürer

The 4-by-4 magic square shown
in Figure 1.7 comes from the
upper right-hand corner of this
engraving done in 1514. Notice
that this date appears in the
magic square. This is a special
magic square that has some
special properties. A few are
listed here; see if you can find
any more:
1. The corners add up to the
   magic number, 34.
2. The center squares
   (10, 11, 6, 7) add up to 34.
3. Opposite pairs of squares
   (3, 2, 15, 14) add up to 34.
4. Slanting squares (2, 8, 15, 9)
   add up to 34.
5. The sum of the first eight
   numbers equal the sum of the
   second eight numbers.

There is also a method that can be used to form only odd-ordered magic
squares. If the size is 3 by 3, work with groups of three numbers; if it is 5 by 5,
work with groups of five numbers, and so on.

Step 1. Write 1 in the middle of the top row. Fill in successive counting numbers moving diagonally upward to the right. If that position is off the
magic square, enter the number at the opposite end of the next row or
column. This has been done for the entries 2 and 4:

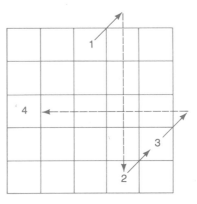

Step 2. Continue the process described in Step 1 until you complete a group of
numbers (three numbers for a 3-by-3 square, five numbers for a 5-by-5
square, seven numbers for a 7-by-7 square, and so on); in our example,
the group of numbers is complete after the number 5. Then move down
one box to begin the next group of five numbers.

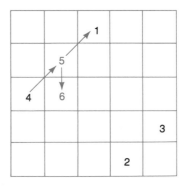

Step 3. Continue the process described in Steps 1 and 2. When a position is
already occupied, enter the number in the space immediately below the
last one filled. The space that follows the upper right-hand corner is
the lower left-hand corner.

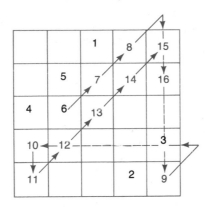

Now, 16 begins the next new group of 5, so it is placed below 15 following the rule from Step 2.

Step 4. Continue the process until the magic square is complete. Check to see that the completed array is indeed a magic square by adding up each column and row, and the corner-to-corner diagonals.

17	24	1	8	15
23	5	7	14	16
4	6	13	20	22
10	12	19	21	3
11	18	25	2	9

Example 2: Complete a 5-by-5 magic square starting with 23.

Solution:

39	46	23	30	37
45	27	29	36	38
26	28	35	42	44
32	34	41	43	25
33	40	47	24	31

The magic sum is 175.

𝔥istorical 𝔑ote

Benjamin Franklin was
interested in magic squares,
but he often complained of
the amount of time they took.
Howard Eves, in his delightful
book *In Mathematical Circles,*
quotes Dr. Franklin as saying
that the time consumed with
these pastimes was time "which I
still think I might have employed
more usefully."

FIGURE 1.10   IXOHOXI

Benjamin Franklin was fond of magic squares and invented the one shown in
Figure 1.9. In this magic square, not only do the rows, columns, and diagonals
add up to 260, but the bent diagonals (indicated by the broken lines) also add up
to 260. The four corners plus the four middle boxes total 260. All four-box
subsquares total 130. So do any four numbers equidistant from the center point.

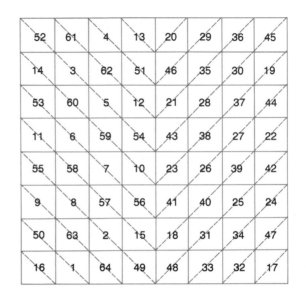

FIGURE 1.9   Benjamin Franklin's magic square

One final magic square that is rather interesting is called IXOHOXI and is
shown in Figure 1.10. It is a magic square whose sum is 19,998. It is interesting
because it is still a magic square when it is turned upside down and also when it
is reflected in a mirror.

Problem Set 1.2

1.   Translate the Lo-shu magic square into modern symbols.

Use rotations of the magic squares in Figure 1.7 to fill in the blanks in Problems 2–4.

2. •

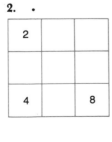

3.

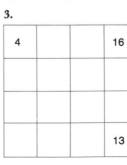

4.

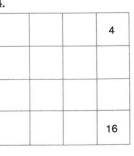

Use reflections of the magic squares in Figure 1.7 to fill in the blanks in Problems 5–7.

5.

6		2

6.

4			1

7.

13			16

Use rotations or reflections of the magic squares in Figure 1.7 to fill in the blanks in Problems 8–10.

8.

1			13
4			

9.

1			4
13			

10.

13			1
16			

Use patterns to fill in the blanks in the magic squares in Problems 11–13.

11.

8	1	6
	7	
	2	

12.

2		
		3
6		

13.

21	7		18
10		15	
14	12	11	17
9	19		

Use the method shown in this section to construct the magic squares in Problems 14–19.

14.

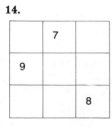

	7	
9		
		8

15.

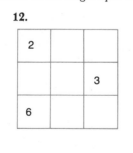

	1	

16.

	43	

1	48	31	50	33	16	63	18
30	51	46	3	62	19	14	35
47	2	49	32	15	34	17	64
52	29	4	45	20	61	36	13
5	44	25	56	9	40	21	60
28	53	8	41	24	57	12	37
43	6	55	26	39	10	59	22
54	27	42	7	58	23	38	11

THE EULER
MAGIC SQUARE

The mathematician Leonhard Euler also worked with magic squares. In the square above, each vertical and horizontal row adds up to 260, with each halfway point totaling 130. Adding extra magic for chess players, a knight starting at box 1 would hit all 64 boxes in numerical order.

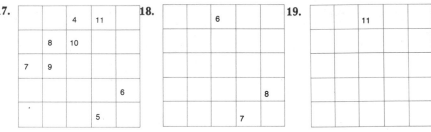

17. **18.** **19.**

20. Make up a standard 7-by-7 magic square.

21. Make up a 7-by-7 magic square starting with a 3.

22. Make up a standard 9-by-9 magic square.

Mind Bogglers

Magic triangles are arrays of consecutive numbers arranged in such a way that the sum of the integers on each side is a magic constant.

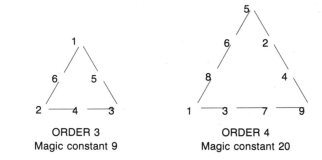

ORDER 3
Magic constant 9

ORDER 4
Magic constant 20

Find the magic triangles indicated in Problems 23–26.

23. Order 3; magic constant 10.   **24.** Order 3; magic constant 11.

25. Order 3; magic constant 12.   **26.** Order 4; magic constant 28.

A 3-by-3 magic cube is an arrangement of 27 numbers that form a cube as shown in the figure:

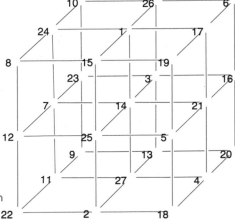

A 3-by-3 magic cube. The sum of each row, column, pillar, and diagonal is 42.

27. Form another 3-by-3 magic cube using a rotation.

28. Form another 3-by-3 magic cube using a reflection.

29. Consider the square shown in the margin.
Circle any number.
Cross out all the numbers in the same row and column.
Circle any remaining number and cross out all the numbers in the same row and column.
Repeat; circle any remaining number and cross out all the numbers in the same row and column.
Circle the remaining number.
The sum of the circled numbers is 48. Why?

10	7	8	11
14	11	12	15
13	10	11	14
15	12	13	16

30. a. Fill in the squares in the figure shown in the margin by adding the numbers at the head of the rows and columns.
    b. Add all the numbers at the head of the columns and rows ($6 + 5 + 7 + 9 + 4 + 10 + 2 + 4$).
    c. Carry out the instructions given in Problem 29 for the completed square.
    d. Now explain why the trick works.

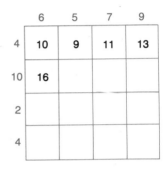

Problems for Individual Study

31. *Magic squares.* Write a short paper about the construction of magic squares. Are there any general methods for finding magic squares with even order?

    References: William H. Benson and Oswald Jacoby, *New Recreations with Magic Squares* (New York: Dover Publications, 1976).

    John Fults, *Magic Squares* (La Salle, Ill.: Open Court, 1974).

    Martin Gardner, "Mathematical Games Department," *Scientific American,* January 1976, pp. 118–122.

    Diayo Sawada, "Magic Squares: Extensions into Mathematics," *The Arithmetic Teacher,* March 1974, pp. 183–188.

    Horace Williams, "A Note of Magic Squares," *The Mathematics Teacher,* October 1974, pp. 511–513.

32. *Art with magic squares.* A process for producing an artistic pattern using magic squares is described in an article "An Art-Full Application Using Magic Squares" by Margaret J. Kenney. (*The Mathematics Teacher,* January 1982, pp. 83–89.) A design can be made from a magic square as follows:

    1. Select a square array of dots whose row size is equivalent to the order of the magic square.
    2. Assign each dot the corresponding number of the magic square.
    3. Connect the numbered dots by straight line segments in numerical order.
    4. Shade in alternating regions.

This process is completed for the *Melancolia* magic square, as shown in the margin.

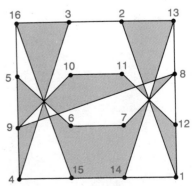

Read the indicated article and design some original magic square art. The following are some examples taken from the article.

1.3 Inductive Reasoning

By studying numerical relationships that exhibit patterns, we can learn much about mathematics and have a lot of fun doing so.

A very familiar pattern is found in the ordinary "times" tables. By pointing out patterns, teachers can make it easier for children to learn some of their multiplication tables. For example, consider the multiplication table for 9s:

$$1 \times 9 = 9 \qquad\qquad 6 \times 9 = 54$$
$$2 \times 9 = 18 \qquad\qquad 7 \times 9 = 63$$
$$3 \times 9 = 27 \qquad\qquad 8 \times 9 = 72$$
$$4 \times 9 = 36 \qquad\qquad 9 \times 9 = 81$$
$$5 \times 9 = 45 \qquad\qquad 10 \times 9 = 90$$

What patterns do you notice? You should be able to see many number relationships by looking at the totals. For example, notice that the sum of the digits to the right of the equality is 9 in all the examples $(1 + 8 = 9, \quad 2 + 7 = 9, \quad 3 + 6 = 9, \quad$ and so on). Will this always be the case for multiplication by 9? (Consider $11 \times 9 = 99$. The sum of the digits here is 18. However, notice the result if you add the digits of 18.) Do you see any other patterns? Can you explain why they "work"? This pattern of adding after multiplying by 9 generates a sequence of numbers: 9, 9, 9, ... We will call this the *nine pattern*.

Example 1: Find the *eight pattern*.

Solution: Consider the successive products of 8:

$$8 \times 1 = 8$$
$$8 \times 2 = 16; \quad 1 + 6 = 7$$
$$8 \times 3 = 24; \quad 2 + 4 = 6$$
$$8 \times 4 = 32; \quad 3 + 2 = 5$$
$$8 \times 5 = 40; \quad 4 + 0 = 4$$

$$8 \times 6 = 48; \quad 4 + 8 = 12; \quad 1 + 2 = 3$$
$$8 \times 7 = 56; \quad 5 + 6 = 11; \quad 1 + 1 = 2$$
$$8 \times 8 = 64; \quad 6 + 4 = 10; \quad 1 + 0 = 1$$
$$8 \times 9 = 72; \quad 7 + 2 = \quad 9$$
$$8 \times 10 = 80; \quad 8 + 0 = \quad 8$$
$$8 \times 11 = 88; \quad 8 + 8 = 16; \quad 1 + 6 = 7$$

The eight pattern is: $8, 7, 6, 5, 4, 3, 2, 1, 9, 8, 7, 6, 5, \ldots$

Example 2: Verify that the *seven pattern* is:

$$7, 5, 3, 1, 8, 6, 4, 2, 9, 7, 5, 3, 1, \ldots$$

Complicated arithmetic problems can sometimes be solved by using patterns. Given a difficult problem, a mathematician will often try first to solve a simpler, but similar, problem. For example, suppose we wish to compute the following number:

$$10 + 123456789 \times 9$$

Instead of doing a lot of arithmetic, let's study the following pattern:

$$2 + 1 \times 9$$
$$3 + 12 \times 9$$
$$4 + 123 \times 9$$

Do you see the next entry in this pattern?

Using Polya's strategy, we begin by working these easier arithmetic problems: Thus, we begin with $2 + 1 \times 9$. You can work left to right: $2 + 1 \times 9 = 3 \times 9 = 27$. Or you can do multiplication first: $2 + 1 \times 9 = 2 + 9 = 11$. Why 11? Why would we ever suggest doing multiplication first? To clarify this process, consider another type of example. Suppose you sold one $2 soccer benefit ticket on Monday and three $4 tickets on Tuesday. What is the total amount collected?

$$\begin{pmatrix} \text{SALES ON} \\ \text{MONDAY} \end{pmatrix} + \begin{pmatrix} \text{SALES ON} \\ \text{TUESDAY} \end{pmatrix} = \begin{pmatrix} \text{TOTAL} \\ \text{SALES} \end{pmatrix}$$
$$2 + 3 \times 4$$

Correct: Multiplication first $2 + 12 = 14$
Incorrect: Left to right $2 + 3 \times 4 = 5 \times 4 = 20$

In mathematics, it is proper to adopt the following agreement.

Order of Operations

If operations are mixed, you should do multiplication *first* (from left to right) and *then* addition and subtraction from left to right.

Thus, the correct result for $2 + 1 \times 9$ is 11. Also,

$$3 + 12 \times 9 = 3 + 108$$
$$= 111$$

and

$$4 + 123 \times 9 = 4 + 1107$$
$$= 1111$$

Do you see a pattern in the answers as well as in the sequence of problems? Can you predict the answer to the next numbers in the pattern:

$$5 + 1234 \times 9 = ?$$

It is five 1s: 11,111. Continue computations like this until you see a pattern, and then predict what the number

$$10 + 123456789 \times 9$$

represents.

The type of reasoning just used—first observing patterns and then predicting answers for complicated problems—is an example of what is called **inductive reasoning.** It is a very important method of thought and is sometimes called the **scientific method.** It involves reasoning from particular facts or individual cases to a general *conjecture*—a statement you think may be true. That is, a generalization is made on the basis of some observed occurrences. The more individual occurrences we observe, the better able we are to make a correct generalization. Peter in the *B.C.* cartoon makes the mistake of generalizing on the basis of a single observation. We must keep in mind that an inductive conclusion is always tentative and may have to be revised on the basis of new evidence. For example, consider this pattern:

$$1 \times 9 = 9$$
$$2 \times 9 = 18$$
$$3 \times 9 = 27$$
$$\cdot$$
$$\cdot$$
$$\cdot$$
$$10 \times 9 = 90$$

You might have made the conjecture that the sum of the digits of the answer to any product involving a 9 is always 9. Certainly the first ten examples substantiate this speculation. But the conjecture is suspect because it is based on so few cases. It is shattered by the very next case, wherein $11 \times 9 = 99$.

On the other hand, we can make a prediction about the exact time of sunrise and sunset tomorrow. This is also an example of inductive reasoning, since our prediction is based on a large number of observed cases. Thus there is a very high probability that we will be successful in our prediction.

Although mathematicians often proceed by inductive reasoning to formulate new ideas, they are not content to stop at the "probable" stage. So they often formalize their predictions into theorems and then try to prove these theorems deductively. Proof and deductive reasoning will be discussed in Chapter 2. Right now, however, consider some additional patterns in Problem Set 1.3 to see if you can reason inductively and formulate some conjectures.

Problem Set 1.3

Perform the operations in Problems 1–10 by using the proper order of operations.

1. $7 + 2 \times 6$ **2.** $8 + 3 \times 2$ **3.** $4 + 6 \times 3$

4. $3 \times 5 + 2 \times 5$ **5.** $6 \times 4 + 3 \times 4$ **6.** $2 \times 8 + 2 \times 7$

7. $1 + 3 \times 2 + 4 + 3 \times 6$ **8.** $3 + 6 \times 2 + 8 + 4 \times 3$

9. $10 + 5 \times 2 + 6 \times 3$ **10.** $4 + 3 \times 8 + 6 + 4 \times 5$

11.

Problem	Answer
$9 \times 1 - 1$	$= 8$
$9 \times 21 - 1$	$= 188$
$9 \times 321 - 1$	$= 2888$
$9 \times 4321 - 1$	$= 38888$

    a. What is the next problem in this sequence?
    b. What is the next answer in this sequence?
    c. Use patterns to predict the answer to

$$9 \times 987654321 - 1$$

12.

Problem	Answer
123456789×9	$= 1111111101$
123456789×18	$= 2222222202$
123456789×27	$= 3333333303$

    a. What is the next problem in this sequence?
    b. What is the next answer in this sequence?
    c. Use patterns to predict the answer to

$$123456789 \times 81$$

13. Compute:

    a. $1 \times 142,857$ **b.** $2 \times 142,857$ **c.** $3 \times 142,857$
    d. $4 \times 142,857$ **e.** $5 \times 142,857$ **f.** $6 \times 142,857$
    g. Can you describe a pattern?

14. Compute:

    a. 11×11 **b.** 111×111
    c. 1111×1111 **d.** Can you describe a pattern?

15. Using the pattern in Problem 12, predict the following *without* direct calculation:

    a.   123456789 × 45   **b.**   123456789 × 54

    c.   123456789 × 63   **d.**   123456789 × 72

16. Find a pattern illustrated in the figures below. What do you think the next figure would look like?

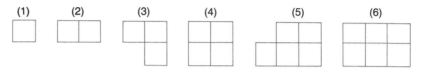

17. Find a pattern illustrated in the figures below. What do you think the next figure would look like?

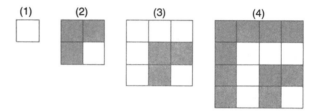

Problems 18–22 are modeled after Examples 1 and 2 in this section. See if you can find the requested patterns.

18. Six pattern   **19.** Five pattern

20. Four pattern   **21.** Three pattern

22. Two pattern

23. Use patterns to find 9 + 123456789 × 9.

24. Use patterns to find 9 + 123456789 × 8.

25. Find (9 × 987654321) − 1 without doing direct multiplication.

26. What is the sum of the first 25 consecutive odd numbers?

27. What is the sum of the first 50 consecutive odd numbers?

28. What is the sum of the first 100 consecutive odd numbers?

29. **a.** Construct a right triangle very carefully, and measure the two acute angles. What is the sum of their measures?

    b. Repeat the above experiment. Are your answers the same?

    c. Make a conjecture about the sum of the measures of the acute angles in any right triangle.

30. **a.** Draw a circle and any diameter in that circle. Now choose any point P on the circumference of the circle, and then draw the triangle with the end points of the diameter and P as the vertices. What is the measure of the angle P?

    b. Repeat this experiment choosing a different point on the circle.

    c. Repeat this experiment by drawing a circle with a different diameter.

    d. Can you make a conjecture about the size of angle P?

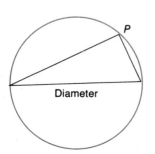

Mind Bogglers

31. How many squares are there in the figure in the margin?

Hint:

has 1 square

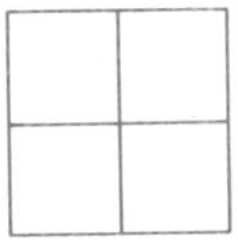

has 4 1-by-1 squares
1 2-by-2 square   TOTAL: 5

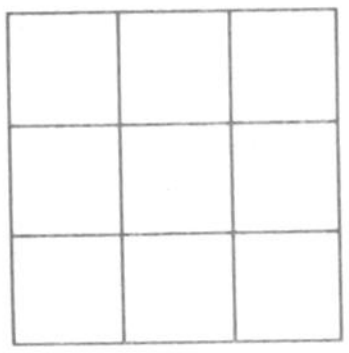

has 9 1-by-1 squares
4 2-by-2 squares
1 3-by-3 square   TOTAL: 14

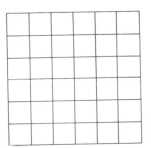

How many squares?

32. Fill in the following chart based on these painted cubes:

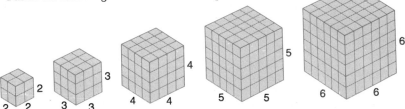

	Length of an edge	Number of cubes	Number of cubes with three painted faces	Number of cubes with two painted faces	Number of cubes with one painted face	Number of cubes with no painted faces
a.	2	8				
b.	3	27				
c.	4					
d.	5					
e.	6					
f.	7					
	. . .					
g.	10					

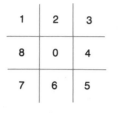

Problems for Individual Study

33. *Two-dimensional tic-tac-toe.* Develop a winning strategy for playing tic-tac-toe. Label the squares in the figure in the margin. You must consider all possible games; look for patterns, and use inductive reasoning.

34. *Three-dimensional tic-tac-toe.* Develop a strategy for playing three-dimensional tic-tac-toe by using inductive reasoning. Label the squares as shown in the figure. An article that will help you to do this is "Three-Dimensional Tic-Tac-Toe," by Gene Mercer and John Kolb, in *The Mathematics Teacher,* vol. LXIV, no. 2 (February 1971). This article also poses five questions; see if you can answer any of them.

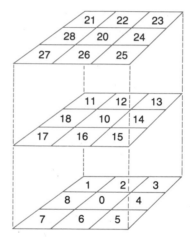

1.4 Mathematical Patterns

Patterns are sometimes used as part of an IQ test. People who have scored at or above the 98th percentile on a standardized IQ test can join an organization called Mensa. The test on the next page is one of their "quickie" tests. You might want to see how well you can do.

It used to be thought that an IQ test measured "innate intelligence" and that a person's IQ score was fairly constant. Today, it is known that this is not the case. IQ test scores can be significantly changed by studying the types of questions asked. For example, let's focus on one type of question on these tests.

Number Sequence

A **number sequence** or **progression** is a collection of numbers arranged in order so that there is a first term, a second term, a third term, and so on.

ARE YOU A GENIUS?

EACH problem is a series of some sort—that is, a succession of either letters, numbers or drawings—with the last item in the series missing. Each series is arranged according to a different rule and, in order to identify the missing item, you must figure out what that rule is.

Now, it's your turn to play. Give yourself a maximum of 20 minutes to answer the 15 questions. If you haven't finished in that time, stop anyway. In the test problems done with drawings, it is always the top row or group that needs to be completed by choosing one drawing from the bottom row.

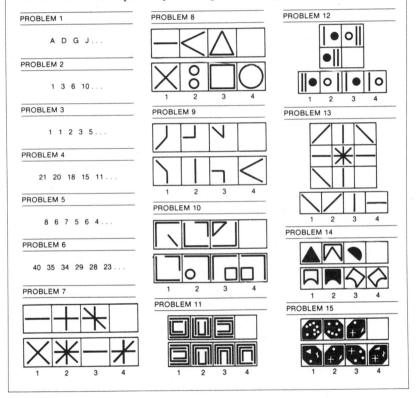

PROBLEM 1

A D G J . . .

PROBLEM 2

1 3 6 10 . . .

PROBLEM 3

1 1 2 3 5 . . .

PROBLEM 4

21 20 18 15 11 . . .

PROBLEM 5

8 6 7 5 6 4 . . .

PROBLEM 6

40 35 34 29 28 23 . . .

A sequence may stop at a particular place, or it may continue indefinitely. Sequences are arranged from left to right, with each number, or term, separated by a comma. In this section we'll look at three categories of progressions.

Example 1:   What is the pattern for the sequence 1, 4, 7, 10, 13, . . . ? Fill in the next three terms.

Solution: There is a difference of 3 between each two terms. That is, 3 is added to each term in order to find the next term. The next three terms are 16, 19, and 22.

The *Peanuts* cartoon illustrates another example of a geometric progression. Charlie Brown is supposed to send a copy of the letter to six of his friends. Let's

If the difference between each two terms of a sequence is the same, it is called the **common difference**. The common difference in Example 1 is 3. If a sequence has a common difference, it is called an **arithmetic sequence**.

Example 2:   Use inductive reasoning to tell which of the given sequences are arithmetic. If the sequence is arithmetic, give the common difference and the next term.

a. 6, 9, 12, 15, . . .   **b.** 66, 72, 78, 84, . . .
c. 4, 11, 18, 24, . . .   **d.** 113, 322, 531, 740, . . .

Solution

a. Check the differences: $9 - 6 = 3$ $12 - 9 = 3$ $15 - 12 = 3$
$$6, \qquad 9, \qquad 12, \qquad 15, \ldots$$

The common difference is 3; the next term is found by adding 3: $15 + 3 = 18$.

b. Check the differences: $72 - 66 = 6$ $78 - 72 = 6$ $84 - 78 = 6$
$$66, \qquad 72, \qquad 78, \qquad 84, \ldots$$

The common difference is 6; the next term is $84 + 6 = 90$.

c. Check the differences: $11 - 4 = 7$ $18 - 11 = 7$ $24 - 18 = 6$
$$4, \qquad 11, \qquad 18, \qquad 24, \ldots$$

There is no common difference, so this sequence is not arithmetic. This example illustrates a common mistake among students: that is, not carefully checking *all* the differences to see that they are the same.

d. Check the differences:

$322 - 113 = 209$ $531 - 322 = 209$ $740 - 531 = 209$
$$113, \qquad 322, \qquad 531, \qquad 740, \ldots$$

The common difference is 209; the next term is $740 + 209 = 949$.

Example 3 illustrates the second category of sequences that we will consider in this book.

Example 3:   What is the pattern for the sequence 2, 4, 8, 16, 32, . . . ? Fill in the next three terms.

Solution: This is *not* an arithmetic sequence because there is no common difference. However, if each term is *divided* by the preceding term, the results are the same. This number found by division is called a **common ratio**. If a sequence has a common ratio, then the sequence is called a **geometric sequence**. Additional terms of a geometric sequence can be found by multiplying successive terms by the common ratio. For this example the common ratio is 2, so the next three terms are

$$32 \times 2 = 64$$
$$64 \times 2 = 128$$
$$128 \times 2 = 256$$

Example 4: Tell which of the given sequences are geometric. If the sequence is geometric, give the common ratio and the next term.

a. $1, 3, 9, 27, \ldots$ **b.** $3, 12, 48, 182, \ldots$
c. $4, 20, 100, 500, \ldots$ **d.** $5, 50, 500, 5000, \ldots$

Solution

a. Check the ratios: $\quad 3 \div 1 = 3 \quad 9 \div 3 = 3 \quad 27 \div 9 = 3$
$$1, \qquad 3, \qquad 9, \qquad 27, \ldots$$

The common ratio is 3; the next term is found by multiplying by 3: $27 \times 3 = \mathbf{81}$.

b. Check the ratios: $\quad 12 \div 3 = 4 \quad 48 \div 12 = 4 \quad 182 \div 48 \neq 4$
$$3, \qquad 12, \qquad 48, \qquad 182, \ldots$$

There is no common ratio, so this sequence is not geometric.

c. Check the ratios: $\quad 20 \div 4 = 5 \quad 100 \div 20 = 5 \quad 500 \div 100 = 5$
$$4, \qquad 20, \qquad 100, \qquad 500, \ldots$$

The common ratio is 5; the next term is $\quad 500 \times 5 = \mathbf{2,500}$.

d. Check the ratios: $\quad 50 \div 5 = 10 \quad 500 \div 50 = 10 \quad 5000 \div 500 = 10$
$$5, \qquad 50, \qquad 500, \qquad 5000, \ldots$$

The common ratio is 10; the next term is **50,000**.

In this book we will classify sequences that are not arithmetic or geometric as *neither*. To find the pattern, begin by looking for a common difference; next look for a common ratio; then look for some other pattern.

Example 5: Classify the following sequences as arithmetic, geometric, both, or neither. If it is arithmetic, give the common difference; if it is geometric, give the common ratio; and, if it is neither, describe the pattern. Give the next term for each sequence.

a. $15, 30, 60, 120, \ldots$ **b.** $15, 30, 45, 60, \ldots$ **c.** $15, 30, 45, 75, \ldots$
d. $15, 20, 26, 33, \ldots$ **e.** $3, 3, 3, 3, \ldots$ **f.** $15, 30, 90, 360, \ldots$
g. $15, 30, 120, 960, \ldots$ **h.** $A, C, F, J, O, \ldots$

Solution

a. Differences:
$$30 - 15 = 15 \qquad 60 - 30 = 30 \qquad \qquad \textit{Not} \text{ arithmetic}$$
$$15, \qquad 30, \qquad 60, \qquad 120, \ldots$$
Ratios: $\quad 30 \div 15 = 2 \quad 60 \div 30 = 2 \quad 120 \div 60 = 2 \qquad$ Common ratio is 2

This is a geometric sequence; the next term is $\quad 120 \times 2 = \mathbf{240}$.

see how many people become involved in a very short time if everyone carries out the directions in the letter.

1st mailing:	6
2nd mailing:	36
3rd mailing:	216
4th mailing:	1296
5th mailing:	7776
6th mailing:	46,656
7th mailing:	279,936
8th mailing:	1,679,616
9th mailing:	10,077,696
10th mailing:	60,466,176
11th mailing:	362,797,056

After 11 mailings, more letters will have been sent than there are people in the United States! The number of letters in only two more mailings would exceed the population of the world.

b. Differences:

$30 - 15 = 15$ $45 - 30 = 15$ $60 - 45 = 15$ Common difference is 15

15,   30,   45,   60, . . .

This is an arithmetic sequence; the next term is $60 + 15 = \mathbf{75}$.

c. Differences: 15   15   30   *Not* arithmetic

15, 30,   45, 75, . . .

Ratios: 2   not 2   *Not* geometric

The pattern is to add successive terms. This is a fairly common way of formulating a sequence:

$15 + 30 = 45$; $30 + 45 = 75$; $45 + 75 = \mathbf{120}$ is the next term.

d. Differences: 5   6   7   *Not* arithmetic

15,   20,   26, 33, . . .

Ratios: $20 \div 15 \neq 26 \div 20$ *Not* geometric

There is a pattern in the differences; that is, the differences form an arithmetic sequence, and the next difference is 8. Thus, the next term is $33 + 8 = \mathbf{41}$.

e. Differences: 0   0   0   Arithmetic

3, 3, 3, 3, . . .

Ratios: 1   1   1   Geometric

This pattern is both arithmetic and geometric.

f. Differences: 15   60   270   *Not* arithmetic

15,   30,   90,   360, . . .

Ratios: $30 \div 15 \neq 90 \div 30$ *Not* geometric

There is a pattern of the ratios; the ratios are 2, 3, and 4. Thus, it seems reasonable to suppose that the next *ratio* in this sequence is 5; this means that the next term in the given sequence is $360 \times 5 = \mathbf{1{,}800}$

g. Differences: 15   90   840   *Not* arithmetic

15, 30, 120,   960, . . .

Ratios: 2   4   8   *Not* geometric

There is a geometric pattern in the ratios; the next ratio is 16, so the next term is $960 \times 16 = \mathbf{15{,}360}$

h. Not all patterns need be patterns of numbers, as this example illustrates. It is, of course, not arithmetic or geometric, but notice that after A there is one letter (B) omitted; then C; then two letters omitted; then F; then three letters omitted; J; four letters omitted; O; so the next gap should have five letters omitted; thus, the next letter in the pattern is **U**

The pattern of adding successive terms, illustrated by Example 5c, suggests a particular famous sequence that has a name. The sequence

$$1, 1, 2, 3, 5, 8, 13, 21, 34, 55, 89, 144, \ldots$$

is called the **Fibonacci sequence.** The numbers of this sequence are called **Fibonacci numbers,** and they have interested mathematicians for centuries—at least since the 13th century, when Leonardo Fibonacci wrote *Liber Abaci,* which discussed the advantages of Hindu-Arabic numerals over Roman numerals. In this book, he had a problem that related Fibonacci numbers to the birth patterns of rabbits. Let's see what he did. The problem was to find the number of rabbits alive after a given number of generations. Suppose a pair of rabbits will produce a new pair of rabbits in their second month and thereafter will produce a new pair every month. The new rabbits will do exactly the same. Starting with one pair, how many pairs will there be in 10 months if no rabbits die?

To solve this problem, we could begin by direct counting.

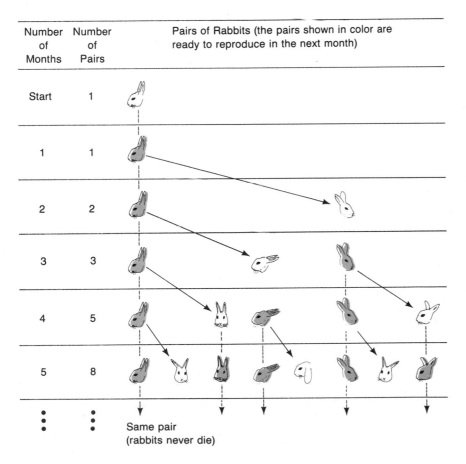

Number of Months	Number of Pairs	Pairs of Rabbits (the pairs shown in color are ready to reproduce in the next month)
Start	1	
1	1	
2	2	
3	3	
4	5	
5	8	
⋮	⋮	Same pair (rabbits never die)

Instead of continuing in this fashion, Leonardo looked for a pattern:

$$1, 1, 2, 3, 5, 8, 13, ?$$

Do you see a pattern for this sequence? (For each term, consider the sum of the two preceding terms.)

Using this pattern, Leonardo was able to compute the number of rabbits alive after ten months (it is the tenth term after the first 1), which is 89. (Continuing with the above sequence: ..., 8, 13, 21, 34, 55, 89, ...) He could also compute the number of rabbits after the first year or after any other interval. Without a pattern, the problem would indeed be a difficult one.

Many mathematicians, such as Verner Hoggatt and Brother Brosseau, have devoted large portions of their careers to the study of Fibonacci numbers. There is even a mathematical society called the Fibonacci Association, which publishes *The Fibonacci Quarterly* and operates a Fibonacci Bibliographical and Research Center.

You may wonder why there is so much interest in the Fibonacci sequence. Some of its applications are far-reaching. It has been used in botany, zoology, physical science, business, economics, statistics, operations research, archeology, fine arts, architecture, education, psychology, sociology, and poetry.* An example of Fibonacci numbers in nature is illustrated by a sunflower. The seeds are arranged in spiral curves as shown in Figure 1.11. If we count the number of clockwise spirals (13 and 21 in this example), they are successive terms in the Fibonacci sequence. This is true of all sunflowers and, indeed, of the seed head of any composite flower such as the daisy or aster.

Closely related to the Fibonacci numbers are the Lucas numbers (after the 19th century French mathematician Lucas):

$$1, 3, 4, 7, 11, 18, 29, \underline{\quad}$$

Can you fill in the blank? [*Hint:* This sequence begins differently but is generated in the same way as the Fibonacci sequence.]

What properties of Lucas or Fibonacci numbers can you discover? These numbers can also be found in Pascal's triangle. Can you find the Fibonacci numbers embedded in Pascal's triangle?

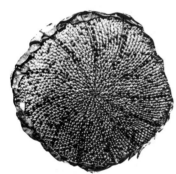

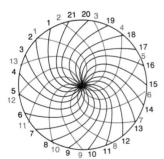

FIGURE 1.11 The arrangement of the pods (phyllotaxy) of a sunflower

Problem Set 1.4

In Problems 1–8, tell which of the given sequences are arithmetic. If the sequence is arithmetic, give the common difference and the next term.

1. 41, 45, 49, 53, ...   **2.** 8, 18, 28, 38, ...

3. 19, 25, 31, 37, ...   **4.** 17, 24, 31, 37, ...

5. 34, 51, 68, 84, ...   **6.** 83, 102, 121, 142, ...

7. 119, 122, 125, 128, ...   **8.** 119, 222, 325, 428, ...

*For more information you can write to the Fibonacci Association, San Jose State University, San Jose, CA 95114 or refer to Verner Hoggatt's *Fibonacci and Lucas Numbers* (Boston: Houghton Mifflin, 1969).

In Problems 9–16, tell which of the given sequences are geometric. If the sequence is geometric, give the common ratio and the next term.

9. 2, 6, 18, 54, . . .

10. 2, 12, 72, 432, . . .

11. 1, 4, 16, 64, . . .

12. 3, 18, 108, 654, . . .

13. 6, 12, 24, 96, . . .

14. 3, 15, 45, 215, . . .

15. 5, 55, 555, 5555, . . .

16. 1, 10, 100, 1000, . . .

Classify the sequences in Problems 17–28 as arithmetic, geometric, both, or neither. If the sequence is arithmetic, give the common difference; if it is geometric, give the common ratio; and if it is neither, describe the pattern. Give the next term.

17. 2, 5, 8, 11, 14, . . .

18. 1, 2, 1, 1, 2, 1, 1, 1, 2, 1, 1, 1, . . .

19. 1, 2, 4, 7, 11, 16, . . .

20. 1, 3, 4, 7, 11, 18, 29, . . .

21. 3, 6, 12, 24, 48, . . .

22. 5, 15, 45, 135, 405, . . .

23. 100, 99, 97, 94, 90, . . .

24. 1, 4, 9, 16, 25, . . .

25. 2, 5, 2, 5, 5, 2, 5, 5, 5, . . .

26. 5, 15, 25, 35, . . .

27. A, A, B, B, C, Ɔ, D, . . .

28. A, C, E, G, I, K, . . .

29. What are the Fibonacci numbers?

30. Write out the first ten terms of the Lucas sequence.

Mind Bogglers

Fill in the missing terms in Problems 31–35.

31. 225, 625, 1225, 2025, _____

32. 1, 8, 27, 64, 125, _____

33. 8, 5, 4, 9, 1, _____

34. A, E, F, H, I, K, L, M, N, _____

35. dog, three, hippopotamus, twelve, lion, four, tiger, _____

36. The number of blocks in each row of the pattern shown in the following figure is arithmetic:

Row 1:   1
Row 2:   6
Row 3:   11
Row 4:   16

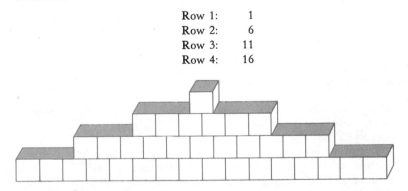

How many blocks in this stack?

How many blocks are there in row 10 if the same pattern is continued?

37. The total number of blocks shown in the block figure for the preceding problem is 34. Use patterns to find the number of blocks needed to build 10 rows of a similar stack.

38. The chain letter in the *Peanuts* cartoon in this section showed how many letters are to be sent for each mailing:

Actually, the total number of letters sent is the number of letters in a particular mailing *plus* the number of letters previously sent.

Thus, 1st mailing: 6
       2nd mailing: 6 + 36 = 42
       3rd mailing: 6 + 36 + 216 = 258

Use patterns to determine the total number of letters sent in five mailings of the chain letter.

39. Repeat Problem 38 for ten mailings.

40. *A visual sequence.* Customizing old automobiles is a very big pastime these days. In fact, enthusiasts have formed thousands of car clubs in Los Angeles, Phoenix, and elsewhere in the West and Southwest.

   Unfortunately, the person who souped up the car shown on the facing page overdid it on the chrome and accessories, and has been suspended from his club for an appalling lack of taste. Each step in the remodeling is shown, starting with the original automobile in the picture labeled j. Can you put the fourteen other pictures in correct sequence?

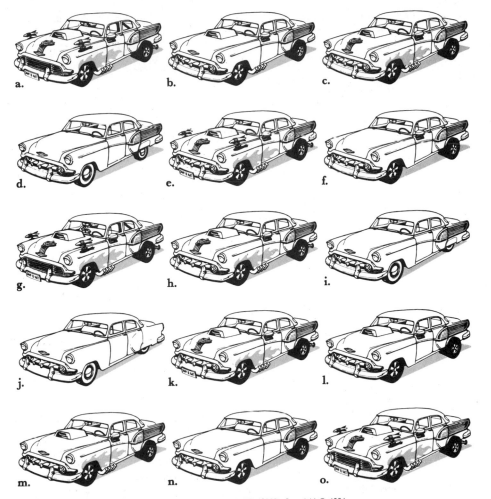

a.   b.   c.

d.   e.   f.

g.   h.   i.

j.   k.   l.

m.   n.   o.

Reprinted from GAMES Magazine (515 Madison Avenue, New York, NY 10022). Copyright © 1981.

41. *Fibonacci magic trick:*

    a. The magician asks the audience for any two numbers (you can limit it to the counting numbers between 1 and 10 to keep the arithmetic manageable).

    b. Add these to obtain a third number.

    c. Add the second and third numbers to obtain a fourth.

    d. Continue until ten numbers are obtained.

    e. Ask the audience to add the ten numbers, while you instantly give the sum.

EXAMPLE:

	Numbers		Solution
a.	5	(1)	n
	9	(2)	m
b.	14	(3)	$n + m$
c.	23	(4)	$n + 2m$
d.	37	(5)	$2n + 3m$
	60	(6)	$3n + 5m$
	97	(7)	$5n + 8m$
	157	(8)	$8n + 13m$
	254	(9)	$13n + 21m$
	411	(10)	$21n + 34m$
e.	1067		$55n + 88m$
			$= 11(5n + 8m)$

The trick depends on the magician's ability to multiply quickly and mentally by 11. Consider the following pattern of multiplication by 11:

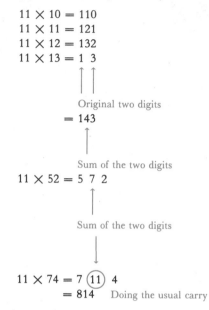

$$11 \times 10 = 110$$
$$11 \times 11 = 121$$
$$11 \times 12 = 132$$
$$11 \times 13 = 1 \quad 3$$

Original two digits

$$= 143$$

Sum of the two digits

$$11 \times 52 = 5 \quad 7 \quad 2$$

Sum of the two digits

$$11 \times 74 = 7 \enspace (11) \enspace 4$$
$$= 814 \quad \text{Doing the usual carry}$$

For the example in the margin on page 35,

$$11 \times 97 = 9 \enspace (16) \enspace 7$$
$$= 1067$$

Explain how this trick works, and why I called it a Fibonacci magic trick.

Problem for Individual Study

42. *Fibonacci numbers.* Write a short paper about Fibonacci numbers. You might like to check with *The Fibonacci Quarterly,* particularly "A Primer on the Fibonacci Sequence," Parts I and II, in the February and April 1963 issues. The articles were written by Verner Hoggatt and S. L. Basin. See also a booklet by Verner Hoggatt, Jr., *Fibonacci and Lucas Numbers,* published in the Houghton Mifflin Mathematics Enrichment Series, 1969. An interesting article for teachers, "Fibonacci Numbers and the Slow Learner," by James Curl, appeared in the October 1968 issue (vol. 6, no. 4) of *The Fibonacci Quarterly.*

*1.5 A Golden Pattern

In the last section, the Fibonacci sequence

$$1, 1, 2, 3, 5, 8, 13, 21, 34, 55, 89, 144, \ldots$$

was introduced. Suppose we consider the ratios of the successive terms of this sequence:

*This is an optional section.

Historical Note

Leonardo Fibonacci (about 1175–1250) was also known as Leonardo da Pisa. Because his father was a customs manager,

$$\frac{1}{1} = 1.000 \qquad \frac{2}{1} = 2.000$$

$$\frac{3}{2} = 1.500 \qquad \frac{5}{3} \approx 1.667$$

$$\frac{8}{5} = 1.600 \qquad \frac{13}{8} = 1.625$$

$$\frac{21}{13} \approx 1.616 \qquad \frac{34}{21} \approx 1.619$$

$$\frac{55}{34} \approx 1.618 \qquad \frac{89}{55} \approx 1.618$$

If you continue finding these ratios, you will notice that the sequence oscillates about a number approximately equal to 1.618.

Suppose we repeat this same procedure with another sequence, this time choosing *any* two nonzero numbers, say 4 and 7. Construct a sequence by adding terms as was done with the Fibonacci sequence:

$$4, 7, 11, 18, 29, 47, 76, 123, 199, 322, \ldots$$

Next, form the ratios of the successive terms:

$$\frac{7}{4} = 1.750; \quad \frac{11}{7} \approx 1.571; \quad \frac{18}{11} \approx 1.636; \quad \frac{29}{18} \approx 1.611;$$

$$\frac{47}{29} \approx 1.621; \quad \frac{76}{47} \approx 1.617; \quad \frac{123}{76} \approx 1.618; \quad \frac{199}{123} \approx 1.618$$

These ratios are oscillating about the same number! There would seem to be something interesting about this number, which is called the **golden ratio** and is sometimes denoted by τ ($\tau \approx 1.6180339885$). This number has been given a special name and symbol because of its frequency of occurrence.

Many everyday rectangular objects have a length-to-width ratio of about 1.6:1 as illustrated in Figure 1.12. Psychologists have tested individuals to determine the rectangles they find most pleasing; the results are those rectangles whose length-to-width ratios are near the golden ratio. Such rectangles are called **golden rectangles.**

Fibonacci visited a number of Eastern and Arabic cities, where he became interested in the Hindu-Arabic numeration system we use today. In Europe at that time, Roman numerals were used for calculation, so Fibonacci wrote *Liber Abaci*, in which he strongly advocated the use of the Hindu-Arabic numeration system. In the book *In Mathematical Circles,* Howard Eves tells the following story: "Fibonacci sometimes signed his work with the name *Leonardo Bigollo*. Now *bigollo* has more than one meaning; it means both 'traveler' and 'blockhead.' In signing his work as he did, Fibonacci may have meant that he was a great traveler, for so he was. But a story has circulated that he took pleasure in using this signature because many of his contemporaries considered him a blockhead (for his interest in the new numbers), and it pleased him to show these critics what a blockhead could accomplish."

From *In Mathematical Circles* by Howard Eves (Boston: Prindle, Weber, & Schmidt, 1969), p. 91.

FIGURE 1.12   A 15 oz box of Kellogg's Sugar Corn Pops is 30 cm $\times$ 19 cm, for a ratio of 1.6. A 1 lb box of C & H sugar is 17 cm $\times$ 10 cm, for a ratio of 1.7.

THE GOLDEN RATIO IN NATURE

1. Begin with any square and draw a quarter circle as shown.

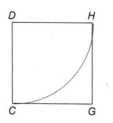

2. Form a golden rectangle (as shown in Figure 1.13).

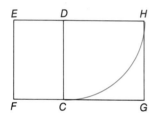

3. Draw a square within the new rectangle *CDEF*; draw a semicircle as shown.

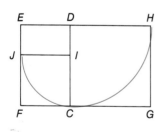

4. Repeat the process.

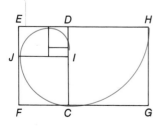

A golden rectangle is easy to construct with a straightedge and a compass. Begin with any square (*CDHG* in Figure 1.13). Divide the square into two equal parts (line *AB* in Figure 1.13). Draw an arc, with the center at *A* and a radius of *AC*, so that it intersects the extension of side *AD*; label this point *E*. Now side *EF* can be drawn. The resulting rectangle *EFGH* is a golden rectangle; *CDEF* is also a golden rectangle.

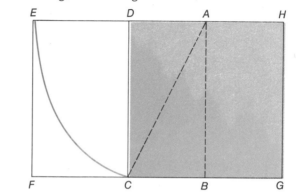

FIGURE 1.13 Constructing a golden rectangle *EFGH*. Another golden rectangle is *CDEF*.

This golden ratio is important to artists in a technique known as "dynamic symmetry." Albrecht Dürer, Leonardo da Vinci, Georges Seurat, George Bellows, and Pieter Mondriaan all studied the golden rectangle as a means of creating dynamic symmetry in their work (see Figure 1.14). In fact, Mondriaan is said to have approached every canvas in terms of the golden ratio.

FIGURE 1.14 *Bathers* (1883–1884) by the French impressionist Georges Seurat. Three successive golden rectangles are shown. Can you find others?

Many studies of the human body itself involve the golden ratio (remember $\tau \approx 1.62$). Figure 1.15 shows a drawing of an idealized athlete.

5. The resulting curve is called a logarithmic spiral and resembles a chambered nautilus.

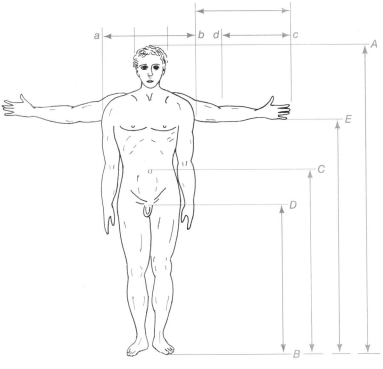

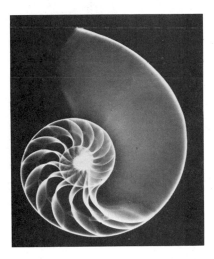

FIGURE 1.15   Proportions of the human body

Let $AC =$ Distance from the top of the head to the navel
$CB =$ Distance from the navel to the floor
$AB =$ Height

Then,

$$CB \div AC \approx \tau \qquad \text{and} \qquad AB \div CB \approx \tau$$

Also, let $ab =$ Shoulder width
$bc =$ Arm length

Then,

$$bc \div ab \approx \tau$$

See if you can find other ratios on the human body approximating τ. Figure 1.16 (page 40) shows a study of the human face by da Vinci, in which the rectangles approximate golden rectangles.

This chapter is concerned with the nature of inductive reasoning, and the golden ratio can provide the point of departure for a wide range of problems. The following problem set will give you an opportunity to pursue some of these directions by using inductive reasoning.

The dimensions of the Parthenon at Athens have a length-to-width ratio that almost exactly equals the golden ratio.

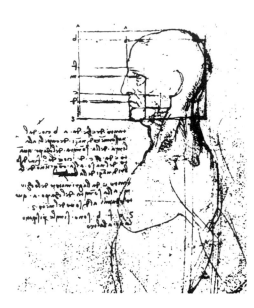

Dynamic symmetry of a human face. A study of golden rectangles by Leonardo da Vinci.

FIGURE 1.16

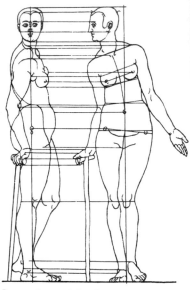

Proportions of the human body by Albrecht Dürer.

David (1501–1504) by Michelangelo illustrates many golden ratios.

Problem Set 1.5

The numbers in the Fibonacci sequence are called Fibonacci numbers, and mathematicians use the notation F_1 to represent the first Fibonacci number, F_2 for the second, F_3 for the third, and so on. Thus,

$$F_1 = 1, \quad F_2 = 1, \quad F_3 = 2, \quad F_4 = 3, \quad F_5 = 5, \quad F_6 = 8, \ldots$$

A table of the first forty Fibonacci numbers is shown in Table 1.1. Use this table to find the numbers requested in Problems 1–10.

1. **a.** F_7 **b.** F_{16} **c.** F_{35} **d.** F_{11}

2. **a.** F_{10} **b.** F_8 **c.** F_{38} **d.** F_{24}

3. **a.** F_9 **b.** F_{19} **c.** F_{31} **d.** F_{27}

4. **a.** F_{12} **b.** F_{15} **c.** F_{39} **d.** F_{26}

5. **a.** F_{13} **b.** F_{18} **c.** F_{33} **d.** F_{25}

6. **a.** F_{41} **b.** F_{42}

7. **a.** $F_1 =$ _____   **b.** $F_1 + F_3 =$ _____
   c. $F_1 + F_3 + F_5 =$ _____   **d.** $F_1 + F_3 + F_5 + F_7 =$ _____

8. **a.** $F_2 =$ _____   **b.** $F_2 + F_4 =$ _____
   c. $F_2 + F_4 + F_6 =$ _____   **d.** $F_2 + F_4 + F_6 + F_8 =$ _____

9. **a.** $F_1 + F_2 =$ _____   **b.** $F_1 + F_2 + F_3 =$ _____
   c. $F_1 + F_2 + F_3 + F_4 =$ _____   **d.** $F_1 + F_2 + F_3 + F_4 + F_5 =$ _____

10. **a.** $F_1 + F_2 + F_3 =$ _____   **b.** $F_2 + F_3 + F_4 =$ _____
    c. $F_3 + F_4 + F_5 =$ _____   **d.** $F_4 + F_5 + F_6 =$ _____

11. If $F_1 \times F_2$ is represented by F_1 $F_2 \times F_3$ by F_3

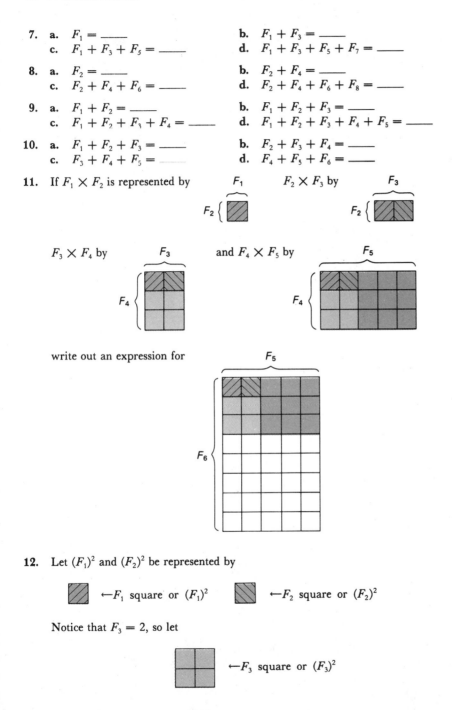

  $F_3 \times F_4$ by F_3 and $F_4 \times F_5$ by F_5

write out an expression for F_5

12. Let $(F_1)^2$ and $(F_2)^2$ be represented by

    ← F_1 square or $(F_1)^2$ ← F_2 square or $(F_2)^2$

Notice that $F_3 = 2$, so let

    ← F_3 square or $(F_3)^2$

How would you represent $(F_4)^2$?

13. Repeat Problem 12 for $(F_5)^2$.

TABLE 1.1   The First Forty
Fibonacci Numbers

n	F_n
1	1
2	1
3	2
4	3
5	5
6	8
7	13
8	21
9	34
10	55
11	89
12	144
13	233
14	377
15	610
16	987
17	1,597
18	2,584
19	4,181
20	6,765
21	10,946
22	17,711
23	28,657
24	46,368
25	75,025
26	121,393
27	196,418
28	317,811
29	514,229
30	832,040
31	1,346,269
32	2,178,309
33	3,524,578
34	5,702,887
35	9,227,465
36	14,930,352
37	24,157,817
38	39,088,169
39	63,245,986
40	102,334,155

Using Figure 1.15, find the requested measurements using your own body as the model.

14. $CB \div AC$

15. $AB \div CB$

16. $AD \div AE$

17. $bc \div ab$

18. $ac \div bc$

19. $dc \div bd$

20. a. Pick any two nonzero numbers. Add them to obtain a third number. Construct a sequence of numbers in the Fibonacci fashion by adding the two previous terms to obtain a new term.

    b. Form the ratios of the successive terms, and show that after a while they oscillate around the golden ratio.

21. Repeat Problem 20 for two other nonzero numbers.

22. On a planet far, far away, Luke finds himself in a strange building with hexagon-shaped rooms. In his search for the princess, he always moves to an adjacent room and always in a southerly direction.

    a. How many paths are there to room 1? room 2? room 3? room 4?

    b. How many paths are there to room 10?

    c. How many paths are there to room n?

23. Draw a logarithmic spiral by constructing golden rectangles, as shown in Figure 1.13. Use graph paper and begin with a square with sides measuring 34 units.

Mind Bogglers

24. Make a conjecture about $F_2 + F_4 + \cdots + F_{2k}$ using Problem 8.

25. Make a conjecture about $F_1 + F_3 + F_5 + \cdots + F_{2k-1}$ using Problem 7.

26. Make a conjecture about $F_1 + F_2 + \cdots + F_k$ using Problem 9.

Problem for Individual Study

27. Do some research on the length-to-width ratios of the packaging of common household items. Form some conclusions. Find some examples of the golden ratio in art. Do some research on dynamic symmetry.

    References: Philip J. Davis and Reuben Hersh, *The Mathematical Experience* (Boston: Houghton Mifflin Company, 1981), pp. 168–171.

               H. E. Huntley, *The Divine Proportion: A Study in Mathematical Beauty* (New York: Dover Publications, 1970).

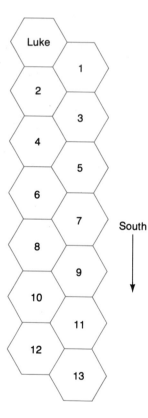

Chapter 1 Outline

  I. Problem Solving

    A. Polya's problem-solving model

      1. Understand the problem.

      2. Develop a plan.

      3. Carry out the plan.

      4. Examine the solution.

    B. Number of paths on a map

      1. Counting

      2. Pascal's triangle

The answers to all of the questions on the chapter tests are given in the back of the book. Check yourself before taking a test in class.

Chapter 1 Test

1. In your own words describe Polya's problem-solving model.

2. A chessboard consists of 64 squares, as shown in the figure in the margin. The rook can move one or more squares horizontally or vertically. Suppose that a rook is in the upper left-hand corner of a chessboard. In how many ways can the rook move to the lower right-hand corner? Assume that the rook always moves toward its destination. [*Hint:* Turn the chessboard so that it forms a triangle with the rook at the top.]

A chessboard problem

*Optional section.

3. Consider the magic square shown here.

21	11	2	6	25
4	18	9	12	22
16	7	13	19	10
23	14	17	8	3
1	15	24	20	5

  a. Fill in the blanks in Figure a using a rotation.
  b. Fill in the blanks in Figure b using a reflection.
  c. Fill in the blanks in Figure c using either a rotation or a reflection.

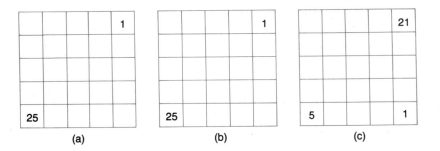

      (a)   (b)   (c)

4. Complete a 5-by-5 magic square by starting with a 2. Compare your answer with the Problem 3 figure in the margin. Is it a rotation of a reflection, or is it a different magic square?

5. Describe the process of inductive reasoning in your own words.

6. Compute 111,111,111 × 111,111,111. Do not use direct multiplication; show all your work.

7. Simplify:

  a. $6 + 2 \times 3$ **b.** $5 \times 100 + 4 \times 10 + 3 \times 1$

8. Complete the given sequences.

  a. 10,000, 1,000, 100, _____
  b. 42, 32, 22, _____
  c. 98, 87, 76, _____

9. **a.** One of the sequences in Problem 8 is a geometric progression. Which one is it?
    b. What is its common ratio?

***10.** **a.** What is F_{10}?
b. How are the numbers F_1, F_2, F_3, F_4, . . . related to τ?

Bonus Question

11. Suppose you multiply

$$9 \times 9 \times 9 \times \cdots \times 9$$

so that there are 1,000 factors. What is the last digit of the product?

*Requires optional Section 1.5.

Date	Cultural History
500 BC	500 BC: Pindar's *Odes*
	480 BC: Siddhartha, the Buddha, delivers his sermons in Deer Park
400 BC	
	323 BC: Alexander the Great completes his conquest of the known world
300 BC	
	218 BC: Hannibal crosses the Alps
200 BC	200 BC: Rosetta Stone engraved
100 BC	100 BC: Birth of Julius Caesar
	20 BC: Virgil's *Aeneid*
0	0: Birth of Christ
100 AD	
200 AD	200 AD: Goths invade Asia Minor
300 AD	324 AD: Founding of Constantinople
400 AD	400 AD: Augustine's *Confessions*
	476 AD: Fall of Rome
500 AD	500 AD: First plans of the Vatican Palace in Rome
600 AD	610 AD: Mohammed's vision
	697 AD: Northern Irish submit to Roman Catholicism
700 AD	

Armored elephant such as those used by Hannibal

St. Augustine

The Koran

Mathematical History

500 BC: Sulvasūtras—Pythagorian numbers
450 BC: Zeno—paradoxes of motion
425 BC: Theodorus of Cyrene—
 irrational numbers
380 BC: Plato's Academy—logic
340 BC: Aristotle—deductive logic
323 BC: Euclid—geometry, perfect numbers

Aristotle and Plato,
Raphael

230 BC: Sieve of Eratosthenes
225 BC: Archimedes—circle, pi, curves, series

180 BC: Hypsicles—number theory

Archimedes

60 BC: Geminus—parallel postulate

50 AD: Negative numbers used in China
75 AD: Heron—measurements, roots,
 surveying

100 AD: Nichomachus—number theory

150 AD: Ptolemy—trigonometry

200 AD: Mayan calendar

250 AD: Diophantus—number theory,
 algebra

Mayan calendaric
numbering

300 AD: Pappus, *Mathematical Collection*

410 AD: Hypatia of Alexandria—first woman
 mentioned in history of mathematics

480 AD: Tsu Ch'ung-chi approximates
 π as $\frac{355}{113}$

Hypatia

628 AD: Brahmagupta—algebra, astronomy

710 AD: Bede—calendar, finger arithmetic

Finger symbols

2

The Nature of Sets and Deductive Reasoning

. . . the two great components of the critical movement, though distinct in origin and following separate paths, are found to converge at last in the thesis: Symbolic Logic is Mathematics, Mathematics is Symbolic Logic, the twain are one.

C. J. Keyser

2.1 Sets

A flock of birds is an everyday example of the mathematical concept of a set.

A fundamental concept in mathematics—and in life for that matter—is the sorting of objects into certain similar groupings. Every language has an abundance of words that mean "a collection" or "a grouping." For example, we speak of a *herd* of cattle, a *flock* of birds, a *school* of fish, a track *team,* a stamp *collection,* and a *set* of dishes. All these grouping words serve the same purpose, and in mathematics we use the word **set** to refer to any collection of objects.

It should be pointed out that "set" is what is called an undefined term. We do not define "set," since an attempt to define it would be circular or would require that we accept other undefined terms.

Set: A collection of objects

Collection: An accumulation

Accumulation: Collection or pile or heap

Pile: A heap

Heap: A pile

The fact that we do not define "set" does not keep us from having an intuitive grasp on how to use the word; it simply means that we do not formally define the concept "set." The objects in a set are called **members** or **elements** of the set and are said to *belong to* or *be contained in* the set.

The concept of sets and set theory is attributed to Georg Cantor (1845–1918), a great German mathematician. Cantor's ideas were not adopted as fundamental underlying concepts in mathematics until the 20th century, but today the idea of set is used to unify and explain nearly every other concept in mathematics. Sets are specified by **description** or by **roster.** In the roster method, sets are specified by listing the elements and enclosing those elements between braces: { }.

If S is a set, we write $a \in S$ if a is a member of the set S, and $b \notin S$ if b is not a member of the set S. For example, let C be the set of cities in California. If a represents the city of Anaheim and b stands for the city of Berlin, we would write

$$a \in C \qquad \text{and} \qquad b \notin C$$

Two special sets are necessary to consider. The first is the set that contains every element under consideration, and the second is the set that contains no elements.

Definition

The **empty set** contains no elements and is denoted by { } or $\varnothing$. The **universal set** contains all the elements under consideration in a given discussion and is denoted by U.

For example, if $U = \{1,2,3,4,5,6,7,8,9\}$, then all sets that we would be considering would have elements only among the elements of U. No set could contain the number 10, since 10 is not included in what has been agreed, at the outset of the discussion, to be the universe. For every problem, a universal set must be specified or implied, and it must remain fixed for that problem. However, when a new problem is begun, a new universal set can be specified.

To deal effectually with sets it is necessary to understand certain relationships among sets.

Definition

A set A is a **subset** of a set B, denoted by $A \subseteq B$, if every element of A is an element of B.

Historical Note

Georg Cantor (1845–1918) was the originator of set theory and of the study of transfinite numbers. When he first published his paper on the theory of sets in 1874, it was very controversial because it was innovative and because it differed from current mathematical thinking. One of Cantor's former teachers, Leopold Kronecker (1823–1891), was particularly strong in his criticism not only of Cantor's work but also of Cantor himself. Cantor believed that "The essence of mathematics lies in its freedom," but, in the final analysis, he was not able to withstand the pressure; he underwent a series of mental breakdowns and died in a mental hospital in 1918.

Consider the following sets:

$$U = \{1,2,3,4,5,6,7,8,9\}$$
$$A = \{2,4,6,8\}$$
$$B = \{1,3,5,7\}$$
$$C = \{5,7\}$$

Now, A, B, and C are subsets of the universal set U (all sets we consider are subsets of the universe, by definition of the universal set). $C \subseteq B$, since every element in C is also an element of B.

However, C is not a subset of A. To substantiate this claim, we must show that there is some member of C that is not a member of A. In this case, we merely note that $5 \in C$ and $5 \notin A$.

Do not confuse the notions of "element" and "subset." That is, 5 is an *element* of C, since it is listed in C; $\{5\}$ is a *subset* of C but is not an element since we do not find $\{5\}$ contained in C—if we did, C might look like this: $\{5,\{5\},7\}$.

Consider all possible subsets of the set C:

$\{5\}$ This is a subset, since $5 \in C$.

$\{7\}$ This is a subset, since $7 \in C$.

$\{5,7\}$ This is a subset, since both 5 and 7 are elements of C.

$\{\ \}$ This is a subset, since all of its elements belong to C. Stated a different way, if it were not a subset of C, then we would have to find an element of $\{\ \}$ that is not in C. Since we could not find such an element, we say that the empty set is a subset of C (and, in fact, of every set). Sometimes we name $\{\ \}$ by using the symbol $\varnothing$. That is, $\varnothing = \{\ \}$.

We see that there are four subsets of C but that C has only two elements. Certainly, then, we must be careful to distinguish between a subset and an element—remember that 5 and {5} mean different things.

The subsets of *C* can be classified into two categories, called *proper* and *improper,* as shown in the margin. Every set has but one improper subset, and that is the set itself. Verify that set *A* has 15 proper subsets and 1 improper subset. The improper subset of *A* is {2,4,6,8}. Some of the proper subsets are the following:

$$\{\ \}, \quad \{2\}, \quad \{4\}, \quad \{6\}, \quad \{8\}, \quad \{2,4\}, \quad \{2,6\}, \quad \{2,8\}, \quad \dots$$

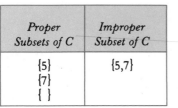

Proper Subsets of C	Improper Subset of C
{5} {7} { }	{5,7}

Definition

Sets *A* and *B* are equal, denoted by $A = B$, if they have exactly the same elements.

A useful way to depict relationships between sets is to let the universal set be represented by a rectangle, with the proper sets in the universe represented by circular or oval-shaped regions, as shown in Figure 2.1.

These figures are called **Venn diagrams,** after John Venn (1834–1923). The Swiss mathematician Leonhard Euler (1707–1783) also used circles to illustrate principles of logic, so sometimes these diagrams are called **Euler circles.** However, Venn was the first person to use them in a general way.

We can also illustrate other relationships between two sets: *A* and *B* may have no elements in common, in which case they are **disjoint,** or they may be overlapping sets having some elements in common.

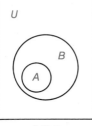

$A \subseteq B$

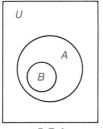

$B \subseteq A$

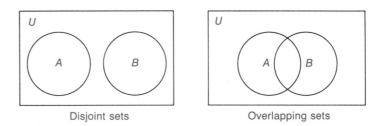

Disjoint sets   Overlapping sets

FIGURE 2.2 Venn diagrams for disjoint and overlapping sets

Sometimes we are given two sets *X* and *Y*, and we know nothing about the way they are related. In this situation we draw a general figure, such as the one shown in Figure 2.3 on page 51.

Notice that these circles divide the universe into four disjoint regions. When we draw the sets in this manner, we do not mean to imply that the only possibility is overlapping sets.

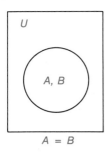

$A = B$

FIGURE 2.1 Venn diagrams for subset and equal relationships

If $X \subseteq Y$, then region 2 is empty.

If $Y \subseteq X$, then region 4 is empty.

If $X = Y$, then regions 2 and 4 are both empty.

If X and Y are disjoint, then region 3 is empty.

Three common operations are performed on sets: union, intersection, and complementation.

Union is an operation for sets A and B in which a set is formed that consists of all the elments that are in A or B or both. The symbol for the operation of union is $\cup$, and we write $A \cup B$.

FIGURE 2.3   Venn diagram showing two general sets

Definition

The **union** of sets A and B, denoted by $A \cup B$,   is the set consisting of all elements of A or B or both.

Example 1:   Let $U = \{1,2,3,4,5,6,7,8,9\}$
$A = \{2,4,6,8\}$
$B = \{1,3,5,7\}$
$C = \{5,7\}$

a. $A \cup C = \{2,4,5,6,7,8\}$.   That is, the union of A and C is that set consisting of all elements in A or in C or in both.

b. $B \cup C = \{1,3,5,7\}$.   Notice that, even though the elements 5 and 7 appear in both sets, they are listed only once. That is, the sets $\{1,3,5,7\}$ and $\{1,3,5,5,7,7\}$ are equal (exactly the same).

c. $A \cup B = \{1,2,3,4,5,6,7,8\}$.

d. $(A \cup B) \cup \{9\} = \{1,2,3,4,5,6,7,8,9\} = U$.   Here we are considering the union of three sets. However, the parentheses indicate the operation that should be performed first. Notice also that, because the solution $\{1,2,3,4,5,6,7,8,9\}$ has a name, we write down the name rather than the set. That is, $(A \cup B) \cup \{9\} = U$ is the simpler representation.

We can use Venn diagrams to illustrate union. In Figure 2.4, we first shade A and then shade B. *The union is all parts that have been shaded at least once.* A second operation for sets is called intersection.

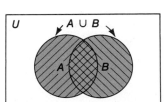

FIGURE 2.4   Venn diagram showing the union of two sets (shaded region)

Definition

The **intersection** of sets A and B, denoted by $A \cap B$,   is the set consisting of all elements common to A and B.

Example 2: Let $U = \{a,b,c,d,e\}$
$$A = \{a,c,e\}$$
$$B = \{c,d,e\}$$
$$C = \{a\}$$
$$D = \{e\}$$

a. $A \cap B = \{c,e\}$.   That is, the intersection of A and B is that set consisting of elements in both A and B.

b. $A \cap C = C$, since $A \cap C = \{a\}$ and $\{a\} = C$; we write down the name for the set that is the intersection.

c. $B \cap C = \varnothing$, since B and C have no elements in common (they are disjoint).

d. $(A \cap B) \cap D = \{c,e\} \cap \{e\}$
$$= \{e\}$$

The parentheses tell us which operation to do first; $\{e\}$ is the set of elements common to the sets $\{c,e\}$ and $D = \{e\}$.

Intersection of sets can also be easily shown using a Venn diagram, as shown in Figure 2.5. To find the intersection of two sets A and B, first shade A and then shade B; *the intersection is all parts shaded twice.*

Complementation is an operation on a set that must be performed in reference to the universal set.

FIGURE 2.5 Venn diagram showing the intersection of two sets (shaded region)

Definition

The **complement** of a set A, denoted by $\overline{A}$, is the set of all elements in U that are not in the set A.

Example 3: Let $U = \{\text{People in California}\}$
$$A = \{\text{People who are over 30}\}$$
$$B = \{\text{People who are 30 or under}\}$$
$$C = \{\text{People who own a car}\}$$

a. $\overline{C} = \{\text{Californians who do not own a car}\}$.

b. $\overline{A} = B$ and $\overline{B} = A$.

c. $\overline{U} = \varnothing$ and $\overline{\varnothing} = U$.

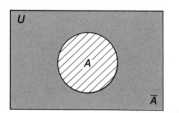

FIGURE 2.6 Venn diagram showing the complement of a set A (shaded region)

Complementation, too, can be shown using a Venn diagram. The unshaded part of Figure 2.6 shows the complement of A. In a Venn diagram, *the complement is everything in U that is not in the set under consideration* (in this case, everything not in A).

Problem Set 2.1

1. Explain the difference between the terms *element of a set* and *subset of a set*. Give examples.

2. Explain the difference between the terms *subset* and *proper subset*.

3. Explain the difference between *universal set* and *empty set*.

4. In your own words, define *union, intersection,* and *complement*. Illustrate with examples.

List all subsets of each set given in Problems 5–10.

5. $\{m,y\}$ 6. $\{4,5\}$ 7. $\{y,o,u\}$

8. $\{3,6,9\}$ 9. $\{m,a,t,h\}$ 10. $\{1,2,3,4\}$

Perform the given set operations in Problems 11–20. Let $U = \{1,2,3,4,5,6,7,8,9,10\}$

11. $\{2,6,8\} \cup \{6,8,10\}$ 12. $\{2,6,8\} \cap \{6,8,10\}$

13. $\{1,2,3,4,5\} \cap \{3,4,5,6,7\}$ 14. $\{1,2,3,4,5\} \cup \{3,4,5,6,7\}$

15. $\{2,5,8\} \cup \{3,6,9\}$ 16. $\{2,5,8\} \cap \{3,6,9\}$

17. $\overline{\{2,8,9\}}$ 18. $\overline{\{1,2,5,7,9\}}$

19. $\overline{\{8\}}$ 20. $\overline{\{6,7,8,9,10\}}$

Let $U = \{1,2,3,4,5,6,7\}$ $A = \{1,2,3,4\}$ $B = \{1,2,5,6\}$ $C = \{3,5,7\}$

List all the members of each of the sets in Problems 21–29.

21. $A \cup B$ 22. $A \cup C$ 23. $B \cup C$

24. $A \cap B$ 25. $A \cap C$ 26. $B \cap C$

27. $\overline{A}$ 28. $\overline{B}$ 29. $\overline{C}$

Draw Venn diagrams for each of the relationships in Problems 30–35.

30. $X \cup Y$ 31. $X \cup Z$ 32. $X \cap Y$

33. $X \cap Z$ 34. $\overline{X}$ 35. $\overline{Y}$

Mind Bogglers

36. a. List all possible subsets of the set $A = \varnothing$.
     b. Repeat part a for $B = \{1\}$.
     c. Repeat part a for $C = \{1,2\}$.
     d. Repeat for $D = \{1,2,3\}$.
     e. Repeat for $E = \{1,2,3,4\}$.
     f. Look for a pattern. Can you guess how many subsets the set $F = \{1,2,3,4,5\}$ has?

37. Using Problem 36, make a conjecture about the number of subsets that can be formed from a set consisting of n elements.

THE EMPTY SET

Several years ago a fellow student showed me an interesting quotation from Don Marquis' *Archy and Mehitabel* (The Merry Flea). Only recently I found it again, and it seems quite appropriate when first introducing the idea of the empty set. Mr. Marquis writes:

"What is there in a vacuum to make one afraid?" said the flea. "There is nothing in it," I said, "and that is what makes one afraid to contemplate it. A person can't think of a place with nothing at all in it without going nutty, and if he tries to think that nothing is something after all, he gets nuttier."

Archy's analysis of this situation is not much different than that of some students I've known. The following little quiz also proves to add to their bewilderment.

True—False

1. _____ $\varnothing \in \varnothing$
2. _____ $\varnothing \subseteq \varnothing$
3. _____ $\varnothing \in \{\varnothing\}$
4. _____ $\varnothing = \varnothing$
5. _____ $\varnothing = \{\varnothing\}$
6. _____ $\varnothing \subseteq \{\varnothing\}$
7. _____ $\varnothing \subseteq A$
     for all sets A
8. _____ $\varnothing = 0$
9. _____ $\varnothing = \{0\}$
10. _____ $\varnothing \subseteq \{0\}$

Answers: 1, 5, 8, 9 are false.

Robert L. Walter
The University of South Dakota
Vermillion, South Dakota

2.2 Combined Operations with Sets

In the last section we defined union, intersection, and complementation of sets. However, the real payoff for studying these relationships comes when dealing with combined operations or with several sets at the same time.

Example 1: Illustrate $\overline{A \cup B}$ using a Venn diagram.

Solution: This is a combined operation. *First,* find $A \cup B$, and *then* find the complement.

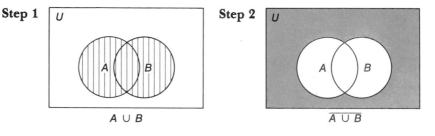

These steps are generally combined into one diagram:

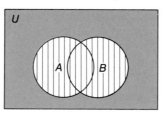

The answer is shown as the shaded color portion; the lines show the intermediate steps. In your own work you will generally find it easier to show the *final answer* (the part shaded in color here) using a second color (pen or highlighter pen).

Example 2: Illustrate $\overline{A} \cup \overline{B}$ using a Venn diagram.

Solution: Compare with Example 1. Here, *first* find $\overline{A}$ and $\overline{B}$, and *then* find the union.

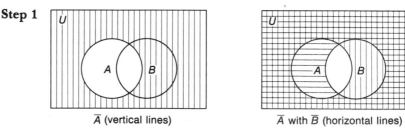

Step 2

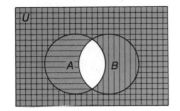

$\overline{A} \cup \overline{B}$ (shaded portion; the union is
all parts that have horizontal
or vertical lines or both)

Your work will show these steps with one diagram.

Notice that $\overline{A \cup B} \neq \overline{A} \cup \overline{B}$. If they were equal, the final *shaded* color portions of the Venn diagrams from Examples 1 and 2 would be the same.

Example 3: Prove $\overline{A \cup B} = \overline{A} \cap \overline{B}$.

Step 1. Draw a Venn diagram for the left side of the equal sign.

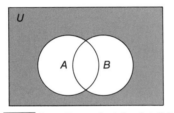

$\overline{A \cup B}$ (see Example 1 for details)

Step 2. Draw a separate Venn diagram for the right side of the equal sign.

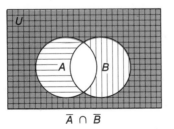

$\overline{A} \cap \overline{B}$

Detail: $\overline{A}$; vertical lines
$\overline{B}$; horizontal lines
$\overline{A} \cap \overline{B}$

      The intersection of horizontal and vertical lines is the shaded portion—it con-
sists of all parts having both horizontal and vertical lines.

Step 3. Compare shaded portions of the two Venn diagrams. They are the same, so we have proved $\overline{A \cup B} = \overline{A} \cap \overline{B}$.

Not only was Augustus De
Morgan a distinguished
mathematician and logician, but
in 1845 he suggested the slanted
line we sometimes use when
writing fractions, such as 1/2 or
2/3. There is a story told about
the time when De Morgan was
explaining to an actuary what
the probability was that a certain
group would be alive at a given
time, and he used a formula
involving π. Recall that π is
defined as the ratio of the
circumference of a circle to its
diameter. The actuary was
astonished and responded: "That
must be a delusion. What can a
circle have to do with the number
of people alive at a given time?"
Of course, actuaries today are
more sophisticated. The number
π is used in a wide variety of
applications. The photo shows
the frontispiece of a book written
by De Morgan in 1838.

The result proved in Example 3 is called **De Morgan's Law.** In the problem
set you are asked to prove the second part of De Morgan's Law.

De Morgan's Laws

$$\overline{X \cup Y} = \overline{X} \cap \overline{Y} \qquad \overline{X \cap Y} = \overline{X} \cup \overline{Y}$$

The *order* of operations for sets is left to right, except if there are parentheses. Operations within parentheses are performed first, as shown by Example 4.

Example 4:   Illustrate $(A \cup C) \cap \overline{C}$ with a Venn diagram.

Solution

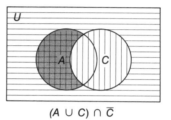

$$(A \cup C) \cap \overline{C}$$

Detail: First draw $A \cup C$ (vertical lines).
Then draw $\overline{C}$ (horizontal lines).
The result is the intersection of the vertical and horizontal parts and is
the shaded portion.

Sometimes you will be asked to consider the relationships among three sets.
The general Venn diagram for this is shown in Figure 2.7. Notice that three
sets divide the universe into eight regions. Can you number each part?

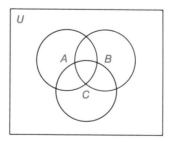

FIGURE 2.7   Three general sets

Example 5:   Using Figure 2.7 as a guide, shade in each of the following sets:

a. $A \cup B$ **b.** $A \cap C$ **c.** $B \cap C$ **d.** $\overline{A}$
e. $\overline{A \cup B}$ **f.** $A \cap B \cap C$ **g.** $A \cup B \cup C$

Solution

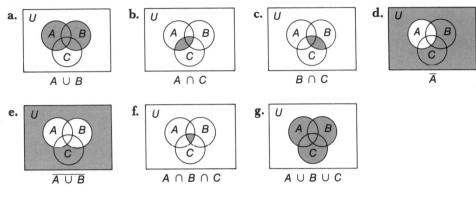

a. $A \cup B$ b. $A \cap C$ c. $B \cap C$ d. $\overline{A}$

e. $\overline{A \cup B}$ f. $A \cap B \cap C$ g. $A \cup B \cup C$

Example 6:   Draw a Venn diagram for $A \cup (B \cap C)$.

Solution: The Venn diagram is shown at the right.

Detail:

Step 1. Parentheses first (vertical lines)
Step 2. Set A (horizontal lines)
Step 3. The union is all parts that show either vertical lines or horizontal lines, or both—this part is shaded.

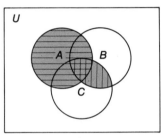

$A \cup (B \cap C)$

Example 7:   Draw a Venn diagram for $\overline{A \cup B} \cap C$.

Solution

Detail:

Step 1. $\overline{A \cup B}$ first (vertical)
Step 2. C (horizontal)
Step 3. The intersection is all parts that show both vertical and horizontal lines—this part is shaded.

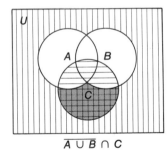

$\overline{A \cup B} \cap C$

Venn diagrams can sometimes be used for survey problems. Suppose a survey indicates that 45 students are taking mathematics and 41 are taking English. How many students are taking math or English? At first, it might seem that all we need to do is add 41 and 45, but such is not the case, as you can see by looking at Figure 2.8a.

Suppose, further, that there are 12 students taking both math and English. In this case we can fill in the number in the intersection first and finish up by using subtraction for the other sections, as shown in Figure 2.8b. There are $33 + 12 + 29 = 74$ students enrolled in mathematics or English.

For three sets, the situation is a little more involved, as illustrated by Exam-

To find out how many students are taking math and English, we need to know the number in the intersection.

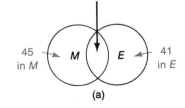

(a)

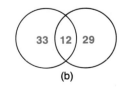

(b)

FIGURE 2.8

ple 8. Remember, the overall procedure is to fill in the number in the innermost set first and work your way out through the Venn diagram by using subtraction.

Example 8: A survey of 100 randomly selected students gave the following information:

45 students are taking mathematics.

41 students are taking English.

40 students are taking history.

15 students are taking math and English.

18 students are taking math and history.

17 students are taking English and history.

7 students are taking all three.

a. How many are taking only mathematics?
b. How many are taking only English?
c. How many are taking only history?
d. How many are not taking any of these courses?

Solution: Draw a Venn diagram and fill in the various regions. Let

$$M = \{\text{Persons taking mathematics}\}$$
$$E = \{\text{Persons taking English}\}$$
$$H = \{\text{Persons taking history}\}$$

Step 1. Fill in the innermost section first—namely, $M \cap E \cap H$; there are 7 in this subset.

Step 2. Fill in the other inner portions by subtraction.

$E \cap H$ is given as 17, but 7 have previously been accounted for, so an additional 10 are added to the Venn diagram (Figure 2.9).

$M \cap H$ is given as 18; fill in an additional 11 members.

$M \cap E$ is given as 15; fill in 8.

Step 3. Fill in the other regions of the Venn diagram.

H is given as 40, but 28 have previously been accounted for, so we need an additional 12 members.

E is given as 41; we need to fill in 16 members.

M is given as 45; we need 19 members.

Step 4. Add all of the numbers in the diagram and see that 83 members have been accounted for. Since 100 students were surveyed, we see that 17 are not taking any of the three courses. We now have all the answers to the questions directly from the Venn diagram: **a.** 19, **b.** 16, **c.** 12, **d.** 17.

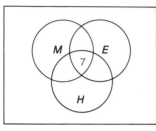

Step 1

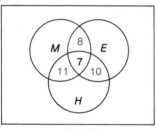

Step 2

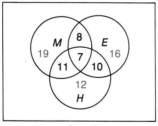

Step 3

FIGURE 2.9   Survey problem

Problem Set 2.2

Draw a Venn diagram to illustrate each relationship given in Problems 1–15.

1. $A \cup \overline{B}$
2. $\overline{A} \cup B$
3. $\overline{A} \cap C$

4. $\overline{B \cap C}$
5. $\overline{A \cap C}$
6. $\overline{A \cup B}$

7. $\overline{B \cup C}$
8. $A \cap (B \cup C)$
9. $A \cup (B \cup C)$

10. $\overline{(A \cup B) \cup C}$
11. $(A \cap B) \cap (A \cap C)$
12. $(A \cap B) \cup (A \cap C)$

13. $\overline{(A \cap B) \cup C}$
14. $A \cap \overline{B} \cup C$
15. $\overline{A \cup B} \cup C$

Let $U = \{1,2,3,4,5,6,7,8,9,10\}$
$A = \{2,4,6,8\}$
$B = \{5,9\}$
$C = \{2,5,8,9,10\}$

List all the members of each set in Problems 16–24.

16. $(A \cup B) \cup C$
17. $(A \cap B) \cap C$
18. $A \cup (B \cap C)$

19. $A \cap (\overline{B} \cup C)$
20. $\overline{A} \cap (B \cup C)$
21. $A \cap \overline{B} \cup C$

22. $\overline{A \cap (B \cup C)}$
23. $\overline{A} \cup (B \cap C)$
24. $\overline{A \cup (B \cap C)}$

In Problems 25–29, use Venn diagrams to prove or disprove each expression.

25. $\overline{A \cup B} = \overline{A} \cup \overline{B}$
26. $\overline{A} \cap \overline{B} = \overline{A \cup B}$
27. $(A \cup B) \cup C = A \cup (B \cup C)$
28. $A \cup (B \cup C) = (A \cup B) \cup (A \cup C)$
29. $A \cap (B \cup C) = (A \cap B) \cap (A \cap C)$

30. Prove De Morgan's Law: $\overline{X \cap Y} = \overline{X} \cup \overline{Y}$.

For Problems 25–29, draw a Venn diagram for the left side of the equation and another for the right side. If the final shaded portions are the same, then you have proved the result. If the final shaded portions are not identical, then you have disproved the result.

31. In a recent survey of 100 women, the following information was gathered:

59 use shampoo A.

51 use shampoo B.

35 use shampoo C.

24 use shampoos A and B.

19 use shampoos A and C.

13 use shampoos B and C.

11 use all three.

Let $A = \{$Women who use shampoo $A\}$; $B = \{$Women who use shampoo $B\}$; and $C = \{$Women who use shampoo $C\}$. Use a Venn diagram to show how many women are in each of the eight possible categories.

32. Matt E. Matic was applying for a job. To see if he could handle the job, the personnel manager sent him out to poll 100 people about their favorite types of TV shows. His data were as follows:

59 preferred comedies.

38 preferred variety shows.

42 preferred serious drama.

18 preferred comedies or variety programs.

12 preferred variety or serious drama.

16 preferred comedies or serious drama.

7 preferred all types.

2 didn't like any TV shows.

If you were the personnel manager, would you hire Matt on the basis of this survey?

33. In an interview of 50 students,

12 liked Proposition 8 and Proposition 13

18 liked Proposition 8 but not Proposition 2

4 liked Proposition 8, Proposition 13, and Proposition 2

25 liked Proposition 8

15 liked Proposition 13

10 liked Proposition 2 but not Proposition 8 or Proposition 13

1 liked Proposition 13 and Proposition 2 but not Proposition 8

a. Of those surveyed, how many did not like any of the three propositions?
b. How many liked Proposition 8 and Proposition 2?
c. Show the completed Venn diagram.

Mind Bogglers

34. If $A \subseteq B$, describe:

a. $A \cup B$ **b.** $A \cap B$ **c.** $\overline{A} \cup B$ **d.** $A \cap \overline{B}$

35. The letter shown on the facing page says "Cross out in order the letters that are common to both [sets].... This leaves TBMY EPEA EVIL ...." Describe this operation using set-operation symbols.

Let $U = \{A,B,E,H,I,K,L,M,N,O,R,S,T,V,Y\}$
       $\mathcal{A} = \{T,H,E,B,R,A,S,S,M,O,N,K,E,Y\}$
       $\mathcal{B} = \{S,E,E,H,E,A,R,S,P,E,A,K,N,O,E,V,I,L\}$

Problem for Individual Study

36. A set is either a member of itself or not a member of itself. For example, suppose in the small California town of Ferndale it is the practice of many of the men to be shaved by the barber. Now, the barber has a rule that has come to be known as The

---

JAY E. COSSEY
2021 LARCHWOOD AVENUE
WILMETTE, ILLINOIS 60091

28 January 1974

Advertising Manager
Heublein, Inc.
Hartford, Conn. 06101

Dear Sir:

Re your recent advertisements concerning the Brass
Monkey --- your conjectures are interesting, but misguided.
The agent's name was Pete Yale and you could have found
him sitting at table number 6 at the Brass Monkey any night.

You were right about the coaster being the tip-off,
except that you went at it the wrong way. Take the name
THE BRASS MONKEY and the motto SEE HEAR SPEAK NO EVIL, and
cross out in order the letters that are common to both:

T H̸E̸ E̸R̸A̸S̸S̸ M̸O̸N̸K̸E̸Y

S̸E̸E̸ H̸E̸A̸R̸ S̸P E A K̸ N̸O̸ E V I L

This leaves TBMY EPEA EVIL, which you arrange in three rows:

T B M Y

E P E A   and then draw an X through the rows:

E V I L

Using the letters in each section, you get PETE YALE –
BM (Brass Monkey) VI (Roman numeral 6).

Pete had a sense of humor, as you might guess. The letters
BM are also the term commonly used on maps for "bench mark".

Sincerely yours,

Barber's Rule: *He shaves those men and only those men who do not shave themselves.* The question is: Does the barber shave himself? If he does shave himself, then, according to The Barber's Rule, he does not shave himself. On the other hand, if he does not shave himself, then, according to The Barber's Rule, he shaves himself.

This problem of paradoxes in set theory has been studied by many famous mathematicians, including Zermelo, Frankel, von Neumann, Bernays, and Poincaré. These studies have given rise to three main schools of thought concerning the foundations of mathematics. Do some investigation of these paradoxes and these schools of thought.

Historical Note

About the time that Cantor's work began to gain acceptance, certain inconsistencies began to appear. One of these inconsistencies, called *Russell's paradox,* is discussed in Problem 36.

2.3 Deductive Reasoning

In the first chapter we studied patterns that allow us to form conjectures using inductive reasoning. Remember, inductive reasoning is the process of forming conjectures that are based on a number of observations of specific instances. The conjectures may be probable, but they are not necessarily true. For example, based on our observations of planetary motion, we may predict that tomorrow the sun will rise at 6:03 A.M. Although it is very probable that this prediction is correct, it is not absolutely certain.

There is, however, a type of reasoning that produces *certain* results. For example, consider the following argument.

Statements 1, 2, and 3 are called an *argument*.

1. If you take the *Times*, then you are well informed.
2. You take the *Times*.
3. Therefore, you are well informed.

If you accept statements 1 and 2 as true, then you *must* accept 3 as true. Statements 1 and 2 are called the **hypotheses** or **premises** of the argument, and 3 is called the **conclusion.** Such reasoning is called **deductive reasoning,** and, if the conclusion follows from the hypotheses, the reasoning is said to be **valid.**

The purpose of this chapter is to build a logical foundation to aid you not only in your study of mathematics and other subjects but also in your day-to-day contact with others.

Logic is a method of reasoning that accepts no conclusions except those that are inescapable. This is possible because of the strict way in which every concept is defined. That is, everything must be defined in a way that leaves no doubt or vagueness in meaning. Nothing can be taken for granted, and dictionary definitions are not usually sufficient. For example, in English one often defines a sentence as "a word or group of words stating, asking, commanding, requesting, or exclaiming something; conventional unit of connected speech or writing, usually containing a subject and predicate, beginning with a capital letter, and ending with an end mark."* In symbolic logic, we use the word *statement* to refer to a declarative sentence.

Historical Note

Logic began to flourish during the classical Greek period. Aristotle (384–322 B.C.) was the first person to systematically study the subject, and he and many other Greeks searched for universal truths that were irrefutable. The logic of this period, referred to as Aristotelian logic, is still used today and is based on the syllogism, which we will study later in this chapter.

Definition

A **statement** is a sentence that is either true or false but not both true and false.

If the sentence is a question or a command, or if it is vague or nonsensical, then it cannot be classified as true or false; thus, we would not call it a statement. For example:

***Webster's New World Dictionary*, College Edition, World Publishing Company, 1960.

1. School starts on Friday the 13th.
2. $5 + 6 = 16$.
3. Fish swim.
4. Mickey Mouse is president.

All of these are statements, since they are either true or false. But consider:

5. Go away!
6. What are you doing?
7. This sentence is false.

Historical Note

These are not statements by our definition, since they cannot possibly be either true or false.

Difficulty in simplifying arguments may arise because of their length, the vagueness of the words used, the literary style, or the possible emotional impact of the words used. Consider the following two arguments.

1. If George Washington was assassinated, then he is dead. Therefore, if he is dead, he was assassinated.
2. If you use heroin, then you first used marijuana. Therefore, if you use marijuana, then you will use heroin.

Logically these two arguments are exactly the same, and both are **invalid** forms of reasoning. Nearly everyone would agree that the first is invalid, but many people see the second as valid. The reason here is the emotional appeal of the words used.

To avoid these difficulties, and to be able to simplify a complicated logical argument, we set up an *artificial symbolic language*. This procedure was first suggested by Leibniz in his search for a *universal characteristic* to unify all of mathematics. What we will do is invent a notational shorthand. We denote simple statements with letters such as p, q, r, s, . . . and then define certain connectives. The problem, then, is to *translate the English statements into symbolic form, simplify the symbolic form,* and then *translate the simpler form back into English statements*.

The problem of length can be avoided by considering only simple statements connected by certain well-defined operators, such as *not, and, or, neither . . . nor, if . . . then, unless, because,* and so on.

A compound statement is formed by combining simple statements with operators. Because of our basic definition of a statement, we see that the **truth value** of any statement is either true (T) or false (F).

The truth value of a compound statement will depend only on the truth values of its component parts. It is not sufficient to assume that we know the meanings of the operators *and, or, not,* and so on, even though they may seem obvious and simple. The strength of logic is that it does not leave any meanings to chance or to individual interpretation. In defining the truth values of these words, we will, however, try to conform to common usage.

If p and q represent two simple statements, then "p and q" is the compound

The second great period for logic came with the use of symbols to simplify complicated logical arguments. This was first done when the great German mathematician Gottfried Leibniz (1646–1716), at the age of 14, attempted to reform Aristotelian logic. He called his logic the *universal characteristic* and wrote, in 1666, that he wanted to create a general method in which truths of reason would be reduced to a calculation so that errors of thought would appear as computational errors.

However, the world took little notice of Leibniz' logic, and it wasn't until George Boole (1815–1864) completed his book *An Investigation of the Laws of Thought* that logic entered its third and most important period.

Boole considered various mathematical operations by separating them from the other commonly used symbols. This idea was popularized by Bertrand Russell (1872–1970) and Alfred North Whitehead (1861–1947) in their monumental *Principia Mathematica*. In this work, they began with a few assumptions and three undefined terms and built a system of symbolic logic. From this, they then formally developed the theorems of arithmetic and mathematics.

TABLE 2.1 Definition of Conjunction

p	q	$p \wedge q$
T	T	T
T	F	F
F	T	F
F	F	F

statement using the operator called **conjunction.** The word *and* is symbolized by $\wedge$. For example,

*I have a penny **and** a quarter in my pocket.*

When will this compound statement be true? There are four possibilities:

1. I have a penny. I have a quarter.
2. I have a penny. I do not have a quarter.
3. I do not have a penny. I have a quarter.
4. I do not have a penny. I do not have a quarter.

Let p and q represent two simple statements as follows:

p: I have a penny.

q: I have a quarter.

The four possibilities can now be shown:

	p	q
1.	T	T
2.	T	F
3.	F	T
4.	F	F

If p and q are both true, we would certainly say that the compound statement is true; otherwise, we would say it is false. Thus, we define "$p \wedge q$" according to Table 2.1. The common usage of the word *and,* used when we discussed the intersection of sets in Section 2.1, conforms to this technical definition of conjunction.

It is worth noting that statements p and q need not be related. For example,

Fish swim and Neil Armstrong was on the moon.

is a true compound statement.

The operator *or,* denoted by $\vee$, is called **disjunction.** The meaning of this simple word is ambiguous, as we can see by considering the following examples.

1. I have a penny or a quarter in my pocket.
2. Ted is speaking in New York or in California at 7:00 P.M. tonight.*

If

p: I have a penny in my pocket

q: I have a quarter in my pocket

*We have used a literary shorthand here. These statements are to be interpreted as "I have a penny in my pocket, or I have a quarter in my pocket" and "Ted is speaking in New York at 7:00 P.M. tonight, or Ted is speaking in California at 7:00 P.M. tonight." We will use this type of shorthand throughout the chapter if no ambiguity results.

> *n*: Ted is speaking in New York at 7:00 P.M. tonight
>
> *c*: Ted is speaking in California at 7:00 P.M. tonight

what do we mean by each of the examples?

Statement 1 may mean:

> I have a penny in my pocket.
>
> I have a quarter in my pocket.
>
> I have both a penny and a quarter in my pocket.

Statement 2 may mean:

> Ted is speaking in New York at 7:00 P.M. tonight.
>
> Ted is speaking in California at 7:00 P.M. tonight.

It does *not* mean that he will do both.

These statements illustrate different usages of the word *or*. Now we are forced to select a single meaning for the operator, so we will choose a definition that will conform to the first example. That is, "*p* or *q*" means "*p* or *q*, perhaps both." Thus, we define "$p \vee q$" according to Table 2.2.

In logic, the "*p* or *q*, perhaps both" that has been defined by Table 2.2 is called the **inclusive or.** The second meaning of the word *or* is "*p* or *q*, but not both." This is called the **exclusive or,** and in this book it will be translated by the words "either . . . or." That is, in this book we write the preceding examples as:

> I have a penny or a quarter in my pocket.
>
> Either Ted is speaking in New York or in California at 7:00 P.M. tonight.

The word *or* that we used in Section 2.1 to define the union of sets conforms to the inclusive use of the word *or* discussed here.

The operator *not*, denoted by $\sim$, is called **negation.** Table 2.3 serves as a straightforward definition of negation. The negation of a true statement is false, and the negation of a false statement is true.

Example 1: If *t*: Otto is telling the truth, then

$\sim t$: Otto is not telling the truth.

We also can translate $\sim t$ as "It is not the case that Otto is telling the truth." You must be careful when negating statements containing the words *all, none,* or *some*. For example, write the negation of:

> *All students have pencils.*

Did you write "No students have pencils" or "All students do not have pencils"? The given statement "All students have pencils" is false, but "No students have pencils" and "All students do not have pencils" are also false.

Remember, if a statement is false, then its negation must be true. The correct negation is

Part of the difficulty in translating English into logical statements is the inaccuracy of our language. For example, how can "fat chance" and "slim chance" mean the same thing?

TABLE 2.2   Definition of Disjunction

p	*q*	$p \vee q$
T	T	T
T	F	T
F	T	T
F	F	F

TABLE 2.3   Definition of Negation

p	$\sim p$
T	F
F	T

Not all students have pencils

or

At least one student does not have a pencil

or

It is not the case that all students have pencils

or

Some students do not have pencils

In mathematics, the word *some* is used to mean "at least one." Table 2.4 gives some of the common negations. In the following examples write the negation of each statement.

	Statement	*Negation*

TABLE 2.4 Negation of *All*, *Some*, and *No*

Statement	*Negation*
All	Some . . . not
Some	No
Some . . . not	All
No	Some

Example 2: All people have compassion. → Some people do not have compassion.

Example 3: Some animals are dirty. → No animal is dirty.

Example 4: Some students do not take Math 10. → All students take Math 10.

Example 5: No students are enthusiastic. → Some students are enthusiastic.

We will consider the negation of compound statements in the next section.

Working with logical arguments requires that we be able to translate from English into symbols and from symbols back into English. For example, the statement

Bonnie and Clyde are cute

can be translated as follows:

Let   *b*: Bonnie is cute

   c: Clyde is cute

Then we have

$b \wedge c$:   Bonnie is cute and Clyde is cute.

Thus, $b \wedge c$ is the symbolic statement.

Parentheses are used to indicate the order of operations. Thus,

$\sim(n \wedge c)$ is read "It is not the case that *n* and *c*."

$\sim(n \wedge c)$ means the negation of the statement "*n* and *c*"

whereas

$\sim n \wedge c$ is read "not *n* and *c*."

$\sim n \wedge c$ means the negation of "*n*" and the statement "*c*"

Suppose that n is T and c is F. What is the truth value of the compound statement $(n \vee c) \wedge \sim(n \wedge c)$?

$$(n \vee c) \wedge \sim(n \wedge c)$$
$$(\text{T} \vee \text{F}) \wedge \sim(\text{T} \wedge \text{F})$$
$$\text{T} \quad \wedge \quad \sim\text{F}$$
$$\text{T} \quad \wedge \quad \text{T}$$
$$\text{T}$$

Thus, the compound statement is true.

Example 6: Test the truth value of $(p \wedge q) \vee (p \vee r)$ when p is T, q is F, and r is F.

Solution

$$(p \wedge q) \vee (p \vee r)$$
$$(\text{T} \wedge \text{F}) \vee (\text{T} \vee \text{F})$$
$$\text{F} \quad \vee \quad \text{T}$$
$$\text{T}$$

Use Tables 2.1, 2.2, and 2.3. For example, T $\vee$ F is found on the second line of Table 2.2 and, as shown on that line, is replaced by T.

The statement is true. This result does *not* depend on the particular statements p, q, and r. As long as p is T, q is F, and r is F, the result will be the same—namely, T.

The procedure requires that we translate not only from English into logical symbols but also from symbols into English. For example, suppose

p: I eat spinach,

q: I am strong.

Translate the statements in the following examples into words.

Example 7: $p \wedge q$ I eat spinach, and I am strong.

Example 8: $\sim p$ I do not eat spinach.

Example 9: $\sim(p \wedge q)$ It is not the case that I eat spinach and am strong.

Example 10: $\sim p \wedge q$ I do not eat spinach, and I am strong.

Example 11: $\sim(\sim q)$ I am not not strong, *or* (if we assume that "not strong" is the same as "weak") I am not weak. In the next section we will show that $\sim(\sim q)$ is the same as q, so this could also be translated as I am strong.

Suppose that p is T and q is F. For Examples 12–16, find the truth value for the translations given in Examples 7–11.

Example 12: $p \wedge q$
$\quad\quad\quad\quad\quad$ T $\wedge$ F
$\quad\quad\quad\quad\quad\quad$ F $\quad\quad\quad$ Thus the statement is false.

Example 13: $\sim p$
$\quad\quad\quad\quad\quad$ $\sim$T
$\quad\quad\quad\quad\quad\quad$ F $\quad\quad\quad$ Thus the statement is false.

Example 14: $\sim(p \wedge q)$
$\quad\quad\quad\quad\quad$ $\sim$(T $\wedge$ F)
$\quad\quad\quad\quad\quad\quad$ $\sim$F
$\quad\quad\quad\quad\quad\quad$ T $\quad\quad\quad$ Thus the statement is true.

Notice that Examples 14 and 15 prove that $\sim(p \wedge q)$ and $\sim p \wedge q$ do not mean the same thing.

Example 15: $\sim p \wedge q$
$\quad\quad\quad\quad\quad$ $\sim$T $\wedge$ F
$\quad\quad\quad\quad\quad$ F $\wedge$ F
$\quad\quad\quad\quad\quad\quad$ F $\quad\quad\quad$ The statement is false.

Notice from Example 16 that $\sim(\sim q)$ has the same truth value as q.

Example 16: $\sim(\sim q)$
$\quad\quad\quad\quad\quad$ $\sim(\sim$F)
$\quad\quad\quad\quad\quad\quad$ $\sim$T
$\quad\quad\quad\quad\quad\quad$ F $\quad\quad\quad$ The statement is false.

Problem Set 2.3

According to our definition, which of the examples in Problems 1–4 are statements?

1. **a.** Hickory, Dickory, Dock, the mouse ran up the clock.
$\quad$ **b.** $3 + 5 = 9$.
$\quad$ **c.** Is John ugly?
$\quad$ **d.** John has a wart on the end of his nose.

2. **a.** March 13, 1984, is Monday.
$\quad$ **b.** Division by zero is impossible.
$\quad$ **c.** Logic is not as difficult as I had anticipated.
$\quad$ **d.** $4 - 6 = 2$.

3. **a.** $6 + 9 \neq 7 + 8$.
$\quad$ **b.** Thomas Jefferson was the twenty-third president.
$\quad$ **c.** Sit down and be quiet!
$\quad$ **d.** If wages continue to rise, then prices will also rise.

4. **a.** John and Mary were married on August 3, 1979.
$\quad$ **b.** $6 + 12 \neq 10 + 8$.

c. Do not read this sentence.

d. Do you have a cold?

5. Define conjunction.   **6.** Define disjunction.   **7.** Define negation.

Write the negation of each statement in Problems 8–10.

8. a. All mathematicians are ogres.   **b.** Some integers are negative.

   c. Some people do not pay taxes.   **d.** No even integers are divisible by 5.

9. a. All counting numbers are divisible by 1.   **b.** Some apples are rotten.

   c. Some integers are not odd.   **d.** No triangles are squares.

10. a. All women are intelligent.   **b.** Some rectangles are squares.

   c. Some rectangles are not squares.   **d.** No nice people are dangerous.

11. Let p represent the statement "Prices will rise" and q the statement "Inflation will be controlled." Translate each of the statements into symbols.

   a. Prices will rise, or inflation will not be controlled.

   b. Prices will rise, but inflation will not be controlled.

   c. Prices will rise, and inflation will not be controlled.

   d. Prices will not rise, and inflation will be controlled.

12. Assume that prices rise and inflation is not controlled. Under these assumptions, which of the statements in Problem 11 are true?

13. Let p represent the statement "Paul is peculiar" and q the statement "Paul likes to read mathematics textbooks." Translate each of the following statements into words.

   a. $p \land q$ **b.** $\sim p \land q$ **c.** $p \lor \sim q$ **d.** $\sim p \lor \sim q$

14. Assume that p is T and q is T. Under these assumptions, which of the statements in Problem 13 are true?

15. Let r represent a statement having a truth value F, and let s and t each represent statements having truth values T. Find the truth value of each statement.

   a. $(r \lor s) \lor t$ **b.** $(r \land s) \land \sim t$

   c. $r \land (s \lor t)$ **d.** $(r \land s) \lor (r \land t)$

16. Find the truth value when p and q are each T and r is F.

   a. $(p \lor q) \land r$ **b.** $(p \land q) \land \sim p$

   c. $p \land (q \lor r)$ **d.** $(p \land q) \lor (p \land r)$

17. Find the truth value when p is F and q and r are each T.

   a. $(p \lor q) \lor (r \land \sim q)$ **b.** $\sim(\sim p) \lor (p \land q)$

   c. $(p \land q) \lor (p \land \sim r)$ **d.** $(\sim p \lor q) \land \sim p$

18. Find the truth value when p is T and q and r are each F.

   a. $\sim(p \land q)$ **b.** $(\sim p \lor q) \land \sim p$

   c. $(p \land \sim q) \lor (p \land q)$ **d.** $(p \lor q) \land [(r \land \sim p) \land (q \lor \sim p)]$

Signpost in the Bronx, New York

19. Does the *B.C.* cartoon in the margin illustrate inductive or deductive reasoning? Explain your answer.

20. Does the following story illustrate inductive or deductive reasoning? Explain your answer.

> ### SIMPLE DEDUCTION
>
> The old fellow in charge of the checkroom in a large hotel was noted for his memory. He never used checks or marks of any sort to help him return hats to their rightful owners.
>
> Thinking to test him, a frequent hotel guest asked him as he received his headgear,
>
> "Sam, how do you know this is my hat?"
>
> "I don't, sir," was the calm response.
>
> "Then why did you give it to me?" asked the guest.
>
> "Because," said Sam, "it's the one you gave me, sir."
>
> —*Lucille J. Goodyear*

Translate the statements in Problems 21–25 into symbols. For each simple statement, be sure to indicate the meanings of the symbols you use. Answers are not unique.

21. W. C. Fields is eating, drinking, and having a good time.

22. Sam will not seek and will not accept the nomination.

23. Jack will not go tonight and Rosamond will not go tomorrow.

24. Fat Albert lives to eat and does not eat to live.

25. The decision will depend on judgment or intuition, and not on who paid the most.

In Problems 26–31, find the truth value when p is T and q and r are each F.

26. $(p \lor q) \land \sim(p \lor \sim q)$ 27. $(p \land \sim q) \lor (r \land \sim q)$

28. $\sim(\sim p) \land (q \lor p)$ 29. $(r \land p) \lor (q \land r)$

30. $\sim(r \land q) \land (q \lor \sim q)$

31. $(q \lor \sim q) \land [(p \land \sim q) \lor (\sim r \lor r)]$

Mind Bogglers

32. What is the largest amount of money you can have in coins and not be able to make change for a dollar? (Silver dollars are not allowed.)

33. Smith received the following note from Melissa: "Dr. Smith, I wish to explain that I was really joking when I told you that I didn't mean what I said about reconsidering my decision not to change my mind." Did Melissa change her mind or didn't she?

34. Their are three errers in this item. See if you can find all three.

2.4 The Conditional

The connectives *and, or,* and *not* introduced in the last section are called the **fundamental operators.** To go beyond these and consider additional operators we need a device known in logic as a **truth table.** A truth table shows how the truth values of compound statements depend on the fundamental operators. Tables 2.1, 2.2, and 2.3 from the last section should be memorized. They are summarized here in Table 2.5.

Historical Note

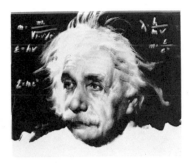

In regard to the real nature of scientific truth, Einstein said "As far as the laws of mathematics refer to reality, they are not certain, and as far as they are certain, they do not refer to reality." Albert Einstein (1879–1955) was one of the intellectual giants of the 20th century. He was a shy, unassuming man who was told as a child that he would never make a success of anything. However, he was able to use the tools of logic and mathematical reasoning to change our understanding of the universe.

TABLE 2.5   Truth Table of the Fundamental Operators

p	q	Conjunction "and" $p \wedge q$	Disjunction "or" $p \vee q$	Negation "not" $\sim p$	$\sim q$
T	T	T	T	F	F
T	F	F	T	F	T
F	T	F	T	T	F
F	F	F	F	T	T

Example 1:   Construct a truth table for the compound statement:

*Alfie did **not** come last night **and** he did **not** pick up his money.*

Solution: Let

p: Alfie came last night,

q: Alfie picked up his money.

Then the statement can be written $\sim p \wedge \sim q$. To begin, list all the possible combinations of truth values for the simple statements p and q.

p	q	
T	T	
T	F	
F	T	
F	F	

Insert the truth values for $\sim p$ and $\sim q$.

p	q	$\sim p$	$\sim q$	
T	T	F	F	
T	F	F	T	
F	T	T	F	
F	F	T	T	

Finally, insert the truth values for $\sim p \wedge \sim q$.

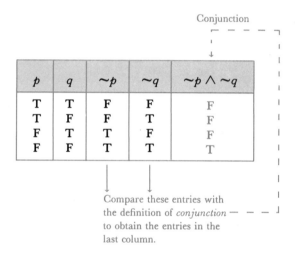

Conjunction

p	q	$\sim p$	$\sim q$	$\sim p \wedge \sim q$
T	T	F	F	F
T	F	F	T	F
F	T	T	F	F
F	F	T	T	T

Compare these entries with
the definition of *conjunction*
to obtain the entries in the
last column.

The only time the compound statement is true is when *both p* and *q* are false.

Example 2:   Construct a truth table for $\sim(\sim p)$.

Solution

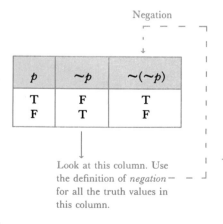

Negation

p	$\sim p$	$\sim(\sim p)$
T	F	T
F	T	F

Look at this column. Use
the definition of *negation*
for all the truth values in
this column.

Notice that $\sim(\sim p)$ and p have the same truth values. If two statements have the same truth values, one can replace the other in any logical expression. This means that the double negative of a statement is the same as the original statement.

Law of Double Negation

$\sim(\sim p)$ may be replaced by p in any logical expression.

There is a story about a logic professor who was telling her class that a double negative is known to mean a negative in some languages and a positive in others (as in English). She continued by saying that there is no spoken language in which a double positive means a negative. Just then she heard from the back of the room a sarcastic "Yeah, yeah."

Example 3: Construct a truth table to determine when the following statement is true.

$$\sim(p \wedge q) \wedge [(p \vee q) \wedge q]$$

Solution: We begin as before and move from left to right, focusing our attention on only two columns at a time (refer to Table 2.5 to find the correct entries).

A p	B q	C $p \wedge q$	D $\sim(p \wedge q)$	E $p \vee q$	F $(p \vee q) \wedge q$	G $\sim(p \wedge q) \wedge [(p \vee q) \wedge q]$
T	T	T	F	T	T	F
T	F	F	T	T	F	F
F	T	F	T	T	T	T
F	F	F	T	F	F	F

Step 1. Attend to the parentheses first. Look at columns A and B along with the definition of conjuction to fill in the entries in column C.

Step 2. Use column C and the definition of negation to fill in the entries of column D.

Step 3. Use columns A and B and the definition of disjunction to fill in the entries of column E.

Step 4. Use columns E and B and the definition of conjunction to fill in the entries of column F.

Step 5. Use columns D and F, which are the left and right sides of the final operation. By using these columns and the definition of conjunction, we obtain the entries of column G.

We can now use truth tables to prove certain useful results and to introduce some additional operators. The first one we'll consider is called the **conditional**. The statement "if p, then q" is called a *conditional statement*. It is symbolized

THERE'S TOO MUCH UNCERTAINTY ABOUT OUR RELATIONSHIP.

if

by $p \rightarrow q$; p is called the **antecedent,** and q is called the **consequent.** There are several ways of using a conditional, as illustrated by the following examples.

1. We can use "if–then" to indicate a *logical* relationship—one in which the consequent follows logically from the antecedent.
    If $\sim (\sim p)$ has the same truth value as p, then p can replace $\sim (\sim p)$.
2. We can use "if–then" to indicate a causal relationship.
    If John drops that rock, then it will land on my foot.
3. We can use "if–then" to report a decision on the part of the speaker.
    If John drops that rock, then I will hit him.
4. We can use "if–then" when the consequent follows from the antecedent by the very definition of the words used.
    If John drives an Oldsmobile, then John drives a car.
5. Finally, we can use "if–then" to make a *material implication.* There is no logical, causal, or definitional relationship between the antecedent and consequent; we use the expression simply to convey humor or emphasis:
    If John got an A on that test, then I'm a monkey's uncle.
    The consequent is obviously false, and the speaker wishes to emphasize that the antecedent is also false.

Our task is to try to devise a definition of the conditional that will apply in all these different types of if–then statements. We'll approach the problem by asking under what circumstances a given conditional would be false. Let's consider another example.

Suppose I make you a promise: "If I receive my check tomorrow, then I will pay you the $10 that I owe you." If I keep my promise, we say the statement is true; if I don't, then it is false.
Let

   p: I receive my check tomorrow,

   q: I will pay you the $10 that I owe you.

We symbolize the promise by $p \rightarrow q$. There are four possibilities:

	p	q	
1.	T	T	*I receive my check tomorrow, and I pay you the $10.* In this case, the promise, or conditional, is true.
2.	T	F	*I receive my check tomorrow, and I do not pay you the $10.* In this case, the conditional is false, since I did not fulfill my promise.
3.	F	T	*I do not receive my check tomorrow, but I pay you the $10.* In this case, I certainly didn't break my promise, so the conditional is true.
4.	F	F	*I do not receive my check tomorrow, and I do not pay you the $10.* Here, again, you could not say I broke my promise, so the conditional is true.

No doubt I came into some unexpected good fortune.

Actually the promise was not tested for case 4, since I didn't receive my check. Assume the principle of "innocent until proved guilty." The only time that I will have broken my promise is in case 2.

The test for a conditional is to determine when it is false. In symbols,

$p \rightarrow q$ is false whenever $p \wedge \sim q$ is true.   Case 2

Or,

$p \rightarrow q$ is true whenever $\sim(p \wedge \sim q)$ is true.

Let's construct a truth table for $\sim(p \wedge \sim q)$.

p	q	$\sim q$	$p \wedge \sim q$	$\sim(p \wedge \sim q)$
T	T	F	F	T
T	F	T	T	F
F	T	F	F	T
F	F	T	F	T

We use this truth table to define the conditional $p \rightarrow q$ as shown in Table 2.6. The following examples illustrate the definition of the conditional, as shown in Table 2.6.

TABLE 2.6 Definition of Conditional

p	q	$p \rightarrow q$
T	T	T
T	F	F
F	T	T
F	F	T

Example 4:   Case 1: T → T
If 7 < 14, then 7 + 2 < 14 + 2.

This is a true statement, since both component parts are true.

Example 5:   Case 2: T → F
If 7 + 5 = 12, then 7 + 10 = 15.

This is a false statement, since the antecedent is true but the consequent is false.

Example 6:   Case 3: F → T
If you have bad breath, then George Washington was president.

This is true, since the consequent is T (whenever the consequent is true the conditional is true). This example shows that the conditional, in mathematics, does not mean that there is any cause-and-effect relationship. *Any* two statements can be connected with the connective of conditional and the result must be T or F.

Example 7:   Case 4: F → F
If 16 = 8, then 8 = 4.

This is a true statement, since both component parts are false.

The *if* part of an implication need not be stated first. All the following statements have the same meaning:

Conditional translation:	*Example*
If p, then q	If you are 18, then you can vote.
q, if p	You can vote, if you are 18.
p, only if q	You are 18 only if you can vote.
$p \subseteq q$	p is a subset of q.

Some statements, although not originally written as a conditional can be put into if–then form. For example, "All ducks are birds" can be rewritten as "If it is a duck, then it is a bird." Thus we add one more form to the list:

All p are q	All 18-year-olds can vote.

This statement appeared on the 1982 federal income tax form.

Example 8:   Translate the following sentence into symbolic form. *If you do not itemize deductions and line 16 is under \$15,000, do not complete lines 17 and 18.*

Step 1. Isolate the simple statements and assign them variables.

> Let d: you itemize deductions
>
> u: line 16 is under \$15,000
>
> s: you complete line 17
>
> e: you complete line 18

Your choice of variables is, of course, arbitrary, but you must be careful not to let variables represent compound statements.

Step 2. Rewrite the sentence, making substitutions for the variables.

$$\text{If } \sim d \text{ and } u, \text{ then } \sim(s \wedge e).$$

Step 3. Complete the translation to symbols.

$$(\sim d \wedge u) \rightarrow \sim(s \wedge e)$$

Related to a conditional $p \rightarrow q$ are other statements, which we will now define.

Definition

Given the conditional $p \rightarrow q$, we define:

1. The *converse:* $q \rightarrow p$
2. The *inverse:* $\sim p \rightarrow \sim q$
3. The *contrapositive:* $\sim q \rightarrow \sim p$

Not all these statements are equivalent in meaning.

Example 9: Write the converse, inverse, and contrapositive of the statement:
If it is a 280ZX, then it is a car.

Let p: It is a 280ZX

q: It is a car

The statement is written $p \rightarrow q$.

Converse: $q \rightarrow p$	If it is a car, then it is a 280ZX.
Inverse: $\sim p \rightarrow \sim q$	If it is not a 280ZX, then it is not a car.
Contrapositive: $\sim q \rightarrow \sim p$	If it is not a car, then it is not a 280ZX.

TABLE 2.7

p	q	$\sim p$	$\sim q$	Statement $p \rightarrow q$	Converse $q \rightarrow p$	Inverse $\sim p \rightarrow \sim q$	Contrapositive $\sim q \rightarrow \sim p$
T	T	F	F	T	T	T	T
T	F	F	T	F	T	T	F
F	T	T	F	T	F	F	T
F	F	T	T	T	T	T	T

Refer to Table 2.7. Notice that the contrapositive and the original statement always have the same truth values, as do the converse and the inverse. Thus we state the following general principle:

Notice that, if the conditional is true, the converse and inverse are not necessarily true. However, the contrapositive is always true if the original conditional is true.

Law of Contraposition ·

A conditional may always be replaced by its contrapositive without having its truth value affected.

Example 10: Given $p \rightarrow \sim q$, write its converse, inverse, and contrapositive.

Solution: Statement: $p \rightarrow \sim q$ *Converse:* $\sim q \rightarrow p$

Inverse: $\sim p \rightarrow q$ *Contrapositive:* $q \rightarrow \sim p$

We have replaced the double negative $\sim(\sim q)$ by q in the inverse and contrapositive. We make this simplification whenever possible.

Assuming that the statement is true, the contrapositive is also true. Consider the following:

p: You obey the law.

q: You will go to jail.

The converse and inverse are
not necessarily true.

Statement: $p \rightarrow \sim q$	If you obey the law, then you will not go to jail.
Contrapositive: $q \rightarrow \sim p$	If you go to jail, then you did not obey the law.
Converse: $\sim q \rightarrow p$	If you do not go to jail, then you obey the law.
Inverse: $\sim p \rightarrow q$	If you do not obey the law, then you will go to jail.

Problem Set 2.4

> • How are you going to teach
> logic in a world where every-
> body talks about the sun set-
> ting, when it's really the hori-
> zon rising?
>
> —*Cal Craig*

Construct a truth table for the statements given in Problems 1–24.

1. $\sim p \vee q$ 2. $\sim p \wedge \sim q$ 3. $\sim (p \wedge q)$

4. $\sim r \vee \sim s$ 5. $\sim(\sim r)$ 6. $(r \wedge s) \vee \sim s$

7. $p \wedge \sim q$ 8. $\sim p \vee \sim q$ 9. $(\sim p \wedge q) \vee \sim q$

10. $(p \wedge \sim q) \wedge p$ 11. $(\sim p \vee q) \wedge (q \wedge p)$ 12. $(p \vee q) \vee (p \wedge \sim q)$

13. $p \vee (p \rightarrow q)$ 14. $p \rightarrow (\sim p \rightarrow q)$ 15. $(p \wedge q) \rightarrow p$

16. $\sim p \rightarrow \sim (p \wedge q)$ 17. $(p \wedge q) \wedge (p \rightarrow \sim q)$ 18. $(p \rightarrow p) \rightarrow (q \rightarrow \sim q)$

19. $(p \rightarrow \sim q) \rightarrow (q \rightarrow \sim p)$ 20. $(p \rightarrow q) \rightarrow (\sim q \rightarrow \sim p)$

21. $[p \wedge (p \vee q)] \rightarrow p$ 22. $(p \wedge q) \wedge \sim r$

23. $[(p \vee q) \wedge \sim r] \wedge r$ 24. $[p \wedge (q \vee \sim p)] \vee r$

25. Let p: $2 + 3 = 5$
         q: $12 - 7 = 5$
     Tell which of the statements in Problems 1–3 are true.

26. Let r: Man pollutes the environment
          s: Man will survive
     Translate each of the statements in Problems 4–6 into words, and tell under which
     conditions each will be true.

27. Write each of the following statements in the if–then format.

     a.   Everything happens to everybody sooner or later if there is time enough.
          (G. B. Shaw)
     b.   We are not weak if we make a proper use of those means which the God of
          Nature has placed in our power. (Patrick Henry)
     c.   A useless life is an early death. (Goethe)
     d.   All work is noble. (Thomas Carlyle)
     e.   Everything's got a moral if only you can find it. (Lewis Carroll)

28. Translate the statements of Problem 27 into symbolic form. Indicate the letters
     used for each simple statement.

29. First decide whether the simple statements are true or false. Then state whether the
     given compound statement is true or false.

a. If $5 + 10 = 16$, then $15 - 10 = 3$.
b. The moon is made of green cheese only if Mickey Mouse is president.
c. If $1 + 1 = 10$, then the moon is made of green cheese.
d. If Alfred E. Newman was elected president in 1980, then F. I. Knight was chosen as his chief mathematician.
e. $3 \cdot 2 = 6$ if and only if water runs uphill.

Write the converse, inverse, and contrapositive of the statements in Problems 30–35.

30. If you break the law, then you will go to jail.

31. I will go on Saturday if I get paid.

32. If you brush your teeth with Smiles toothpaste, then you will have fewer cavities.

33. $\sim p \rightarrow \sim q$ **34.** $\sim r \rightarrow t$ **35.** $\sim t \rightarrow \sim s$

Mind Bogglers

36. If the Roman god Jupiter could do anything, could he make an object that he could not lift?

37. Decide about the truth or falsity of the following: "If wishes were horses, then beggars could ride."

38. A man is about to be electrocuted but is given a chance to save his life. In the execution chamber are two chairs, labeled 1 and 2, and a jailer. One chair is electrified; the other is not. The prisoner must sit in one of the chairs, but, before doing so, he may ask the jailer one question, to which the jailer must answer yes or no. The jailer is a consistent liar or else a consistent truthteller, but the prisoner does not know which. Knowing that the jailer either deliberately lies or faithfully tells the truth, what question should the prisoner ask?

Hint 1: Let p be the proposition "Number 1 is the hot seat" and let q be the proposition "You are telling the truth."

Hint 2: Since I would like you to try working the problem without using this hint, I'll make you read it backward: "?hturt eht gnillet era uoy fi ylno dna fi taes toh eht si 1 rebmun taht eurt ti sI"

Problems for Individual Study

39. a. Explain the difference between the words *necessary* and *sufficient*.
    b. How do they relate to the conditional?
    c. Put the following statement into if–then format: "The presence of oxygen is a necessary condition for combustion."
    d. Put the following statement into if–then format: "Combustion is a sufficient condition for the presence of oxygen."
    e. Put the following statements into if–then format:
      i. p is a necessary condition for q.
      ii. p is a sufficient condition for q.
    f. Now put the statements in parts b–e into the form of a Euler diagram.

40. Sometimes statements p and q are described as contradictory, contrary, or inconsistent. Consult a logic text, and then define these terms using truth tables.

This is a sign found in Vermont.

2.5 Additional Connectives

In the last section we took great care to point out that a statement $p \rightarrow q$ and its converse $q \rightarrow p$ do not have the same truth values. However, it may be the case that $p \rightarrow q$ *and also* $q \rightarrow p$. In this case we write

$$p \leftrightarrow q$$

and call this operator the **biconditional.** To determine the truth values of the biconditional, we construct a truth table for $(p \rightarrow q) \wedge (q \rightarrow p)$ as shown in Table 2.8. (Why?)

TABLE 2.8

p	q	$p \rightarrow q$	$q \rightarrow p$	$(p \rightarrow q) \wedge (q \rightarrow p)$
T	T	T	T	T
T	F	F	T	F
F	T	T	F	F
F	F	T	T	T

This leads us to define the biconditional so that it is true only when both p and q are true or when both p and q are false (that is, whenever they have the same truth values). (See Table 2.9.)

In mathematics, $p \leftrightarrow q$ is translated in several ways, all of which have the same meaning:

TABLE 2.9   Definition of Biconditional

p	q	$p \leftrightarrow q$
T	T	T
T	F	F
F	T	F
F	F	T

---

1. p if and only if q.
2. q if and only if p.
3. If p then q, and conversely.
4. If q then p, and conversely.

Example 1:   Rewrite the following as one statement:

1. If a polygon has three sides, then it is a triangle.
2. If a polygon is a triangle, then it has three sides.

Solution: A polygon is a triangle if and only if it has three sides.

We will define triangles and polygons in Chapter 7 (you do not need to use those definitions here).

For a given statement, the set of logical possibilities for which the given statement is true is called its *truth set*. If its truth set is identical to its universal

set, then the statement is a logically true statement and is called a **tautology.**
This means that a compound statement is a tautology if you obtain only Ts on a
truth table.

Example 2:   Is $(p \lor q) \to (\sim q \to p)$ a tautology?

Solution

p	q	$p \lor q$	$\sim q$	$\sim q \to p$	$(p \lor q) \to (\sim q \to p)$
T	T	T	F	T	T
T	F	T	T	T	T
F	T	T	F	T	T
F	F	F	T	F	T

Thus, it is a tautology.

If a conditional is a tautology, as in Example 2, then it is called an **implication** and is symbolized by $\Rightarrow$. That is, Example 2 can be written

$$(p \lor q) \Rightarrow (\sim q \to p)$$

The implication symbol $p \Rightarrow q$ is pronounced "p implies q."

A biconditional statement $p \leftrightarrow q$ that is also a tautology is a **logical equivalence,** written $p \Leftrightarrow q$ and read "p is logically equivalent to q."

Example 3:   Show $(p \to q) \Leftrightarrow \sim(p \land \sim q)$.

Solution

p	q	$p \to q$	$\sim q$	$p \land \sim q$	$\sim(p \land \sim q)$	$(p \to q) \leftrightarrow \sim(p \land \sim q)$
T	T	T	F	F	T	T
T	F	F	T	T	F	T
F	T	T	F	F	T	T
F	F	T	T	F	T	T

Since all possibilities are true, it is a logical equivalence and we write

$$(p \to q) \Leftrightarrow \sim(p \land \sim q)$$

We can use the idea of logical equivalence to write two previously stated laws:

Law of Double Negation: $\sim(\sim p) \Leftrightarrow p$

Law of Contraposition: $(\sim q \rightarrow \sim p) \Leftrightarrow (p \rightarrow q)$

You will be asked to prove these in Problem Set 2.5. Other important laws are developed in the next section. Two additional laws, however, will be considered in this section.

De Morgan's Laws

$$\sim(p \vee q) \Leftrightarrow \sim p \wedge \sim q$$
$$\sim(p \wedge q) \Leftrightarrow \sim p \vee \sim q$$

As you will see in Section 2.7 there is a strong tie between set theory and symbolic logic, so it is no coincidence that there are De Morgan's Laws in both set theory and here again in symbolic logic. We will prove the first one and leave the second for the problem set. Proof (by truth table):

p	q	$p \wedge q$	$\sim(p \wedge q)$	$\sim p$	$\sim q$	$\sim p \vee \sim q$	$\sim(p \wedge q) \leftrightarrow \sim p \vee \sim q$
T	T	T	F	F	F	F	T
T	F	F	T	F	T	T	T
F	T	F	T	T	F	T	T
F	F	F	T	T	T	T	T

Since the last column is all Ts, we can write

$$\sim(p \wedge q) \Leftrightarrow \sim p \vee \sim q$$

De Morgan's Laws can be used to write the negation of a compound statement using conjunction and disjunction as shown by Examples 4 and 5.

Example 4:   Write the negation of the compound statement: John went to work or he went to bed.

Solution

Let w: John went to work

        b: John went to bed

Then, the symbolic statement is: $w \vee b$.
   Negation: $\sim(w \vee b) \Leftrightarrow \sim w \wedge \sim b$ De Morgan's Law
   Translation: John did not go to work and he did not go to bed.

Example 5: Write the negation of the compound statement: Alfie didn't come last night and didn't pick up his money.

Solution

  Let p: Alfie came last night
       q: Alfie picked up his money

Then the symbolic statement is: $\sim p \wedge \sim q$.
   Negation: $\sim(\sim p \wedge \sim q) \Leftrightarrow \sim(\sim p) \vee \sim(\sim q)$ De Morgan's Law
                        $\Leftrightarrow p \vee q$ Law of Double Negation
   Translation: Alfie came last night or he picked up his money.

Sometimes it is necessary to find the negation of a conditional, $p \rightarrow q$. In Problem Set 2.5 (Problem 31) you are asked to prove that:

$$(p \rightarrow q) \Leftrightarrow \sim p \vee q$$

Thus, the negation of $p \rightarrow q$ is the negation of $\sim p \vee q$:

      $\sim(\sim p \vee q) \Leftrightarrow \sim(\sim p) \wedge \sim q$ De Morgan's Law
                $\Leftrightarrow p \wedge \sim q$ Law of Double Negation

The negation of $p \rightarrow q$ is $p \wedge \sim q$.

Example 6: The officer said to the detective, "If John was at the scene of the crime, then he knows that Jean could not have done it."
   "No, Colombo," answered the detective, "that is not correct. John was at the scene of the crime and Jean did it." Was the detective's statement a correct negation of Colombo's conditional statement?

Solution

  Let p: John was at the scene of the crime
       q: Jean committed the crime

Then, a symbolic statement is: $p \rightarrow \sim q$.
   Negation: $p \wedge \sim(\sim q)$
          $p \wedge q$ Law of Double Negation
   Translation: John was at the scene of the crime and Jean committed the crime. The detective's negation was correct.

Occasionally, we encounter other operators, and it's necessary to formulate precise definitions of these additional operators. As Table 2.10 shows, they are all defined in terms of our previous operators.

Historical Note

Galileo Galilei (1564–1642) is best known for his work in astronomy, but he should be remembered as the father of modern science in general, and of physics in particular. Galileo believed mathematics was the basis for science. Using nature as his teacher, he reasoned deductively and described the qualities of matter as quantities of mathematics. He laid the foundation of physics that Newton would build upon almost 50 years later (see the Historical Note on page 336).

TABLE 2.10   Additional Operators

p	q	Either p or q defined as $(p \lor q) \land \sim(p \land q)$	Neither p nor q defined as $\sim(p \lor q)$	p unless q defined as $\sim q \to p$	p because q defined as $(p \land q) \land (q \to p)$	No p is q defined as $p \to \sim q$
T	T	F	F	T	T	F
T	F	T	F	T	F	T
F	T	T	F	T	F	T
F	F	F	T	F	F	T

Example 7:   Show that the statement "p because q" is equivalent to conjunction.

Solution: Use a truth table: "p because q" means $(p \land q) \land (q \to p)$ and conjunction means $p \land q$.

p	q	$(p \land q) \land (q \to p)$	$p \land q$	$(p \land q) \land (q \to p) \leftrightarrow p \land q$
T	T	T	T	T
T	F	F	F	T
F	T	F	F	T
F	F	F	F	T

From Table 2.10 ⎯⎯⎯⎯⎯⎯⎯⎯⎯⎯ From Table 2.5

The last column is all Ts, so we can write

$$(p \land q) \land (q \to p) \leftrightarrow p \land q$$

Problem Set 2.5

Use truth tables in Problems 1–6 to determine whether the given compound statement is a tautology.

1. $(p \land q) \lor (p \to \sim q)$ **2.** $(p \lor q) \land (q \to \sim p)$ **3.** $(\sim p \to q) \to p$

4. $(\sim q \to p) \to q$ **5.** $(p \land q) \leftrightarrow (p \lor q)$ **6.** $(p \to q) \leftrightarrow (\sim p \lor q)$

7.   Verify the results of the truth table for the operator *unless*.

8.   Verify the results of the truth table for the operator *neither . . . nor*.

9. Verify the results of the truth table for the operator *either . . . or*.

10. Verify the results of the truth table for the operator *because*.

11. Show that the definition for *neither . . . nor* could also be $\sim p \wedge \sim q$.

Translate the statements of Problems 12–19 into symbols. For each simple statement, be sure to indicate the meanings of the symbols you use.

12. Neither smoking nor drinking is good for your health.

13. I will not buy a new house unless all provisions of the sale are clearly understood.

14. I cannot go with you because I have a previous engagement.

15. No man is an island.

16. Either I will invest my money in stocks or I will put it in a savings account.

17. Be nice to people on your way up 'cause you'll meet 'em on your way down. (Jimmy Durante)

18. No man who has once heartily and wholly laughed can be altogether irreclaimably bad. (Thomas Carlyle)

19. If by the mere force of numbers a majority should deprive a minority of any clearly written constitutional right, it might, in a moral point of view, justify revolution. (Abraham Lincoln)

Write the negation of the compound statements in Problems 20–27.

20. Jane went to Macy's or Sears.

21. Paul went out for baseball or soccer.

22. Tim is not here and he is not at home.

23. Sally is not on time and she missed the boat.

24. If I can't go with you, then I'll go with Bill.

25. If you're out of Schlitz, you're out of beer.

26. If $x + 2 = 5$, then $x = 3$.

27. If $x - 5 = 4$, then $x = 1$.

28. Prove the Law of Double Negation.

29. Prove the Law of Contraposition.

30. Prove De Morgan's Law: $\sim(p \wedge q) \Leftrightarrow \sim p \vee \sim q$

31. Prove $(p \rightarrow q) \Leftrightarrow (\sim p \vee q)$

Mind Bogglers

32. One day in a foreign country I met three politicians. Now, all of the politicians of this country belonged to one of two political parties. The first was the Veracious Party, consisting of persons who could tell only the truth. The other party, called the Deceit Party, consisted of persons who were chronic liars. I asked these politi-

cians to which party they belonged. The first said something I did not hear. The second remarked "He said he belonged to the Veracious Party." The third said "You're a liar!" To which party did the third politician belong?

Translate the statements in Problems 33–35 into symbolic form.

33. Either Alfie is not afraid to go, or Bogie and Clyde will have lied.

34. Either Donna does not like Elmer because Elmer is bald, or Donna likes Frank and George because they are handsome twins.

35. Either neither you nor I am honest, or Hank is a liar because Iggy did not have the money.

Problem for Individual Study

36. Between now and the end of the course, look for logical arguments in newspapers, periodicals, and books. Translate these arguments into symbolic form. Turn in as many of them as you can find. Be sure to indicate where you found each argument.

2.6 The Nature of Proof in Problem Solving

"I THINK YOU SHOULD BE MORE EXPLICIT HERE IN STEP TWO."

One of mathematics' greatest strengths is its concern with the logical proof of its propositions. Any logical system must start with some undefined terms, definitions, and postulates or axioms. We have seen several examples of each of these. From here, other assertions can be made. These assertions are called theorems, and they must be proved using the rules of logic. In this book, we are concerned not so much with "proving mathematics" as with investigating the nature of mathematics. It seems appropriate, then, to speak of the nature of proof, since the idea of proof does occupy a great portion of a mathematician's time.

In Chapter 1 we discussed inductive reasoning. Experimentation, guessing, and looking for patterns are all part of inductive reasoning. After a conjecture or generalization has been made it needs to be proved using the rules of logic. Once a conjecture has been proved, it is called a **theorem.** Often, several years will pass from the conjectural stage to the final proved form. Certain definitions must be made, and certain axioms or postulates must be accepted. Perhaps the proof of the conjecture will require the results of some previous theorems.

Two of the major types of reasoning used by mathematicians are direct and indirect reasoning.

Direct reasoning is the drawing of a conclusion from two premises. The following example, called a **syllogism,** illustrates direct reasoning.

$p \rightarrow q$	If you receive an A on the final, then you will pass the course.
p	You receive an A on the final.
q	You pass the course.

This argument consists of two *premises,* or *hypotheses,* and a *conclusion;* the argument is valid if

$$[(p \rightarrow q) \wedge p] \rightarrow q$$

is always true. By means of Table 2.11 we can see that it is always true, so the argument is valid and we can write:

$$[(p \rightarrow q) \wedge p] \Rightarrow q$$

TABLE 2.11   Truth Table for Direct Reasoning

p	q	$p \rightarrow q$	$(p \rightarrow q) \wedge p$	$[(p \rightarrow q) \wedge p] \rightarrow q$
T	T	T	T	T
T	F	F	F	T
F	T	T	F	T
F	F	T	F	T

The pattern of argument is illustrated as follows:

Major premise:	$p \rightarrow q$
Minor premise:	p
Conclusion:	$\therefore q$

DIRECT REASONING

This reasoning is also sometimes called *modus ponens, law of detachment,* or *assuming the antecedent.*

Example 1: If you play chess, then you are intelligent.
You play chess.

$\therefore$ You are intelligent.

Three dots ($\therefore$) are used to mean "therefore."

Example 2: If you are a logical person, then you will understand this example.
You are a logical person.

$\therefore$ You understand this example.

We say that these arguments are *valid,* since we recognize them as direct reasoning.

The following syllogism illustrates what we call **indirect reasoning.**

$p \rightarrow q$ If you receive an A on the final, then you will pass the course.
$\sim q$ You did not pass the course.

$\therefore \sim p$ $\therefore$ You did not receive an A on the final.

We can prove this is valid by direct reasoning as follows:

$$[(\sim q \rightarrow \sim p) \wedge \sim q] \rightarrow \sim p$$

Also, since $\sim q \rightarrow \sim p$ is logically equivalent to the contrapositive $p \rightarrow q$, we have

$$[(p \rightarrow q) \wedge \sim q] \Rightarrow \sim p$$

You can also prove indirect reasoning is valid by using a truth table, and you are asked to do this in the problem set. The pattern of argument is illustrated as follows:

INDIRECT REASONING

This method is also known as reasoning by *denying the consequent* or as *modus tollens*.

Major premise:	$p \rightarrow q$
Minor premise:	$\sim q$
Conclusion:	$\therefore \sim p$

Example 3:   If the cat takes the rat, then the rat will take the cheese.
                 The rat does not take the cheese.

$\therefore$ The cat does not take the rat.

Example 4:   If you received an A on the test, then I am Napoleon.
                 I am not Napoleon.

$\therefore$ You did not receive an A on the test.

Sometimes we must consider some extended arguments. Transitivity allows us to reason through several premises to some conclusion.
The argument form is given as follows:

Premise:	$p \rightarrow q$
Premise:	$q \rightarrow r$
Conclusion:	$\therefore p \rightarrow r$

Table 2.12 shows the proof for transitivity.

Since transitivity is always true, we may write

$$[(p \rightarrow q) \wedge (q \rightarrow r)] \Rightarrow (p \rightarrow r)$$

TABLE 2.12   Truth Table for Transitivity

p	q	r	$p \rightarrow q$	$q \rightarrow r$	$p \rightarrow r$	$(p \rightarrow q) \wedge (q \rightarrow r)$	$[(p \rightarrow q) \wedge (q \rightarrow r)] \rightarrow (p \rightarrow r)$
T	T	T	T	T	T	T	T
T	T	F	T	F	F	F	T
T	F	T	F	T	T	F	T
T	F	F	F	T	F	F	T
F	T	T	T	T	T	T	T
F	T	F	T	F	T	F	T
F	F	T	T	T	T	T	T
F	F	F	T	T	T	T	T

Example 5:   If you attend class, then you will pass the course. $p \rightarrow q$
                    If you pass the course, then you will graduate. $q \rightarrow r$
                    ∴If you attend class, then you will graduate. $\therefore p \rightarrow r$

Transitivity can be extended so that a chain of several if–then sentences is connected together. For example, we could continue:

Example 6:   If you graduate, then you will get a good job.
                    If you get a good job, then you will meet the right people.
                    If you meet the right people, then you will become well known.

                    ∴If you attend class, then you will become well known.

The poem "For Want of a Nail a Kingdom Was Lost" used this type of reasoning to conclude: If a nail is lost, then a kingdom is lost. (See Problem 16 in Problem Set 2.6.)

Several of these argument forms may be combined into one argument. Remember:

---

1. **Translate into symbols.**
2. **Simplify the symbolic argument.** You might replace a statement by its contrapositive, use direct or indirect reasoning, or use transitivity.
3. **Translate the conclusion back into words.**

Example 7: Form a valid conclusion using all these statements:

1. If I receive a check for \$500, then we will go on vacation.
2. If the car breaks down, then we will not go on vacation.
3. The car breaks down.

Solution: First, change into symbolic form:

1. $c \rightarrow v$ where c = I receive a \$500 check
2. $b \rightarrow \sim v$ v = We will go on vacation
3. b b = Car breaks down

Next, simplify the symbolic argument. In this example, we must rearrange the premises.

2. $b \rightarrow \sim v$ Second premise
1. $c \rightarrow v$ First premise
3. b

Now $c \rightarrow v$ is the same as $\sim v \rightarrow \sim c$ (replace the statement by its contrapositive). That is, if we replace (1) by (1') $\sim v \rightarrow \sim c$, we obtain the following argument.

Notice that, when presenting the argument, we state a reason for each step. With a little practice we'll be able to give reasons for all our arguments.

2. $b \rightarrow \sim v$ Second premise
1'. $\underline{\sim v \rightarrow \sim c}$ Contrapositive of first premise
     $\therefore b \rightarrow \sim c$ Transitive
3. $\underline{b\quad\quad\quad}$ Third premise
     $\therefore \sim c$ Direct reasoning

Finally, we translate the conclusion back into words: "I did not receive a check for \$500."

Problem Set 2.6

Name the type of reasoning illustrated by the arguments given in Problems 1–9.

1. If I inherit \$1,000, I will buy you a cookie.
   I inherit \$1,000.
   Therefore, I will buy you a cookie.

2. If a^2 is even, then a must be even.
a is odd.
Therefore, a^2 is odd.

Hint: Assume that a is a counting number. If a counting number is odd, then it is not even.

3. All snarks are fribbles.
All fribbles are ugly.
Therefore, all snarks are ugly.

Hint: Rewrite the premises in if–then form.

4. If you understand a problem, it is easy.
The problem is not easy.
Therefore, you do not understand the problem.

5. If I don't get a raise in pay, I will quit.
I don't get a raise.
Therefore, I quit.

6. If Fermat's Last Theorem is ever proved, then my life is complete.
My life is not complete.
Therefore, Fermat's Last Theorem is not proved.

7. All mathematicians are eccentrics.
All eccentrics are rich.
Therefore, all mathematicians are rich.

8. No students are enthusiastic.
You are enthusiastic.
Therefore, you are not a student.

9. If you like beer, you'll like Bud.
You don't like Bud.
Therefore, you don't like beer.

10. Using symbolic form, state the following:

a. Direct reasoning   **b.** Indirect reasoning   **c.** Transitivity

11. Prove indirect reasoning using a truth table.

In Problems 12–28 form a valid conclusion using all the statements for each argument.

12. If you can learn mathematics, then you are intelligent.
If you are intelligent, then you understand human nature.

13. If I am idle, then I become lazy.   **14.** All trebbles are frebbles.
I am idle.   All frebbles are expensive.

15. If we interfere with the publication of false information, we are guilty of suppressing the freedom of others. We are not guilty of suppressing the freedom of others.

16. If a nail is lost, then a shoe is lost.
If a shoe is lost, then a horse is lost.
If a horse is lost, then a rider is lost.
If a rider is lost, then a battle is lost.
If a battle is lost, then a kingdom is lost.

𝕳𝖎𝖘𝖙𝖔𝖗𝖎𝖈𝖆𝖑 𝕹𝖔𝖙𝖊

Problems 25–28 are patterned
after the English mathematician
Charles Dodgson's (1832–1898)
famous logic problems. Dodgson,
probably better known by his
pseudonym, Lewis Carroll, was
the author of *Alice in Wonderland*
and *Through the Looking Glass*
as well as numerous mathematics
textbooks. Queen Victoria
was said to have been so taken by
his children's books that she
requested copies of every book he
had written. Imagine her surprise
when she received a pile
of mathematics books! Dodgson
stammered, and he felt most
at ease with children, which
accounts for his interest in
children's books. The fascinating
thing about his children's books
is the fact that they can also be
appreciated from a mathematical
standpoint. For example,
Through the Looking Glass is
based on a game of chess, and, in
Alice in Wonderland, Alice's size
and proportions form a closed set
of projective transformations.

Hint for Problem 29: You do not
need to use all the premises.

17. If you climb the highest mountain, you will feel great.
If you feel great, then you are happy.

18. If $b = 0$, then $a = 0$. $a \neq 0$.

19. If $a \cdot b = 0$, then $a = 0$ or $b = 0$. $a \cdot b = 0$.

20. If I eat that piece of pie, I will get fat.
I will not get fat.

21. If we win first prize, we will go to Europe.
If we are ingenious, we will win first prize.
We are ingenious.

22. If I can earn enough money this summer, I will attend college in the fall.
If I do not participate in student demonstrations, then I will not attend college.
I earned enough money this summer.

23. If I am tired, then I cannot finish my homework.
If I understand the material, then I can finish my homework.

24. If you go to college, then you will get a good job.
If you get a good job, then you will make a lot of money.
If you do not obey the law, then you will not make a lot of money.
You go to college.

25. All puppies are nice.
This animal is a puppy.
No nice creatures are dangerous.

26. Babies are illogical.
Nobody is despised who can manage a crocodile.
Illogical persons are despised.

27. Everyone who is sane can do logic.   **28.** No ducks waltz.
No lunatics are fit to serve on a jury.   No officers ever decline to waltz.
None of your sons can do logic.   All my poultry are ducks.

29. If the butler was present, he would have been seen; if he was seen, he would have
been questioned. If he had been questioned, he would have replied; if he had
replied, he would have been heard. But the butler was not heard. If the butler
was neither seen nor heard, then he must have been on duty; if he was on duty,
then he must have been present. Therefore, the butler was questioned. Is the conclusion valid?

30. There is a logic game available called WFF 'N PROOF, which I highly recommend. It is suitable for all levels of maturity. In its advertising, the following
statements are made:

> **1.** There are three numbered statements in this box.
> **2.** Two of these numbered statements are not true.
> **3.** The average increase in I.Q. scores of those people who learn to play
> WFF 'N PROOF is more than 20 points.

Is statement 3 true?

Mind Bogglers

31. If Alice shouldn't, then Carol would. It is impossible that the following statements can both be true at the same time:

a. Alice should.
b. Betty couldn't.

If Carol would, then Alice should and Betty could.
Therefore, Betty could. Is the conclusion valid?

32. In the story in the *Buz Sawyer* cartoon, Lucille is able to figure out the boys' fish story. The solution has been blacked out. See if you can determine how many fish were caught by Rosco, Jake, and Elmo.

33. Form a valid conclusion using all of the following statements:
No kitten that loves fish is unteachable.
No kitten without a tail will play with a gorilla.
Kittens with whiskers always love fish.
No teachable kitten has green eyes.
No kittens have tails unless they have whiskers.

34. Form a valid conclusion using all of the following statements:

When I work a logic problem without grumbling, you may be sure it is one that I can understand.

These problems are not arranged in regular order, like the problems I am used to.

No easy problem ever makes my head ache.

I can't understand problems that are not arranged in regular order, like those I am used to.

I never grumble at a problem unless it gives me a headache.

35. Form a valid conclusion using all of the following statements:

Every idea of mine that cannot be expressed as a syllogism is really ridiculous.

None of my ideas about rock stars is worth writing down.

No idea of mine that fails to come true can be expressed as a syllogism.

I never have any really ridiculous idea that I do not at once refer to my lawyer.

All my dreams are about rock stars.

I never refer any idea of mine to my lawyer unless it is worth writing down.

36. I found the following article in the *Reader's Digest*. Answer the questions asked in the article.

WHODUNIT?

A brain-busting puzzle for mystery fans

Condensed from "Murder Ink"   Perpetrated by Dilys Winn

Boggs has been found dead in the club lounge, his wine poisoned. Four men, seated on a sofa and two chairs in front of the fireplace as shown above, are discussing the foul deed. Their names are Howell, Scott, Jennings and Wilton. They are, not necessarily respectively, a general, schoolmaster, admiral and doctor.

(a) The waiter pours a glass of whiskey for Jennings and a beer for Scott.

(b) In the mirror over the fireplace, the general sees the door close behind the waiter. He turns to speak to Wilton, next to him.

(c) Neither Howell nor Scott has any sisters.

(d) The schoolmaster is a teetotaler.

(e) Howell, who is sitting in one of the chairs, is the admiral's brother-in-law. The schoolmaster is next to Howell on his left.

Suddenly a hand moves stealthily to put something in Jennings' whiskey. It is the murderer. No one has left his seat and no one else is in the room.

What is the profession of each man, where is he sitting, and who is the murderer?

*2.7 Logic Circuits

In this section we will apply logic to the theory of switching circuits. For our purposes we will assume that all switches are two-state devices; that is, they are either *on* or *off*. Schematically, a switch is represented as shown in Figure 2.10. We call one state "on" and the other state "off"; the switch will be either on or off and will never be in the middle. However, sometimes it is drawn in the middle (as shown at the bottom of Figure 2.10) to indicate that at any one time it might be either on or off.

Suppose we connect a light with this switch so that the light is *on* when the switch is *on* and *off* when the switch is *off*. A light is schematically drawn as ◯. Since we need a power source, a battery, ⊣|⊢ , is shown at the bottom. The light will light whenever a circuit is completed. Notice that the circuit is complete when *p* is switched on and not complete when *p* is off. (See Figure 2.11.)

Since these switches can assume either one of two states, *on* or *off*, and since our logical statements can assume either one of two states, *true* or *false*, it seems natural to define a relationship between a *switch* and a *logical statement*. Electronic circuits, which we will call *logical gates*, are devices with one or more inputs and one output. We will consider three of these:

1. Negation circuit or NOT-gate
2. Series circuit or AND-gate
3. Parallel circuit or OR-gate

These circuits can be used to illustrate some of the logical principles we have been discussing. First, however, we must make the following agreements. We have been considering switches that are on or off and a light that is on or off. Let us call switch *p* a logical proposition that is either true or false. If *p* is *true*, switch *p* will be represented as an *on-switch*. If *p* is *false*, switch *p* will be *off*. Similarly, we will let the light be the truth value of the simple or compound statements. Now let's consider some logical circuits. Recall the truth table for negation. We see that the truth value for ~*p* is the opposite of the truth value for *p*. In terms of circuits, this means that, when switch *p* is on, the light must be off, and, if *p* is off, the light is on. (You probably have some "negation switches" in your home.)

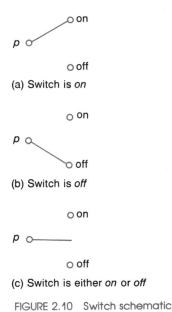

(a) Switch is *on*

(b) Switch is *off*

(c) Switch is either *on* or *off*

FIGURE 2.10 Switch schematic

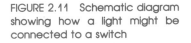

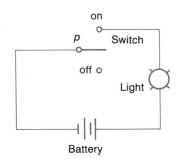

FIGURE 2.11 Schematic diagram showing how a light might be connected to a switch

Negation

p	~*p*
T	F
F	T

The electrical circuit for negation is shown in Figure 2.12.

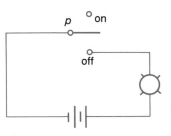

FIGURE 2.12 Negation circuit

*This is an optional section.

Conjunction

p	q	$p \wedge q$
T	T	T
T	F	F
F	T	F
F	F	F

Conjunction is a little more difficult since two switches, p and q, are involved. The truth value of the light should indicate that it is true if and only if p and q are both true, as shown in the truth table for conjunction. Switches p and q must be connected *in series,* so that the circuit is completed if and only if both p and q are on (like Christmas lights that all go out if one goes out). (See Figure 2.13.)

FIGURE 2.13 Series (conjunction) circuit

Disjunction

p	q	$p \vee q$
T	T	T
T	F	T
F	T	T
F	F	F

The truth value of $p \vee q$ is such that the light is on whenever either p or q is on. That is, p and q must be connected *in parallel,* so that the circuit is completed if either p or q is on (like Christmas lights that stay lit even if one light burns out). Study Figure 2.14, and then try to predict when the light will be on.

It is often necessary to combine the circuits of Figures 2.12, 2.13, and 2.14 into more complicated circuits that can be used to represent more complicated compound statements. When we do this, the schematics can become a little unmanageable. Therefore we will invent new notational symbols to stand for entire circuits. The first of these is called the *NOT-gate.* Instead of drawing the circuit shown in Figure 2.12, we will simply represent the entire circuit as follows:

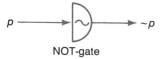

NOT-gate

Notice that the NOT-gate has a single input (proposition p) and a single output (proposition $\sim p$). The input p stands for the p-switch in Figure 2.12. The output $\sim p$ stands for the light in Figure 2.12. That is, current will flow out of the gate in those cases in which the light of Figure 2.12 is on and will not flow out of the gate in those cases in which the light is off.

The AND- and OR-gates, on the other hand, have two switches, p and q. These are symbolized as two inputs, p and q. The single output stands for the

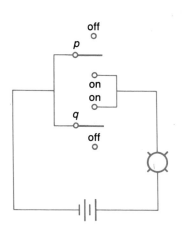

FIGURE 2.14 Parallel (disjunction) circuit

light. These gates are represented symbolically as shown in the margin.

Logic circuits can be constructed to simulate logical truth tables. For example, suppose we wish to find the truth values for $\sim(p \lor q)$. We would design the following circuit (remember that parentheses indicate the operation to be performed first):

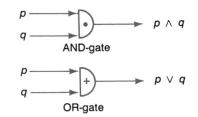

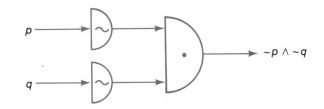

Example 1:   Design a circuit for $\sim p \land \sim q$.

Solution

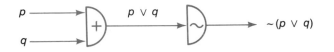

Example 2:   Design a circuit that will find the truth values for $(p \lor q) \land q$.

Solution: We use gates to symbolize $(p \lor q) \land q$:

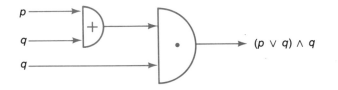

Next we must have some way of determining when the output values of $(p \lor q) \land q$ are true and when they are false. We can do this by again connecting a light to the circuit. When $(p \lor q) \land q$ is true, the light should be on; when it is false, the light should be off. We make this connection as follows:

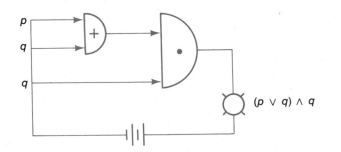

You may have noticed the relationship among sets, logic, and circuits. These relationships are summarized in Table 2.13.

TABLE 2.13

	Sets	*Logic*	*Circuits*
Elements	Sets: $A, B, C, \ldots$	Propositions: $p, q, r, \ldots$	Switches
Operations	Intersection: ∩	Conjunction: ∧	Series circuit:
	Union: ∪	Disjunction: ∨	Parallel circuit:
	Complement: $\overline{A}$	Negation: $\sim p$	Negation circuit:

Problem Set 2.7

1. Consider Figure 2.12. In your own words, explain when the light will be on and when it will be off.

2. What do the following circuit symbols mean?

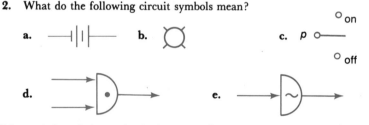

Using switches, design a circuit that would find the truth values for each of the statements in Problems 3–11.

3. $p \wedge q$ 4. $p \vee q$ 5. $\sim p \wedge q$

6. $p \wedge \sim q$ 7. $\sim(p \vee q)$ 8. $\sim(p \wedge q)$

9. $p \rightarrow q$ 10. $q \rightarrow p$ 11. $p \rightarrow \sim q$

12. Recall from Section 2.4 that $p \rightarrow q$ is defined to be $\sim p \vee q$. Design a circuit for $p \rightarrow q$ by using AND-gates and NOT-gates.

Using AND-gates, OR-gates, and NOT-gates, design a circuit that will find the truth values for the statements in Problems 13–21.

13. $p \wedge \sim q$ 14. $p \wedge (q \vee r)$ 15. $(p \wedge q) \vee (p \wedge r)$

16. $\sim[(p \wedge q) \vee r]$ 17. $\sim(\sim p \vee q)$ 18. $(\sim p \wedge q) \vee (p \vee \sim q)$

19. $\sim p \rightarrow q$ 20. $p \rightarrow \sim q$ 21. $\sim q \rightarrow \sim p$

22. The cost of a circuit is often a factor when working with switching circuits. For example, the circuit you constructed in Problem 12 by using the definition of the

conditional required three gates. Construct a truth table for $p \rightarrow q$ and $\sim p \vee q$. What do you notice about the truth values of the result? Design a circuit for $p \rightarrow q$ by using only one OR-gate and one NOT-gate.

Mind Bogglers

23. **a.** Suppose an engineer designs the following circuit: $\sim\{\sim[(p \vee q) \wedge \sim p] \wedge \sim p\}$. What will the circuit look like?

b. If each gate costs 2¢ and a company is going to manufacture one million items using this circuit, how much will the circuit cost?

c. Use truth tables to find the following.
   i. $\sim\{\sim[(p \vee q) \wedge \sim p] \wedge \sim p\}$
   ii. $p \vee q$
   iii. What do you notice about the final column of each of these truth tables?

d. Draw a simpler circuit that will output the same values as those in part a.

e. If the company is going to manufacture one million items using this simpler circuit, how much will the circuit cost? How much will the company save by using this simpler circuit rather than the more complicated one?

24. Alfie, Bogie, and Clyde are the members of a senate committee. Design a circuit that will output *yea* or *nay* depending on the way the majority of the committee members voted.

25. Suppose that the senate committee of Problem 24 has five members. Design a circuit that will output the result of the majority of the committee's vote.

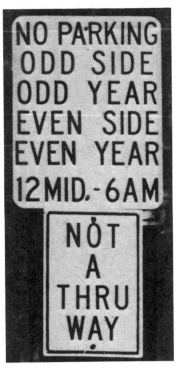

NO PARKING ODD SIDE ODD YEAR EVEN SIDE EVEN YEAR 12 MID. - 6 AM

NOT A THRU WAY

HOW ODD!

Many drivers must throw up their hands in utter confusion when they come upon this wacky sign in Medford, Mass. Local police say the sign is really easy to decipher: You can't park on the odd side of the street—the side in which the house numbers are odd—from midnight to 6 a.m. in an odd-numbered year such as 1985. Likewise, parking on the even side is taboo in an even year. The reason for this zany arrangement is to keep one side clear of snow for emergency vehicle use.

Chapter 2 Outline

I. Sets
   A. Specifying sets
      1. Description
      2. Roster
      3. Braces
   B. Special sets
      1. Empty set
      2. Universal set
   C. Subsets
      1. Proper
      2. Improper
      3. Equal sets
   D. Venn diagrams
      1. Disjoint sets
      2. Overlapping sets
      3. General diagram for two sets
   E. Set operations
      1. Union
      2. Intersection
      3. Complementation

II. Combined Operations with Sets
   A. Venn diagrams
   B. Proofs using Venn diagrams
   C. De Morgan's Laws
   D. General diagram for three sets; combined operations
   E. Survey problems

III. Deductive Reasoning
   A. Statements
   B. Conjunction
   C. Disjunction
   D. Negation
   E. Truth value of compound statements
   F. Translations

IV. The Conditional
   A. Truth tables
   B. Law of double negation
   C. Conditional
      1. Definition

      2. Antecedent and consequent   **G.** Other operators
      3. Connectives   **1.** Either . . . or
         a. p only if q **2.** Neither . . . nor
         b. all p is q **3.** Unless
    D. Converse, inverse, and   **4.** Because
      contrapositive   **5.** No p is q
    E. Law of contraposition   **VI.** The Nature of Proof
V. Additional Connectives   **A.** Direct reasoning
    A. Biconditional   **B.** Indirect reasoning
      1. If and only if   **C.** Transitivity
      2. If p then q, and conversely   **D.** Simplifying arguments
    B. Tautology   ***VII.** Logic Circuits
    C. Implication   **A.** Switches
    D. Logical equivalence   **B.** Negation circuit
    E. De Morgan's Laws   **C.** Series circuit
    F. Negation of compound   **D.** Parallel circuit
      statements

Chapter 2 Test

1. Let $U = \{1,2,3,4,5,6,7,8,9,10\}$
      $A = \{1,3,5,7,9\}$
      $B = \{2,4,6,9,10\}$
  Find:

    a. $A \cup B$ **b.** $A \cap B$ **c.** $\overline{B}$
    d. $A \cup \overline{B}$ **e.** $\overline{A} \cap B \cup A)$

2. Indicate whether the following are true or false:

    a. $\varnothing \in \varnothing$ **b.** $\varnothing \subseteq \varnothing$ **c.** $\varnothing = \varnothing$
    d. $\varnothing = \{\varnothing\}$ **e.** $\varnothing = 0$

3. Prove or disprove:

    a. $A \cup (B \cap C) = (A \cup B) \cap C$
    b. $(A \cup B) \cap C = (A \cap C) \cup (B \cap C)$

4. Does the story in the margin illustrate inductive or deductive reasoning? Explain your answer.

5. Complete the following truth table:

p	q	$p \wedge q$	$p \vee q$	$\sim p$	$\sim q$	$p \rightarrow q$	$p \leftrightarrow q$
T	T						
T	F						
F	T						
F	F						

Optional section.

Q: What has 18 legs and catches flies?
A: I don't know, what?
Q: A baseball team. What has 36 legs and catches flies?
A: I don't know that either.
Q: Two baseball teams. If the United States has 100 senators, and each state has 2 senators, what does . . .
A: I know this one!
Q: Good. What does each state have?
A: Three baseball teams!

6. a. Construct a truth table for $\sim(p \wedge q)$.
   b. Show $\sim(p \wedge q) \Leftrightarrow (\sim p) \vee (\sim q)$.

7. Consider the statement "If I don't go to college, then I will not make a lot of money." Write the converse, inverse, and contrapositive.

8. a. State and prove the principle of direct reasoning.
   b. Give an example of indirect reasoning.
   c. Is the following valid?

$$p \rightarrow q$$
$$\underline{\sim q}$$
$$\therefore \sim p$$

   Can you support your answer?

9. Form a valid conclusion using all the statements for each argument.

   a. No person going to Hawaii fails to get a tan.
   Mr. Gent went to Hawaii.
   b. If you want peace, prepare for war.
   You do not prepare for war.
   c. If we go on a picnic, then we must pack a lunch.
   If we pack a lunch, then we must buy bread.
   If we buy bread, then we must go to the store.
   d. If I attend to my duties, I will be rewarded.
   If I am lazy, I will not be rewarded.
   I am lazy.
   e. All organic food is healthy.
   All artificial sweeteners are unhealthy.
   No prune is nonorganic.

*10. a. Design a circuit to find the truth values of $\sim p \vee \sim q$.
   b. Using AND-gates, OR-gates, and NOT-gates, design a circuit for $\sim[(p \vee q) \wedge q]$.

Bonus Question

11. *Liar problem.* On the South Side, one of the mob had just knocked off a store. Since the Boss had told them all to lay low, he was a bit mad. He decided to have a talk with the boys.

From the boys' comments, can you help the Boss figure out who committed the crime? Assume that the Boss is telling the truth.

*Requires optional Section 2.7.

Date	Cultural History
800	800: Charlemagne crowned emperor of Holy Roman Empire 832: Utrecht Psalter 843: Beginning of Carolingian dynasties

Charlemagne

| 850 | 871: Alfred the Great Schism of the Church |

Alfred the Great

| 900 | 900: Vikings discover Greenland Reign of Otto I |

Viking ship

St. John the Baptist, Reni

| 950 | 950: Beginning of the Dark Ages

990: Development of systematic musical notation

993: First canonization of saints |

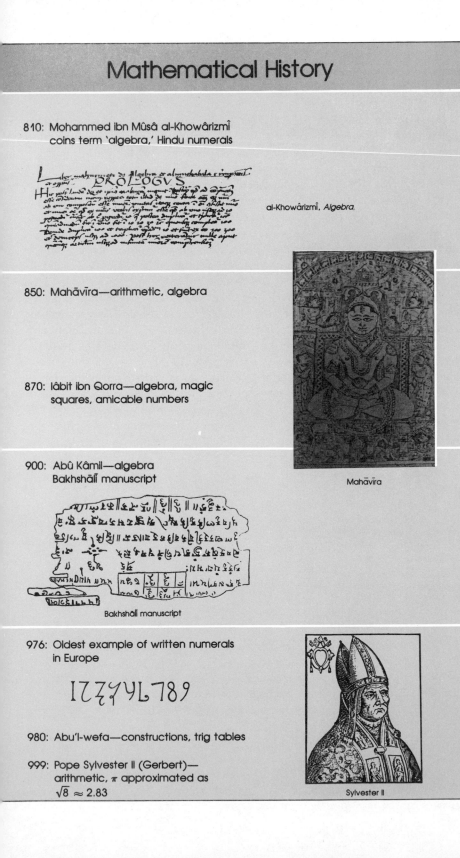

Mathematical History

810: Mohammed ibn Mûsâ al-Khowârizmî coins term 'algebra,' Hindu numerals

al-Khowârizmî, *Algebra*,

850: Mahāvīra—arithmetic, algebra

870: Iâbit ibn Qorra—algebra, magic squares, amicable numbers

Mahāvīra

900: Abû Kâmil—algebra
Bakhshālī manuscript

Bakhshālī manuscript

976: Oldest example of written numerals in Europe

980: Abu'l-wefa—constructions, trig tables

999: Pope Sylvester II (Gerbert)—arithmetic, π approximated as $\sqrt{8} \approx 2.83$

Sylvester II

3

The Nature of Calculators and Computers

To err is human, but to really foul things up requires a computer.

Anonymous

3.1 History of Computing Devices

"It says, 'don't fold, spindle, or mutilate.'"

In his book *Future Shock,* Alvin Toffler divided humanity's time on earth into 800 lifetimes. The 800th lifetime, in which we now live, has produced more knowledge than the previous 700 combined, and this has been made possible because of computers.

Most people still think of computers in the way they were portrayed in the 1950s—as giant brains full of wires, chips, and integrated circuits with the potential for turning us into second-class citizens. It is astounding that computers, which influence our lives to a considerable extent every day, can remain misunderstood by the majority of our population. Even people who venture into one of the new computer stores for information may feel intimidated and overwhelmed by all the new terminology.

To function intelligently in our society, we need to understand how and why computers influence our lives. Every educated person should know something about computers and, if possible, have some experience with one. As computers diligently carry out assigned and generally tedious tasks, people are free to do creative thinking. It is mathematics, not computation, that is fascinating. Prolonged computations lead to weariness, errors, and a distaste for arithmetic. It is not surprising that we have developed mechanical devices to perform our calculations.

The first "aid" to arithmetic computations is finger counting. It has the advantages of low cost and instant availability. You are familiar with addition and subtraction on your fingers, but here is a method of multiplication by nine. Place both hands as shown.

To multiply 4 × 9, simply bend the fourth finger from the left.

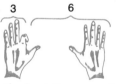

The answer is read as 36, the bent finger serving to distinguish between the first and second digits in the answer.

What about a two-digit number times 9? There is a procedure, provided the tens digit is smaller than the ones digit. For example, to multiply 36 × 9, separate the third and fourth fingers from the left (as shown), since 36 has 3 tens.

Next bend the sixth finger from the left, since 36 has 6 units. Now the answer can be read directly from the fingers. The answer is 324.

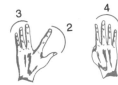

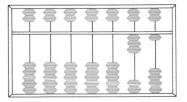

FIGURE 3.1   Abacus. The number shown is 31.

Numbers can also be represented by stones, slip knots, or beads on a string. These devices eventually evolved into the abacus (Figure 3.1). Abacuses were used thousands of years ago and are still used. In the hands of an expert, they rival even mechanical (but not electronic) calculators in speed. An abacus consists of rods containing sliding beads, four or five in the lower section and one or two in the upper. One in the upper equals five in the lower, and two in the upper equals one in the lower section of the next higher denomination. The abacus is useful today both for figuring business transactions and for teaching mathematics to youngsters. One can actually "see" the "carry" in addition and the "borrow" in subtraction (see Problem Set 3.1, Problem 13).

The 17th century saw the beginnings of modern calculating machines. John Napier invented a device in 1624 that was similar to a multiplication table with movable parts (see Figure 3.2 and Problem 14 in the problem set), but the first real step was taken in 1642 by Blaise Pascal.

Historical Note

John Napier (1550–1617) invented these rods, or, as they were sometimes called, *Napier's bones*. Napier also invented logarithms, which led to the development of the slide rule (shown in Figure 3.7). Napier considered himself a theologian and not a mathematician. In 1593 he published a commentary on the Book of Revelation and argued that the Pope at Rome was the anti-Christ. He also predicted that the end of the world would come near the end of the 17th century. Napier believed his reputation would rest solely on this work, but instead we remember him because of what he considered his minor mathematical diversions.

FIGURE 3.2   Napier's rods (1624)

Historical Note

Blaise Pascal (1623–1662)

Pascal was mentioned earlier in reference to Pascal's triangle. Born in 1623, he was a brilliant child and by the age of 13 was introduced to many of the intellectuals of France. He studied physics, mathematics, and religion. However:

the young Pascal was almost lost to history. At the age of one, he became seriously ill with either tuberculosis or rickets. The doctors found that the sight of water made him hysterical, and he would often throw tantrums when he saw his father and mother together.

Medicine was still quite primitive, and witchcraft was looked upon as the doer of physical harm. Blaise's illness was diagnosed as a sorcerer's spell. To remove the curse, it was necessary to find the witch, and the town turned against a poor old woman living on the charity of Mrs. Pascal and a frequent visitor to the Pascal house. This pathetic woman was under constant persecution. Although she believed herself innocent, to relieve this endless pressure, she threw herself before Étienne Pascal and promised to reveal all if he would not have her hanged.

Her story was full of mysticism. She said that she had earlier asked Blaise's father, a lawyer, to defend her against an unjust suit, and, in retaliation for his denying her request, she had put a death spell on the child. Although the charm could not be withdrawn, she said it could be transferred to bring about some other animal's death. A horse was first offered, but it was determined that a cat would be less costly. Descending the stairs with a cat in hand, she met two friars, who scared the old lady, and in her fright she threw the animal out of a window. Unlike the story of nine lives, this cat hit the pavement and died. Time was to be the healer, and Blaise did eventually recover.

When Pascal was 19, he began to develop a machine that was supposed to add long columns of figures (see Figure 3.3). The machine was essentially like those of today. He built several versions, and, since all were unreliable, he considered his project a failure; but with the machine he introduced basic principles that are used in modern mathematical calculators.

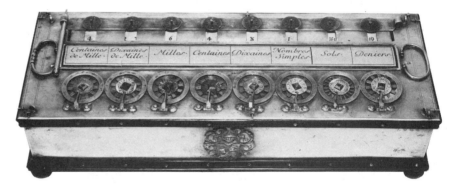

FIGURE 3.3 Pascal's calculator (1642)

The next advance in calculating devices came from Germany in about 1672, when the great mathematician Gottfried Leibniz studied Pascal's calculators, made improvements on them, and drew up plans for a mechanical calculator (see Figure 3.4). In 1695 a machine was actually built, but the resulting calculator was also unreliable. Yet Leibniz:

almost captured one of the most important aspects of modern-day computing but failed to see its application to mathematics. The binary system, so important in present computers, was envisioned by Leibniz, but he saw it in terms of religious significance. He saw God represented by 1 and nothing represented by 0. Leibniz used the binary scale as proof that God had created the world (1) out of nothing (0). In fact, he used the binary system as proof to convert the emperor of China to accept a God who could create a universe out of nothing.

𝕳istorical 𝕹ote

Gottfried Leibniz (1646–1716)

FIGURE 3.4   Leibniz' reckoning machine (1695)

The next advance came, oddly enough, from the weaving industry. In about 1725 a man named Basile Bouchon invented an endless loop of *punched paper tape* to aid in the weaving of designs with the hand loom, and in 1728 a Frenchman, M. Falcon, used *punched paper cards* for the same task. The device was employed on a Jacquard loom (see Figure 3.5). Although it is generally called the Jacquard card, it was invented by Falcon and was the forerunner of modern punched cards. Falcon took a card and punched holes in it where he wanted the needles to be directed; to change the pattern, he needed only to change the card.

(a) Jacquard loom with cards (1801)   **(b) Detail of one of the punched cards**

FIGURE 3.5

Charles Babbage (1792–1871)

Punched cards led to the idea that the arithmetic functions of a calculating machine could be "programmed" in advance. This concept was proposed by the eccentric Englishman Charles Babbage (1792–1871), who was years ahead of his time. His calculating machine was grandiose, with thousands of gears, shafts, ratchets, and counters. He called it his "difference engine" (see Figure 3.6).

(a) Difference engine (1812)   **(b) Analytic engine (1851)**

FIGURE 3.6   Babbage's calculating machines

Four years later Babbage had still not built his difference engine, but he had designed a much more complicated machine—one capable of accuracy to 20 decimal places—that, however, could not be completed since the technical knowledge to build it was not far enough advanced. This machine, the "Analytic Engine," was the forerunner of modern computers. In fact, International Business Machines Corporation has built an Analytic Engine based completely on Babbage's design but using, of course, modern technology—and it works perfectly.

Babbage continued for 40 years trying to build his Analytic Engine, always seeking financial help to subsidize his project. However, his eccentricities often got him into trouble. For example, he hated organ-grinders and other street musicians, and for this reason people ridiculed him. He was a no-nonsense man and would not tolerate inaccuracy. Were he willing to compromise his ideas of perfection with the technical skill then available, he might have developed a workable machine. Yet his desire for precision was not limited to his work. After Lord Tennyson wrote "The Vision of Sin," Babbage sent this note to the poet:

Sir,

In your otherwise beautiful poem there is a verse which reads
"Every minute dies a man,
Every minute one is born."

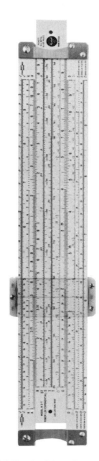

FIGURE 3.7   A slide rule

It must be manifest that if this were true, the population of the world would be at a standstill. In truth the rate of birth is slightly in excess of that of death. I would suggest that in the next edition of your poem you have it read—

"Every moment dies a man,
Every moment $1\frac{1}{16}$ is born."

Strictly speaking, this is not correct; the actual figure is so long that I cannot fit it into one line, but I believe that the figure $1\frac{1}{16}$ will be sufficiently accurate for poetry.

I am, Sir, yours, etc.

Babbage's Analytic Engine was all mechanical. Modern calculating machines are powered electrically and can perform additions, subtractions, multiplications, and divisions. For raising to a power or taking roots, the device used for many years was a slide rule (see Figure 3.7), which was also invented by Napier. The answers given by a slide rule are only visual approximations and do not have the precision that is often required. Electronic pocket calculators, which perform all these operations quickly and accurately, represent the latest advance in calculators. Since they are so useful, as well as inexpensive and readily available, we will discuss their use in the next section.

The devices discussed thus far in this section would all be classified as calculators. With some of the new programmable calculators, the distinction between a calculator and a computer is less well defined than it was in the past. The first generally recognized automatic-sequence electromechanical computer was the Mark I, developed by Howard Aiken at Harvard in 1944. The first fully electronic digital computer was built in 1946 at the University of Pennsylvania and was called ENIAC (Electronic Numerical Integrator and Calculator). This and other first-generation computers used vacuum tubes, which took up a large amount of space, consumed enormous amounts of electricity, and were not very efficient. The ENIAC, for example, filled a space 30 × 50 feet, weighed 30 tons, had 18,000 tubes, and used enough electricity to operate three 150 kilowatt radio stations.

In the second generation of computers, tubes were replaced by transistors, which reduced space and power requirements (see Figure 3.8). The advantage of the transistor was that it did away with the separate materials (carbon, ceramics, tungsten, and so on) used in fabricating components and physically joined the materials into one structure (usually silicon). Nevertheless, the problem remained of soldering individual transistors and diodes into a circuit.

The problem of individually connecting the transistors was solved by assembling several transistors, diodes, resistors, and capacitors onto a single circuit board that could be plugged into a computer (TV manufacturers now use this type of circuitry).

The next thing designers began to worry about was speed. For an electrical pulse to travel through 8 inches of wire, it takes one-billionth of a second, or a **nanosecond.** (Light travels about 1 foot in 1 nanosecond.) To us this seems like a small amount of time, but for the computer designer, it is long enough to cause real problems.

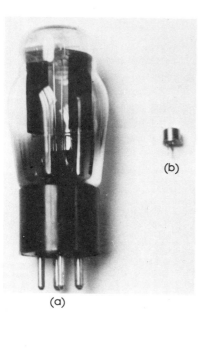

(a)

(b)

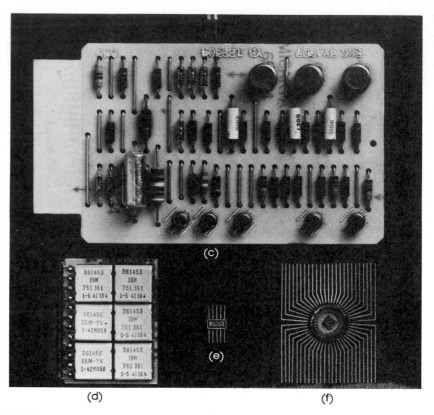

(c)

(d)   (f)

(e)

FIGURE 3.8   The evolution of computer hardware: *First Generation:* Vacuum tube, 1946 (a). *Second Generation:* Transistor, 1952 (b). Circuit board, 1960 (c); the one shown incorporates 45 devices. Solid state logic, 1963 (d); the one shown provides for 240 devices (40 in each tiny can). *Third Generation:* Integrated circuit, 1965 (e, f); the first (e) contains 91 transistors and resistors; (f) contains 864 devices. The integrated circuit in Figure 3.9 contains thousands of devices. *Fourth Generation:* Microcomputers, 1975. *Fifth Generation and beyond:* The evolution of computer devices is changing so rapidly that there is no longer a serious attempt to catalogue the evolution of computer devices by generation.

FIGURE 3.9   International Business Machines integrated circuit. A typical integrated circuit like this contains many individual circuits, each containing transistors, diodes, and resistors. The density of an integrated circuit like this one might exceed 90,000 devices per square inch.

More compact circuits were designed, and transistors, diodes, and resistors were all formed on a single chip of silicon called an **integrated circuit.** This microminiaturization of components led to what we call the third generation of computers (see Figures 3.8 and 3.9).

Today's computers are built to perform millions of operations per second. The state of the industry is advancing so rapidly that textbooks are outdated before they can be published. According to the 1982 edition of the *Guinness Book of World Records,* the fastest computer is the Cray-1, which can operate at a rate of 200 *million* results per *second.* With fourth-generation computers, we saw a rapid decrease in the size and cost of computers. *Minicomputers* are generally a cubic meter or less in size and cost under $10,000. *Personal computers,* such as the Apple IIe, TRS-80, IBM PC, and Atari 800, are all personal desktop computers that are designed for individual use. The machine, or computer itself, is referred to as **hardware.** A personal computer might sell for under $1,000, but many extras, called **peripherals,** can increase the price by several thousand dollars. **Software** refers to programs which tell the computer how to carry out a specific task. Programs are discussed in Sections 3.3–3.6.

FIGURE 3.10   Apple IIe and Apple III personal computers. (Copyright 1982, Apple Computer, Inc. Used by permission of Apple Computer, Inc., 20525 Mariani Ave., Cupertino, CA 95014.)

What does the future hold? A design goal set by International Business Machines Corporation is to fit an entire computer, including all its memory, into a box 8 centimeters by 8 centimeters by 10 centimeters. That is less than half the volume of this book! Computers will no longer consist of transistors or other semiconductor components but will instead consist of superconductors called Josephson-junction devices. This new generation of computers will shrink the time necessary to execute a single command from 50 nanoseconds (for present-day computers) to about 4 nanoseconds. A computer based on Josephson-junction switches could be as powerful as an IBM model 370/168—a very powerful machine—yet fit into a cube 6 inches on a side!

This ruler is 8 cm long.
Imagine a computer in a box
8 cm × 8 cm × 10 cm.

Problem Set 3.1

1. Use finger multiplication to do the following calculations:

    a. 3 × 9   **b.** 7 × 9   **c.** 27 × 9   **d.** 48 × 9   **e.** 56 × 9

𝔥𝔦𝔰𝔱𝔬𝔯𝔦𝔠𝔞𝔩 𝔑𝔬𝔱𝔢

Brian Josephson (1940–   ) was
only 33 years old when he won
the Nobel Prize in physics for
his work in superconductivity.
Superconductivity, according
to Josephson, occurs when you
cool a metal, and its electrical
resistance falls toward a lower
limit. In many pure metals all
electrical resistance vanishes at
a crucial transition temperature,
at which time it enters the
superconducting state. At that
time it will sustain a current
indefinitely.

2. Describe some of the computing devices that were used before the invention of the electronic computer.

3. In what way was Jacquard's automated loom related to computers?

4. What do we mean when we refer to first-, second-, third-, and fourth-generation computers?

5. What is a nanosecond?

6. Distinguish between hardware and software.

7. It has been said that "computers influence our lives increasingly every year, and the trend will continue." Do you see this as a benefit or a detriment to humanity? Explain your reasons.

8. In the next section we'll discuss the use of electronic calculators. Much has been written about the merits of using these calculators in the elementary grades. Do you see calculators as a benefit or a detriment to mathematical education? Explain your answer.

9. In his article "Toward an Intelligence beyond Man's" (*Time,* February 20, 1978), Robert Jastrow claims that by the 1990s computer intelligence will match that of the human brain. He quotes Dartmouth president John Kemeny, a leading mathematician and computer scientist, as saying that he "sees the ultimate relation between man and computer as a symbiotic union of two living species, each completely dependent on the other for survival." He calls the computer a new form of life. Do you agree or disagree with the hypothesis that a computer could be "a new form of life"?

10. What do you mean by "thinking" and "reasoning"? Try to formulate these ideas as clearly as possible, and then discuss the following question: "Can computers think or reason?"

Mind Boggler

11. The following are some basic definitions for "thinking." *According to each of the given definitions,* answer the question "Can computers think?"

    a. To remember
    b. To subject to the process of logical thought
    c. To form a mental picture of
    d. To perceive or recognize
    e. To have feeling or consideration for
    f. To create or devise
    g. To have the ability to learn

Problems for Individual Study

12. A method of finger calculation called Chisanbop has created a great deal of interest since it was demonstrated on the Johnny Carson show. It is described by one of its developers, Edwin Lieberthal, and the Director of the Psychological Research Laboratory, William Lamon, in "Chisanbop Finger Calculation" (*California MathematiCs Journal,* vol. 3, no. 2, October 1978, pp. 2–10). Another source of informa-

tion about Chisanbop is Chisanbop Enterprises, Mount Vernon, NY 10550. Write a report about this finger calculation method.

13. *Abacus.* Write a short paper or prepare a classroom demonstration on the use of an abacus. Build your own device as a project.

    Reference: Martin Gardner, "Mathematical Games," *Scientific American,* January 1970.

14. Build a working model of Napier's rods.

15. Write a report on the history of the computer chip.

    Reference: "The Chip," *National Geographic,* October 1982.

16. Several classics from literature, as well as contemporary movies and books, deal with attitudes toward computers and automated machinery. For example, HAL in *2001: A Space Odyssey* tries to take command of the spacecraft, and PROTEUS IV in *Demon Seed* even attempts to procreate a living child with a human "bride." Books such as *Frankenstein, Brave New World,* and *1984* precede computers but offer warnings about future supremacy of technology over humans. In light of these words and the recent developments in computer science, what moral responsibility do you think scientists must take for their creations?

    References: Arthur C. Clarke, *2001: A Space Odyssey* (New York: New American Library, 1968).

    Aldous Huxley, *Brave New World* (New York: Harper & Row, 1932).

    George Orwell, *1984* (New York: Harcourt Brace Jovanovich, 1949).

    Mary Shelley, *Frankenstein* (New York: Macmillan, 1970).

17. Alan Turing devised a test for "thinking" in the 1950s. In his now famous paper, Turing asked, "What would we ask a computer to do before we would say that it could think?" Turing devised what is now known as the "Turing Test." Describe this test and comment on whether you think a computer will ever be able to pass the test.

    Reference: Peter Kugel, " 'Can Computers Think?' Part II: What Is Thinking?" *Creative Computing,* September 1975, pp. 104–109.

3.2 Pocket Calculators

In the last few years pocket calculators have been one of the fastest growing items in the United States. There are probably two reasons for this increase in popularity. Most people (including mathematicians!) don't like to do arithmetic, and a good four-function calculator can be purchased for under $20. (You might even find one on sale for around $10.)

Calculators are classified according to their ability to do different types of problems, as well as by the type of logic they use to do the calculations. The problem of selecting a calculator is further compounded by the multiplicity of

Historical Note

The following is an interesting anecdote about Napier from Howard Eves' *In Mathematical Circles:*

> It is no wonder that Napier's remarkable ingenuity and imagination led some to believe he was mentally unbalanced and others to regard him as a dealer in the black art. Many stories, probably unfounded, are told in support of these views. Thus there was the time he announced that his coal black rooster would identify to him which one of his servants was stealing from him. He put his rooster in a box in a darkened room and instructed the servants to enter one by one and to place a hand on the rooster's back. Napier assured his servants that his rooster would expose the culprit to him at the completion of these performances. Now, unknown to the servants, Napier had coated the bird's back with lampblack. The innocent servants, having nothing to fear, did as they were bidden, but the guilty one decided to protect himself by not touching the bird. In this way he was exposed, for he was the only servant to return from the darkened room with clean hands.

A pocket calculator is necessary for working the problems in this section.

brands from which to choose. Therefore, choosing a calculator and learning to use it require some sort of instruction.

The different levels of calculators are distinguished primarily by their price.

1. *Four-function calculators (under $10)*. These calculators have a keyboard consisting of the numerals and the four arithmetic operations, or functions: addition $\boxed{+}$, subtraction $\boxed{-}$, multiplication $\boxed{\times}$, and division $\boxed{\div}$.

2. *Four-function calculators with memory ($10–$20)*. Usually no more expensive than four-function calculators, these offer a memory register, $\boxed{M}$, $\boxed{STO}$, or $\boxed{M^+}$. The more expensive models may have more than one memory register. Memory registers allow you to store partial calculations for later recall. Some checkbook models will even remember the total when they are turned off.

3. *Scientific calculators ($20–$50)*. These calculators have additional mathematical functions, such as square root $\boxed{\sqrt{}}$, trigonometric $\boxed{SIN}$, $\boxed{COS}$, $\boxed{TAN}$, and logarithmic $\boxed{LOG}$. Depending on the particular brand, a scientific model may have other keys as well.

4. *Special-purpose calculators ($40–$400)*. Special-use calculators for business, statistics surveying, medicine, and even gambling and chess are available.

5. *Programmable calculators ($50–$600)*. These calculators allow the insertion of different cards that "remember" the sequence of steps for complex calculations.

For most nonscientific purposes, a four-function calculator with memory will be sufficient for everyday usage. If you anticipate taking several mathematics and/or science courses, you'll find a scientific calculator to be a worthwhile investment.

There are essentially three types of logic used by these calculators: arithmetic, algebraic, and RPN. The order of operations tells us to perform multiplication before addition, so that the correct value for

$$2 + 3 \times 4$$

is 14 (multiply first). An algebraic calculator "knows" this fact and will give the correct answer, whereas an arithmetic calculator will simply work from left to right to obtain the incorrect answer 20. Therefore, if you have an arithmetic-logic calculator, you will need to be careful about the order of operations. Some arithmetic-logic calculators provide parentheses $\boxed{(}$ $\boxed{)}$ so that operations can be grouped as in

but then you must remember to insert the parentheses.

The last type of logic is RPN. A calculator using this logic is characterized by $\boxed{ENTER}$ or $\boxed{SAVE}$ keys and does not have an equal key $\boxed{=}$. With an RPN calculator the operation symbol is entered after the numbers have been entered. These three types of logic can be illustrated by the problem $2 + 3 \times 4$.

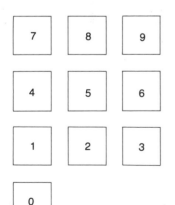

FIGURE 3.11 Standard calculator keyboard. Additional keys depend on brand and model.

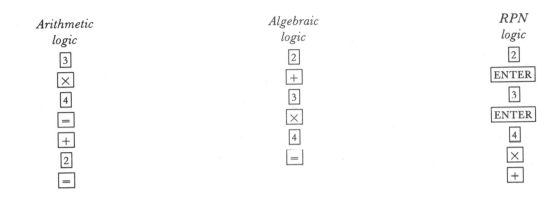

Arithmetic logic

3
×
4
=
+
2
=

Algebraic logic

2
+
3
×
4
=

RPN logic

2
ENTER
3
ENTER
4
×
+

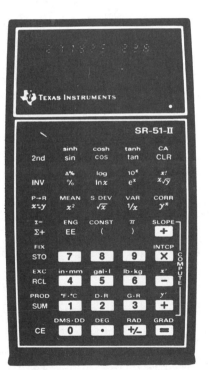

(a) The Sharp calculator is an example of a calculator with arithmetic logic.

(b) This Texas Instruments SR-51-II is an example of a calculator with algebraic logic.

(c) The Hewlett-Packard calculator uses RPN logic.

FIGURE 3.12 Types of calculators. When you're purchasing a calculator, try this test problem: 2 + 3 × 4 = If the display shows 20, it is an arithmetic-logic calculator. If the answer is 14 (the correct answer), then it is an algebraic-logic calculator. If the problem requires an ENTER or SAVE key, then it is an RPN-logic calculator.

Regardless of the type of logic your calculator uses, it is a good idea to check your owner's manual for each type of problem, since many different brands of calculators are on the market. There may be slight differences in how each brand handles a particular problem.

In this book, we will illustrate the examples using algebraic logic, and, if you have a calculator with RPN logic, you can use your owner's manual to change the examples to RPN. We will also indicate the keys to be pushed by drawing boxes around the numbers and operational signs as shown above, and we'll use only the keys on a four-function calculator with memory. Numerals for calculator display will be designated as

$$0123456789$$

You should work each of these problems on your own calculator. If you don't obtain exactly the same display as shown in the examples, consult your owner's manual to find out how your calculator differs from the one used in these examples.

Example 1:   Addition: $14 + 38$

Be sure to turn your calculator on, or clear the machine if it is already on. A clear button is designated by $\boxed{\text{C}}$, and the display will show 0 after the clear button is pushed.

Solution
Press: $\boxed{\text{ON}}$ $\boxed{1}$ $\boxed{4}$ $\boxed{+}$ $\boxed{3}$ $\boxed{8}$ $\boxed{=}$
Display: 0.   1.   14.   14.   3.   38.   52.

You will need to check these steps every time you use your calculator, but after a while it becomes automatic. We will not remind you of this on each example.

Your calculator should also have a button labeled $\boxed{\text{CE}}$. This means *clear entry* and is used if you make a mistake keying in a number and don't want to start over with the problem. For example, if you want $2 + 3$ and push

$$\boxed{2} \quad \boxed{+} \quad \boxed{4}$$

you can then push

$$\boxed{\text{CE}} \quad \boxed{3} \quad \boxed{=}$$

Some calculators have a button $\boxed{\text{CE/C}}$, which combines the *clear* and *clear entry* keys.

to obtain the correct answer. This is especially helpful if you are in the middle of a long problem.

Example 2:   Addition with decimals: $14.6 + 38.9 + 6 + 12.817$

Solution
Press:

$\boxed{1}$ $\boxed{4}$ $\boxed{.}$ $\boxed{6}$ $\boxed{+}$ $\boxed{38.9}$ $\boxed{+}$ $\boxed{6}$ $\boxed{+}$ $\boxed{12.817}$ $\boxed{=}$

Display:

1.   14.   14.   14.6   14.6   38.9   53.5   6.   59.5   12.817   72.317

For purposes of notation, we'll show each number in a single box, which will mean you key in one numeral at a time, as shown here. From now on, this will be indicated by

$\boxed{14.6}$ 14.6

Notice that every time you push $\boxed{+}$, a subtotal is shown in the display. Also notice that you do not need to key-in the decimal point for a whole number (such as 6 in Example 2).

Example 3: Mixed operations: $4 + 3 \times 5 - 7$

Solution

Press: $\boxed{4}$ $\boxed{+}$ $\boxed{3}$ $\boxed{\times}$ $\boxed{5}$ $\boxed{-}$ $\boxed{7}$ $\boxed{=}$

Display: 4. 4. 4. 3. 5. 19. 7. 12.

If you have an algebraic-logic calculator, your machine will perform the correct order of operations. If it is an arithmetic-logic calculator, it will give the incorrect answer of 28.

Example 4: Multiplication with decimals: $\$4.95 \times 6$

Solution

Press: $\boxed{4.95}$ $\boxed{\times}$ $\boxed{6}$ $\longleftarrow$ $\boxed{=}$

Display: 4.95 4.95 6. 29.7

Notice that you do not need to push zeros or even a decimal point.

The result is $29.70.

Notice from Example 4 that your calculator may not display zeros to the right of the last nonzero digit in the answer.

There is also a limit to the accuracy of your calculator. You may have a calculator with a 6-, 8-, 10-, or 12-digit display. You can test the accuracy with the following example.

Example 5: Division: $2 \div 3$

Solution

Press: $\boxed{2}$ $\boxed{\div}$ $\boxed{3}$ $\boxed{=}$

Display: 2. 2. 3. .66666667

There may be some discrepancy between this answer and the one you obtain on your calculator. Some machines will not round the answer as shown here but will show the display

.66666666

Some calculators will simply show an overflow and won't accept larger numbers. Every calculator has some upper limit beyond which it will not function. You'll need to find out what this limit is for your machine.

Example 6: Sometimes problems can be inverted so that the answer can be easily checked. For example, to determine an ancient Arab proverb, calculate

$$5 \times .9547 + 2 \times 353.$$

Solution: If you have correctly handled the order of operations your display should be

710.7735

Now turn the display upside-down to read the answer:

SELL OIL

The answer is 710.7735, but the check is found by turning the calculator over to see if it forms a word.

Problem Set 3.2

1.   Briefly describe the three different types of calculator logic.

Work Problems 2–11 on a calculator. After completing the calculation, use the question or statement in the margin to check your answer by turning your calculator over.

2. What is the opposite of low?

3. You step on them.

4. Where do you live on the pipeline?

5. How do you look taller?

6. What is above your feet?

7. What is on your feet?

8. How do you like your calculator?

9. How are you feeling?

10. How will you raise the money?

11. This is fun!

2. 14×351

3. 218×263

4. $3478.06 + 2256.028 + 1979.919 + .00091$

5. $57,300 + .094 + 32.27 + 2.09 + .0074$

6. $(1979 \times 1356) \div 452$

7. $(515 \times 20,600) \div 200$

8. $498 \times 6925 + 1245 \times 1662$

9. $182 + 52 \times 91$

10. $(.14)(19) + (197)(25.08)(19)$

11. $1 \div (.005 + .02)$

12.   In 1975 *Godfather II* won the Academy Award for the best picture. A little calculator exercise will tell you why. Suppose the studio paid $5,210 for each of 42 advertisements (multiply 5,210 by 42), gave 212 parties costing an average of $2,824 (add the product of 212 and 2,824), and spent $8,784.70 for postage (add that to previous total). The reason for the picture's success becomes apparent when you divide this result by the number of voting members of the academy (150) and turn the calculator over. What you see is why the godfather always wins.

13.   Superman is faster than a speeding bullet, so put 4,000 feet per second (the rate of a speeding bullet) into your calculator. He is more powerful than a locomotive, so add the average weight of a locomotive (438,000 pounds). He is able to leap tall buildings with a single bound, so add 1,472 (the height of the Empire State Building). Now send mild-mannered Clark Kent into a phone booth (add 67,242 for the number of phone booths in Metropolis), and push the equal-sign button. If you turn your calculator over, you will see what a red-faced Superman says when Lois Lane catches him in the phone booth without his cape.

14.   What do Congress and belly dancers have in common? Multiply the prime number 2,417 by the number of months in the year, divide by the number of letters in the word *congress,* and then multiply by the number of letters in "George Washington." Turn the machine around to read the answer. For greater precision, add .7956 and subtract .1776.

15.   Zerah Colburn (1804–1840) toured America when he was 6 years old in order to display his calculating ability. He could instantaneously give the square and cube roots of large numbers. It is reported that it took him only a few seconds to find

Historical Note

From time to time there have been people who have had the ability to perform in their heads calculations that would rival today's electronic calculators. Some of them are described in Problems 15–17. Most calculating prodigies toured the country to display their abilities. Some were idiot savants, who were brilliant in handling immense numbers and stupid in everything else (for example, see Jedidiah Buxton in Problem 16). Others had powers that fell off as they grew older (see Zerah Colburn in Problem 15), while

$$8 \times 8 \times 8 \times \cdots \times 8$$

$$\underbrace{}_{\text{16 factors}}$$

Use your calculator to *help you* find this number.

16. Jedidiah Buxton (1707–1772) never learned to write and was described as slow witted, but given any distance he could tell you the number of inches, and given any length of time he could tell you the number of seconds. If he listened to a speech or a sermon, he could tell the number of words or syllables in it. It reportedly took him only a few moments to mentally calculate the number of cubic inches in a right-angle block of stone 23,145,789 yards long, 5,642,732 yards wide, and 54,465 yards thick. Estimate this answer on your calculator. You will use scientific notation because of the limitations of your calculator, but remember Jedidiah gave the *exact* answer by working the problem in his head.

17. George Bidder (1806–1878) not only possessed exceptional power at calculations but also went on to obtain a good education. He could give immediate answers to problems of compound interest and annuities. One question he was asked was "If the moon is 123,256 miles from the earth and sound travels at the rate of 4 miles per minute, how long would it be before the inhabitants of the moon could hear the battle of Waterloo?" By calculating *mentally,* he gave the answer in less than one minute! Use your calculator to give the answer in days, hours, and minutes, to the nearest minute.

18. *Favorite number trick.* Enter your favorite counting number from 1 to 9. Multiply by 259; then multiply this result by 429. What is your answer? Try it for three different choices.

19. *Magic .618.* Pick any two numbers (5) and (9), and add them to obtain a third number (14). Add the sum (14) to the second number (9) to obtain a fourth (23). Continue to add in this fashion until you have obtained 20 numbers. Divide the last two numbers (in either order), and write down the first three digits after the decimal point. Repeat for another two starting numbers, and compare the final results.

20. Problem 19 is related to the golden ratio introduced in Section 1.5. Use a calculator to find F_n/F_{n-1} for several n's. Make a conjecture (guess) about this ratio for very large values of n.

21. Pick any three-digit number, and repeat the digits to make a six-digit number. Divide by 7, then 11, and then 13. What is the result? Repeat for three different numbers.

22. *Factorial* is an operation of repeated multiplication: $n!$ (pronounced "n factorial") means
$$n(n-1)(n-2) \cdots 3 \cdot 2 \cdot 1$$
where $1! = 1$ and $0! = 1$. For example,
$$5! = 5 \cdot 4 \cdot 3 \cdot 2 \cdot 1$$
$$= 120$$

Find the following:

a. 4!   b. 6!   c. 8!   d. 10!   e. $\dfrac{10!}{8!}$

others were bright and had many mental skills (for example, see George Bidder in Problem 17). Another brilliant calculating prodigy was Truman Safford (1836–1901), who was an astronomer but never exhibited his talents publicly. When he was 10 years old he was asked to find the surface area of a regular pyramid with a slant height of 17 ft and a pentagon-shaped base with sides of 33.5 ft. He gave the answer, 3,354.5558 sq ft, in two minutes.

For Problem 20, you can use Table 1.1 on page 41.

EXAMPLE FOR PROBLEM 21:

123

123,123

$$\begin{array}{r} 17,589 \\ 7 \overline{)123,123} \\ \vdots \\ \vdots \end{array}$$

Remember, your calculator does not give exact answers. For example,

$$\frac{1}{3} = .333\ldots$$

If you find

$$\boxed{1} \ \boxed{\div} \ \boxed{3}$$

on *any* calculator, it will give you an approximate answer correct to a certain number of significant digits. That is,

$$\frac{1}{3} \neq .3333333333$$

If the exact decimal representation for $\frac{1}{3}$ is .333... and not .33333333, find the exact decimal representation for each of the numbers given in Problems 23–28.

23. $\dfrac{1}{6}$ **24.** $\dfrac{1}{11}$ **25.** $\dfrac{1}{7}$ **26.** $\dfrac{1}{41}$ **27.** $\dfrac{1}{271}$ **28.** $\dfrac{1}{37}$

29. Use your calculator to write each fraction as a decimal.

    a. $\dfrac{1}{7}$ **b.** $\dfrac{2}{7}$ **c.** $\dfrac{3}{7}$ **d.** $\dfrac{4}{7}$ **e.** $\dfrac{5}{7}$ **f.** $\dfrac{6}{7}$

30. Use Problem 29 to describe a pattern.

Mind Bogglers

31. Use your calculator to find

    a. $\dfrac{1}{17}$ **b.** $\dfrac{2}{17}$ **c.** $\dfrac{3}{17}$ **d.** $\dfrac{4}{17}$

    e. If the pattern of parts a–d is the same as that described in Problem 30, can you guess the entire decimal representation for $\frac{1}{17}$ even though it exceeds the accuracy of the calculator?

32. Explain the following pattern.

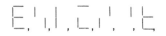

    Fill in the next entry in this pattern.

These problems are called alphametrics. The rules for constructing these problems state:

1. Each letter must stand for a unique digit, and no digit is represented by more than one letter.
2. The leftmost letter of any word may not equal zero.
3. Each numerical combination is a word in the English language.

Many cryptic-arithmetic problems can be found in *The Journal of Recreational Mathematics.*

Cryptic arithmetic. Cryptic arithmetic is a mathematical puzzle in which letters have been replaced by the digits of numbers. Replace each letter by a digit (the same digit for the same letter throughout; different digits for different letters), and the arithmetic will be performed correctly. See if you can do these puzzles:

33.   SEND
    + MORE
    ————
    MONEY

34.   DAD
    SEND
    + MORE
    ————
    MONEY

35.   THIS
    IS
    + VERY
    ————
    EASY

3.3 Introduction to Computers

Computers are used for a variety of purposes. **Data processing** involves large collections of data on which relatively few calculations need to be made. For example, a credit card company could use the system to keep track of the balances of its customers' accounts. **Information retrieval** involves locating and displaying material from a description of its content. For example, a police department may want information about a suspect or a realtor may want listings that meet the criteria of a client. **Pattern recognition** is the identification and classification of shapes, forms, or relationships. For example, a chess player will examine the chessboard carefully to determine a good next move. Finally, a computer can be used for **simulation** of a real situation. For example, a proto-type of a new spacecraft can be tested under various circumstances to determine its limitations. The computer carries out these tasks by accepting information, performing mathematical and logical operations with the information, and then supplying the results of the operations as new information. Every computing system is made up of five main components: input; storage or memory; central processing unit (CPU) or accumulator or arithmetic unit; control; and output. (See Figure 3.13.)

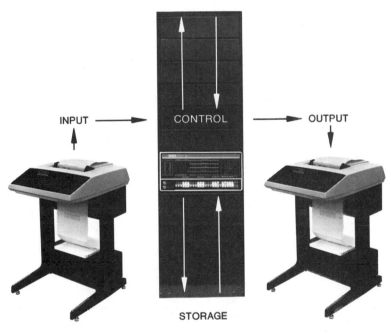

FIGURE 3.13   Relationships among the five main components of a digital computer

Input refers to the procedure of entering data or instructions into a com-puter. The most common method used by students is on a **CRT keyboard** (see Figure 3.14a); CRT stands for Cathode Ray Tube display. You type in your instructions or messages manually, just as on a typewriter, and the words or numbers appear on a video screen. The computer then responds on the video screen. This response is called **output.** Another common output device is a **line printer,** which can output a great deal of data at a rate that is much faster than the human eye can read.

Another method of inputting data is with the familiar **punched card** (see Figure 3.14b). Punched cards are prepared on a machine called a **keypunch,** which has a keyboard similar to a typewriter and can produce new cards or duplicate old ones. Many computers read mark-sense cards, which are like punched cards but do not need to have the holes punched into the cards.

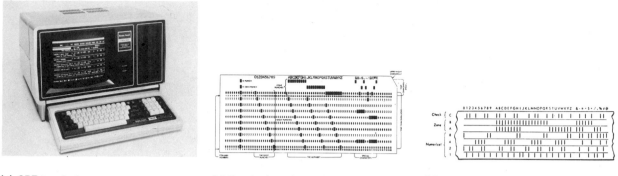

(a) CRT terminal   (b) Punched card   (c) Magnetic tape

FIGURE 3.14   Common methods of inputting information into a computer

Magnetic tape is another way to store or enter information into a computer (see Figure 3.14c). This is a very fast method of inputting information into a computer. Many home computers use ordinary cassette tape for this purpose.

Some computers have unconventional input devices, as shown in Figure 3.15. The General Motors advertisement illustrates a "seeing" computer, the UPI newspaper article talks of a "speaking" computer, and the third article mentions a "hearing" computer.

The **storage unit** stores information until it is needed for some purpose. Storage might be visualized as a long corridor in a hotel. Each door is numbered, and behind each door are a limited number of "slots" to hold information. Each compartment (door) is capable of holding one piece of information or one of the instructions in a program.

ADDRESS 1	INFORMATION
ADDRESS 2	INFORMATION
ADDRESS 3	IS
ADDRESS 4	STORED
ADDRESS 5	IN
ADDRESS 6	EACH
ADDRESS 7	NUMBERED
ADDRESS 8	LOCATION

The number associated with each of these slots is called the location or **address.** The information inside the box is called a **word.** During the process of solving a problem, the storage unit contains:

1. A set of instructions that specify the sequence of operations required to complete the problem
2. The input data

SCIENTIST SAYS COMPUTERS THAT TALK A REALITY

PASADENA (UPI)—Talking computers are a reality, according to a scientist at the California Institute of Technology.

Computers can now be taught to speak, including such niceties as stressing important words and pausing for punctuation, said Dr. John R. Pierce, a communications expert and former executive director of Bell Laboratories communications sciences division.

"Computers that imitate human speech with a remarkable degree of accuracy already are a reality." Pierce said Monday in a lecture at Cal Tech.

Noriko Umeda, a Japanese linguist, has succeeded in synthesizing human speech from a computer equipped with a pronouncing dictionary in its memory system, Pierce said.

Some computers are now equipped to give voice responses in some circumstances, by triggering a tape recorded message. But teaching the computer itself to talk leads to greater accuracy—because the computer will no longer repeat a mistake made by a human announcer—and "economy, rapidity and flexibility," Pierce said.

"Tape recorded answers have a very limited range and only give a few responses. A computer would have a large body of information stored in it alphabetically. The computer would be able to put these words together in an intelligible form which could answer questions." Pierce said.

Although talking computers are still imperfect, the idea is now workable, he said.

Talking computers can give instructions to workmen, he said, and could be linked to a telephone to provide advice services, such as counseling housewives on what dishes could be prepared from leftovers.

COMPUTERS TO 'HEAR' COMMANDS

By Edward O'Brien
Newhouse News Service

WASHINGTON—Pentagon research sometimes wanders into fields far removed from the battlefield. Now Defense officials predict confidently that in two or three years they will have developed computers that take instruction from a human voice.

"The use of natural spoken English as an input language to computers will revolutionize the effectiveness and utility of computer systems," Stephen Lukasik, director of defense Advanced Research Projects, has reported to Congress.

The study is at the midpoint and has produced computer systems that understand continuous speech with a vocabulary of about 100 words. This is much more sophisticated than anything done by government or private industry.

The next 900 words will come more easily because of progress already made in teaching the computers to break speech into separate words, find individual sounds within words, and use clues provided by the vocabulary, grammar, and subject matter.

"Speech is the most natural means for a person to express himself, and we will be able to integrate the computer much more closely to our daily workings if we can develop means for it to understand natural speech," Lukasik said.

"Fast typists (feeding information into computers) can normally type 60 to 70 words per minute, while non-typists may be able to achieve only six to 10 words.

"Almost everybody, however, talks in excess of 100 words per minute, and no special manual skill or equipment is required . . . with voice access, we can expect computers to be as easy to use as telephones are today."

Voice control of computers, however, may not be the ultimate.

A computer that can see gets its first real job.

 General Motors Research Laboratories
Warren, Michigan 48090

A "seeing" computer developed by the General Motors Research Laboratories has recently become the first of its kind to go to work on a U.S. automotive production line. The employer: GM's Delco Electronics Division.

FIGURE 3.15 Examples of news articles about computer input/output devices that simulate human senses

𝕳istorical 𝕹ote

You are no doubt aware of the special numerals on the bottom of your bank checks.

0 1 2 3 4 5 6 7 8 9
| : " ; ∎

These are called MICR (Magnetic Ink Character Recording) numerals and are printed in magnetic ink so they can be read mechanically. Although they may look futuristic, modern-day computers do not use MICR, which was designed in the late '50s, but instead use an alphabet called OCR (Optical Character Recognition).

ABCDEFGHIJKLM
NOPQRSTUVWXYZ
0123456789
.,:;=+/?'"
⅄♪⊣$%|&*{}∎

Another modern optical scanner has come to your supermarket, using what is called a bar code. All packages have or will have a bar code:

All the grocery clerk needs to do is move the bar code over a sensing device, and the correct amount is charged and the tax recorded. The computer keeps track of the store inventory by noting that the item has been purchased.

3. The intermediate results for as long as they are required in the development of the problem
4. The final result before it is sent to the output unit

In practice, each number, letter of the alphabet, punctuation mark or other symbol, and special computer instruction is "coded" using only two symbols (represented as 0 and 1). This is called a *binary coding* and was chosen because it assumes only two states (on–off, hole–no hole, right–left, and so on). (See Figure 3.16.)

FIGURE 3.16   Two-state device. Light bulbs serve as a good example of two-state devices. The 1 is symbolized by "on," and the 0 is symbolized by "off."

Ordinary numerals		Possible binary code	Ordinary letters		Possible binary code
0	=	000 000	A	=	100 000
1	=	000 001	B	=	100 001
2	=	000 010	C	=	100 010
3	=	000 011	⋮		
⋮			S	=	110 011
			T	=	110 100
			⋮		
			SPACE	=	111 111

Each numeral, letter, or other keyboard symbol is represented by a six-digit binary code. Each digit of this code is called a bit, so this is an example of a six-bit code. For example, the numeral "3" is internally stored in a computer as 000 011, and a space is coded as 111 111. If you wanted to store the information "3 cats" in a computer, you would need six locations or addresses:

Address numbers	Information	
1000	000011	← Code for 3
1001	111111	← Code for space
1010	100010	← Code for C
1011	100000	← Code for A
1100	110100	← Code for T
1101	110011	← Code for S

Notice that an address number has nothing to do with the contents of that address. The early computers stored these individual bits (digits 0 or 1) with switches, vacuum tubes, or cathode ray tubes. In the "3 cats" example, each word contains 6 bits. In practice, however, a word may contain 36 or more bits, which allows the use of much larger numbers. That is, a location might look like the following:

Address 00101 | 000111110000000110110110000000001101 |

The binary code is part of what is known as the **binary numeration system.** It is similar to the decimal system, except that instead of ten symbols there are only two (0 and 1). The following code is established. Look for a pattern.

Decimal:	0	1	2	3	4	5	6	7
	↕	↕	↕	↕	↕	↕	↕	↕
Binary:	0	1	10	11	100	101	110	111

Decimal:	8	9	10	11	12	13	14	15
	↕	↕	↕	↕	↕	↕	↕	↕
Binary:	1000	1001	1010	1011	1100	1101	1110	1111

Decimal:	16	17	. . .	31	32	. . .
	↕	↕		↕	↕	
Binary:	10000	10001	. . .	11111	100000	. . .

In order to prevent confusion between decimal and binary numerals, when the context does not make it clear which is being represented, a subscript "two" is used on the binary numeral. For example, if you see 101, it could mean "one hundred one" in decimal or it could mean "one zero one base two" (the decimal number five) in binary. In such a case we write 101 for "one hundred one" and 101_{two} for the binary numeral.

To change a number from binary to decimal, it is necessary only to write the number out in "expanded form" to see what it means in decimal. Notice that a 1 in a certain column has a different meaning from a 1 in another column.

Example 1:

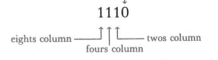

This means: 1 eight + 1 four + 1 two + 0 units

$$= (1 \times 8) + (1 \times 4) + (1 \times 2) + 0$$
$$= 8 + 4 + 2$$
$$= 14$$

Example 2:

$$10111_{two} = (1 \times 16) + (0 \times 8) + (1 \times 4) + (1 \times 2) + 1$$
$$= 16 + 4 + 2 + 1$$
$$= 23$$

To change a number from decimal to binary, it is necessary to find how many units (1 if odd, 0 if even), how many twos, how many fours, and so on are contained in that given number. This can be accomplished by repeated division by 2.

Example 3:   Change 47 to a binary numeral.

0 r. 1		There is one thirty-two
2 $\overline{)1}$ r. 0		There is zero sixteen
2 $\overline{)2}$ r. 1		There is one eight
2 $\overline{)5}$ r. 1		There is one four
2 $\overline{)11}$ r. 1		There is one two
2 $\overline{)23}$ r. 1		There is one unit
Start here → 2 $\overline{)47}$		

(Work up / Read down)

If you read down the remainders, you obtain the binary numeral 101111, which represents

one thirty-two + zero sixteen + one eight + one four + one two + one unit

$$= \quad 32 \quad + \quad 0 \quad + \quad 8 \quad + \quad 4 \quad + \quad 2 \quad + \quad 1$$
$$= \quad 47$$

Numbers, letters, and other symbols, therefore, are written using some code. These binary bits form the code that is used to make up words that are then stored in various addresses. These addresses are then kept in some memory device.

There are two kinds of primary memory devices used in computers today: ROM (read-only memory) and RAM (random-access memory). ROM is for information that is permanently stored in the computer and is not alterable. RAM is memory that can be changed as needed. A RAM chip today can store 64 kilobits (65,536 characters, or about 10,000 words of English text). The RAM capacity of a computer can be expanded by adding extra memory boards, or modules. The standard medium for extra memory storage is the **floppy disk,** which is a flexible disk of Mylar plastic 5¼ or 8 inches in diameter. These disks provide information storage in localized areas of a thin magnetic film that is coated onto a nonmagnetic supporting surface. Information is stored in the form of tiny magnetized spots in the magnetic film. The magnetic film and the read/write head move in relation to each other to bring a storage site into position for writing or reading information.

The **central processing unit (CPU)** is the part of the computer in which addition, subtraction, multiplication, or other operations are carried on. It is also called the *accumulator* or *arithmetic unit*. The CPU keeps a running tally of whatever is put into it. This is accomplished with logic circuits, which enable it to do these operations. In a personal computer, the CPU is a microprocessor (a single integrated circuit on a chip; see Problem 15 of Problem Set 3.1). In this book we are not concerned with *how* these operations occur; we simply assume that the computer "knows how" to carry out these operations.

The **control unit** of the computer directs the flow of data to and from the various components of the system. It obtains instructions, one at a time, from the storage unit and causes the CPU to carry out the operations.

When data instructions are read into the computer, control will usually direct them to storage. Both the data and the instructions can be placed anywhere in storage, although they are usually put apart from each other and kept in some predetermined sequence.

In solving a problem, the computer is directed to start at the first instruction of the program stored at a given location. Once the first instruction has been executed, the control circuits usually get the next instruction from the next higher storage position.

" HOLD IT, FELLAS. THERE'S BEEN A BREAKTHROUGH AND THAT ENTIRE UNIT IS BEING REPLACED WITH THIS "

Problem Set 3.3

1. What is meant by an input/output device?

2. What is the function of the control part of a computer?

3. What does the CPU in a computer do?

4. What is meant by computer memory?

5. Name three methods for inputting information into a computer.

6. Explain what we mean by "address," "word," and "bit" when we speak of computers.

*7. *GIGO.* There is a slogan among computer operators: *garbage in, garbage out.* This phrase reflects the fact that a computer will do exactly what it is told to do. Consider the following humorous story.

   A very large computer system had been built for the military. It was built and staffed by the best computer people in the country.

This lion was made entirely of computer components.

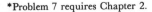
*Problem 7 requires Chapter 2.

"The system is now ready to answer questions," said the spokesperson for
the project.

A four-star general bit off the end of a cigar, looked whimsically at his
comrades and said, "Ask the machine if there will be war or peace."

The machine replied: YES.

"Yes *what*?" bellowed the general.

The operator typed in this question, and the machine answered: YES SIR.

In terms of what you know about logic from Chapter 2, why did the computer
answer the first question YES correctly?

8.   Give examples of:

   a. Data processing   **b.** Information retrieval
   c. Pattern recognition   **d.** Simulation

9.   What decimal number is represented in Figure 3.16, page 124?

10.   What decimal number is represented by these light bulbs?

   a.   **b.**

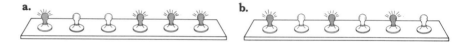

Write each of the numbers given in Problems 11–19 as a decimal numeral.

11. 1101_{two} **12.** 1001_{two} **13.** 1011_{two}

14. 11101_{two} **15.** 10111_{two} **16.** 11011_{two}

17. 1100011_{two} **18.** 1110111_{two} **19.** 10111000_{two}

Write each of the numbers given in Problems 20–31 as a binary numeral.

20. 13   **21.** 15   **22.** 35   **23.** 46   **24.** 51   **25.** 63

26. 64   **27.** 256   **28.** 128   **29.** 615   **30.** 795   **31.** 803

Mind Bogglers

32.   The *binary numeration system* has only two symbols, 0 and 1, which makes counting in the system simple. Arithmetic is particularly easy in the binary system, since
the only arithmetic "facts" one needs are the following:

Addition		
+	0	1
0	0	1
1	1	10

Multiplication		
×	0	1
0	0	0
1	0	1

Perform the indicated operations.

a. 1101_{two} **b.** 11011_{two} **c.** 10110_{two}
   $+ 1100_{two}$ $- 10110_{two}$ $\times 101_{two}$

33. The *octal numeration system* refers to the base eight system, which uses the symbols 0, 1, 2, 3, 4, 5, 6, 7. Describe a process for converting octal numbers to decimal and decimal numerals to octal.

Programmers often use octal numerals (see Problem 33) instead of binary numerals because it is easy to convert from binary to octal directly. Table 3.1 shows this conversion. A binary number 1 1 0 1 1 1 0 1 1 1 is separated into three-digit groupings by starting with the right end of the number and supplying leading zeros at the left if necessary:

$$001 \; 101 \; 110 \; 111$$

The binary groups are then replaced by their octal equivalents:

$$001_{two} = 1_{eight} \qquad 101_{two} = 5_{eight} \qquad 110_{two} = 6_{eight} \qquad 111_{two} = 7_{eight}$$

and the binary number is converted to its octal equivalent,

$$1 5 6 7$$

Conversely, an octal numeral can be expanded to a binary numeral using the same table of equivalents

$$5307_{eight} = 101 \; 011 \; 000 \; 111_{two}$$

Use this information for Problems 34–36.

TABLE 3.1 Octal-Binary Equivalence

Octal	Binary
0	000
1	001
2	010
3	011
4	100
5	101
6	110
7	111

34. Convert to the binary system.

   a. 5_{eight} **b.** 6_{eight} **c.** 14_{eight} **d.** 624_{eight} **e.** 473_{eight}

35. Convert to the binary system.

   a. 7045_{eight} **b.** 3062_{eight} **c.** 5700_{eight}
   d. 04320_{eight} **e.** 00200_{eight}

36. Convert to the octal system.

   a. 101_{two} **b.** 100_{two} **c.** 011_{two}
   d. $000 \; 000 \; 111 \; 111 \; 101 \; 000_{two}$ **e.** $100 \; 000 \; 000 \; 101 \; 110 \; 111_{two}$

37. *Magic cards.* You will need to use a pair of scissors for this problem. Construct the following cards:

Card 1

Card 2

Card 3

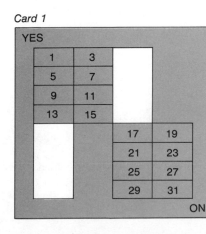

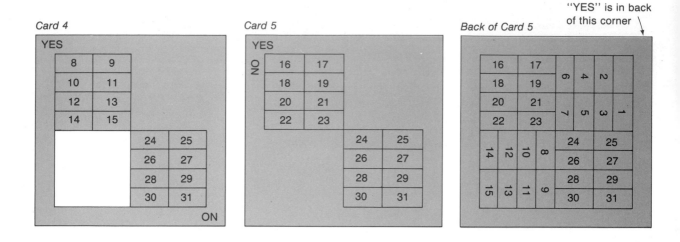

Cut out those parts of the cards that are white. Ask someone to select a number between 1 and 31. Hold the cards face up on top of one another, and ask the person if his or her number is on the card. If the answer is yes, place the card on the table face up with the word *yes* at the upper left corner. If the answer is no, place the card face up with the word *no* at the upper left corner. Repeat this procedure with each card, placing the cards on top of one another on the table. After you have gone through the five cards, turn the entire stack over. The chosen number will be seen. Explain why this trick works.

Problems for Individual Study

38. The locations given in the marginal note about computer vending machines are in California, near the home of the author. Locate the facility nearest your home where anyone can gain access to a computer. Visit the facility and prepare a report.

39. Find out what local, state, and federal governments have stored in their computers about you and your family. Find out what you can see and what others can see. This will provide you with an interesting intellectual journey, if you wish to take it.

40. Build and operate a simple analog or digital computer. It is not as difficult as you might imagine. See, for example, *Popular Electronics,* January 1975.

3.4 BASIC Programming

To get a computer to operate, a complete set of step-by-step instructions must be provided. These sets of instructions are known as *programs,* and the process of setting them up is known as **programming.**

Each step or instruction for the solution of a problem must be provided for the computer in detail. For example, if a teacher asks a pupil to add 7 and 4, the student will generally say 11; the programmer, however, not only must ask the

COMPUTER VENDING MACHINES?

You are all familiar with vending machines dispensing everything from soda pop to pantyhose. But did you know that you can also buy computer time on a vending machine? In 1972 a coin-operated minicomputer was installed in a library in Monterey, California. In Menlo Park, California, at People's Computing Center you can rent a computer terminal for $3 an hour. You can use it to write programs, play games, do personal calculations, or help with your math homework. At the Lawrence Hall of Science in Berkeley, California, you can use a computer for $1 an hour. Perhaps one day in the not-too-distant future a computer vending machine may be as accessible as a cigarette vending machine is today.

Computer Career Opportunities
Honeywell Corporation

Future career opportunities in the rapidly growing world of computers seem to be practically limitless. These opportunities may be direct, as in the case of those who manufacture and operate computers, or indirect, as in the case of businessmen, scientists, and others who use computer systems.

Increasingly great numbers of skilled personnel will be needed by the computer industry itself:

Designers and manufacturers of systems

Engineers and scientists for research and development

Sales personnel skilled in marketing methods

Systems analysts to analyze and meet special requirements of customers

Programmers who prepare programs to meet customers' needs

Computer operators to run systems

Personnel for clerical and data preparation jobs

Managers of computer operations

Management interpreters of computer systems, needs, opportunities

Specialists in areas such as business, science, education, and government

Interdisciplinarians—those who can understand and meet the needs of persons from varied professions united on mutual projects

More than a thousand colleges and universities in the U.S.A. and Canada, according to a recent survey, now offer courses in the computer sciences and data processing. Computer usage is being taught in many high schools and even in some grammar schools. Many independent training schools exist for high school and college graduates.

The use of remote terminals, that connect to a central computer system sometimes thousands of miles away, is becoming commonplace. Industry experts say it's only a matter of time and cost reduction before the use of household terminals, for a variety of purposes ranging from information services to entertainment, becomes as ordinary as the use of the telephone.

Economists predict that by the end of the century, or earlier, the computer industry and directly associated industries will be the largest American business.

computer to add 7 and 4 but also must ask it to indicate the answer.

To write a program, we should carry out the following five steps:

1. Analyze the problem and break it down into its basic parts.
2. Prepare a flowchart or write out a description of a process for carrying out the task.
3. Put the flowchart or description into a language the machine can "understand."
4. Test the program to see if it does what it is supposed to do.
5. **Debug** the program, if necessary. This means that you remove any deficiencies in the program.

A **flowchart** is an outline that lists in sequence the steps to be performed. The directions are usually listed within squares, circles, or other geometric figures (as shown in Figure 3.17). The direction, or flow, of the process is shown by arrows.

• Edwin S. Cohen, Assistant Secretary of the Treasury for tax policy, marvels over the prowess of the Internal Revenue Service in using computers. Still, he was disturbed to find that the computers sent him two sets of 1040 forms, one to his current address and the other to his Charlottesville, Va., home. Cohen wasn't eager to pay his taxes twice, so he asked the IRS computer people to fix things up. Sorry, Cohen says he was told, there was no way to correct the error because it was impossible for it to have happened.

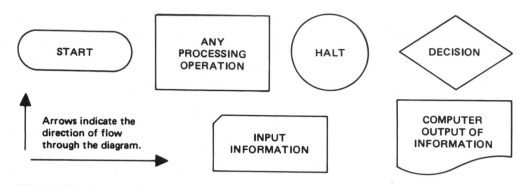

FIGURE 3.17   Common flowchart symbols

The flowchart provides a way of checking the overall analysis of a process or problem, and, when completed, it also provides a way for describing the process to someone else. Using the flowchart as a guide, the details of the program are easier to write.

Flowcharts can direct us in other types of activities besides writing computer programs. For example, the General Information Leaflet of the General Library of the University of California, Davis, uses the flowchart shown in Figure 3.18 to explain how to find library materials. Note that no detail should be overlooked in a flowchart.

Example 1:   Follow the directions given by the following flowchart.

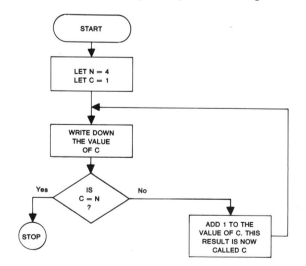

Solution

START; $N = 4$, $C = 1$.

Write down the value of C: 1

Is $C = N$?   No,   ($C = 1$ and $N = 4$),   so add 1 to the value of C:
$C = 1 + 1 = 2$.   Write down the value of C: 2

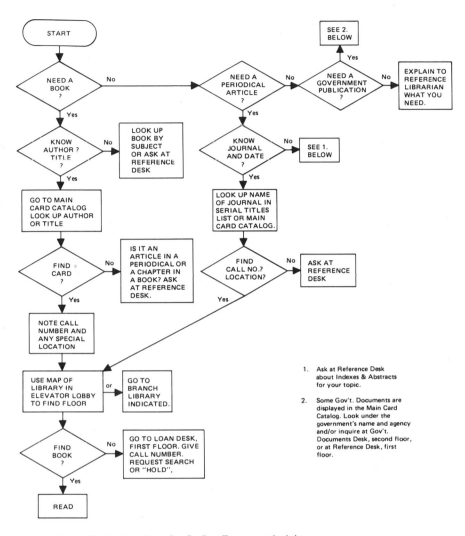

FIGURE 3.18 The basic steps for finding library materials

Is $C = N$? No, ($C = 2$ and $N = 4$), so add 1 to the value of C:
    $C = 2 + 1 = 3$. Write down the value of C: 3

Is $C = N$? No, ($C = 3$ and $N = 4$), so add 1 to the value of C:
    $C = 3 + 1 = 4$. Write down the value of C: 4

Is $C = N$? Yes ($C = 4$ and $N = 4$), so stop.

The output is shown above in color; it is: 1
                                              2
                                              3
                                              4

A **square number** is a number that can be written as a counting number times itself.

Factored form of a square number ——→ ⎫ ⎧ ——— Square number

$$1 \times 1 = 1$$
$$2 \times 2 = 4$$
$$3 \times 3 = 9$$
$$4 \times 4 = 16$$
$$5 \times 5 = 25$$
$$\vdots$$

Sometimes the factored form of a square number is written using an **exponent of 2** as follows:

1^2 is pronounced "one squared" and is 1

2^2 is pronounced "two squared" and is 4

3^2 is pronounced "three squared" and is 9

Definition of N^2

The number N^2, pronounced "N squared" means

$$N \times N$$

Example 2:   Draw a flowchart that will direct someone to write down the square numbers up to 50.

Solution

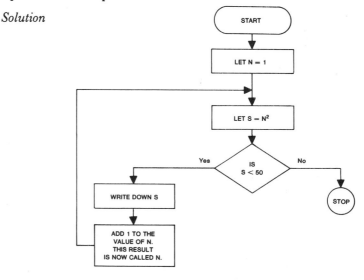

After preparing a flowchart, the next step is to write a program. Computers use many different languages. Some widely used languages and the computers that use them are shown in Table 3.2.

TABLE 3.2 Languages Available for Some Popular Computers

Language \ Computer	APPLE IIe	ATARI	TRS-80	PET/ CBM	IBM/ PC
BASIC	X	X	X	X	X
PASCAL	X	X	X		X
APL			X		X
FORTRAN	X		X		X
COBOL	X		X		X
C, TINY C	X		X	X	
FORTH	X	X	X		X
LIST	X	X			X
PILOT	X	X			X
LOGO	X		X		X

Many of the names for various computer languages started out as acronyms. For example, BASIC (Beginner's All-Purpose Symbolic Instruction Code), FORTRAN (FORmula TRANslator), and COBOL (COmmon Business Oriented Language).

In order to communicate with a computer, you need to learn one of these languages. We'll concentrate on a very popular conversational language called **BASIC.** BASIC consists of short, easy-to-learn commands that are very similar to normal English. We will talk to the computer on a CRT keyboard, and the computer will respond on the screen of that same device.

A BASIC program is made up of a sequence of statements called **commands.** Different commands instruct the computer to carry out different processes. We'll discuss some of these commands in this and the following sections. In order to tell a computer the order of commands, **line numbers** are used. That is, each command in a program is preceded by a counting number (called its *line number*) followed by a space. When running a program, the computer will execute the commands in the order of ascending line numbers, regardless of the order in which they are typed on the CRT. Thus, we usually use the line numbers 10, 20, 30 (rather than 1, 2, 3) so that we can insert new statements to debug or modify the program at some later time.

You are now ready for a simple program. Every computer has some sort of a log-in procedure. These procedures vary from machine to machine and facility to facility, and they must be kept relatively secret so that not just anyone can use the computer. You will need to find out the log-in procedure for the computer or terminal you will be using. All the examples in this book will assume that you have correctly followed the log-in procedure for your particular computer.

Suppose you wish to have the computer type the phrase "I love you." In order to do this, you will use a command called PRINT.

JOIN
STAB
SOCIETY TO
ABOLISH BASIC

Glossary of Computer Terms

Address: Designation of the location of data within internal memory or on a magnetic disk or tape.

BASIC: Beginner's All-Purpose Symbolic Instruction Code, a higher-level language (see below) common to most microcomputers.

Bit: Binary Digit, the smallest unit of data storage with a value of either 0 or 1, thereby representing whether a circuit is open or closed.

Bug: An error in the design or makeup of a computer program (software bug) or a hardware component of the system (hardware bug).

Byte: The fundamental block of data that can be processed by a computer. In most microcomputers, a byte is a group of eight adjacent bits and the rough equivalent of one alphanumeric.

Command: A single instruction to prompt the computer to perform a specific predefined operation.

Compiler: An operating system program that converts an entire program written in a higher-level language into machine language before the program is executed.

Computer: A device which, under the direction of a program, can process data, alter its own program instructions, and perform computations and logical operations without human intervention.

Computer literacy: A catch-all phrase to describe one's ability to understand computer technology, its operation, applications, and the implications of its use.

Console: A device (e.g., a CRT terminal) used to control and monitor the operation of a computer.

CPU: Central Processing Unit, the primary section of the computer that contains the memory, logic, and arithmetic procedures necessary to process data and perform computations. The CPU also controls the functions performed by the input, output, and memory devices.

CRT terminal: Cathode Ray Tube terminal, an I/O device that uses a screen to display data and a keyboard to input data.

Cursor: Indicator (often flashing) on a video display to designate where the next character generated or input will be placed.

Debug: The organized process of testing for, locating, and correcting errors within a program.

Disk drive: A mechanical device that uses the rotating surface of a magnetic disk for the high-speed transfer and storage of data.

Documentation: The written description of a piece of software or hardware.

DOS: Disk Operating System, the set of programs that allows both the user and the computer to communicate with a disk drive.

Floppy disk: Storage medium, which is a flexible platter (8 or $5\frac{1}{4}$ inches in diameter) of mylar plastic coated with a magnetic material. Data are represented on the disk by electrical impulses.

Flowchart: A diagram illustrating the logical sequence of steps for the solution of a problem, an algorithm, or the flow of a computer program.

GIGO: Garbage In, Garbage Out, an old axiom regarding the use of computers.

Hard copy: The output produced on paper by a printer, usually in human-readable form.

Hardware: The physical components (mechanical, magnetic, electronic) of a computer system.

Higher-level language: A computer programming language (e.g., BASIC, PASCAL, LOGO) that approaches the syntax of English and is both easier to use and to learn than machine language. It is also not system-dependent.

Input device: Component of a system that allows the entry of data on a program into the computer's memory.

Integrated circuit: The plastic or ceramic body that contains a chip and the leads connecting it to other components.

Interactive: Software that allows continuous two-way communication between the user and the computer.

Interface: The electronics necessary for a computer to communicate with a peripheral.

K: The symbol represents 1024 (or 2^{10}). For example, 48K bytes of memory is the same as 48×1024 or 49,152 bytes of memory.

Keyboard: Typewriter-like device that allows the user to input data into the computer.

Peripheral: A device, such as a printer, that is connected to a computer and is operated by the computer.

RAM: Random-Access Memory, or memory where each location is uniformly accessible, and is often used for the storage of a program and the data being processed.

ROM: Read-Only Memory, or memory that cannot be altered either by the user or a loss of power. In microcomputers, the ROM usually contains the operating system and system programs.

Software: The routines, programs, and associated documentation in a computer system.

Subroutine: A collection of related instructions designed both to perform a specific task and to be called or used by another program routine.

Syntax error: The breaking of a rule governing the structure of the programming language being used.

Terminal: A device that allows the user to communicate with a computer.

Time-sharing: A technique that allows several users to appear to have simultaneous access to a single computer.

Word processing: The process of creating, modifying, deleting, and formatting textual material.

This is a counting number; it is the line number
of the command; this is your (the programmer's) choice.

10   PRINT "I LOVE YOU."

To prevent confusion, computers
use Ø for "zero" and O for "oh."

Whatever is enclosed in quotation marks will be printed by the computer.

Command

20   END

The end of the program is designated by END (yes, even one this short).

This program, along with another example, is shown below as it would look on the CRT.

Program A	*Program B*
10 PRINT "I LOVE YOU."	10 PRINT "I AM ROBBIE THE ROBOT."
20 END	20 PRINT "I LOVE YOU."
	30 PRINT "GOODBYE"
	40 END

Now that we've written a program, the next step is to run the program. The RUN command causes the computer to begin the program. When you tell the computer to RUN, it will begin at the lowest line number and execute in consecutive order all commands in the program. A RUN command is called a **system command** and is not preceded by a line number. A system command is a command that is not stored in the memory, but is executed immediately after pressing the return button.

Example 3: Run Programs A and B given above.

a. Suppose Program A is input. Type RUN and return (which we'll denote by RUN↩), and the computer will respond with

I LOVE YOU.

b. Suppose Program B is input.

RUN↩

I AM ROBBIE THE ROBOT.
I LOVE YOU.
GOODBYE

Notice that each new line number generates a new line of output.

When you need to use a computer to do arithmetic operations, you will use * for multiplication, / for division, and ↑2 for squaring a number, as shown in Table 3.3. The up arrow (↑) on a computer is also used for the general operation of exponentiation, which we will discuss in Section 4.3.

TABLE 3.3 Mathematical Symbols in BASIC

Symbol		Math Notation	BASIC
↑	Squaring	3^2	3↑2
*	Multiplication	$3 \cdot 4$	3 * 4
/	Division	$3 \div 4$ or $\frac{3}{4}$	3/4
+	Addition	$3 + 4$	3 + 4
—	Subtraction	$3 - 4$	3 — 4

Example 4: Write a program to calculate $5 \times 3 - 6$.

Solution
```
10  PRINT 5*3 - 6
20  END
```

Example 5: The formula for the area of a circle is $A = \pi r^2$, where π is approximately 3.1416. Use this formula to calculate the area of a circle with radius 6.3.

Solution
```
10  PRINT 3.1416 * 6.3↑2
20  END
```

Historical Note

In 1962 an Atlas-Agena rocket blasted off from Cape Kennedy; it was to have been the first U.S. spacecraft to fly by Venus. When it was about 90 miles above the earth, it had to be destroyed. The reason for the failure was later determined to have been due to a programmer who left out a hyphen when writing the flight plan. That missing hyphen cost the United States about $18,500,000.

Debugging means checking and revising your program to make sure it does what it is supposed to do. The programs in this section will be easy to debug, but if your program is more complex and involves several decisions and options, it will be much more difficult to check.

There are two types of mistakes. The first is a logic error, which is an error in your thinking. A flowchart may help you overcome this type of error if the problem is complicated. The second type of error is a coding error, which is a clerical error. The historical note gives an example of an $18 million coding error.

In the next section, we will introduce the BASIC commands you will use when writing a program. We will also write some simple programs.

Problem Set 3.4

1. Discuss the steps to follow when writing a program.

2. What is a flowchart?

3. Draw the symbol used in a flowchart to denote:

    a. Start   **b.** Stop   **c.** A decision   **d.** Input data
    e. Output data

4. Tell what each of the following BASIC operation symbols mean.

   a. —   **b.** *   **c.** /   **d.** +   **e.** ↑

5. In your own words, describe what is meant by debugging a program.

6. Consider the following program.

```
10  PRINT "GOOD MORNING."
20  PRINT "CAN YOU DO THIS PROBLEM CORRECTLY?"
30  END
```

   Without using a computer, explain what the computer will type after you give a RUN⤸ command.

7. Suppose you have just written the program shown in Problem 6. Then you type

```
25  PRINT 2+3
35  END
```

   Without using a computer, explain what the computer will type after you give a RUN⤸ command.

In Problems 8–15, write each BASIC expression in ordinary algebraic notation.

8. **a.** `4*X + 3`   **b.** `(5/4)*Y + 14↑2`

9. **a.** `35*X↑2 - 13*X + 2`   **b.** `6*X - 7`

10. **a.** `5*X↑2 - 3*X + 4`   **b.** `(X - 5)*(2*X + 4)↑2`

11. **a.** `6.29↑2 - 7`   **b.** `13*X↑2 + (15/2)`

12. `4*(X↑2 + 5)*(3*X↑2 - 3)↑2`

13. `(16.34/12.5)*(42.1↑2 - 64)`

14. `(13/2)*X↑2 + (15/4)*X - 17`

15. `(5/2)*X + 135/2`

In Problems 16–23 write each expression in BASIC notation.

16. **a.** $\dfrac{2}{3}x^2$ **b.** $3x^2 - 17$

17. **a.** $5y^2 - 6x^2 + 11$ **b.** $14y^2 + 12x^2 + 3$

18. **a.** $12(x^2 + 4)$ **b.** $\dfrac{15x + 7}{2}$

19. **a.** $(5 - x)(x + 3)^2$ **b.** $6(x + 3)(2x - 7)^2$

20. $(x + 1)(2x - 3)(x^2 + 4)$ 21. $(2x - 3)(3x^2 + 1)$

22. $\dfrac{1}{4}x^2 - \dfrac{1}{2}x + 12$ 23. $\dfrac{2}{3}x^2 + \dfrac{1}{3}x - 17$

Write a simple BASIC program to carry out the tasks given in each of Problems 24–29.

24. Say "I'm a happy computer."

A hairdresser created a pouf
So high that it looked like a
   goof!
He said, while a-tugging,
"Madame, I am debugging
This artless and curlycued
   roof!"

25. Say "I will demonstrate my computational skill."

26. Calculate $\dfrac{53 + 87}{2}$.

27. Calculate $(1 + .08)^2$.

28. Calculate $5,000(1 + .06)^2$.

29. Calculate $\dfrac{23.5^2 - 5(61.1)}{2}$.

30. Follow the directions shown in the flowchart. Is the answer predicted in the flowchart correct?

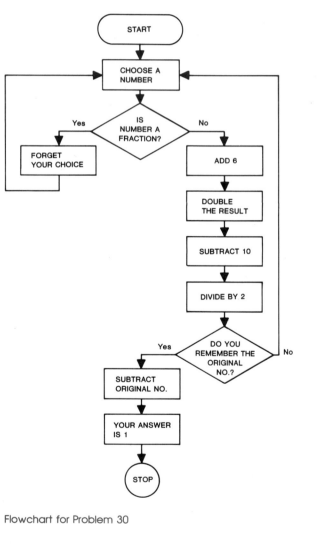

Flowchart for Problem 30

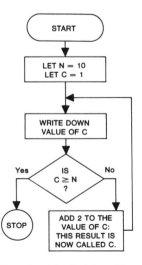

Flowchart for Problem 31

Computers and Music

To the performer of STALKS AND TREES AND DROPS AND CLOUDS:

You need two sets of instruments.

Set A, for STALKS AND TREES: Seven non-reverberating instruments, dry, dead, explosive, whipping, rattling, wooden, etc. Preferably, but not necessarily, just one of each of seven very different kinds. The seven symbols for Set A, and the number of times each appears in the score:

Set B, for DROPS AND CLOUDS: Six reverberating instruments, resounding, ringing, vibrating, sizzling, noisy, whispering, etc. Preferably, but not necessarily, several of each of six very different kinds. The six symbols for Set B, and the number of times each appears in the score:

During the first preparatory stage (see General Preface) you have to determine the particular kind of instrument, which each symbol will represent throughout the piece.

The example of music composition by computer shown here depicts instructions to the performer and a part of the score from Herbert Brun's *Stalks and Trees and Drops and Clouds* (1967). The complete score consists of 31 pages, and it would last about $7\frac{1}{2}$ minutes in performance. It was composed on an IBM 7094 computer.

In another effort, two Southwestern College professors used an IBM 360/22 computer to compose music. John Bibbo, mathematics and computer programming instructor at the college, assisted Dr. Victor Saucedo, music instructor, in a computerized approach to musical composing. Dr. Saucedo was asked to present his composition *RAN. I.X. For Solo Clarinet and Tape* at the Western Regional Conference of the American Society of University Composers held at Fullerton State University in November 1976.

31. Follow the directions shown in the flowchart on the facing page.

32. Write a flowchart to describe how to add two numbers on an algebraic-logic calculator.

33. Write a flowchart that will tell someone how to add two numbers.

34. Prepare a flowchart to describe the process of taking attendance by calling roll. You must take into account the possibilities of both tardy and absent persons.

Mind Bogglers

35. It has been said that anyone who does not know how to program a computer is functionally illiterate. Do you agree or disagree? Give some arguments to support your answer. Project this question into the future; do you think it will ever be true?

36. Arrange four checkers as shown. Try to interchange the checkers so the black ones are at the left and the red ones at the right.

Rules:
   i.   You may move only one checker at a time.
   ii.   You may jump over only one checker.
   iii.   Black checkers may be moved only to the left; red checkers may be moved only to the right.
   iv.   No two checkers may occupy the same space at the same time.
What is the minimum number of moves required to complete the game?

Problems for Individual Study

37. Generalize the results of Problem 36. That is, look for a pattern, and try to predict the minimum number of moves required to complete the game with n black checkers and n red checkers, using $2n + 1$ squares. See "Mathematizing 'Frogs': Heuristics, Proof, and Generalization in the Context of a Recreational Problem," by William Higginson, *Mathematics Teacher*, October 1981, pp. 505–515.

38. The shortage of computer programmers has never been greater. In 1980 there were 150,000 persons employed as programmers, and there were openings for at least 50,000 more. Check out the job opportunities for programmers in your area.

3.5 Communicating with a Computer

In this section, we'll introduce some of the more common BASIC commands and write some programs. If you have access to a computer, it will be helpful to try running some of these programs.

    BASIC treats all numbers as decimal numbers; if the input number does not have a decimal, the computer assumes a decimal point after the last numeral. Also, if the answer to an arithmetic operation is $\frac{1}{3}$, the computer would write this as .3333333333. We also use variables in BASIC, which are formed by a single letter or by a letter followed by a digit.

LIST COMMAND

To find out what is in the computer's core memory at any stage, we use a LIST command, which tells the computer to list everything stored in its core. The following are variations of the list command:

LIST n,   This command causes the computer to list line n only.

LIST n, m This command causes the computer to list lines n through m, inclusive.

LIST   This command causes the computer to list the entire program in its memory.

These list commands are system commands.

END COMMAND

The END command signifies that the program is complete. If a program has a command with a higher line number, then those commands will not be carried out. The program will terminate upon reaching an END command. An END command is optional in Applesoft BASIC.*

PRINT COMMAND

As you saw in the last section, the PRINT command carries out several functions. The first is to perform arithmetical calculations on the computer. For example, suppose we type the following:

```
10 PRINT 7+4
50 END
```

Notice that the END command tells the computer that the program is complete at this point. Now, if we give the computer a RUN command, the computer will output the answer.

```
10 PRINT 7+4
50 END

RUN
11
```

*BASIC has a number of variations, most of which are minor and depend on the computer being used. You should check with your instructor for possible variations in the system you are using.

lyzed objectively within minutes. It also gives flexibility so we can gather data daily on the patient's treatment process, do future follow-ups, and receive valuable feedback."

The actual treatment is determined after a staff evaluation of the computer printouts. The mental health staff then enters its daily reports into the computer producing a running account of each patient's progress.

"This system was developed to account for the accessibility and success of our treatment programs," Johnson said.

As you learn about programming, keep in mind some of the deception that is carried out in the name of computers. People seem to think that "If a computer did it, then it must be correct." This is false, and the fact that a computer is involved has no effect on the validity of the results. Have you ever been told that

"The computer requires that . . ."

"There is nothing we can do; it is done by computer."

"The computer won't permit it."

What they really mean is that the *program* doesn't permit it or that they don't want to write the program to permit it.

Several computations can be done with one program. Each new line in the program will require that the answer be on a new line. For example,

```
10  PRINT 55+11
20  PRINT 55*11
30  PRINT 55-11,55/11
50  END
```

will cause the computer to output the answers for $55 + 11$ and $55*11$ on successive lines, but the answers for $55 - 11$ and $55/11$ will be recorded on the same line:

```
10  PRINT 55+11
20  PRINT 55*11
30  PRINT 55-11,55/11
50  END

RUN
 66
 605
 44            5
```

Remember, a single PRINT statement can contain more than one expression, but the results will be printed on the same line (up to the permissible width of the page). If the output for one line is too long, the end of it is automatically printed on the next line. If a PRINT statement contains more than one expression, those expressions are separated by commas or semicolons.

The PRINT command can also be used for spaces in the output. For example, the command PRINT followed by no variables or numerals will simply generate a line feed, that is, skip a line.

```
10  PRINT 1+2
20  PRINT 2+3
30  PRINT
40  PRINT
50  PRINT 4+5
60  END

RUN
 3
 5


 9
```

In the last section we also used the PRINT command to type messages. For example, suppose we would like the computer to tell us "Hello." This can be accomplished by giving a PRINT command followed by the word enclosed in quotation marks.

```
10  PRINT "HELLO"
20  END
```

A run of this program would simply cause the computer to say "HELLO."

```
RUN
HELLO
```

Let's look at this use of the PRINT command more carefully by studying the following program:

```
10  PRINT "4+7=", 4+7
20  END
```

What is the difference between "4 + 7 =" and 4 + 7 in the program? The "4 + 7 =" causes the computer to type out

The difference is the quotation marks.

$$4+7=$$

and does no arithmetic. The second part, 4 + 7, tells the computer to do the arithmetic (which causes the computer to type out 11).

```
RUN
4+7=        11
```

LET COMMAND

We assign values to locations by using a LET command. For example, we might say

```
10  LET A=2
20  LET B=5
30  LET P=3.1416
40  PRINT A,B
50  PRINT P
60  PRINT A+B
70  PRINT
80  PRINT P*B↑2
100  END
```

We usually write flowcharts for more involved programs, but the flowchart for this program is shown in Figure 3.19. Notice that each instruction on the flowchart is translated into a separate line in the program.

In programming, the statement

```
10  LET A=2
```

is *not* the same as the command

```
10  LET 2=A
```

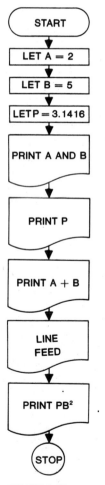

FIGURE 3.19

The latter would cause an error statement. For example, suppose $A = 2$ and $B = 5$ (lines $1\emptyset$ and $2\emptyset$ of the preceding program), and then later in the program we write:

```
90   LET A=B
```

This will cause the value of A to change from 2 to 5, and B will remain unchanged. On the other hand, if we instead used

```
90   LET B=A
```

then B will change in value from 5 to 2 and A will be unchanged.

INPUT COMMAND

Suppose we wish to compute the amount of money present after one year for various interest rates. The simple interest formula is

$$A = P(1 + rt)$$

where P = Principal (amount invested)
$\qquad r$ = Interest rate (written as a decimal)
$\qquad t$ = Time (in years)
$\qquad A$ = Amount present (after t years)

For purposes of this example, we will let $P = \$10,000$ and $t = 1$. Since we must have some way of telling the computer the value of r without rewriting the program each time, a BASIC command called INPUT will be used. It is used to supply data while the program is running. When the computer receives an INPUT command, it will type out a question mark, ?, and wait for the operator to input the data. When the operator is finished inputting data, the return key ($\circlearrowright$) must be pressed. Consider the following BASIC program (the flowchart is shown in Figure 3.20):

```
10   INPUT R
20   LET A=10000*(1+R)
30   PRINT A
40   END
```

When a RUN$\circlearrowright$ command is given, the computer will type

```
                    ?
```

The operator must then type the value for r (three runs of the program are shown for interest rates 5%, 12%, and 18%).

```
RUN
?  .05        Notice that interest rates are input as decimals.
   10500


RUN
?  .12
   12000
```

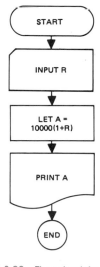

FIGURE 3.20 Flowchart for $A = P(1 + rt)$ where $P = \$10,000$ and $t = 1$.

```
RUN
? .18
  18000
```

The programmer may find it helpful to remind those people using the program what is being asked for with the question mark, so the program can be modified as follows (notice that even the spaces must be included inside the quotation marks):

```
 4  PRINT "I WILL CALCULATE THE AMOUNT PRESENT FROM A $10,000 ";
 6  PRINT "INVESTMENT AFTER"
 8  PRINT "1 YEAR AT SIMPLE INTEREST. WHAT IS THE INTEREST RATE";
24  PRINT
26  PRINT "THE AMOUNT PRESENT IS $";
```

Combining these steps with the original, we get:

```
 4  PRINT "I WILL CALCULATE THE AMOUNT PRESENT FROM A $10,000 ";
 6  PRINT "INVESTMENT AFTER"
 8  PRINT "1 YEAR AT SIMPLE INTEREST. WHAT IS THE INTEREST RATE";
10  INPUT R
20  LET A=10000*(1+R)
24  PRINT
26  PRINT "THE AMOUNT PRESENT IS $";
30  PRINT A
40  END
```

This second version of the program has several nonessentials from a programming standpoint, but it makes the computer much more conversational.

```
RUN
I WILL CALCULATE THE AMOUNT PRESENT FROM A $10,000 INVESTMENT AFTER
1 YEAR AT SIMPLE INTEREST. WHAT IS THE INTEREST RATE? .05

THE AMOUNT PRESENT IS $ 10500

READY
```

Notice the semicolon at the end of lines 4, 8, and 26. It means that there should be *no line feed* when going to the next line of the program.

It is possible to input several variables at once. Suppose we wish to input not only R but also P, as in the command

```
INPUT P, R
```

We have the following variation:

```
10  INPUT P, R
20  LET A=P*(1+R)
30  PRINT A
40  END
```

When we run this program, two variables must be input after the question mark.

Can you add PRINT commands to this program to make it more conversational?

```
10  INPUT P, R
20  LET A=P*(1+R)
30  PRINT A
40  END

RUN
? .5000,.07
   5350

RUN
? 30000,.075
   32250

RUN
? 30000,.09
   32700
```

This first run of the program is calculating the simple interest formula for one year at 7% with $5,000 invested.

Can you interpret what is being calculated in the second two runs of this program?

Problem Set 3.5

1. Give an example of each of the three uses for the PRINT command.

2. What is the significance of the semicolon at the end of a line beginning with a PRINT command?

3. Name a command that allows you to put information into the computer.

4. You have fed the following program into a computer:

```
10  PRINT 6+5
20  PRINT 6-5
30  PRINT 6*5, 6/3
40  END
```

What will the computer type if you give a RUN command?

5. You have fed the following program into a computer:

```
10  PRINT "6+5=",6+5
20  PRINT "6-5=",6-5
30  PRINT "6*5=",6*5,"6/3=",6/3
40  END
```

What will the computer type if you give a RUN command?

6. Compare the following program with the one in Problem 5.

```
10  PRINT "6+5=";6+5
20  PRINT "6-5=";6-5
```

```
30  PRINT "6*5=";"6/3=";6/3
40  END
```

What is the difference in output?

7. You have fed the following program into a computer:

```
10  PRINT 4+5
20  PRINT "HELLO, ";
30  PRINT "I LIKE YOU."
40  PRINT "4+5=";4+5
50  END
```

What will the computer type if you give a RUN⤶ command?

8. You have fed the following program into a computer:

```
10  PRINT 5+10
20  PRINT 5*10
30  PRINT 10/5, 5/10
40  END
```

What will the computer type if you give a RUN⤶ command?

9. You have fed the following program into a computer:

```
10  PRINT 7+12
20  PRINT 7-12, 7*12
30  END
```

What will the computer type if you give a RUN⤶ command?

10. Find the error in the following program:

```
10  PRINT "I WILL COMPUTE (2*X↑2)(X+1)
20  INPUT X
30  PRINT (2*X↑2)*(X+1)
40  END
```

11. Find the error in the following program:

```
10  INPUT X
20  PRINT (X↑2-3)(2*X+1)
30  END
```

12. You have fed the following program into a computer:

```
10  LET A=2
20  PRINT "WORKING"
30  LET B=3
40  PRINT A+B
50  END
```

What will the computer type if you give a RUN⤶ command?

13. You have fed the following program into a computer:

```
10  PRINT "I WILL CALCULATE IQ."
20  PRINT "WHAT IS MENTAL AGE";
```

COMPUTERS IN MEDICINE

Computers are becoming an important tool in medicine and medical research. For example, at the Texas Institute for Rehabilitation in Houston, scientists are using an IBM computer to produce highly accurate, three-dimensional measurements of the human body for studies of problems ranging from spinal deformities in children to weight loss in astronauts. Overlapping photographs of the body are taken simultaneously, and then a plotting device identifies as many as 40,000 reference points, which are then fed into the computer. With a computer-driven plotter, a contour map of the body (like the one shown here) is then drawn to assist doctors in their diagnosis and treatment.

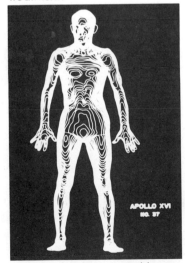

Body contour maps provide one means of studying weight losses experienced by astronauts during prolonged space flights. Such maps enable scientists to compare body volume before and after the flights.

```
30  INPUT M
40  PRINT "WHAT IS CHRONOLOGICAL AGE";
50  INPUT C
60  PRINT
70  PRINT"IQ IS ";
80  PRINT (M/C)*100
```

What will the computer type if the program is run for a mental age of 12 and a chronological age of 10?

14. What will the computer type if you RUN⤸ the following program for a 4 by 5 rectangle?

```
 10  PRINT "I WILL CALCULATE THE PERIMETER AND ";
 20  PRINT "AREA OF A RECTANGLE."
 30  PRINT "WHAT IS THE LENGTH";
 40  INPUT L
 50  PRINT "WHAT IS THE WIDTH";
 60  INPUT W
 70  PRINT "THE PERIMETER IS ";
 80  PRINT 2*(L+W);
 90  PRINT "AREA IS ";
100  PRINT L*W
110  END
```

15. Write a BASIC program to ask for a value for x, and then use this value to evaluate $5x^2 + 17x - 128$. Also, show the flowchart.

16. Write a BASIC program to ask for a value of x, and then use this value to evaluate $12.8x^2 + 14.76x + 6\frac{1}{3}$. Also, show the flowchart.

17. Rewrite the following program in four steps so that the output is unchanged.

```
10  LET R=2
20  LET P=3.1416
30  LET A=P*R↑2
40  PRINT "A=";A
50  END
```

18. Make three different choices for x, y, z, and show the output.

```
10  INPUT X,Y,Z
20  LET M=(X+Y+Z)/3
30  PRINT "X=";X,"Y=";Y,"Z=";Z,"M=";M
40  PRINT "TRY THIS AGAIN FOR SOME OTHER VALUES."
50  END
```

19. Suppose you are considering a job with a company as a sales representative. You are offered $100 per week plus 5% commission on sales. Write a program that will compute your weekly salary if you input the total amount of sales. Be sure to include a flowchart.

20. A manufacturer of sticky widgets wishes to determine a monthly cost of manufacturing on an item that costs $2.68 per item for materials and $14 per item for labor.

MACHINE TRANSLATION

In the 1960s there was a big push to use computers to do foreign-language translation. Computers were supplied with a small bilingual dictionary with the corresponding words in the two languages. It soon became apparent that word-for-word translation was virtually useless. The addition of a dictionary of phrases brought only marginal improvement.

These translators can be tested by translating English to Russian and then Russian back to English. Hopefully, one should end up with about the same as one started. Using this method, the maxim, "Out of sight, out of mind" ended up as "The person is blind, and is insane."

Another example was, "The spirit is willing but the flesh is weak." It was translated as "The wine is good but the meat is raw."

Needless to say, computer translation is presently used very little. And it is doubtful that it will be useful in the near future, even though there are now calculators that will translate individual words from one language to another.

The company also has fixed expenses of $6,000 per month (this includes advertising, taxes, plant facilities, and so on). Write a program that will compute monthly cost if you input the number of items manufactured. Be sure to include a flowchart.

Mind Bogglers

Problems 21 and 22 require that you have access to a computer.

21. *Computer problem.* Think of any counting number. Add 9. Square. Subtract the square of the original number. Subtract 61. Multiply by 2. Add 24. Subtract 36 times the original number. Take the square root. What is the result?

 a. Show algebraically why everyone who works the problem correctly gets the same answer.
 b. Write a BASIC program to solve the puzzle. In BASIC, the square root of a number N is found by giving the command SQR(N).

22. *Computer problem.* Write a BASIC program for obtaining the pattern shown.

```
XXXXXX   XXXXXX   XXXXXX   XXXXXX   XXXXXX
XX   X   X        X    X   X        X
XXXXXX   XXXXXX   XXXXXX   X        XXXXXX
XX       X        X    X   X        X
XX       X        X    X   X        X
XX       XXXXXX   X    X   XXXXXX   XXXXXX
```

*3.6 Programming Repetitive Processes

Repetition of arithmetic operations is a tedious task that can be assigned to computers. Suppose we wish to add the first 100 counting numbers:

$$1 + 2 + 3 + 4 + \cdots + 97 + 98 + 99 + 100$$

We could write a program similar to the ones of the previous section, but these methods would still be tedious. Instead, we will use a **loop**, which repeats a sequence of operations. The loop is eventually completed (we hope!), and the sequence of instructions proceeds to the stopping point.

The simplest command that sets up a loop is called the GOTO command. (GOTO is pronounced as two words: "go to.") The command transfers the control to a line number other than the next higher one. The flowchart in Figure 3.21 illustrates the GOTO command. This command can cause the computer to go into a loop, as shown by the program below:

```
10   LET N=1
20   PRINT N
30   LET N=N+1
40   GOTO 20
50   END
```

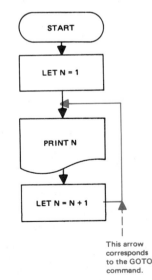

FIGURE 3.21 Flowchart illustrating a GOTO command

*This is an optional section.

Computer Poetry

The four computer-written poems at the right were produced by Margaret Masterman and Robin McKinnon Wood using human–machine interaction at the Cambridge Language Research Unit. The program is written in the TRAC language.

"Only once in every generation is there a computer that can write poetry like this."

©DATAMATION

This poem is a translation of a computer-written German poem. It was done at Stuttgart's Technical College from a program written by Professor Max Bense. Professor Bense has been trying to discover if it is possible to formulate problems in esthetics mathematically.

THE POEM

The joyful dreams rain down
The heart kisses the blade of grass
The green diverts the tender lover
Far away is a melancholy vastness
The foxes are sleeping peacefully
The dream caresses the lights
Dreamy sleep wins an earth
Grace freezes where this glow dallies
Magically the languishing shepherd
dances.

1 POEM   eons deep in the
          ice
          I paint all time in
          a whorl
          bang the sludge
          has cracked

2 POEM   eons deep in the
          ice
          I see gelled time
          in a whorl
          pffftt the sludge
          has cracked

3 POEM   all green in the
          leaves
          I smell dark
          pools in the trees
          crash the moon
          has fled

4 POEM   all white in the
          buds
          I flash snow
          peaks in the
          spring
          bang the sun has
          fogged

If we type RUN⤸, this program will continue forever typing out the successive counting numbers. Can you follow the steps of this program? The loop is set up using the GOTO statement. The problem is, however, how to "get out" of the loop. To do this, we introduce some additional commands.

The **READ** and **DATA** commands allow you to input data into the program itself rather than with the INPUT command we discussed in the last section. This command is illustrated in Example 1.

Example 1: Consider the following program and determine the computer output if you give a RUN⤸ command.

```
        10  PRINT "NUMBER","NUMBER SQUARED"
        20  READ N
        30  LET S=N↑2
        40  PRINT N,S
        50  GOTO 20
        60  DATA 5,6,7,8,198
        70  END
```

Transfers back to line 20. The second time through, it will read the next value in the DATA statement, namely, 6.

The first time the computer "sees" N it will read the first value under the DATA statement (5 in this example).

Solution: The output will be:

```
        RUN⤸

        NUMBER    NUMBER SQUARED
          5            25
          6            36
          7            49
          8            64
        198          39204


        OUT OF DATA LINE 20
```

The computer continues to loop until all the data are processed.

A second, and perhaps more sophisticated, way of ending a loop is to use an **IF-THEN** or **IF-GOTO** command. These commands (whichever one your computer uses) give the computer its basic decision-making ability. Consider the following example:

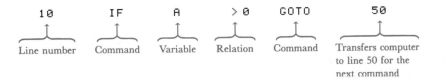

10	IF	A	> 0	GOTO	50
Line number	Command	Variable	Relation	Command	Transfers computer to line 50 for the next command

TABLE 3.4 Relation Symbols
That Can Be Used with
IF-THEN Commands

Math Symbol	BASIC Symbol
=	=
>	>
≥	>=
<	<
≤	<=
≠	<>

Some different relations that can be used are shown in Table 3.4. On a flow-chart, conditional transfers are symbolized by a diamond-shaped box, as shown in Figure 3.22.

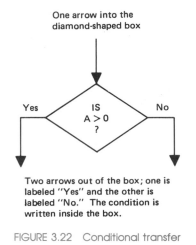

One arrow into the
diamond-shaped box

Yes   IS
           A > 0
            ?   No

Two arrows out of the box; one is
labeled "Yes" and the other is
labeled "No." The condition is
written inside the box.

FIGURE 3.22 Conditional transfer

Example 2: Consider the following program and determine the computer output if you give a RUN↵ command.

```
10   PRINT "INPUT TWO NUMBERS, AND I WILL TELL YOU WHETHER THEIR ";
20   PRINT "PRODUCT IS "
30   PRINT "POSITIVE, NEGATIVE, OR ZERO."
40   PRINT "WHAT ARE THE NUMBERS";
50   INPUT A, B
60   PRINT
70   PRINT "THE PRODUCT IS ";
80   IF A*B=0 GOTO 150
90   IF A*B>0 GOTO 200
100  IF A*B<0 GOTO 250
150  PRINT "ZERO."
160  GOTO 300
200  PRINT "POSITIVE."
210  GOTO 300
250  PRINT "NEGATIVE."
300  END
```

Solution: Here are some sample outputs for this program:

```
RUN
INPUT TWO NUMBERS, AND I WILL TELL YOU WHETHER THEIR PRODUCT IS
POSITIVE, NEGATIVE, OR ZERO.
WHAT ARE THE NUMBERS? 4,-5
```

Computer Graphics

Computers can be programmed to create a wide vari- both two and three dimensions. They can also be
ety of designs. They can draw charts and maps in used to simulate art.

```
THE PRODUCT IS NEGATIVE.

READY

RUN
INPUT TWO NUMBERS, AND I WILL TELL YOU WHETHER THEIR PRODUCT IS
POSITIVE, NEGATIVE, OR ZERO.
WHAT ARE THE NUMBERS? -4,-5

THE PRODUCT IS POSITIVE.
```

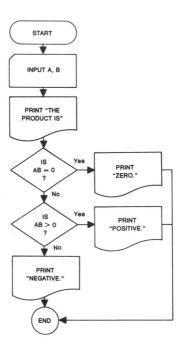

FIGURE 3.23 Flowchart for the example illustrating the IF-THEN or IF-GOTO command (Example 2)

```
RUN
INPUT TWO NUMBERS, AND I WILL TELL YOU WHETHER THEIR PRODUCT IS
POSITIVE, NEGATIVE, OR ZERO.
WHAT ARE THE NUMBERS? 0,123

THE PRODUCT IS ZERO
```

The flowchart for this program is shown in Figure 3.23.

Example 3:   We began this section by asking for a program to find the sum of the first 100 counting numbers. Write this BASIC program.

Solution

```
10   LET S=1
20   LET T=1
30   LET T=T+1
40   LET S=S+T
50   IF T<100 GOTO 30
60   PRINT S
70   END

RUN

5050
```

Problem Set 3.6

1.   What is a loop?

2.   What procedure can be used to terminate a loop?

3.   Consider the following program:

```
10    PRINT "H";
20    PRINT "I";
30    PRINT " ";
40    GOTO 70
50    PRINT "END";
60    GOTO 110
70    PRINT "F";
80    PRINT "R";
90    PRINT "I";
100   GOTO 50
110   PRINT ".";
120   END
```

What will the computer type if you give a RUN⤸ command?

4. Consider the following program:

```
 10 PRINT "T";
 20 PRINT "H";
 30 GOTO 100
 40 PRINT "IS ";
 50 PRINT "C";
 60 PRINT "O";
 70 GOTO 120
 80 PRINT "ECT";
 90 GOTO 140
100 PRINT "IS ";
110 GOTO 40
120 PRINT "RR";
130 GOTO 80
140 PRINT ".";
150 END
```

What will the computer type if you give a RUN↩ command?

5. The program in Problem 3 is inefficient from a programmer's standpoint, since the same output can be generated by using only two commands. Refine the program in Problem 3 by writing a new program that is more efficient. By "refine" we mean write a better program that will generate the same output.

6. The program in Problem 4 is inefficient from a programmer's standpoint, since the same output can be generated by using only two commands. Refine the program in Problem 4 by writing a new program that is more efficient. By "refine" we mean write a better program that will generate the same output.

7. Suppose you are given the following program:

```
10 LET A=7
20 IF A>100 GOTO 80
30 IF A<1 GOTO 60
40 PRINT "A IS BETWEEN 1 AND 100 INCLUSIVE."
50 GOTO 90
60 PRINT "A IS LESS THAN 1."
70 GOTO 90
80 PRINT "A IS GREATER THAN 100."
90 END
```

What will the computer type if you give a RUN↩ command?

8. Repeat Problem 7 for $A = 107$.

9. Repeat Problem 7 for $A = -10$.

10. Write a program that will compare two numbers, A and B, and then print the larger number.

11. Consider the following program and determine the computer output if you give a RUN↩ command.

erties you can own," he said. "It takes a large amount of capital to make any money from them. They do not have a high efficiency rating."

"The best properties are the orange ones, followed by the red if the player has a fair amount of money or, if not, the light blue ones."

"If you have both the red and the orange you are practically unbeatable," he said.

Railroads are fine early in the game but are not of great value later, Treinish said.

His program showed that Reading Railroad is the most frequently landed on property but that Shortline is the least visited. Chance is landed on more often than any other spot on the board.

He has programmed the computer to play the game, not only with people but with other computers. It will buy and sell, handle money, roll the dice and catch people cheating. And it really knows how to trade.

"The computer would immediately reject a deal if it was going to be cheated. If it accepts your deal then you've probably been cheated," Treinish said Thursday.

Treinish started working on his program about a year ago because he was a Monopoly fan and wanted to know what the probabilities of success were with different property purchases. He had already taught himself some computer language and was using computers to do his homework in physics, math and chemistry.

"I was looking for a project to enter in the Westinghouse Talent Search," he said, and his program has made him one of 300 semi-finalists.

```
10   PRINT "NUMBER","NUMBER SQUARED"
20   READ N
30   LET S=N↑2
40   PRINT N,S
50   GOTO 20
60   DATA 1,2,4,3
70   END
```

12. Repeat Problem 11 where line 60 is changed to: DATA 0,5,4,2.

13. Consider the following program and determine the computer output if you give a RUN⤸ command.

```
10   READ N
20   LET A=2*N+3
30   PRINT N,A
40   GOTO 20
50   DATA 1,5,2,4
60   END
```

14. Repeat Problem 13 where line 50 is changed to: DATA 3,6,8,5.

15. Consider the following program and determine the computer output if you give a RUN⤸ command.

```
10   LET N=1
20   LET B=N↑2
30   PRINT B
40   IF N=3 GOTO 70
50   LET N=N+1
60   GOTO 20
70   END
```

16. Consider the following program and determine the computer output if you give a RUN⤸ command.

```
10   LET M=1
20   LET C=3*M-2
30   PRINT C
40   IF M=4 GOTO 70
50   LET M=M+1
60   GOTO 20
70   END
```

17. The old song "100 Bottles of Beer on the Wall" is quite tedious because of the lyrics. Write a computer program that will generate an output:

```
100  BOTTLES OF BEER ON THE WALL,
 99  BOTTLES OF BEER ON THE WALL,
 98  BOTTLES OF BEER ON THE WALL,

              ⋮

  2  BOTTLES OF BEER ON THE WALL,
  1  BOTTLE OF BEER ON THE WALL,
     THAT'S ALL FOLKS!
```

18. Write a BASIC program that will print out the first one hundred numbers backwards:

$$100, 99, 98, 97, \ldots, 3, 2, 1$$

19. Write a BASIC program to output the first 100 Fibonacci numbers. Recall that Fibonacci numbers are numbers in the pattern

$$1, 1, 2, 3, 5, 8, 13, 21, \ldots$$

20. Write a BASIC program that will print out the square numbers up to 50. A flowchart is given in Example 2 of Section 3.4.

21. Write a BASIC program that will print out the first 50 square numbers.

Computers and Architecture

When architectural drawings require a large amount of repetition, computers can be used quite effectively. This computer drawing shows the furniture, telephone, and electrical layouts for a floor of the Sears Tower in Chicago. Drawings were made for four separate plans for each of the building's 110 stories.

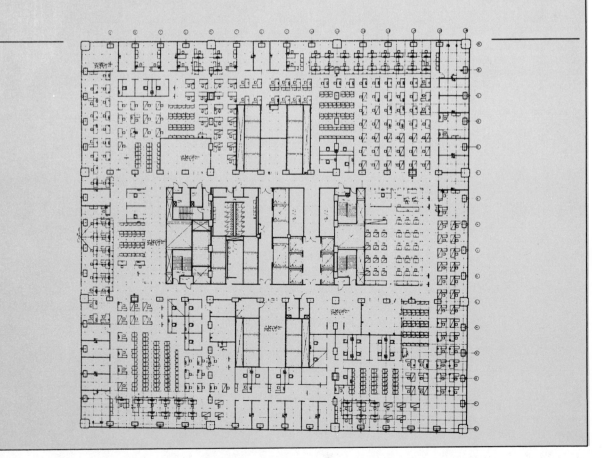

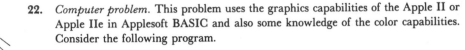

Mind Boggler

Problem 22 requires that you have access to a computer.

22. *Computer problem.* This problem uses the graphics capabilities of the Apple II or Apple IIe in Applesoft BASIC and also some knowledge of the color capabilities. Consider the following program.

```
10   GR : BLUE=6 : PINK=11
20   FOR Y=0 TO 39
30   FOR X=0 TO 39
40   COLOR=PINK
50   IF X>Y THEN COLOR=BLUE
60   PLOT X,Y
70   NEXT X
80   NEXT Y
90   END
```

a. What will appear when the program is run? Make the following replacements for line 50 and tell what appears when the program is run in each case.

b. `50   IF X+Y>40 THEN COLOR=BLUE`

c. `50   IF X<10 OR Y>30 THEN COLOR=BLUE`

d. `50   IF X<10 AND Y>30 THEN COLOR=BLUE`

e. `50   IF X*Y>100 THEN COLOR=BLUE`

"LET ME OUT!"

Chapter 3 Outline

I. History of Computing Devices
    A. Ancient methods of calculating
        1. Finger multiplication
        2. Abacus
    B. The dawning of computers
        1. Leibniz
        2. Babbage
        3. Jacquard
        4. Mark I, ENIAC
        5. Pocket calculators
    C. Modern computers
II. Pocket Calculators
    A. Types of calculators
        1. Four-function
        2. Four-function with memory
        3. Scientific
        4. Special-purpose
        5. Programmable
    B. Calculator logic
        1. Arithmetic
        2. Algebraic

      3. RPN

   C. Elementary operations

III. Introduction to Computers

   A. Functions of a computer

      1. Data processing

      2. Information retrieval

      3. Pattern recognition

      4. Simulation

   B. The main components of a computer

      1. Input/output

         a. CRT

         b. Punched card

         c. Magnetic tape

         d. Line printer

      2. Storage or memory

         a. Word

         b. Address

         c. Bit

         d. Types of storage

      3. Central processing unit (CPU)

      4. Control

   C. Binary numeration system

      1. Expanded form

      2. Change from binary to decimal

      3. Change from decimal to binary

IV. BASIC Programming

   A. Programming

   B. Flowcharts

   C. Square numbers

   D. BASIC language

      1. Comparison with other languages

      2. Definition

      3. Commands

      4. PRINT command

      5. END command

      6. System commands (for example, RUN command)

      7. Math symbols in BASIC

V. Communicating with a Computer

   A. LIST commands (system commands)

   B. END command

   C. PRINT command

      1. Text

      2. Calculation

      3. Line feed

   D. LET command

   E. INPUT command

*VI. Programming Repetitive Processes
    A. Loops
    B. GOTO command
    C. READ/DATA commands
    D. IF-THEN and IF-GOTO commands

Chapter 3 Test

1. Briefly discuss some of the events leading up to the invention of the computer.

2. Use a calculator to find

    a. $63 + 28.7 - 6.8 \times 7$ **b.** $\dfrac{41.3 + 6.85}{16}$

3. **a.** *Calculator problem.* To find out what to do when the market is dropping, calculate

$$7700 + 7.01 \times 5$$

    and turn your calculator over to read the answer.
    b. *Calculator problem.* The child prodigy Jacques Inaudi (1867–1923) instantly gave the correct answer to the following problem by doing all the arithmetic mentally:

$$\frac{4811^2 - 1}{6}$$

    Use a calculator to find the answer.

4. Name the five main components of a computer, and briefly explain the function of each.

5. Briefly explain what we mean when we speak of computer "programming."

6. Fill in the blanks.

    a. $11101_{two} =$ _____ (decimal)   **b.** $1111011_{two} =$ _____ (decimal)
    c. $12 =$ _____ (binary)   **d.** $52 =$ _____ (binary)

7. **a.** Write 5*(X + 2)↑2 in ordinary mathematical notation.
    b. Write $3(x + 4)(2x - 7)^2$ in BASIC notation.

8. Write a BASIC language program to ask the value of x, and then compute $18x^2 + 10$. Include a flowchart.

---

*Optional section.

*9. Suppose you are given the following BASIC program:

```
 10  PRINT "TH";
 20  PRINT "E";
 30  PRINT " ";
 40  PRINT "M";
 50  PRINT "IS";
 60  PRINT "S";
 70  PRINT "ISS";
 80  GOTO 100
 90  PRINT "H";
100  PRINT "IPP";
110  GOTO 130
120  PRINT "Y";
130  PRINT "I ";
140  PRINT "I";
150  PRINT "S ";
160  PRINT "WET";
170  GOTO 190
180  PRINT "NOODLE";
190  PRINT "."
200  END
```

What will the computer type if you give a RUN↵ command?

10. Write a BASIC program to calculate the areas of various circles; the radii are the input values. (Use 3.1416 for the computer approximation of π.)   The formula for the area of a circle is $A = \pi r^2$.

Bonus Question

11. Write a BASIC program that will print out the decimal representation of the unit fractions from $\frac{1}{1}$ to $\frac{1}{100}$ where the denominators increase by 1.

*Requires optional Section 3.6.

Date	Cultural History
1000	980–1002: Emperor Otto II 1003: Leif Erickson crosses Atlantic to Vinland 1028: School of Chartes 
1050	1050: Normans penetrate England 1054: Macbeth defeated at Dunsinane 1065: Consecration of Westminster Abbey 1096: Start of the First Crusade 
1100	1110: Chinese invent the playing card 1125: Commencement of troubadour music 
1150	1154: Beginning of Plantagenet reign 1165: Maimonides: *Mishneh Torah* 1186: Domesday Book, tax census ordered by William the Conqueror 

Otto II

Crusader

Troubadour, Dürer

Hebrew Book of Law

Mathematical History

1000: Śrīdhara recognizes the importance of the zero

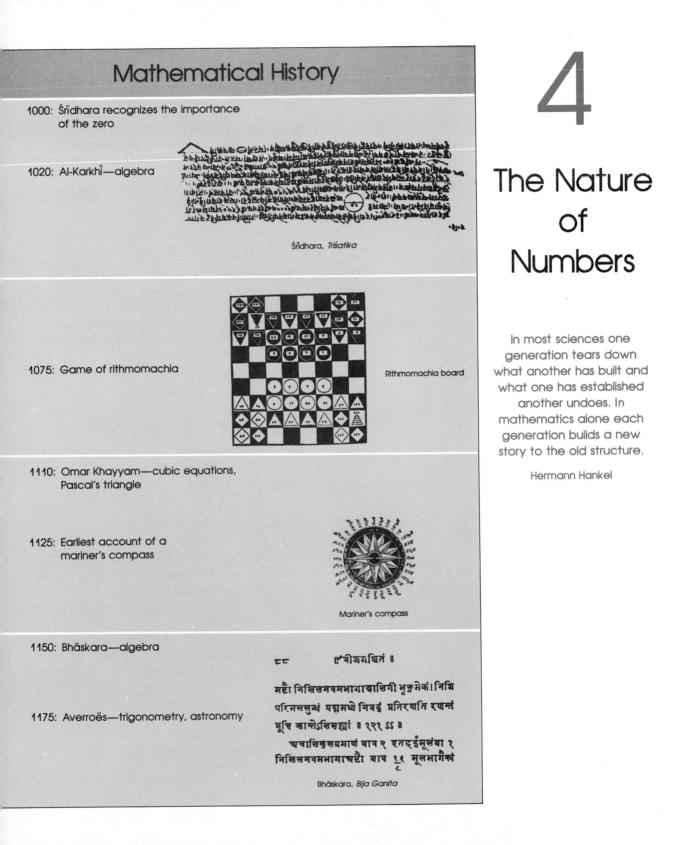

Śrīdhara, *Triśatika*

1020: Al-Karkhî—algebra

1075: Game of rithmomachia

Rithmomachia board

1110: Omar Khayyam—cubic equations, Pascal's triangle

1125: Earliest account of a mariner's compass

Mariner's compass

1150: Bhāskara—algebra

1175: Averroës—trigonometry, astronomy

Bhāskara, *Bija Ganita*

4

The Nature of Numbers

In most sciences one generation tears down what another has built and what one has established another undoes. In mathematics alone each generation builds a new story to the old structure.

Hermann Hankel

4.1 Early Numeration Systems

A numeration system is a set of basic symbols and some rules for making other symbols from them, the purpose of the whole game being the identification of numbers. The invention of a precise and "workable" system is one of the greatest inventions of humanity. It is certainly equal to the invention of the alphabet, which enabled humans to carry the knowledge of one generation to the next. It is simple for us to use the symbol 17 to represent this many objects:

$$XXXXX\ XXXXX$$
$$XXXXX\ XX$$

However, it took us centuries to arrive at this stage of symbolic representation. It wasn't the first and probably won't be the last numeration system to be developed. Here are some of the ways that 17 has been written:

Tally	⊮⊮⊮ ⊮⊮⊮ ⊮⊮⊮ ||
Egyptian	∩ |||||||
Roman	XVII
Mayan	≡
Linguistic	Seventeen
	Siebzehn
	Dix-sept

The concept represented by each of these symbols is the same, but the symbols differ. The concept or idea of "seventeenness" is called a *number;* the symbol used to represent the concept is called a *numeral*. The difference between *number* and *numeral* is analogous to the difference between a person and his or her name, as is illustrated by the *Peanuts* cartoon.

Two of the earliest civilizations known to use numerals were the Egyptian and Babylonian. We shall examine these systems for two reasons. First, it will help us more fully to understand our own numeration (decimal) system; second, it will help us to see how the ideas of these other systems have been incorporated into our system.

Perhaps the earliest type of written numeration system developed was a **simple grouping system.** The Egyptians used such a system by the time of the first dynasty, around 2850 B.C. The symbols of the Egyptian system were part of their hieroglyphics and are shown in Table 4.1.

Any number is expressed by using these symbols additively; each symbol is repeated the required number of times, but with no more than nine repetitions.

TABLE 4.1 Egyptian Hieroglyphic Numerals

Our Numeral (Decimal)	Egyptian Numeral	Descriptive Name
1	I	Stroke
10	∩	Heel bone
100	ϑ	Scroll
1000	✗	Lotus flower
10,000	⌒	Pointing finger
100,000	⌒	Polliwog
1,000,000	✗	Astonished man

Example 1: $12{,}345 = $ ⌒✗✗ϑϑ∩∩/ϑ∩∩|||||

The position of the individual symbols is not important. That is,

$$⌒✗✗{}^{ϑϑ∩∩}_{ϑ∩∩}||||| = ✗✗{}^{ϑϑ}_{ϑ}⌒{}^{∩∩}_{∩∩}|||||$$

$$= |||||⌒✗{}^{ϑϑ}_{∩∩∩∩}✗ϑ$$

The Egyptians had a simple repetitive-type arithmetic. Addition and subtraction were performed by repeating the symbols and by regrouping:

Example 2:

$$\begin{array}{r} 245 \\ +457 \\ \hline \end{array}$$

$$ϑϑ{}^{∩∩}_{∩∩}|||||$$
$$+ \quad ϑϑ∩∩∩|||||$$
$$ϑϑ ∩∩ ||$$
$$\overline{ϑϑϑ∩∩∩}$$
$$ϑϑϑ∩∩∩∩∩ ||||||||||$$

Regroup: $ϑϑϑ∩∩∩∩∩| ||$
$$ϑϑϑ∩∩∩∩∩|$$

Regroup again: $ϑϑϑϑ ||$
$$ϑϑϑ$$

Example 3:

$$\begin{array}{r} 142 \\ - \ 67 \\ \hline \end{array}$$

ϑ ∩ ∩ ||
∩ ∩
∩∩∩ |||||||
∩∩∩

Regroup: ∩∩∩∩∩∩∩ ||||||||||
∩∩∩∩∩∩

$$- \quad ∩∩∩ \qquad ||||||$$
$$∩∩∩$$
$$\overline{∩∩∩∩} \qquad |||||$$
$$∩∩∩$$

Historical Note

There are two important sources for our information about the Egyptian numeration system. The first is the Moscow papyrus, which was written in about 1850 B.C. and contains 25 mathematical problems. The second is the Rhind papyrus. The chronology of this papyrus is interesting. Around 1575 B.C. the scribe Ahmose copied (and most likely edited) material from an older manuscript. Then, 3500 years later this became known as the Rhind Mathematical papyrus. In 1858 A. Henry Rhind bought the papyrus in Luxor. He found it in a room of a small building near the Ramesseum. The Rhind papyrus contains 85 problems and was a type of mathematical handbook for the Egyptians. From these two sources, we conclude that most of the problems were of a practical nature (although *some* were of a theoretical nature).

Historical Note

From the development of the arithmetical calculations of finding two-thirds of any number, the Egyptians were able to determine π with surprising accuracy, to solve first- and second-degree equations, and to sum arithmetic and geometric progressions. They also found the formula for the volume of a frustum of a square pyramid, as well as the formula for the surface area of a hemisphere. They were able to do all this with very few mathematical tools.

Multiplication and division were performed by successions of additions that did not require the memorization of a multiplication table and could easily be done on an abacus-type device (see Problem 33 of Problem Set 4.1).

FIGURE 4.1   This is a portion of the Rhind papyrus showing the area of a triangle.

The Egyptians also used unit fractions $(1/n)$ in their computations. These were indicated by placing the symbol $\bigcirc$ over the numeral for the denominator. Thus

$$\underset{|||}{\bigcirc} = \frac{1}{3} \qquad \underset{||||}{\bigcirc} = \frac{1}{4} \qquad \underset{\cap}{\bigcirc} = \frac{1}{10} \qquad \underset{\underset{|||}{\cap\cap\cap}}{\bigcirc} = \frac{1}{33}$$

The fraction $\frac{1}{2}$ was an exception:

$$\underset{||}{\bigcirc} \quad \text{or} \quad \boxed{} = \frac{1}{2}$$

Fractions that were not unit fractions were represented as sums of unit fractions. For example,

$$\frac{2}{7} \text{ would be expressed as } \frac{1}{4} + \frac{1}{28}$$

Repetitions were not allowed; that is, $\frac{2}{7}$ would not be written as $\frac{1}{7} + \frac{1}{7}$ but could be written as $\frac{1}{28} + \frac{1}{4}$ (called the *decomposition of* $\frac{2}{7}$).

$$\frac{3}{5} \text{ would be expressed as } \frac{1}{2} + \frac{1}{10}$$

$$\frac{2}{99} \text{ would be expressed as } \frac{1}{66} + \frac{1}{198}$$

The fraction $\frac{2}{3}$ was the only nonunit fraction not written as a sum:

$$\text{⍟ or ⍟} = \frac{2}{3}$$

The Rhind papyrus includes a table for all such decompositions for odd denominators from 5 to 101. This papyrus also uses a symbolism for addition and subtraction. Addition was indicated by a pair of legs walking to the left, and subtraction was a pair of legs walking to the right.

The Babylonian numeration system differed from the Egyptian in several respects. Whereas the Egyptian system was a simple grouping system, the Babylonians employed a much more useful **positional system.** Since they lacked papyrus, they used mostly clay as a writing medium, and thus the Babylonian cuneiform was much less pictorial than the Egyptian system. They employed only two wedge-shaped characters, which date from 2000 B.C. and are shown in Table 4.2.

TABLE 4.2 Babylonian Cuneiform Numerals

Our Numeral (Decimal)	Babylonian Numeral
1	▼
2	▼▼
9	▼▼▼▼▼▼▼▼▼
10	⟨
59	⟨⟨⟨▼▼▼▼▼ ⟨⟨ ▼▼▼▼

Historical Note

Sumerian clay tablet

There are many clay tablets available to us (more than 50,000), and the work of interpreting them is still going on. Most of our knowledge of the tablets is less than 50 years old and has been provided by Professors Otto Neugebauer and F. Thureau-Dangin.

Notice from the table that, for numbers 1 through 59, the system is repetitive. However, unlike in the Egyptian system, the position of the symbols was important. The ⟨ *must* appear to the left of any ▼s to represent numbers smaller than 60. For numbers larger than 60, the symbols ⟨ and ▼ are to the left of ⟨, and any symbols to the left of the ⟨ have a value 60 times their original value. That is,

$$\text{▼ ⟨⟨▼▼▼ / ⟨ ▼▼} \qquad \text{means} \qquad (1 \times 60) + 35$$

Example 4: ▼▼▼⟨ ⟨⟨▼▼▼ / ⟨⟨▼▼▼ ▼▼▼ $= (3 \times 60) + 59 = 239$

Example 5: ▼▼▼ / ▼▼ ⟨▼ $= (5 \times 60) + 11 = 311$

Example 6: ⟨⟨▼▼▼▼⟨ ▼▼▼▼▼▼ $= (23 \times 60) + 16 = 1,396$

> **CHOCOLATE USED FOR MATH**
>
> NEW YORK (UPI)—The Aztec Indians, first people known to have consumed chocolate as a food, also used it as a counting unit, the Chocolate Information Council reports. Rather than counting by tens as we do, they used 20 as a median. For example: quantities up to 28 were represented by dots; the next unit was a flag, repeated for amounts up to 400; next, a fir tree, which signified up to 800. The next unit—8,000—was indicated by a sack because each sack of cocoa contained about 8,000 beans.

This system is called a *sexagesimal system* and uses the principle of position. However, the system is not fully positional, since only numbers larger than 60 use the position principle; numbers within each basic 60-group are written by a simple grouping system. A true positional sexagesimal system would require 60 different symbols.

The Babylonians carried their positional system a step further. If any numerals were to the left of the second 60-group, they had the value of 60×60 or 60^2. Thus

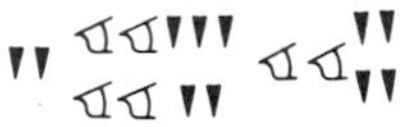

$$= (2 \times 60^2) + (45 \times 60) + 24$$
$$= 7{,}200 + 2{,}700 + 24$$
$$= 9{,}924$$

The Babylonians also made use of a subtractive symbol, . That is, 38 could be written

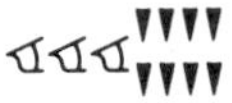

 or

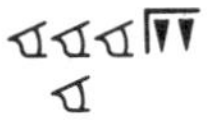

Example 7: Change to a decimal numeral. This number means

$$(3 \times 60) + 20 - 1 = 199$$

Example 8: Change

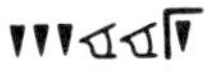

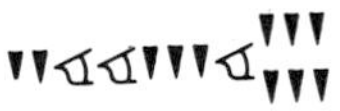

to a decimal numeral. This number means

$$(2 \times 60^2) + (23 \times 60) + 16 = 7{,}200 + 1{,}380 + 16$$
$$= 8{,}596$$

Example 9: Change 1,234 to Babylonian numerals. Now, $60 \times 20 = 1{,}200$, so we write $(20 \times 60) + 34 = 1{,}234$. That is,

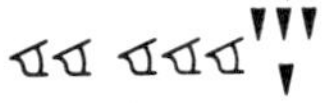

(Note an ambiguity: $(20 \times 60) + 34$ or 54?)

Example 10: Change 4,571 to Babylonian numerals.

$$1 \times 60^2 = 3{,}600$$
$$16 \times 60 = 960$$
$$11 \times 1 = 11$$
$$4{,}571$$

> **MAN WHO WANTS TO BE A NUMBER**
>
> **Minneapolis**
>
> At a time when many people complain of losing their identities to a long string of numbers, Michael Herbert Dengler is waging a fight for just that—he wants to be called 1069.
>
> Dengler, a teacher turned short-order cook, is due in Hennepin county District Court Monday to resume the legal struggle he began four years ago.
>
> The case at one point reached the North Dakota Supreme Court. In rejecting Dengler's plea that court said it was "without precedent in the entire experience of Anglo-American law."

Thus,

$$\text{ᛁ⊲}^{\text{ᛁᛁᛁ}}_{\text{ᛁᛁᛁ}}\text{⊲ᛁ}$$

is the way 4,571 was written.

Although this positional numeration system was in many ways superior to the Egyptian system, it suffered from the lack of a zero or place-holder symbol. For example, how is the number 60 represented? Does ᛁᛁ mean 2 or 61? In Example 9 the value of the number had to be found from the context. (Scholars tell us that such ambiguity can be resolved only by a careful study of the context.) However, in later Babylonia, around 300 B.C., records show that there is a zero symbol, ≨ , and this idea was later used by the Hindus.

Arithmetic with the Babylonian numerals is quite simple, since there are only two symbols. A study of the Babylonian arithmetic will be left for the student (see Problems 21–25 of Problem Set 4.1).

Problem Set 4.1

1. Explain the difference between *number* and *numeral*. Give examples of each.

2. Discuss the similarities and differences of:

   a. A simple grouping system
   b. A positional system
   Give examples of each.

3. **a.** Does ⌡∩∩ represent the same number as ∩∩⌡ ?
   b. Do ᛁ⊲⊲, ⊲ᛁ⊲, and ⊲⊲ᛁ represent the same number?
   Explain your answers.

4. Write 258 and 852:

   a. In Egyptian hieroglyphics
   b. In Babylonian cuneiform symbols

5. Write the following numbers in Egyptian hieroglyphics:

   a. 47   **b.** 75   **c.** 521   **d.** 1,976   **e.** 5,492

6. Write the numbers given in Problem 5 using the Babylonian numeration system.

7. What do you regard as the shortcomings and contributions of the Egyptian numeration system?

8. What do you regard as the shortcomings and contributions of the Babylonian numeration system?

So Dengler, 32, who once taught social studies in a Fargo, N.D., high school, moved to Minneapolis to renew the fight. And until the matter is settled, he's calling himself 1069, pronounced "one zero six nine."

"My friends call me 'One Zero,'" he says.

Dengler says he wants his name changed for personal and philosophic reasons. He says his protracted court battle has resulted in embarrassment—and some practical problems.

Minnesota refused to issue a driver's license to 1069. And Dengler had trouble landing a job. One company that makes copying machines told him, "You come in here with a name. We'll give you a number."

Four years ago, Dengler circulated a memo among fellow teachers and administrators in Fargo, informing them that henceforth he was to be known as 1069. In the subsequent uproar, Dengler resigned.

Dengler says that to understand fully the reasons he wants to change his name, a person must be on his wavelength. Few are, but friends and former colleagues insist he's sincere.

"He was bright, creative and intelligent," said Warren Gullickson, principal at Franklin School in Fargo, where Dengler once taught.

"He's absolutely sincere about this. This is where he's coming from."

Write the numerals in Problems 9–15 in decimal numerals. (Decimal numerals refer to the system used in everyday arithmetic. Thus, 5, 248, and $\frac{1}{3}$ are considered decimal numerals.)

9. a.  ⌒99∩∩∩∩∩ |||||
    b. ⌒99

10. a. ⬡
    b.  ⌒∩

11. a.  ⟨⟨▼▼▼▼
    b.  ▼▼▼▼⟨⟨▼

12. a.  ⟨⟨⟨ ▼▼▼ ⟨⟨
    b. 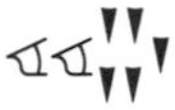 ⟨⟨ ▼▼ ▼▼▼ ▼

13. a.  9∩∩||| ⌐
    b.  ⌒ 9

14. a.  ⌒ ||∩
    b.  ∩∩||||| ⌐

15. a.  ▼▼⟨⟨⟨⟨▼▼
    b.  ▼⟨⟨⟨▼⟨⟨▼

16. Place a check mark in the appropriate columns to indicate the presence of the designated property.

	Egyptian	Babylonian
Grouping system		
Positional		
Repetitive		
Additive		
Subtractive		
Zero symbol		

Perform the indicated operations in Problems 17–24.

17. 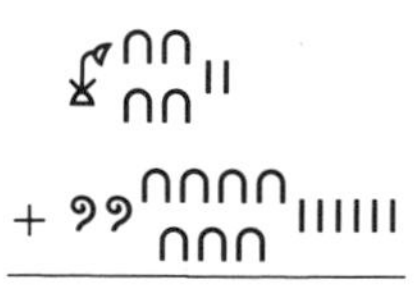
    ⌒∩∩ ∩∩ ||
    + 99∩∩∩∩ ∩∩∩ ||||||
    ──────────────

18. 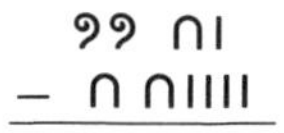
    99 ∩|
    − ∩ ∩|||||
    ──────────────

19. 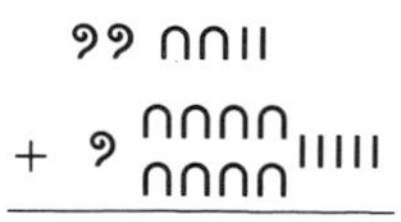
    99 ∩∩||
    + 9 ∩∩∩∩ ∩∩∩∩ |||||
    ──────────────

20. 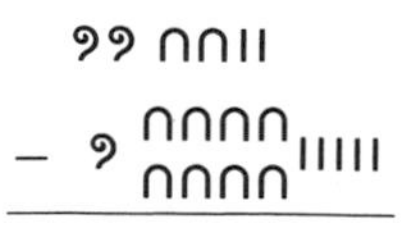
    99 ∩∩||
    − 9 ∩∩∩∩ ∩∩∩∩ |||||
    ──────────────

21. 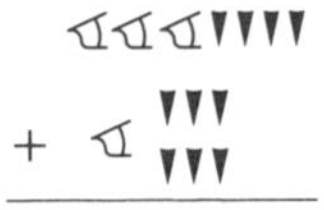 ⟨⟨⟨⟨▼▼▼▼
  + ⟨ ▼▼▼
      ▼▼▼
  ———————

22. 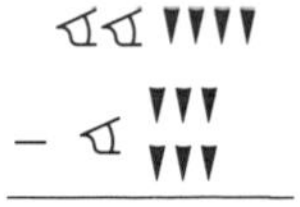 ⟨⟨ ▼▼▼▼
  − ⟨ ▼▼▼
      ▼▼▼
  ———————

23. 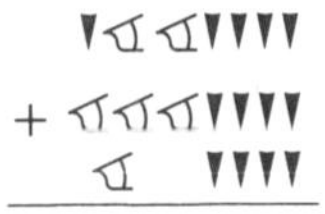 ▼⟨ ⟨▼▼▼▼
  + ⟨▼⟨▼⟨▼▼▼▼
      ⟨ ▼▼▼▼
  ———————————

24. 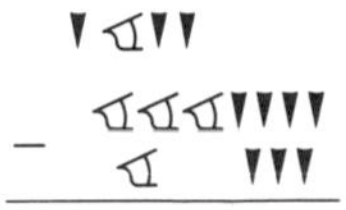 ▼⟨▼▼
  − ⟨⟨⟨⟨▼▼▼▼
      ⟨ ▼▼▼
  ———————————

25. Discuss addition and subtraction for Babylonian numerals. Show examples.

26. One of the problems found on the Rhind papyrus is sometimes translated as follows:

> In each of 7 houses are 7 cats;
> Each cat kills 7 mice;
> Each mouse would have eaten 7 ears of spelt (wheat);
> Each ear of spelt would have produced 7 kehat (half a peck) of grain.
> Query: How much grain is saved by the 7 houses' cats?

Can you provide an answer to the question?

27. Problem 26 taken from the Rhind papyrus, reminds us of an 18th century Mother Goose rhyme:

> As I was going to St. Ives
> I met a man with seven wives.
> Every wife had seven sacks,
> Every sack had seven cats,
> Every cat had seven kits.
> Kits, cats, sacks, and wives.
> How many were there going to St. Ives?

Read this rhyme very carefully, and then answer the question.

Mind Bogglers

28. *Calculator problem.* Another problem on the Rhind papyrus asks: "A quantity and its two thirds and its half and its one seventh together make 33. Find the quantity." The answer given on the papyrus is

$$14 + \frac{1}{4} + \frac{1}{56} + \frac{1}{97} + \frac{1}{194} + \frac{1}{388} + \frac{1}{679} + \frac{1}{776}$$

a. Use a calculator to approximate the answer given on the papyrus in decimal form.

b. Is the answer given on the papyrus correct? Find the exact answer in fractional form.

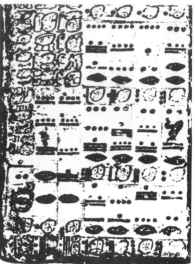

This example of Mayan numerals comes from the Dresden codex. See if you can "decode" any of the numerals.

A unit fraction (a fraction with a numerator 1) is sometimes called an Egyptian fraction. On page 168 we said that the Egyptians expressed their fractions as a sum of distinct (different) unit fractions. How might the Egyptians have written the following fractions?

29. $\dfrac{3}{4}$ **30.** $\dfrac{47}{60}$ **31.** $\dfrac{67}{120}$ **32.** $\dfrac{7}{17}$

Problems for Individual Study

33. Write a paper discussing the Egyptian method of multiplication.

   References: James Newman, *The World of Mathematics,* vol. I (New York: Simon and Schuster, 1956), pp. 170–178.

   Howard Eves, *Introduction to the History of Mathematics,* 3rd ed. (New York: Holt, Rinehart, and Winston, 1969).

34. Write a short paper discussing the development of our numeration system. Refer to places and important dates. You might also like to do research on the Roman, Chinese-Japanese, Greek, and Mayan numeration systems.

35. *Egyptian fractions.* We mentioned that the Egyptians wrote their fractions as sums of unit fractions. Show that every positive fraction less than 1 can be written as a sum of unit fractions.

   Reference: Bernhardt Wohlgemuth, "Egyptian Fractions," *Journal of Recreational Math,* vol. 5, no. 1 (1972), pp. 55–58.

36. Read Chapters 2 and 3 (pp. 9–47) of *A History of Mathematics* by Carl B. Boyer (New York: Wiley, 1968) and report on the Egyptian and Babylonian numeration systems. Answer Problems 7 and 8 in this problem set as part of your report.

37. *Ancient computing methods.* How do you multiply with Roman numerals? What is the scratch system? The lattice method of computation? What changes in our methods of long multiplication and long division are being suggested in some new arithmetic textbooks? How is the abacus used for computation? How are Napier's bones used for multiplication? How did the old computing machines work? How were logarithms invented? Who invented the slide rule?
   Exhibit suggestions: Charts of sample computations by ancient methods; ancient number representations such as pebbles, tally sticks, tally marks in sand, Roman number computations, abaci, Napier's bones, old computing devices.

4.2 Hindu-Arabic Numeration System

The numeration system in common use today has ten symbols, the digits 0,1,2,3,4,5,6,7,8, and 9. The selection of ten digits was no doubt a result of our having ten fingers (digits).

   The symbols originated in India in about 300 B.C. However, because the early specimens do not contain a zero or use a positional system, this numeration system offered no advantage over other systems then in use in India.

   The date of the invention of the zero symbol is not known. The symbol did not originate in India but probably came from the late Babylonian period via the Greek world.

By the year 750 A.D. the zero symbol and the idea of a positional system had been brought to Baghdad and translated into Arabic. We are not certain how these numerals were introduced into Europe, but they likely came via Spain in the 8th century. Gerbert, who later became Pope Sylvester II, studied in Spain and was the first European scholar known to have taught these numerals. Because of their origins, these numerals are called **Hindu-Arabic numerals.** Since ten basic symbols are used, the Hindu-Arabic numeration system is also called the *decimal numeration system,* from the Latin word *decem,* meaning "ten."

FIGURE 4.2 This picture depicts a contest between one man computing with a form of abacus and another man computing with Hindu-Arabic numerals.

Although we now know that the decimal system is very efficient, its introduction met with considerable controversy. Two opposing factions, the "algorists" and the "abacists," arose. Those favoring the Hindu-Arabic system were called algorists, since the symbols were introduced into Europe in a book called (in Latin) *Liber Algorismi de Numero Indorum,* by the Arab mathematician al-Khowârizmî. The word *algorismi* is the origin of our word *algorism.* The abacists favored the status quo—using Roman numerals and doing arithmetic on an abacus. The battle between the abacists and the algorists lasted for 400 years (see Figure 4.2). The Roman Catholic church exerted great influence in commerce, science, and theology. Roman numerals were easy to write and learn, and addition and subtraction with them were easier than with the "new" Hindu-Arabic numerals. Those not using Roman numerals were criticized for using "heathen" numerals. It seems incredible that our decimal system has been in general use only since about the year 1500.

Let's examine the Hindu-Arabic or decimal numeration system a little more closely.

1. It uses ten symbols, called digits: 0,1,2,3,4,5,6,7,8, and 9.
2. Larger numbers are expressed in terms of powers of 10.
3. It is positional.

Now, let's review how we count objects.

/	1
//	2
///	3
////	4
/////	5
//////	6
///////	7
////////	8
booking /////////	9
//////////	

At this point we could invent another symbol, as the Egyptians did, or we could reuse the digit symbols by repeating them or by altering their position.

𝕳istorical 𝕹ote

Hindu 300 B.C.

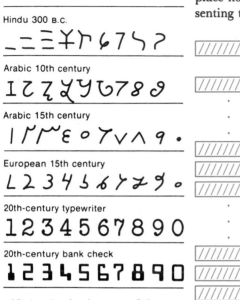

Arabic 10th century

Arabic 15th century

European 15th century

20th-century typewriter

20th-century bank check

Notice the development of these numerals. The symbols were subject to a good deal of change in appearance due to differences in writing, but the invention of the printing press helped standardize them into the form we are familiar with today. Recently, with the advent of computers, we have seen still another form for our numerals, called OCR (see page 124). Is this the end, or will we see other changes in the years ahead?

That is, 10 will mean 1 group of [//////////]. The symbol 0 was invented as a place-holder to show that the 1 here is in a different position from the 1 representing the /.

[//////////] /   11   This means 1 group of [//////////] and 1 extra.

[//////////] //   12   This means 1 group and 2 extra.

[//////////]
[//////////] ////   34   This means 3 groups (let's call each group a "ten") and 4 extra.
[//////////]

[//////////] [//////////]
[//////////] [//////////]
[//////////] [//////////]
[//////////] [//////////]
[//////////] [//////////]

In the last instance we have a group of groups, or ten groups (or ten-tens, if you like). Let's call this group of groups a "10 · 10" or "10^2" or a "hundred." We again use position and repeat the symbol 1 with a still different meaning: 100.

Thus, what does 134 represent? It means we have the following:

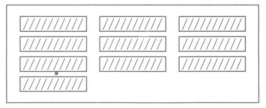

We could denote this more simply by writing;

$$(1 \times 10^2) + (3 + 10) + 4$$

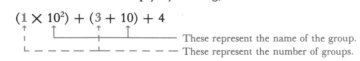

These represent the name of the group.
These represent the number of groups.

This leads us to the name

one hundred, three tens, four ones

which is read "one hundred thirty-four." A number written in this fashion is in **expanded notation.** As we pointed out earlier, mathematicians often look for patterns; so we notice that the expanded notation for 31,452 suggests a pattern:

$$(3 \times 10^4) + (1 \times 10^3) + (4 \times 10^2) + (5 \times 10) + 2$$

Definition of Exponent

For any nonzero number b and any counting number n,

$$b^n = \underbrace{b \times b \times b \times \cdots \times b}_{n \text{ factors}}$$

$$b^0 = 1$$

$$b^{-n} = \frac{1}{b^n}$$

The number b is called the **base** and n is called the **exponent.**

We write $10 = 10^1$ and $1 = 10^0$. Recall that we defined $10^0 = 1$. Then the pattern is complete:

$$(3 \times 10^4) + (1 \times 10^3) + (4 \times 10^2) + (5 \times 10^1) + (2 \times 10^0)$$

A period, called a **decimal point** in the decimal system, is used to separate the fractional parts from the whole parts. The most common base for us will be 10.

$$\frac{1}{10} = 10^{-1} \qquad \frac{1}{100} = 10^{-2}$$

$$\frac{1}{1,000} = 10^{-3} \qquad \text{since} \qquad 10^{-3} = \frac{1}{10^3} = \frac{1}{10 \times 10 \times 10} = \frac{1}{1,000}$$

Example 1: Write 479.352 using expanded notation.

Solution
$$(4 \times 10^2) + (7 \times 10^1) + (9 \times 10^0) + (3 \times 10^{-1}) + (5 \times 10^{-2}) + (2 \times 10^{-3})$$

We often fail to see all the great ideas that are incorporated into this system, and it is difficult to understand the problems inherent in developing a numeration system. In the next section, we'll change the focus from the *numerals* used to the kind of *numbers* existing independently of the particular system used. We will, of course, develop the various *number systems* using Hindu-Arabic *numerals.*

HONG KONG AUCTIONS LICENSE PLATES

HONG KONG (AP)— Movie mogul Sir Run Run Shaw paid about $55,000 for his Rolls Royce in 1979. He paid $66,000 for his license plate—the number 6.

Getting a special license plate in this British colony is an expensive proposition and the demand for them is so high that the government auctions them off.

Sir Run Run wanted the number 6 because he was the sixth-born son in his family. Also the number, when pronounced in Cantonese Chinese, sounds like the word for happiness.

Ruy Alberto da Silva, senior executive officer of licensing, said those who want special plate numbers are mostly Chinese, perhaps a carryover of their superstitious tradition.

Many Chinese consider the numbers 3, 6, and 8 as lucky but 4 is "deadly." They also like round and small numbers.

A license plate bearing CM1, for example, brought the equivalent in Hong Kong dollars of more than $42,000 U.S. early this year—$12,000 more than the current listed price of a Cadillac Seville here and $7,000 more than that of a Mercedes-Benz 280SE.

To help you realize some of these problems, you are asked in Problem 23 to invent your own numeration system.

𝕳istorical 𝕹ote

The Hindus often stated their problems poetically, since the problems were frequently solved just for the fun of it. As an example, consider the following problem adapted from an early Hindu writer and quoted in Howard Eves' *In Mathematical Circles:*

> The square root of half the number of bees in a swarm has flown out upon a jessamine bush; $\frac{8}{9}$ of the swarm has remained behind; one female bee flies about a male that is buzzing within a lotus flower into which he was allured in the night by its sweet odor, but is now imprisoned in it. Tell me, most enchanting lady, the number of bees.

In connection with his problems, the mathematician Brahmagupta said: "These problems are proposed simply for pleasure; the wise man can invent a thousand others, or he can solve the problems of others by the rules given here. As the sun eclipses the stars by his brilliancy, so the man of knowledge will eclipse the fame of others in assemblies of the people if he proposes algebraic problems, and still more if he solves them."

Problem Set 4.2

1. Let b be any nonzero number and n any counting number. Define:

   a. b^n **b.** b^0 **c.** b^{-n}

Write the numbers in Problems 2–10 in decimal notation.

2. **a.** 10^{-4} **b.** 5×10^3

3. **a.** 10^{-2} **b.** 3×10^2

4. **a.** 10^5 **b.** 8×10^{-4}

5. **a.** 10^6 **b.** 7×10^{-3}

6. **a.** $(1 \times 10^4) + (0 \times 10^3) + (2 \times 10^2) + (3 \times 10^1) + (4 \times 10^0)$
   b. $(6 \times 10^1) + (5 \times 10^0) + (0 \times 10^{-1}) + (8 \times 10^{-2}) + (9 \times 10^{-3})$

7. **a.** $(5 \times 10^5) + (2 \times 10^4) + (1 \times 10^3) + (6 \times 10^2) + (5 \times 10^1) + (8 \times 10^0)$
   b. $(6 \times 10^7) + (4 \times 10^3) + (1 \times 10^0)$

8. **a.** $(5 \times 10^5) + (4 \times 10^2) + (5 \times 10^1)$
   $\qquad + (7 \times 10^0) + (3 \times 10^{-1}) + (4 \times 10^{-2})$
   b. $(7 \times 10^6) + (3 \times 10^{-2})$

9. $(3 \times 10^3) + (2 \times 10^1) + (8 \times 10^0)$
   $\qquad + (5 \times 10^{-1}) + (4 \times 10^{-2}) + (6 \times 10^{-3}) + (2 \times 10^{-4})$

10. $(2 \times 10^4) + (6 \times 10^2) + (3 \times 10^0)$
    $\qquad + (4 \times 10^{-1}) + (7 \times 10^{-3}) + (6 \times 10^{-4}) + (9 \times 10^{-5})$

Write the numbers in Problems 11–17 in expanded notation.

11. **a.** 741 **b.** 728,407

12. **a.** 0.06421 **b.** 27.572

13. **a.** 521 **b.** 6,245

14. **a.** 428.31 **b.** 2,345,681

15. **a.** 47.03215 **b.** 100,000.001

16. **a.** .00000527 **b.** 5,245.5

17. **a.** 678,000.01 **b.** 57,285.9361

18. What is the meaning of the 5 in each of these numerals?

    a. 805 **b.** 508 **c.** 0.00567

19. Illustrate the meaning of 123 by showing the appropriate groupings.

20. Illustrate the meaning of 145 by showing the appropriate groupings.

Mind Bogglers

21. Can you find the pattern?

$$0, 1, 2, 10, 11, 12, 20, 21, 22, 100, \ldots$$

22. Notice the following pattern for multiplication by 11:

$$14 \times 11 = 1 \qquad 4 \qquad \text{Original two digits}$$
$$= 1 \quad 5 \quad 4$$

⎣——— Insert the sum of the two digits

$$52 \times 11 = 5 \qquad 2$$
$$= 5 \quad 7 \quad 2$$

Use expanded notation to show why this pattern "works."

23. Invent your own numeration system. You may use the *ideas* of the Egyptian, Babylonian, and Hindu-Arabic numeration system (such as repetitive, positional, and so on), but don't use the specifics, such as the symbols used or the names for those symbols, of these (or any other) systems.

Problem for Individual Study

24. *The role of mathematics in Western civilization.* What are some of the significant events in the development of mathematics? Who are some of the famous people who have contributed to mathematical knowledge? What are some of the amusing anecdotes from the lives of famous mathematicians? How has mathematics been involved in science, politics, military campaigns, architecture, transportation, philosophy, art, music, and literature?

Exhibit suggestions: Charts and models depicting the inventions and contributions of such men as Archimedes, Newton, and Gauss; a time line showing the significant developments in mathematics.

References: E. T. Bell, *Men of Mathematics* (New York: Simon and Schuster, 1937).

Howard Eves, *In Mathematical Circles,* vols. 1 and 2 (Boston: Prindle, Weber, & Schmidt, 1969).

Howard Eves, *Mathematical Circles Revisited* (Boston: Prindle, Weber, & Schmidt, 1971).

Howard Eves, *Mathematical Circles Adieu* (Boston: Prindle, Weber, & Schmidt, 1977).

Virginia Newell et al., *Black Mathematicians and Their Works* (Ardmore: Dorrance & Company, 1980).

4.3 Large and Small Numbers

Exponential notation is useful not only with expanded notation as shown in the last section, but also to represent very large and very small numbers in a form called **scientific notation.**

Definition of Scientific Notation

The *scientific notation* for a number is that number written as a power of 10 times another number x, such that x is between 1 and 10 $(1 \leq x < 10)$.

One of the first times very large numbers were used was by Archimedes in about 250 B.C., when he reportedly computed the number of grains of sand in the universe to be 10^{63}. Archimedes (about 287–212 B.C.) was one of the greatest mathematicians of all times. He once boasted that he could move the earth if he had a place to stand. King Hiero of Syracuse told him to make good his boast by moving a ship (fully loaded) out of dry dock. He did this single-handedly with the use of pulleys. This feat so impressed the king that he involved Archimedes in the defense of Syracuse against the Romans. Archimedes invented catapults, pulleys to raise and smash Roman ships, and devices to set them on fire. The Roman general Marcellus left strict orders to let no harm come to Archimedes during the siege, but nevertheless Archimedes was killed by a Roman soldier.

A very nice arithmetical pattern makes it easy to represent numbers in scientific notation. Consider the following examples, and notice what happens to the decimal point.

$$9.42 \times 10^1 = 94.2$$
$$9.42 \times 10^2 = 942.$$
$$9.42 \times 10^3 = 9420.$$
$$9.42 \times 10^4 = 94{,}200.$$

We find these answers by direct multiplication.

Do you see a pattern? If we multiply 9.42×10^5, how many places to the right will the decimal point be moved?

$$9.42 \times 10^5 = 942{,}000.$$

5 places ⟶

Using this pattern, can you multiply the following *without direct calculation?*

$$9.42 \times 10^{12}$$

This answer is found by observing the pattern, not by directly multiplying.

For small numbers there is a similar pattern:

$$9.42 \times 10^2 = 942.$$
$$9.42 \times 10^1 = 94.2$$
$$9.42 \times 10^0 = 9.42$$

We interpret the zero exponent as "decimal point moves 0 places."

To continue, direct calculation extends the pattern:

$$9.42 \times 10^{-1} = .942$$
$$9.42 \times 10^{-2} = .0942$$
$$9.42 \times 10^{-3} = .00942$$

These numbers are found by direct multiplication. For example,

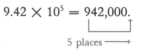

$$9.42 \times 10^{-2} = 9.42 \times \frac{1}{100}$$
$$= 9.42 \times .01$$
$$= .0942$$

Do you see that the same pattern for multiplying by 10 with a negative exponent also holds? Can you multiply 9.42×10^{-6} *without direct calculation?*
The solution is as follows:

$$9.42 \times 10^{-6} = .00000942$$

Moved six places to the left

Example 1: $123{,}600 = 1.236 \times 10^5$

Example 2: $.000035 = 3.5 \times 10^{-5}$

Example 3: $1{,}000{,}000{,}000{,}000 = 10^{12}$

Example 4: $7.35 = 7.35 \times 10^0$ or just 7.35. Notice that a number between 1 and 10 is also in scientific notation, since it can be written as that number times 10^0, as with 7.35×10^0.

Example 5: Light travels at about 186,000 miles/second. One year comprises 31,557,600 seconds. Thus, light travels about

$$5{,}869{,}713{,}600{,}000 \text{ miles}$$

in one year. In scientific notation we write

$$5.87 \times 10^{12} \text{ miles/light-year}$$

Example 6: The estimated age of the earth is about

$$5 \times 10^9 \text{ years}$$

Example 7: The sun loses about

$$4.3 \times 10^9 \text{ kilograms}$$

of solar mass per second.

Example 8: The number of possible combinations of genes that an individual may possess is about

$$10^{9{,}400{,}000{,}000}$$

There is a way of representing scientific notation on calculators and on computers. On a calculator it can be entered by pressing an $\boxed{\text{EE}}$ or $\boxed{\text{EEx}}$ key. On a calculator or on a computer, scientific notation is not represented by using an exponent, so instead of being called scientific notation it is called **floating point form.** When representing very large or very small answers most calculators will automatically output the answers in floating point notation. Examples 9–12 compare the different forms for large and small numbers with which you should be familiar.

	Decimal Notation Fixed Point	Scientific Notation	Calculator Notation		BASIC Notation
Example 9:	745	7.45×10^2	7.45	02	7.45E+2
Example 10:	1,230,000,000	1.23×10^9	1.23	09	1.23E+9
Example 11:	.00573	5.73×10^{-3}	5.73	−03	5.73E−3
Example 12:	.00000 06239	6.239×10^{-7}	6.239	−07	6.239E−7

Notice that the ordinary decimal notation is sometimes called **fixed point form.**

We have been looking at some very large numbers. But just how large is large? Most of us are accustomed to hearing about millions and even billions, but do we really understand the magnitude of these numbers? Would you do a better job than Dennis's parents in the cartoon at explaining "How much is a million?"

A million is a fairly modest number, 10^6. Yet if we were to count one number per second, nonstop, it would take us about 278 hours or approximately $11\frac{1}{2}$ days to count to a million. Not a million days have elapsed since the birth of Christ (a million days is about 2,700 years). A million letters in small type would fill a book of about 700 pages. A million bottle caps placed in one line would stretch about 17 miles (see the photo on page 184).

But the age in which we live has been called the age of billions. The U.S. national budget is measured in terms of billions of dollars (soon it will be a trillion-dollar budget). Just how large is a billion? How long would it take you to count to a billion? To get some idea about how large a billion is, let's compare it to some familiar units.

1. Suppose you give away $1,000 every day. It would take more than 2,700 years to give away a billion dollars.
2. A stack of a billion $1 bills would be more than 59 miles high.
3. At 8% interest, a billion dollars would earn you $219,178.08 interest per day.
4. A billion seconds ago was the bombing of Pearl Harbor.
5. A billion minutes ago Christ was living on earth.
6. A billion hours ago people had not yet appeared on earth.

But a billion is only 10^9, a mere nothing compared with the real giants. There is an old story of a king who, being under obligation to one of his subjects, offered to reward him in any way the subject desired. Being of mathematical mind and modest tastes, the subject simply asked for a chessboard with one grain of wheat on the first square, two on the second, four on the third, and so forth. The old king was delighted with this request because he had a beautiful daughter and had feared the subject would ask for her hand in marriage. However, the king was soon sorry he had granted the request. He needed 2^{63} grains of wheat for the last square alone! Now,

$$2^{63} = 9,223,372,036,854,775,808$$

If we purchase some raw wheat and count the number of grains in one cubic inch, we find that there are about 250 grains per cubic inch. We can then compute (using scientific notation) the amount of wheat needed to equal 2^{63} grains. There are 2,150 cubic inches per bushel, or 537,500 grains of wheat in a bushel. This means that 2^{63} grains of wheat would equal about 17,383,000,000,000 bushels. For the last several years, the yearly output of wheat in the United States has been about 1.25 billion bushels. At this rate of production, it would take the United States almost 14,000 years to satisfy the requirements for the last square of the chessboard! The story goes no further. The chances are that the king lost his temper and the subject lost his head before the sixty-fourth square of the chessboard was reached.

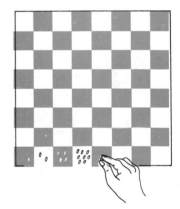

Let's consider a real giant. What is the largest number you can write using exactly three digits? Is it 987? How about 999? Can you think of a larger one? How about 9^{9^9} (certainly larger than 999). Now, 9^{9^9} means $9^{(9^9)}$ or 9 raised to the 9^9 power. Then, $9^9 = 387,420,489$; thus, $9^{9^9} = 9^{387,420,489}$. To compare this with the "little" number from the chessboard problem, 2^{63}, realize that $9^{387,420,489}$ is greater (a lot greater) than $8^{387,420,489} = (2^3)^{387,420,489} = 2^{1,162,261,467}$, which makes our big number 2^{63} look minuscule.

The number 9^{9^9} would take 6 million lines to write. It has 369,693,100 digits, and it begins 428,124,773, ... and ends with 89. That number of bacteria would overflow the Milky Way, and the number of grains of sand of this and every other planet in our solar system would not be as large as this giant.

TABLE 4.3   Some Large Numbers

Million: 1,000,000
Billion: 1,000,000,000
Trillion: 1,000,000,000,000
Quadrillion: 1,000,000,000,000,000 ←——————— 10,000 trillion (1 followed by 16 zeros) is here: It has been estimated that
Quintillion: 1,000,000,000,000,000,000   the total of all copies of all words ever printed (including everything from
Sextillion: 1,000,000,000,000,000,000,000   the Sunday newspapers, novels, textbooks, back to the Gutenburg Bible)
Septillion: 1,000,000,000,000,000,000,000,000   is smaller than this number!
Octillion: 1,000,000,000,000,000,000,000,000,000
Nonillion: 1,000,000,000,000,000,000,000,000,000,000
Decillion: 1,000,000,000,000,000,000,000,000,000,000,000
Undecillion: 1,000,000,000,000,000,000,000,000,000,000,000,000

                                        .
                                        .
                                        .

Vigintillion: 1,000,000,000,000,000,000,000,000,000,000,000,000,000,000,000,000,000,000,000,000,000

For every large number, there is a corresponding small number. If 9^{9^9} is very large, then $\dfrac{1}{9^{9^9}}$ is very small. Therefore, when we speak of finding large numbers, we are also finding small numbers.

These third-graders at Michigan School for the Blind in Lansing decided to find out what a million something is like. One year later they are seen frolicking on 1 million bottle caps. The students found the caps. Their teacher, Mrs. Jackie Taylor, said it was difficult to describe what a million bottle caps look like, but "They smell like a very large brewery."

We have been looking at some very large (and very small) numbers, but there are yet larger numbers. Professor Edward Kasner of Colombia University named the "googol" and "googolplex." A googol is 10^{100}, and a googolplex is 10^{googol}. Thus, as Schroeder in the *Peanuts* strip tells us, a googol is 1 followed by 100 zeros. It has been estimated that, if we tried to write out a googolplex on a single line with a typewriter, it would not fit between the earth and the moon.

Have we reached a limit? Are there still larger numbers? What about the following numbers?

$$googol^{googol} \quad \text{or} \quad googol^{googol^{googol}}$$

These are real giants. Is there a largest number? Clearly not, since, if you think you've found the largest number, you would only have to add 1 to have a larger number. All these numbers are large beyond comprehension, but they are still *finite*. Is it possible to have even larger numbers that are not finite? In the next section, we'll consider the familiar set of numbers called the *natural numbers*. This is a set of numbers that is not finite. It is called an *infinite* set of numbers.

Problem Set 4.3

1. Consider the number 10^6.

    a. What is the common name for this number?
    b. What is the base?
    c. What is the exponent?
    d. According to the definition of exponential notation, what does the number mean?

2. Consider the number 10^{-1}.

    a. What is the common name for this number?
    b. What is the base?
    c. What is the exponent?
    d. According to the definition of exponential notation, what does the number mean?

Write each of the numbers in Problems 3–14 in (a) scientific notation; (b) floating point form on a calculator; (c) floating point form in BASIC.

3. 3,200
4. .0004
5. 5,629

6. 23.79
7. 35,000,000,000
8. 63,000,000

9. .00000 00000 00035
10. .00001
11. .00000 0006

12. googol
13. .00000 00000 68
14. 1,200,300,000

Write each of the numbers in Problems 15–19 in scientific notation.

15. Drawn to the scale shown in the figure in the margin, the distance between Earth and Mars (220,000,000 miles) would be .0000025 in.

16. A thermochemical calorie is about 41,840,000 ergs.

17. The velocity of light in a vacuum is about 30,000,000,000 cm/sec.

18. The wavelength of the orange-red line of krypton 86 is about 6,100 Å.

19. The world's largest library, the Library of Congress, has 59,000,000 items.

Write each of the numbers in Problems 20–35 in fixed point notation.

20. **a.** 7^2 **b.** 2^6
21. **a.** 7.2×10^{10} **b.** 2.1×10^{-3}

22. **a.** 6×10^3 **b.** 4.1×10^{-7}
23. **a.** 6.81×10^0 **b.** 6.9×10^{-5}

24. **a.** 4.56E+3   **b.** 5.8E−7
25. **a.** 4.07E+4   **b.** 7.078E+2

26. **a.** 7.9E+14   **b.** 6.02E−5
27. **a.** 3.217E−6   **b.** 8.89E−11

28. A kilowatt-hour is about 3.6×10^6 joules.

29. A ton is about 9.07×10^2 kilograms.

30. Saturn is about 8.86×10^8 miles from the sun.

31. The mass of the sun is about 3.33×10^5 times the mass of the earth.

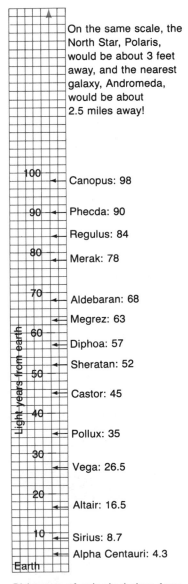

On the same scale, the North Star, Polaris, would be about 3 feet away, and the nearest galaxy, Andromeda, would be about 2.5 miles away!

Canopus: 98
Phecda: 90
Regulus: 84
Merak: 78
Aldebaran: 68
Megrez: 63
Diphoa: 57
Sheratan: 52
Castor: 45
Pollux: 35
Vega: 26.5
Altair: 16.5
Sirius: 8.7
Alpha Centauri: 4.3

Light-years from earth

Distances of selected stars from Earth in light-years

32. The sun develops about 5×10^{23} horsepower per second.

33. The volume of a typical neuron is about 3×10^{-8} cm^3.

34. If the sun were a light bulb, it would be rated at 3.8×10^{25} watts.

35. There are 1.28×10^6 supermarket employees in the United States.

36. The newspaper clipping in the margin column describes a trillion. Write the numbers in the article in scientific notation.

37. In 1973 about 1.6 million all-aluminum cans were returned for recycling, which was enough metal to put storm windows in a half-million homes. Installing the storm windows resulted in a savings of over a half-billion kilowatt-hours for the year—enough electricity to supply the energy for about 63,000 homes. Write the numbers in the problem in scientific notation.

38. In the 17th century, Christian Huygens estimated the star Sirius to be about 4.2×10^{17} cm away. It is really 20 times farther away. Write both these distances without exponents.

39. The newspaper clipping in the margin column says that our atmosphere weighs 5 quadrillion, 157 trillion tons. Write this number in scientific notation.

40. Approximately how high would a stack of 1 million $1 bills be? (There are 233 new $1 bills per inch.)

41. Estimate how many pennies it would take to make a stack 1 inch high. A fistful of pennies and a ruler will help you. Approximately how high would a stack of 1 million pennies be?

42. There are 2,260,000 grains in a pound of sugar. If the U.S. production of sugar in 1977 was 5,523,000 tons, estimate the number of grains of sugar produced in a year in the United States. Use scientific notation. (There are 2,240 lb per ton.)

43. Light travels at 186,282 miles per second. A light-year is the distance that light travels in one year. Approximately how many miles are there in one light-year? [*Hint:* Approximate the speed of light by $1.86 \cdot 10^5$. In one year there are approximately $(6 \cdot 10) \cdot (6 \cdot 10) \cdot (2.4 \cdot 10) \cdot (3.65 \cdot 10^2)$ seconds. This is about $315 \cdot 10^5$ or $3.15 \cdot 10^7$ seconds per year.]

44. Use Problem 43 and the margin figure that goes with Problem 15 to write the distances in miles of the following stars from Earth.

    a. Alpha Centauri   **b.** Sirius   **c.** Phecda

45. **a.** Use the margin figure that goes with Problem 15 to estimate the number of light-years that Polaris is from Earth.
    b. Use Problem 43 to write your answer to part a in miles.

Mind Bogglers

46. If the entire population of the world moved to California and each person was given an equal amount of area, how much space would you *guess* that each person would have (multiple choice):

WHAT IS A TRILLION?

HOW MUCH IS A TRILLION? It is 1,000,000,000,000 or a million millions. A trillion inches is more than 15.8 million miles or half the distance to Venus. The moon is about 230,000 miles, so a trillion *inches* is about 68 round trips to the moon. A trillion seconds is about 81,700 years!

SOVIETS WEIGH THE ATMOSPHERE

MOSCOW (UPI)—The atmosphere weighs 5 quadrillion, 157 trillion tons, the Soviets said in an announcement of great import to scientists and connoisseurs of trivia.

The official Tass news agency said an electronic brain known as the Minsk-22 computer had figured out the total weight of Earth's air cover more precisely than had been done before. "The mass of air enveloping our planet is now estimated at 5 quadrillion, 157 trillion tons," Tass said.

"This calculation is essential for research in cosmonautics, space geodesy and gravimetry," Tass said.

A. 12 sq in. B. 12 sq ft
C. 125 sq ft D. 1250 sq ft
E. 1 sq mi

Now, use the information in the margin to calculate the amount of space. Does your answer to this problem correspond to the guess you made?

47. The total population of the earth is about 3.5×10^9.

    a. If each person has the room of a prison cell (50 sq ft), and if there are about 2.5×10^7 sq ft in a sq mi, how many people could fit into a square mile?

    b. How many square miles would be required to accommodate the entire population of the earth?

    c. If the total land area of the earth is about 5.2×10^7 sq mi, and if all the land area is divided equally, how many acres of land would each person be allocated? (1 sq mi = 640 acres)

48. **a.** Guess what percentage of the world's population could be packed into a cubical box measuring $\frac{1}{2}$ mi on each side. [*Hint:* The volume of a typical person is about 2 cu ft.]

    b. Now calculate the answer to part a using the earth's population as given in Problem 47.

49. If it took one second to write down each digit, how long would it take to write all the numbers from 1 to 1,000,000?

50. Imagine that you have written down the numbers from 1 to 1,000,000. What is the total number of zeros you have recorded?

Problem for Individual Study

51. *Population growth.* The time it takes the population to double is constantly decreasing. The table shows the U.S. population for census years 1800–1980. As we can see from the accompanying figure, the population increases geometrically. Is the same kind of increase true for your own state? Consult an almanac to obtain your information.

POPULATION EXPLOSION?

What would happen if the *entire world population* moved to California?

California

158,600 square miles

World population about
3,500,000,000!

How much space would each person have (make a guess)?

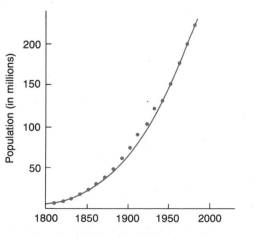

U.S. population for census years 1800–1980

U.S. Population during
Census Years 1800–1980

Year	Population (in millions)
1800	5.3
1810	7.2
1820	9.6
1830	12.8
1840	17.1
1850	23.2
1860	31.4
1870	39.8
1880	50.2
1890	63.0
1900	76.0
1910	92.0
1920	105.7
1930	122.8
1940	131.7
1950	151.3
1960	179.3
1970	203.2
1980	226.5

> • There still remain three studies suitable for free man. Arithmetic is one of them.
>
> *Plato*

4.4 Natural Numbers

The most basic set of numbers used by any society is the set of numbers used for counting:

$$\{1,2,3,4, \ldots \}$$

This set is called the set of counting numbers or **natural numbers.** Let's assume that you understand what the numbers in this set represent and that you understand the operation of addition, $+$. We have been using this set of numbers up to this point in the text.

There are a few self-evident properties of addition for this set of natural numbers. They are called "self-evident" because they almost seem too obvious to be stated explicitly. For example, if we jump into the air, we expect to come back down. That assumption is well founded in experience and is also based on an assumption that jumping has certain undeniable properties. But astronauts have found that some very basic assumptions are valid on earth and false in space. Recognizing these assumptions (properties, axioms, laws, or postulates) is important.

When we add any two natural numbers, we know that we will obtain a natural number as an answer. This "knowing" is an assumption based on experience, but we have actually experienced only a small number of cases for all the possible sums of numbers. The scientist—and the mathematician in particular—is very skeptical about making assumptions too quickly. There is a story about a mathematician riding in a car with a friend. The friend remarked "Oh look, the Joneses have painted their house." The mathematician responded "Yes, at least this side."

The first self-evident property of the natural numbers concerns the order in which they are added. It is called the **commutative property for addition** and states that the order in which two numbers are added makes no difference; that is (if we read from left to right),

$$a + b = b + a$$

The word *commute* can mean to travel back and forth from home to work; this back-and-forth idea can help you remember that the commutative property applies if you read from left to right or right to left.

for any two natural numbers a and b. The commutative property allows us to rearrange numbers; it is called the property of order. Together with the second property, called the associative property, it is used in calculation and simplification.

The **associative property for addition** allows us to group numbers for addition. Suppose you wish to add three numbers, say 2, 3, and 5:

The word *associate* can mean to connect, join, or unite; with this property you associate two of the added numbers.

$$2 + 3 + 5$$

In order to add these numbers we must first add two of them and then add this sum to the third. The associative property tells us that, no matter which two are added first, the final result is the same. If parentheses are used to indicate the numbers to be added first, then this property can be symbolized by

$$(2 + 3) + 5 = 2 + (3 + 5)$$

The parentheses indicate the operations to be performed first.

This associative property for addition holds for *any* three or more natural numbers.

Add the column of numbers in the margin. How long did it take? Five seconds is long enough if you use the associative and commutative properties for addition:

$$(9 + 1) + (8 + 2) + (7 + 3) + (6 + 4) + 5$$
$$= \quad 10 \quad + \quad 10 \quad + \quad 10 \quad + \quad 10 \quad + 5$$
$$= \quad 45$$

9
8
7
6
5
4
3
2
1

However, it takes much longer if you don't rearrange and regroup the numbers:

$$(9 + 8) + (7 + 6 + 5 + 4 + 3 + 2 + 1)$$
$$= (17 + 7) + (6 + 5 + 4 + 3 + 2 + 1)$$
$$= (24 + 6) + (5 + 4 + 3 + 2 + 1)$$
$$= (30 + 5) + (4 + 3 + 2 + 1)$$
$$= (35 + 4) + (3 + 2 + 1)$$
$$= (39 + 3) + (2 + 1)$$
$$= (42 + 2) + 1$$
$$= 44 + 1$$
$$= 45$$

The properties of associativity and commutativity are not restricted to the operation of addition. They are general properties that can be applied to any operation for any given set.

Multiplication is defined as repeated addition.

Definition of Multiplication

$$a \times b \qquad (b \neq 0)$$

means

$$\underbrace{a + a + a + \cdots + a}_{b \text{ addends}}$$

If $b = 0$, then $a \cdot 0 = 0$.

Multiplication is denoted by a cross ($\times$), as in $3 \times a$; a dot, as in $3 \cdot a$; parentheses, as in $3(a)$; or juxtaposition, as in $3a$.

We can now check commutativity and associativity for multiplication in the set of natural numbers.

$$Commutativity: \quad 2 \times 3 \overset{?}{=} 3 \times 2$$
$$Associativity: \quad (2 \times 3) \times 4 \overset{?}{=} 2 \times (3 \times 4)$$

The question mark above the equal sign signifies that we should not assume the conclusion (namely, that they are equal) until we check the arithmetic. Even though we can check these properties for particular natural numbers, it is impossible to check them for *all* natural numbers, so we will accept the following axioms.

For any natural numbers a, b, and c:

Commutative properties

$$Addition: \quad a + b = b + a$$
$$Multiplication: \quad ab = ba$$

Associative properties

$$Addition: \quad (a + b) + c = a + (b + c)$$
$$Multiplication: \quad (ab)c = a(bc)$$

Historical Note

It was difficult to decide where to put the historical note about Leonhard Euler (1707–1783), since Euler's name is attached to every branch of mathematics. He was the most prolific writer on the subject of mathematics, and his mathematical textbooks were masterfully written. His writing was not at all slowed down by his total blindness for the last 17 years of his life. He possessed a phenomenal memory, had almost total recall, and could mentally calculate long and complicated problems. The story of Euler's death is told by Howard Eves in his book *In Mathematical Circles* (Boston: Prindle, Weber & Schmidt, 1969, vol. 2, p. 54):

It is possible to have a set with more than one operation. For example, we can add and multiply in the set of counting numbers. Are there properties in this set that involve both operations? Consider an example.

Example 1:   Suppose you are selling tickets for a raffle, and the tickets cost \$2 each. You sell 3 tickets on Monday and 4 tickets on Tuesday. How much money did you collect?

Solution 1:   You sold a total of $3 + 4 = 7$ tickets, which cost \$2 each, so you collected $2 \cdot 7 = 14$ dollars. That is,

$$2 \times (3 + 4) = 14$$

Solution 2:   You collected $2 \cdot 3 = 6$ dollars the first day and $2 \cdot 4 = 8$ dollars the second day for a total of $6 + 8 = 14$ dollars. That is,

$$(2 \times 3) + (2 \times 4) = 14$$

Since these solutions are equal, we say:

$$2 \times (3 + 4) = (2 \times 3) + (2 \times 4)$$

Do you suppose this would be true if the tickets cost a dollars and you sold b tickets on Monday and c tickets on Tuesday? Then the equation would look like:

$$a \times (b + c) = (a \times b) + (a \times c)$$

or simply

$$a(b + c) = ab + ac$$

This example illustrates the **distributive property.**

Distributive Property for Multiplication over Addition

$$a(b + c) = ab + ac$$

In the set of counting numbers, is addition distributive over multiplication? We wish to check:

$$3 + (4 \times 5) \overset{?}{=} (3 + 4) \times (3 + 5)$$

Checking:

$$3 + (4 \times 5) = 3 + 20 \qquad \text{and} \qquad (3 + 4) \times (3 + 5) = 7 \times 8$$
$$= 23 \qquad\qquad\qquad\qquad\qquad = 56$$

Thus, addition is not distributive over multiplication in the set of counting numbers.

The distributive property can also help to simplify arithmetic. Suppose we wish to multiply 9 by 71. We can use the distributive property to think:

$$9 \times 71 = 9 \times (70 + 1) = (9 \times 70) + (9 \times 1)$$
$$= 630 + 9$$
$$= 639$$

This allows us to do the problem quickly and simply in our heads.

Since these properties hold for the operations of addition and multiplication, we might reasonably ask if they hold for other operations. Subtraction is defined as the opposite of addition.

Definition of Subtraction

$$a - b = x \qquad \text{means} \qquad a = b + x$$

To verify the commutative property for subtraction, we check a particular example:

$$3 - 2 \overset{?}{=} 2 - 3$$

Euler retained vigor and power of mind up to the moment of his death, which occurred in his seventy-seventh year, on September 18, 1783. He had amused himself in the afternoon calculating the laws of ascent of balloons. He then dined with Lexell and his family, and outlined the calculation of the orbit of the recently discovered planet Uranus. A short time later he begged that his grandson be brought in. While playing with the youngster and sipping some tea, he suffered a stroke. His pipe fell from his hand and he uttered, "I die." At that instant, in the words of Condorcet, "Euler ceased to live and calculate."

From the definition of subtraction,

$$2 - 3 = \square$$

means

$$2 = \square + 3$$

so we need to find a number that, when added to 3, gives the result 2.

Now, $3 - 2 = 1$, but $2 - 3$ doesn't even exist in the set of natural numbers. What does it mean to say that a result doesn't exist in the set of natural numbers? It means that, to provide the result of the operation of subtraction for $2 - 3$, we must find a number added to 3 that gives the result 2. But there is *no such natural number.* In mathematics we give a name to such an occurrence. We say that a set is *closed* for an operation if every possible result of performing that operation is a number contained in the given set. Thus, the set of natural numbers is *closed* for addition and multiplication but *not closed* for subtraction.

Notice that we use the words
not closed instead of *open.*

Closure Property

The set of natural numbers is closed for:

1. *Addition,* since $a + b$ is a natural number for any natural numbers a and b
2. *Multiplication,* since ab is a natural number for any natural numbers a and b

---

You need find only *one*
counterexample to show that
a property does not hold.

Example 2:   The set $A = \{1,2,3,4,5,6,7,8,9,10\}$ is *not closed* for the operation of addition, since $7 + 8 = 15$.   Since $15 \notin A$, we see that the set is not closed.

Example 3:   The set $B = \{0,1\}$ is closed for the operation of multiplication, since all possible products are in B.

$$0 \times 0 = 0$$
$$0 \times 1 = 0$$
$$1 \times 0 = 0$$
$$1 \times 1 = 1$$

In order to make sure you understand the properties discussed in this section, this set of problems will focus on the properties of closure, commutativity, associativity, and distributivity rather than the set of natural numbers and operations of addition, multiplication, and subtraction. Since you are so familiar with the set of natural numbers and with these operations, you could probably answer questions about them without much reflection on the concepts involved. Therefore, in Problem Set 4.4, certain new operations (other than addition, multiplication, and subtraction) will be defined. In using a table, rows are horizontal and columns are vertical. Thus, 3×4 on a multiplication table would be found in the third row and fourth column.

Problem Set 4.4

In Problems 1–10, classify each as an example of the commutative property, the associative property, or both.

To distinguish between the commutative and associative properties, remember the following:
1. When the *commutative* property is used, the order in which the elements appear from left to right is changed, but the grouping is not changed.
2. When the *associative* property is used, the elements are grouped differently, but the order in which they appear is not changed.

1. $3 + 5 = 5 + 3$

2. $6 + (2 + 3) = (6 + 2) + 3$

3. $6 + (2 + 3) = 6 + (3 + 2)$

4. $6 + (2 + 3) = (2 + 3) + 6$

5. $6 + (2 + 3) = (6 + 3) + 2$

6. $(4 + 5)(6 + 9) = (4 + 5)(9 + 6)$

7. $(4 + 5)(6 + 9) = (6 + 9)(4 + 5)$

8. $(3 + 5) + (2 + 4) = (3 + 5 + 2) + 4$

9. $(3 + 5) + (2 + 4) = (3 + 5) + (4 + 2)$

10. $(3 + 5) + (2 + 4) = (3 + 4) + (5 + 2)$

11. Does the following Blondie cartoon remind you of the associative or the commutative property? Why?

FIRST, I'LL ASK HIM FOR A TEN-DOLLAR RAISE---IF HE SAYS "NO," I'LL COME DOWN TO FIVE

MR. DITHERS, I THINK A FIVE-DOLLAR RAISE IS IN ORDER FOR ME

GO AWAY

OKAY, THEN I'LL SETTLE FOR TEN

I THINK I SAID THAT WRONG

YOUNG RAYMOND 10-15

12. "Isn't this one just too sweet, dear?" asked the wife as she tried on a beautiful diamond ring.
"No," her husband replied. "It's just too dear, sweet."
Does this story remind you of the associative or the commutative property?

13. Is the operation of putting on your shoes and socks commutative?

14. In the English language the meanings of certain phrases can be very different depending on the association of the words. For example,

(MAN EATING) TIGER

is not the same as

MAN (EATING TIGER)

Decide whether each of the following groups of words is associative:

a. HIGH SCHOOL STUDENT   b. RED FIRE ENGINE
c. SLOW CURVE SIGN   d. TRAVELING SALESMAN JOKE
e. BARE FACTS PERSON   f. BROWN SMOKING JACKET

"How thrilling — what time does it start?"
NATIONAL ENQUIRER

15. Think of three nonassociative word triples as shown in Problem 14.

16. Consider the set $A = \{1,4,7,9\}$ with an operation ＊ defined by the table.

＊	1	4	7	9
1	9	7	1	4
4	7	9	4	1
7	1	4	7	9
9	4	1	9	7

Find each of the following.

a. 7＊4   **b.** 9＊1   **c.** 1＊7   **d.** 9＊9

17. Consider the set $F = \{1, {}^{-}1, i, {}^{-}i\}$ with an operation $\times$ defined by the table.

$\times$	1	${}^{-}1$	i	${}^{-}i$
1	1	${}^{-}1$	i	${}^{-}i$
${}^{-}1$	${}^{-}1$	1	${}^{-}i$	i
i	i	${}^{-}i$	${}^{-}1$	1
${}^{-}i$	${}^{-}i$	i	1	${}^{-}1$

Find each of the following.

a. $({}^{-}1) \times i$ **b.** $i \times i$ **c.** $({}^{-}i) \times ({}^{-}1)$ **d.** $({}^{-}i) \times ({}^{-}i)$

18. Consider the set A and the operation ＊ from Problem 16.

a. Is the set A closed for ＊? Give reasons for your answer.
b. Is ＊ associative for the set A? Give reasons.
c. Is ＊ commutative for the set A? Give reasons.

19. Consider the set F and the operation $\times$ from Problem 17.

a. Is the set F closed for $\times$? Give reasons for your answer.
b. Is $\times$ associative for set F? Give reasons.
c. Is $\times$ commutative for set F? Give reasons.

20. Let A be the process of putting on a shirt; let
        B be the process of putting on a pair of socks; and let
        C be the process of putting on a pair of shoes.
     Let $\emptyset$ be the operation of "followed by."

a. Is $\emptyset$ commutative for $\{A,B,C\}$?
b. Is $\emptyset$ associative for $\{A,B,C\}$?

21. Consider the set of counting numbers and an operation Σ, which means "select the first of the two." That is,

$$4 \Sigma 3 = 4$$
$$3 \Sigma 4 = 3$$
$$5 \Sigma 7 = 5$$
$$6 \Sigma 6 = 6$$

a. Is the set of counting numbers closed for Σ? Give reasons.
b. Is Σ associative for the set of counting numbers? Give reasons.
c. Is Σ commutative for the set of counting numbers? Give reasons.

22. Consider the operation ▼ defined by the table.

a. Find □▼△.   **b.** Find △▼○.
c. Does ○▼□ = □▼○?
d. Does (○▼△)▼△ = ○▼△(△▼△)?
e. Is ▼ a commutative operation? Give reasons.

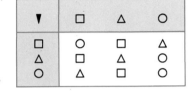

▼	□	△	○
□	○	□	△
△	□	△	○
○	△	□	○

23. Let @ mean "select the smaller number" and Σ mean "select the first of the two." Is Σ distributive over @ in the set of counting numbers?

24. Do the following problems mentally using the distributive property.

a. 6×82 **b.** 8×41 **c.** 7×49 **d.** 5×99 **e.** 4×88

***25.** Is the operation of union for sets a commutative operation? [*Hint:* The example shown in the margin illustrates the question for *one* example, but it doesn't *prove* that union is associative. To prove this result, use Venn diagrams for *any* sets X, Y, and Z.] That is, prove

$$(X \cup Y) \cup Z = X \cup (Y \cup Z)$$

***26.** Is the operation of intersection for sets a commutative operation?

***27.** Is the operation of intersection for sets an associative operation?

28. Is the set of even numbers closed for the operation of addition? Of multiplication?

29. Is the set of odd numbers closed for the operation of addition? Of multiplication?

30. Consider the soldier facing in a given direction (say north). Let us denote "left face" by L, "right face" by R, "about face" by A, and "stand fast" by F. Then we define $H = \{L,R,A,F\}$ and an operation $\emptyset$ meaning "followed by." Thus, $L \emptyset L = A$ means "left face" followed by "left face" is the same as the command "about face." Also, $L \emptyset A = R$
$A \emptyset R = L$
$L \emptyset F = L$
Make a table summarizing all possible movements. Use the headings L, R, A, and F.

31. Using Problem 30, find:

a. $R \emptyset R$ **b.** $A \emptyset A$ **c.** $L \emptyset L$ **d.** $R \emptyset L$ **e.** $L \emptyset A$

*These problems require Chapter 2.

Let $U = \{1,2,3,4,5,6,7,8,9,10\}$
$P = \{1,4,7\}$
$Q = \{2,4,9,10\}$
$R = \{6,7,8,9\}$
Does
$(P \cup Q) \cup R = P \cup (Q \cup R)$?
Check: $(P \cup Q) \cup R$
$= \{1,2,4,7,9,10\} \cup \{6,7,8,9\}$
$= \{1,2,4,6,7,8,9,10\}$
$P \cup (Q \cup R)$
$= \{1,4,7\} \cup \{2,4,6,7,8,9,10\}$
$= \{1,2,4,6,7,8,9,10\}$
Thus, for these sets,

$(P \cup Q) \cup R = P \cup (Q \cup R)$

F means "don't move from your present position." It does not mean "return to your original position."

32. Using Problem 30:

    a. Is H closed for the operation $\varnothing$?

    b. Does $\varnothing$ satisfy the associative property for the set H?

    c. Does $\varnothing$ satisfy the commutative property for the set H?

33. Let S be a set of stairs in a given stairway in some building. Number the stairs in ascending order: $s_1, s_2, s_3, \ldots, s_n$. Then $S = \{s_1, s_2, \ldots, s_n\}$, where the elements of S are individual steps in the stairway. Consider the operation of walking up the staircase one step at a time. If we denote this operation by $\uparrow$, we have

$$s_1 \uparrow s_2 = s_3 \qquad \text{and} \qquad s_2 \uparrow s_3 = s_4$$

meaning that, given two consecutive steps, the result is the next step in ascending order. Similarly, we define a descending operation $\downarrow$:

$$s_5 \downarrow s_4 = s_3 \qquad s_{17} \downarrow s_{16} = s_{15}$$

The steps must be consecutive, so $s_3 \uparrow s_{19}$ or $s_7 \downarrow s_5$ is not defined. If S is a set of stairs in a typical building, the operations of $\uparrow$ and $\downarrow$ are not closed (why?). However, if we consider the set of stairs

$$S = \{s_1, s_2, \ldots, s_{45}\}$$

depicted by the Escher lithograph, is the set S closed for $\uparrow$ and $\downarrow$?

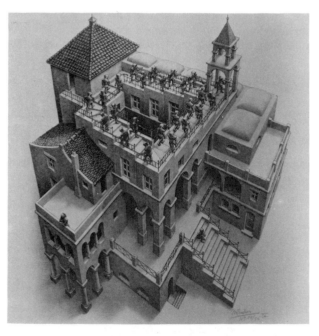

M. C. Escher's *Ascending, Descending*

34. Let $S = \{1,2,3, \ldots ,99,100\}$. Define an operation $\boxtimes$: $a \boxtimes b = 2a + b$, where $+$ refers to ordinary addition. Check the commutative and associative properties.

35. *Computer problem.* In the text, we indicated that the commutative property could be verified by checking all possibilities. If the set is not too big, we can do this on a computer as follows: Let $S = \{1,2,3,4,5,6,7,8,9,10\}$ for the operation of multiplication. We wish to see whether $a \cdot b = b \cdot a$.

```
 10  LET A=1
 20  LET B=1
 30  IF A*B=B*A GOTO 80
 40  PRINT "NOT COMMUTATIVE, SINCE I HAVE FOUND A";
 45  PRINT " COUNTEREXAMPLE WHEN"
 50  PRINT "A= ";A;
 60  PRINT ", AND B= ";B
 70  GOTO 130
 80  LET B=B+1
 90  IF B<11 GOTO 30
100  LET A=A+1
110  IF A<11 GOTO 20
120  PRINT "THE SET IS COMMUTATIVE FOR MULTIPLICATION."
130  END
```

Write a program similar to this one to see if S is commutative for the operation of subtraction.

36. *Computer problem.* As noted in the text, testing the associative property can be a tedious task, since all possibilities must be checked. Let $S = \{1,2,3,4,5,6,7,8,9,10\}$ for the operation of multiplication. We wish to see if $(a \cdot b) \cdot c = a \cdot (b \cdot c)$. We can write a BASIC program as follows:

```
 10  LET A=1 TO 10
 20  LET B=1 TO 10
 30  LET C=1 TO 10
 40  IF (A*B)*C=A*(B*C) GOTO 100
 50  PRINT "NOT ASSOCIATIVE, SINCE I HAVE FOUND A";
 55  PRINT " COUNTEREXAMPLE WHEN"
 60  PRINT "A= ";A;
 70  PRINT ", B= ";B;
 80  PRINT ", AND C= ";C
 90  GOTO 170
100  LET C=C+1
110  IF C<11 GOTO 40
120  LET B=B+1
130  IF B<11 GOTO 30
140  LET A=A+1
150  IF A<11 GOTO 20
160  PRINT "SET IS ASSOCIATIVE FOR MULTIPLICATION."
170  END
```

Write a program to see if S is associative for the operation of addition.

37. *Computer problem.* Check associativity for multiplication in the set $C = \{0,1\}$. Write a program that will print out all the possibilities.

Mind Boggler

38.   *The beer glass puzzle.* The following puzzle by Mel Stover consists of three pieces as shown in the figure.

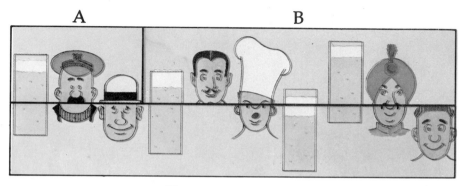

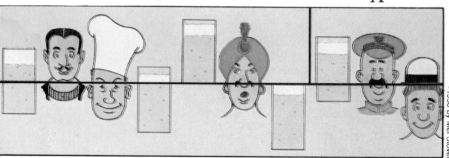

The beer glass puzzle

If we place the pieces together as shown in the top part of the figure, we see 6 men and 4 beers. However, if we *commute* pieces A and B as shown in the bottom part of the figure, we count 5 men and 5 beers! Thus,

$$AB \neq BA$$

Can you explain how a man changes into a beer simply by switching the locations of the pieces?

4.5 Integers

Historically, an agricultural-type society would need only natural numbers, but what about a subtraction like

$$5 - 5 = ?$$

Certainly society would have a need for a number representing $5 - 5$, so a new number, called **zero,** was invented, so that $5 = 5 + 0$ (remember the definition of subtraction). If this new number is annexed to the set of natural numbers, the set is called the set of *whole numbers:*

$$W = \{0,1,2,3,4, \ldots\}$$

Did the Egyptians or Babylonians have such a number?

This one annexation to the existing numbers satisfied society's needs for several thousands of years.

However, as society evolved, the need for bookkeeping advanced, and eventually the need for answers to problems like $5 - 6 = ?$ arose. The question that we need to answer is this: Can we annex new numbers to the set W so that it is possible to carry out *all* subtractions? That is, can we annex elements to W so that it becomes a closed set for subtraction? The numbers that need to be annexed are the opposites of the natural numbers. If we add these to the set W we have the following set:

$$Z = \{\ldots, ^-3, ^-2, ^-1, 0, ^+1, ^+2, ^+3, \ldots\}$$

This is called the set of **integers.** It is customary to refer to certain subsets of Z as follows:

1. Positive integers: $\{^+1, ^+2, ^+3, \ldots\}$
2. Zero: $\{0\}$
3. Negative integers: $\{^-1, ^-2, ^-3, \ldots\}$

Now that we have this new enlarged set of numbers, will we be able to carry out all possible additions, subtractions, and multiplications? Before we answer this question, let's review the processes by which we operate within the set of integers.

To see how to add integers, use the number line shown in Figure 4.3.

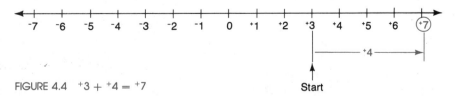

FIGURE 4.3

We shall make moves along the number line. Suppose you have the following sums:

$$^+3 + {}^+4 \qquad \text{and} \qquad ^+3 + {}^-4$$

Start at $^+3$; add $^+4$ by moving four units to the right (Figure 4.4), and add $^-4$ by moving four units to the left (Figure 4.5).

FIGURE 4.4 $^+3 + {}^+4 = {}^+7$ Start

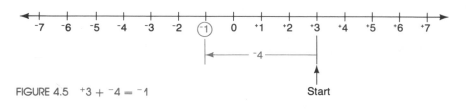

FIGURE 4.5 $^+3 + {}^-4 = {}^-1$

Start

Examples 1–3 illustrate what is called *directed distance*. A three-unit move to the right is different from a three-unit move to the left. However, if you simply want to indicate the distance between two points, you speak of the *undirected distance*.

Example 1: $^-3 + {}^-4 = ?$

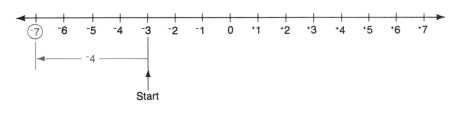

Therefore, $^-3 + {}^-4 = {}^-7$.

Example 2: $^+5 + {}^-6 = ?$

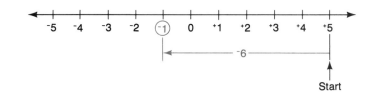

Therefore, $^+5 + {}^-6 = {}^-1$.

Example 3: $^-3 + {}^+4 = ?$

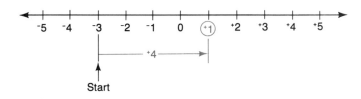

Therefore, $^-3 + {}^+4 = {}^+1$.

Because it is not always convenient or practical to add on a number line, we make some generalizations. We are interested in looking at the numerical value of a quantity, independent of direction or sign. To do this, we define the notion of **absolute value** of a number, which represents the distance of that number from the origin when plotted on a number line.

Definition

The *absolute value* of x is its undirected distance from zero. We use the symbol $|x|$ for absolute value.

Historical Note

Historically, the negative integers were developed quite late. There are indications that the Chinese had some knowledge of negative numbers as early as 200 B.C., and in the 7th century A.D. the Hindu Brahmagupta stated the rules for operations with positive and negative numbers. The Chinese represented negatives by putting them in red (compare with the present-day accountant), and the Hindus represented them by putting a circle or a dot over the number.

However, as late as the 16th century, some European scholars were calling numbers such as $0 - 1$ absurd. In 1545 Cardano (1501–1576), an Italian scholar who presented the elementary properties of negative numbers, called the positive numbers "true" numbers and the negative numbers "fictitious" numbers. However, they did become universally accepted; as a matter of fact, the word *integer* that we use to describe the set is derived from "numbers with integrity."

Example 4: $|^+5|$ is 5   **Example 5:** $|^-5|$ is 5   **Example 6:** $|^-(^-3)| = 3$

Now notice that, if we're adding numbers with the same sign, the result is the same as the sum of the absolute values except for a plus or a minus sign (since their directions on the number line are the same). If we're adding numbers with different signs, their directions are opposite, so the net result is the difference of the absolute values. This can be summarized as follows (see also Figure 4.6, page 202):

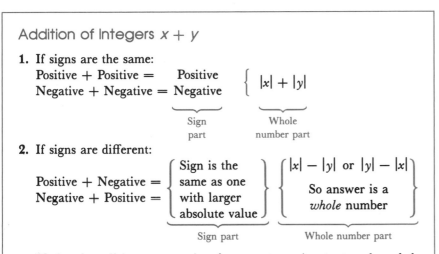

Addition of Integers $x + y$

1. If signs are the same:
   Positive + Positive =   Positive
   Negative + Negative = Negative $\left\{ |x| + |y| \right.$

   Sign part   Whole number part

2. If signs are different:
   Positive + Negative = $\left\{ \begin{array}{l} \text{Sign is the same as one with larger absolute value} \end{array} \right.$ $\left\{ \begin{array}{l} |x| - |y| \text{ or } |y| - |x| \\ \text{So answer is a } whole \text{ number} \end{array} \right.$
   Negative + Positive =

   Sign part   Whole number part

Notice that all integers consist of two parts: a sign part and a whole number part.

Example 7: $^+41 + {}^+13 = {}^+54$ **Example 8:** $^-41 + {}^-13 = {}^-54$

Example 9: $^+41 + {}^-13 = {}^+28$ **Example 10:** $^-41 + {}^+13 = {}^-28$

It is now easy to verify that the set of integers is closed for addition. We know that the whole numbers are closed for addition and that the sum of two integers will be a whole number or the opposite of a whole number and thus will be an integer (this is not proof and must be accepted without proof).

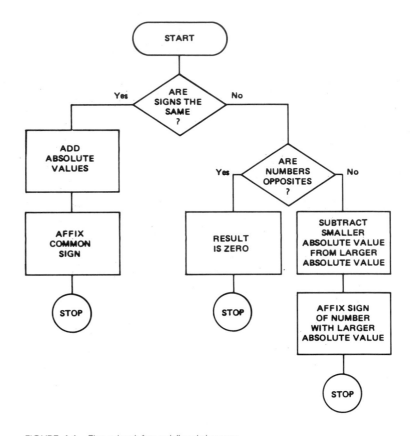

FIGURE 4.6   Flowchart for adding integers

Sometimes we will still use the
positive-integer form for clarity
or emphasis.

Since the set of positive integers is the set of natural numbers, let's agree that whenever we write a natural number 1,2,3,4,... we can also write $^+1, ^+2, ^+3, ^+4, ...$ That is, $1 = ^+1$, $2 = ^+2$, $3 = ^+3, ...$

For the whole numbers, multiplication can be defined as repeated addition, since we say that $4 \cdot 5$ means

$$\underbrace{4 + 4 + 4 + 4 + 4}_{5 \text{ addends}}$$

However, we cannot do this for the integers, since $4 \cdot (^-5)$, or

$$\underbrace{4 + 4 + \cdots + 4}_{^-5 \text{ addends}}$$

doesn't "make sense." We cannot repeat the addition process a negative number of times. Thus, the best we can do is seek out patterns to help us choose a "good" definition for the multiplication of integers.

There are four cases to consider: **1.** Positive · Positive
2. Positive · Negative
3. Negative · Positive
4. Negative · Negative

1. Suppose we wish to multiply two positive integers:

$$^+3 \cdot {}^+4$$

We agreed that $^+3 = 3$ and $^+4 = 4$; therefore,

$$^+3 \cdot {}^+4 = 3 \cdot 4$$

Now 3 and 4 are natural numbers, and we know how to multiply natural numbers. Thus, since $3 \cdot 4 = 12 = {}^+12$, we say that $^+3 \cdot {}^+4 = {}^+12$. We can then conclude:

The product of two positive numbers is a positive number.

2. Next consider

$$^+3 \cdot {}^-4$$

To see how to find this product, we make use of a very important characteristic of mathematics discussed in Section 1.1—the discovery of a new concept by looking at some patterns.

$$^+3 \cdot {}^+4 = {}^+12$$
$$^+3 \cdot {}^+3 = {}^+9$$
$$^+3 \cdot {}^+2 = {}^+6$$

Would you know what to write next? Here it is:

$$^+3 \cdot {}^+1 = {}^+3$$
$$^+3 \cdot 0 = 0$$

What comes next? *Answer this question before reading further.*

$$^+3 \cdot {}^-1 = {}^-3$$
$$^+3 \cdot {}^-2 = {}^-6$$
$$^+3 \cdot {}^-3 = {}^-9$$
$$^+3 \cdot {}^-4 = {}^-12$$
$$\vdots$$

Do you know how to continue? Try building a few more such patterns using different numbers. What did you discover about the product of a positive number and a negative number?

The product of a positive number and a negative number is a negative number.

3. Consider the problem of a negative times a positive:

$$^-3 \cdot {}^+4$$

   Since you know that $^+4 \cdot {}^-3 = {}^-12$, you should also know that $^-3 \cdot {}^+4 = {}^-12$ if we are to retain all the basic properties for numbers that we have previously discussed. That is, we want the commutative property to hold for the multiplication of integers. Therefore:

The product of a negative number and a positive number is a negative number.

4. Consider the final example, the product of two negative integers:

$$^-3 \cdot {}^-4$$

Let's build another pattern.

$$^-3 \cdot {}^+4 = {}^-12$$
$$^-3 \cdot {}^+3 = {}^-9$$
$$^-3 \cdot {}^+2 = {}^-6$$

What would you write next? Here it is:

$$^-3 \cdot {}^+1 = {}^-3$$
$$^-3 \cdot 0 = 0$$

What comes next? *Answer this question before reading further.*

$$^-3 \cdot {}^-1 = {}^+3$$
$$^-3 \cdot {}^-2 = {}^+6$$
$$^-3 \cdot {}^-3 = {}^+9$$
$$^-3 \cdot {}^-4 = {}^+12$$
$$\cdot$$
$$\cdot$$
$$\cdot$$

Thus, as the pattern indicates:

The product of two negative numbers is a positive number.

We can summarize our discussion as follows (see also Figure 4.7):

Multiplication of Integers $x \cdot y$

1. Positive · Positive = Positive
2. Positive · Negative = Negative
3. Negative · Positive = Negative
4. Negative · Negative = Positive

Sign part

$$\underbrace{|x| \cdot |y|}$$
Whole number part

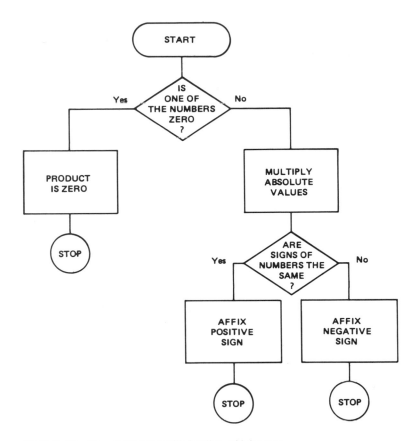

FIGURE 4.7   Flowchart for multiplication of integers

Sirnivasa Ramanujan
(1887–1920)

Ramanujan, who lived only 33 years, was a mathematical prodigy of great originality. He was largely self-taught, but was "discovered" in 1913 by the eminent British mathematician G. H. Hardy. Hardy brought Ramanujan to Cambridge, and in 1918 Ramanujan became the first Indian to become a Fellow of the Royal Society. An often-told story about Hardy and Ramanujan was that when Hardy visited Ramanujan in the hospital, he came in a taxi bearing the number 1729. He asked Ramanujan if there was anything interesting about this number. Without hesitation, Ramanujan said there was: It is the smallest positive integer that can be represented in two different ways as a sum of two cubes:

$$1729 = 1^3 + 12^3$$
$$= 9^3 + 10^3$$

Example 11: $(^+41)(^+13) = {}^+533$ **Example 12:** $(^-41)(^-13) = {}^+533$

Example 13: $(^+41)(^-13) = {}^-533$ **Example 14:** $(^-41)(^+13) = {}^-533$

On a number line, subtraction is "going back" in the opposite direction instead of "going ahead," as in addition. For two positive numbers, the process seems quite natural. For example:

$$^+4 - {}^+3 = {}^+1$$

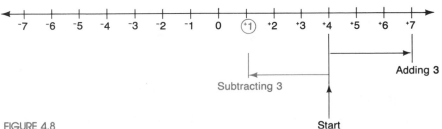

FIGURE 4.8

But what about subtracting negative numbers? Negative already indicates "going back," or to the left. Does subtraction of a negative mean "going ahead," or to the right? Consider the following pattern:

$$^+4 - {}^+4 = 0$$
$$^+4 - {}^+3 = {}^+1$$
$$^+4 - {}^+2 = {}^+2$$
$$^+4 - {}^+1 = {}^+3$$
$$^+4 - 0 = {}^+4$$
$$^+4 - {}^-1 = ?$$
$$^+4 - {}^-2 = ?$$
$$^+4 - {}^-3 = ?$$

It does seem as if smaller numbers in each case should give larger differences. On a number line, $^+4 - {}^-3$ means "going in the opposite direction from $^-3$."

$$^+4 - {}^-3 = 7$$

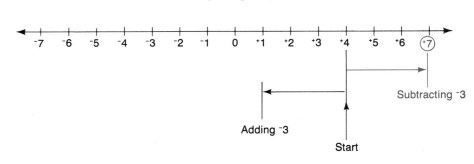

FIGURE 4.9

Guided by these results, we make the following definition for subtraction.

Definition of Subtraction of Integers

$$x - y = x + (^-y)$$

To subtract, add the opposite (of the number being subtracted).

Example 15: $^+41 - {}^+13 = {}^+41 + {}^-13 = {}^+28$
                          └── Add the ──┘   └ Complete the addition of $^+41 + {}^-13$
                             opposite

Example 16: $^+41 - {}^-13 = {}^+41 + {}^+13 = {}^+54$

Example 17: $^-41 - {}^+13 = {}^-41 + {}^-13 = {}^-54$

Example 18: $^-41 - {}^-13 = {}^-41 + {}^+13 = {}^-28$

Let's take an overview of what has been done in the last few sections. We began with the natural numbers, which are closed for addition and multiplication. Next, we defined subtraction and created a situation where it was impossible to subtract some numbers from others. We then "created" a new set (called the *integers*) by adding to the set of whole numbers the opposite of each of its members.

Now, since the subtraction of integers is defined in terms of addition, you can easily show that the integers are closed for subtraction. You are asked to do this in Problem 24 of Problem Set 4.5.

The last of the basic operations is division, which is defined as the opposite operation of multiplication.

Definition of Division

If a, b, and z are integers, where $b \neq 0$,

$$\frac{a}{b} = z \quad \text{means} \quad a = b \cdot z$$

a divided by b is written

$$a \div b \quad \text{or} \quad \frac{a}{b}$$

See also Figure 4.10 on page 208.

Example 19: $^-12 \div 6 = {}^-2$ **Example 20:** $\dfrac{^-18}{2} = {}^-9$

Example 21: $\dfrac{^-10}{^-5} = 2$ **Example 22:** $\dfrac{65}{^-13} = {}^-5$

Notice that we required that $b \neq 0$. Why do we not allow division by zero? We will consider two cases.

1. Division of a nonzero number by zero.

$$a \div 0 \quad \text{or} \quad \frac{a}{0} = x$$

Is there such a number so that $a \div 0$ makes sense? We see that any number x would have to be such that $a = 0 \cdot x$. Now, $0 \cdot x = 0$; so if $a \neq 0$, we see that such a case is impossible. That is, $a/0$ does not exist.

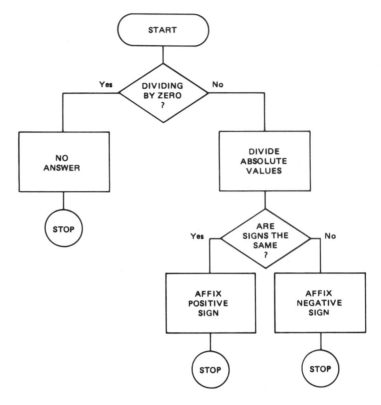

FIGURE 4.10 Flowchart for dividing integers

Abraham Lincoln used a biblical reference (Mark 3:25) to initiate his campaign in 1858; he said, "A house divided against itself cannot stand." I offer a corollary to Lincoln's statement: "A house divided by itself is one (provided that house is not zero, of course)."

2. Division of zero by zero.

$$\frac{0}{0} = x$$

Is there such a number x? We see that *any* x makes this true, since $0 \cdot x = 0$ for all x. But this leads to certain absurdities. If $\frac{0}{0} = 2$ and $\frac{0}{0} = 5$, then $2 = 5$, since numbers equal to the same number are equal to each other.

Is the set of integers closed for division? Certainly one can find many examples in which an integer divided by an integer is an integer. Does this mean that the set of integers is closed? What about $1 \div 2$ or $4 \div 5$? These numbers do not exist in the set of integers; thus, the set is *not* closed for division. Now, as long as a society has no need for such division problems, the question of inventing new numbers will not arise. However, as the need to divide 1 into 2 or more parts arises, some new numbers will have to be invented so that the set will be closed for division. We'll do this in the next section. The problem with inventing such new numbers is that it must be done in such a way that the properties of the existing numbers are left unchanged. That is, closure for addition, subtraction, and multiplication must be retained.

Problem Set 4.5

1. Suppose you are given the following number line:

```
        M   A   T   H       I   S       F   U   N
 ←――+――+――+――+――+――+――+――+――+――+――+――+――+――+――→
   ⁻7  ⁻6  ⁻5  ⁻4  ⁻3  ⁻2  ⁻1   0  ⁺1  ⁺2  ⁺3  ⁺4  ⁺5  ⁺6  ⁺7
```

   a. Explain why a move from H to N is described by $^+7$.
   b. Describe a move from T to F.
   c. Describe a move from U to A.

2. Add the following integers on a number line.

   a. $^+5 + {}^+3$ **b.** $^-5 + {}^+3$ **c.** $^+4 + {}^-7$
   d. $^+3 + {}^+5$ **e.** $^-3 + {}^+5$ **f.** $^-2 + {}^-4$

3. Find the sum.

   a. $7 + {}^-3$ **b.** $^-9 + 5$
   c. $^-10 + {}^-4$ **d.** $162 + {}^-27$
   e. $^-12 + ({}^-4 + {}^-3)$ **f.** $^-5 + ({}^-6 + 10)$

4. Find the sum.

   a. $^-15 + 8$ **b.** $(5 + {}^-7) + (5 + {}^-4)$
   c. $^-14 + 27$ **d.** $42 + {}^-121$
   e. $62 + {}^-62$ **f.** $(6 + {}^-8) + (6 + 8)$

5. Find the difference.

   a. $10 - 7$ **b.** $7 - 10$ **c.** $6 - {}^-4$
   d. $^-5 - {}^-10$ **e.** $5 - {}^-5$ **f.** $7 - {}^-3$

6. Find the difference.

   a. $0 - {}^-15$ **b.** $^-46 - {}^-46$ **c.** $^-7 - {}^-18$
   d. $^-4 - 8$ **e.** $62 - {}^-112$ **f.** $5 - {}^-416$

7. Simplify.

   a. $({}^-31) + ({}^-16)$ **b.** $^-14 + {}^-21$
   c. $^-9 + 16 + {}^-11$ **d.** $^-3 - ({}^-6 - {}^+4)$
   e. $5 + {}^-19 + 15$ **f.** $|^-8| + {}^-8$

8. Simplify.

   a. $^-9 - (4 - 5)$ **b.** $^-({}^-23 + 14)$
   c. $^-7 - (6 - 4)$ **d.** $^-({}^-18) + |^-18|$
   e. $|^-3| - ({}^-({}^-2))$ **f.** $18 - 4 - 12 - {}^-3$

9. Simplify.

   a. $(^+3)(^-6)$ **b.** $(^-5)(^+4)$
   c. $\dfrac{^-4}{^-2}$ **d.** $\dfrac{^-6}{^-3}$
   e. $(^-5)(^+7) - (^-9)$ **f.** $(^-2)(^+3) - (^-8)$

B.C.

WHAT DO YOU REALLY THINK OF ME AS A PERSON?

I THINK YOU'RE NICE.

* GOSH *.... ON A SCALE OF 1 TO 10 WHAT DO YOU THINK OF ME AS A LOVER?

IS IT PERMISSIBLE TO USE NEGATIVE NUMBERS?

by johnny hart

10. Simplify.

    a. $14 \cdot {}^-5$ **b.** ${}^-14 \cdot {}^-5$ **c.** ${}^-5 \cdot (8 - 12)$

    d. $({}^-5 \cdot 8) - 12$ **e.** $\dfrac{{}^-12}{4}$ **f.** $\dfrac{{}^-63}{{}^-9}$

11. Simplify.

    a. $\dfrac{12}{{}^-4}$ **b.** ${}^-6 \cdot \dfrac{14}{{}^-2}$ **c.** $\dfrac{{}^-528}{{}^-4}$

    d. $45 - (25 \cdot {}^-5)$ **e.** $({}^-1)^3$ **f.** $10 \cdot \left(\dfrac{{}^-8}{{}^-2}\right)$

12. Perform the following operations.

    a. $15 - {}^-7$ **b.** ${}^-6 - {}^-6$

    c. ${}^-18 - 5$ **d.** ${}^-8 + 7 + 16$

    e. $14 + {}^-10 - 8 - 11$ **f.** $6 + {}^-8 - 5$

13. Perform the following operations.

    a. $({}^-2)({}^-3) + ({}^-1)({}^+6)$ **b.** $({}^-5)({}^-1) + ({}^-9)({}^+2)$

    c. $({}^-32) \div ({}^-8) - 5 - ({}^-7)$ **d.** $({}^-15) \div ({}^-5) - 4 - ({}^-8)$

    e. $({}^-5)({}^+2) + ({}^-3)({}^-4) - ({}^+6)({}^-7)$

    f. $({}^-3)({}^-7 + 2)$

14. Perform the following operations.

    a. $6 - {}^-2$ **b.** ${}^-5 - {}^-3$

    c. ${}^-15 - {}^-6$ **d.** ${}^-4 + 6 - 8$

    e. $15 - {}^-3 - 4 - 11$ **f.** ${}^-12 + {}^-7 - 10 - 14$

15. Perform the following operations.

    a. $(7)({}^-8)$ **b.** $({}^-5)(15)$

    c. ${}^-5 \cdot {}^-6$ **d.** $\dfrac{{}^-42}{3}$

    e. $({}^-54 \div {}^-9) \div 3$ **f.** ${}^-54 \div ({}^-9 \div 3)$

Parts e and f of Problem 15 provide a counterexample for some property. Can you name this property?

16. Perform the following operations.

    a. $(5)(2)$ **b.** $({}^-6)(14)$ **c.** ${}^-2 \cdot {}^-3$

    d. $\dfrac{34}{{}^-2}$ **e.** $(48 \div {}^-6) \div {}^-2$ **f.** $48 \div ({}^-6 \div {}^-2)$

*17. Draw a Venn diagram showing the following sets.

$$N = \{\text{Natural numbers}\}$$
$$W = \{\text{Whole numbers}\}$$
$$Z = \{\text{Integers}\}$$

*18. What is the intersection of the set of positive integers and the set of negative integers? Is the union of these two sets the set of integers?

*Chapter 2 is required for these problems.

19. **a.** State the commutative property.
    b. Is addition or subtraction commutative in the set of integers?
    c. Is multiplication or division commutative in the set of integers? Support your answers.

20. Explain, in your own words, the difference between $0 \div 5$ and $5 \div 0$.

21. **a.** State the associative property.
    b. Is addition or subtraction associative in the set of integers?
    c. Is multiplication or division associative in the set of integers? Be sure to give reasons for your answers.

22. Perform the indicated operations.

    a. $(42 \div {}^-7) \div {}^-2$
    b. $\dfrac{\frac{42}{-7}}{-2}$

    c. ${}^-56 \div (14 \div 2)$
    d. $\dfrac{\frac{-56}{14}}{2}$

23. Perform the indicated operations.

    a. $({}^-2)^4$
    b. $({}^-1)^{67}$
    c. $({}^-1)^{1972}$
    d. $({}^-1)^{2k}$, where k is any natural number

24. Show that the set of integers is closed for subtraction.

25. Do you think any finite subset of the set of integers will be closed under multiplication?

Mind Bogglers

26. *Four fours.* B.C. has a mental block against fours, as we can see in the cartoon. See if you can handle fours by writing the numbers from 1 to 10 using four 4s for each. Here are the first three completed for you:

    $$\frac{4}{4} + 4 - 4 = 1 \qquad \frac{4}{4} + \frac{4}{4} = 2 \qquad \frac{4 + 4 + 4}{4} = 3$$

27. *Calculator problem.* Multiply 1234567×9999999:

    a. By using a calculator
    b. By looking for a pattern

28. Here is a problem that was proposed by Professor Richard Andree of the University of Oklahoma in 1971 and, to my knowledge, is still unsolved.

Build a sequence as follows:

a. Pick any number—for example, 7.
b. If the last number in the sequence is odd, the next number is found by tripling it and adding 1.
c. If the last number is even, the following number is half of it. For example,

$$7, 22, 11, 34, 17, 52, 26, 13, 40, 20, 10, 5, 16, 8, 4, 2, 1$$

The conjecture (still unproved) is that all such sequences eventually go to 1. Try this for several examples.

Current research on this problem indicates that all numbers less than 10^{40} go to 1.

4.6 Rational Numbers

Historically, the need for a closed set for division came before the need for closure for subtraction. We need to find some number k so that

$$1 \div 2 = k$$

As we saw in Section 4.1, the ancient Egyptians limited their fractions by requiring the numerator to be 1. The Romans avoided fractions by the use of subunits. Feet were divided into inches and pounds into ounces, and a twelfth part of the Roman unit was called an *uncia*.

However, people soon felt the practical need to obtain greater accuracy in measurement and the theoretical need to close the number system with respect to the operation of division. In the set of integers, some divisions are possible:

$$\frac{10}{5}, \frac{-4}{2}, \frac{-16}{-8}, \cdots$$

However, certain others are not:

$$\frac{1}{2}, \frac{-16}{5}, \frac{5}{12}, \cdots$$

Just as we extended the set of natural numbers by creating the concept of opposites, we can extend the set of integers. That is, the number $\frac{5}{12}$ is defined to be that number obtained when 5 is divided by 12. This new set, consisting of the integers as well as the quotients of integers, is called the set of **rational numbers.**

Historical Note

The symbol $\div$ was adopted by John Wallis (1616–1703) and was used in Great Britain and in the United States but not on the European continent, where the colon (:) was used. In 1923 the National Committee on Mathematical Requirements stated: "Since neither $\div$ nor : as signs of division plays any part in business life, it seems proper to consider only the needs of algebra, and to make more use of the fractional form and (where the meaning is clear) of the symbol /, and to drop the symbol $\div$ in writing algebraic expressions."*

*From *Report of the National Committee on Mathematical Requirements under the Auspices of the Mathematical Association of America, Inc.* (1923), p. 81.

Definition

The *rational numbers,* denoted by Q, are the set of all numbers of the form a/b, where a and b are integers and $b \neq 0$.

Notice that a rational number has fractional form. In arithmetic you learned that if a number is written in the form $\frac{a}{b}$ it means $a \div b$ and that a is called the **numerator** and b the **denominator.** Also, $\frac{a}{b}$ is called

a **proper fraction** if $a < b$;

an **improper fraction** if $a > b$; and

a **whole number** if $a = b$.

The fraction $\frac{a}{b}$ is **reduced** if there is no natural number (other than 1) that divides into *both* the numerator and denominator.

In Examples 1–5, reduce the fractions.

Example 1: $\dfrac{8}{16} = \dfrac{1}{2}$ Notice that 8 divides into *both* 8 and 16, so $\frac{8}{16}$ is not reduced. Mentally, carry out the following steps:

$$\frac{8}{16} = \frac{8 \div 8}{16 \div 8} = \frac{1}{2}$$

Example 2: $\dfrac{6}{9} = \dfrac{2}{3}$ Mentally: $\dfrac{6}{9} = \dfrac{6 \div 3}{9 \div 3} = \dfrac{2}{3}$

Example 3: $\dfrac{14}{27}$ This is reduced since the only natural number that divides into both 14 and 27 is 1.

Example 4: $\dfrac{15}{5} = 3$ Mentally: $\dfrac{15}{5} = \dfrac{15 \div 5}{5 \div 5} = \dfrac{3}{1} = 3$

Example 5: $\dfrac{15}{7}$ This is reduced since the only natural number that divides into both 15 and 7 is 1.

In Example 5, you might want to write $\frac{15}{7}$ as $2\frac{1}{7}$, which is certainly acceptable. The form $2\frac{1}{7}$ is called **mixed number** form (it mixes whole numbers and fractions), and it means $2 + \frac{1}{7}$. In this book we do require that fractional answers be written in reduced form, but not necessarily in mixed number form. If the fractions are very complicated, a better procedure for reducing fractions than repeated dividing needs to be developed. This will be done in Chapter 6.

The next problem to solve is the question of closure for the operations. That is, is the set of rational numbers closed for addition, subtraction, multiplication, and division? Let's first review the definitions of these operations.

We are familiar with the idea of adding fractions with common denominators, but what if the given fractions don't have common denominators? You may remember a process for *finding common denominators;* this process is discussed in Chapter 6. But did you know it is possible to define addition of fractions without using the idea of common denominators? Consider the following definition.

Addition of Rational Numbers

If a/b and c/d are rational numbers, then

$$\frac{a}{b} + \frac{c}{d} = \frac{ad + bc}{bd}$$

Remember from Section 4.4 that the notation ad, bc, and bd means multiply the numbers a and d, b and c, and b and d, as shown in Examples 6 and 7.

Example 6: $\dfrac{1}{3} + \dfrac{1}{2} = \dfrac{1 \times 2 + 3 \times 1}{3 \times 2} = \dfrac{5}{6}$

Example 7: $\dfrac{4}{5} + \dfrac{7}{9} = \dfrac{4 \times 9 + 5 \times 7}{5 \times 9} = \dfrac{36 + 35}{45} = \dfrac{71}{45}$

Will this definition satisfy all the needs of a society? Is it possible to add *any* two rational numbers and get a rational number? To show that the addition of two rational numbers always results in a rational number, it is necessary to show that, given any two elements of the set Q, their sum is also an element of the set. For example, suppose we are given two rational numbers x/y and w/z. This means that x and w are integers, and y and z are nonzero integers. We wish to prove that $(xz + wy)/yz$ is a rational number. Now, xz and wy are integers, since the set of integers is closed for multiplication. Also, $xz + wy$ is an integer, by closure for addition in the set of integers. Similarly, yz must be a nonzero integer; thus,

$$\frac{xz + wy}{yz}$$

is an integer divided by a nonzero integer and therefore a member of Q.
   Subtraction is similar to addition.

Subtraction of Rational Numbers

If a/b and c/d are rational numbers, then

$$\frac{a}{b} - \frac{c}{d} = \frac{ad - bc}{bd}$$

We can repeat our preceding discussion for subtraction to show that the rationals are closed for subtraction. The details are left to the student (see Problem 32 in Problem Set 4.6).

Example 8: $\dfrac{2}{3} - \dfrac{^-1}{2} = \dfrac{2 \times 2 - 3 \times (^-1)}{3 \times 2} = \dfrac{4 + 3}{6} = \dfrac{7}{6}$

Example 9: $\dfrac{4}{5} - \dfrac{7}{9} = \dfrac{36 - 35}{45} = \dfrac{1}{45}$

Multiplication of fractions is easier than addition or subtraction. To multiply fractions, simply write each number as a fraction (this may require forms like $\frac{5}{1}$ or improper fractions) and multiply numerators, and then multiply denominators.

Multiplication of Rational Numbers

If a/b and c/d are rational numbers, then

$$\frac{a}{b} \times \frac{c}{d} = \frac{ac}{bd}$$

Example 10: $\dfrac{1}{3} \times \dfrac{2}{5} = \boxed{\dfrac{1 \times 2}{3 \times 5}}$ This step can often be done in your head.

$= \dfrac{2}{15}$

Example 11: $\dfrac{2}{3} \times \dfrac{^-4}{7} = \dfrac{^-8}{21}$

Example 12: $^-5 \times \dfrac{^-2}{3}$

When multiplying a whole number and a fraction, write the whole number as a fraction, and then multiply:

$$\frac{^-5}{1} \times \frac{^-2}{3} = \frac{10}{3}$$

Example 13: $3\frac{1}{2} \times 2\frac{3}{5}$

When multiplying mixed numbers, write the mixed numbers as improper fractions, and then multiply:

$$\frac{7}{2} \times \frac{13}{5} = \frac{91}{10}$$

Example 14: $\dfrac{3}{4} \times \dfrac{2}{3} = \dfrac{6}{12}$ but $\dfrac{6}{12} = \dfrac{6 \div 6}{12 \div 6} = \dfrac{1}{2}$

Notice that the actual multiplication was a wasted step, because the answer was simply divided again in order to reduce it! Therefore, the proper procedure is to cancel common factors before you do the multiplication. That is, to reduce $\frac{6}{12}$, notice that 6 divides into both 6 and 12 and write

These little numbers mean that 6 divides into 6 one time, and 6 divides into 12 twice (6 is the common factor).

$$\dfrac{\overset{1}{\cancel{6}}}{\underset{2}{\cancel{12}}} = \dfrac{1}{2}$$

Example 15: $\dfrac{2}{5} \times \dfrac{3}{4} = \dfrac{\overset{1}{\cancel{2}} \times 3}{5 \times \underset{2}{\cancel{4}}} = \dfrac{3}{10}$

Notice that this is very similar to the original multiplication as stated to the left of the equal sign. You can save a step by canceling with the original product, as shown in the next example.

Example 16: $\dfrac{\overset{1}{\cancel{3}}}{5} \times \dfrac{1}{\underset{1}{\cancel{3}}} = \dfrac{1}{5}$

Example 17: $3\frac{2}{3} \times 2\frac{2}{5} = \dfrac{11}{\underset{1}{\cancel{3}}} \times \dfrac{\overset{4}{\cancel{12}}}{5} = \dfrac{44}{5}$

Now we can see that the set of rational numbers is closed under the operation of multiplication. Suppose x/y and z/w are any two rational numbers. We wish to show that their product is also a rational number. By definition, $(x/y) \times (z/w) = xz/yw$. Now, since x, z, y, and w are integers, with $y \neq 0$ and $w \neq 0$, we know that xz and yw must also be integers and $yw \neq 0$; by the definition of a rational number, xz/yw is a rational number. Therefore, we know that the set is closed for the operation of multiplication.

Just as the integers gave rise to integers when added, multiplied, or subtracted, so all sums, products, and differences of rational numbers are rational numbers. But are the rationals a closed set for division?

Consider the problem

$$\dfrac{2}{3} \div \dfrac{4}{5} = \boxed{}$$

Before considering this problem, look at a more familiar problem,

$$\dfrac{12}{4} = *$$

By the definition of division, we must find some number for * so that

$$4 \times * = 12$$

Similarly, for

$$\frac{2}{3} \div \frac{4}{5} = \boxed{}$$

we must find some number for $\boxed{}$ so that

$$\frac{4}{5} \times \boxed{} = \frac{2}{3}$$

What do we put into the box to get the answer? Do it in two parts. First multiply $\frac{4}{5}$ by $\frac{5}{4}$ to obtain 1. *Then* multiply by $\frac{2}{3}$ to obtain the result:

$$\frac{4}{5} \times \boxed{\frac{5}{4} \times \frac{2}{3}} = \frac{2}{3}$$

Thus,

$$\frac{2}{3} \div \frac{4}{5} = \frac{2}{3} \times \frac{5}{4}$$

which seems to suggest that we "invert the fraction we are dividing by, and then multiply."

Division of Rational Numbers

If a/b and c/d are rational numbers, and $c \neq 0$, then

$$\frac{\dfrac{a}{b}}{\dfrac{c}{d}} = \frac{ad}{bc}$$

Stated in words, to divide fractions, "invert and multiply."

Note the additional requirement that $c \neq 0$. If $c = 0$, then $c/d = 0$, and this is the same as our previous restriction for division—namely, *you can never divide by zero* (remember you are dividing by c/d). However, the rationals are closed for nonzero division, since ad/bc is a rational (can you explain why?).

Example 18: $\quad \dfrac{^-3}{4} \div \dfrac{^-4}{7} = \dfrac{^-3}{4} \times \dfrac{7}{^-4} = \dfrac{^-21}{^-16} = \dfrac{21}{16}$

Example 19: $\dfrac{4}{5} \div \dfrac{7}{9} = \dfrac{4}{5} \times \dfrac{9}{7} = \dfrac{36}{35}$

Having reached this stage in the development of numbers, you might ask if we need any other numbers. We have shown that the set of rationals is closed for addition, subtraction, multiplication, and nonzero division; thus, no new numbers could be formed by applying these operations. Could society advance without any new numbers? Is there a need to invent any more numbers to combine with the set of rationals?

Problem Set 4.6

Completely reduce the fractions in Problems 1–6.

1. **a.** $\dfrac{3}{9}$ **b.** $\dfrac{6}{9}$ **c.** $\dfrac{2}{10}$

2. **a.** $\dfrac{3}{12}$ **b.** $\dfrac{4}{12}$ **c.** $\dfrac{6}{12}$

3. **a.** $\dfrac{14}{7}$ **b.** $\dfrac{38}{19}$ **c.** $\dfrac{92}{2}$

4. **a.** $\dfrac{72}{15}$ **b.** $\dfrac{42}{14}$ **c.** $\dfrac{16}{24}$

5. **a.** $\dfrac{18}{30}$ **b.** $\dfrac{70}{105}$ **c.** $\dfrac{50}{400}$

6. **a.** $\dfrac{140}{420}$ **b.** $\dfrac{150}{1,000}$ **c.** $\dfrac{2,500}{10,000}$

Perform the indicated operations in Problems 7–30.

7. $\dfrac{2}{3} + \dfrac{7}{9}$ 8. $\dfrac{^-5}{7} + \dfrac{4}{3}$ 9. $\dfrac{^-12}{35} - \dfrac{8}{15}$

10. $2^{-1} + 3^{-1} + 5^{-1}$ 11. $\left(\dfrac{4}{5} + 2\right) + \dfrac{^-7}{25}$ 12. $2 + 2^{-1}$

13. $\dfrac{7}{9} - \dfrac{2}{3}$ 14. $\dfrac{4}{7} - \dfrac{^-5}{9}$ 15. $\dfrac{^-3}{5} - \dfrac{^-6}{9}$

16. $\left(\dfrac{3}{10} - \dfrac{7}{100}\right) - \dfrac{9}{1,000}$ 17. $\dfrac{3}{10} - \left(\dfrac{7}{100} - \dfrac{9}{1,000}\right)$

18. $5 - 5^{-1}$ 19. $\dfrac{2}{3} \times \dfrac{5}{7}$

20. $\dfrac{^-1}{8} \times \dfrac{2}{^-5}$ 21. $5 \times \dfrac{6}{7}$

22. $\dfrac{47}{-5} \times \dfrac{-15}{26}$

23. $\left(\dfrac{-5}{7} \times \dfrac{-14}{25}\right) \times \dfrac{15}{-21}$

24. 7×7^{-1}

25. $\dfrac{4}{9} \div \dfrac{2}{3}$

26. $\dfrac{\dfrac{-2}{9}}{\dfrac{6}{7}}$

27. $\dfrac{105}{-11} \div \dfrac{-15}{33}$

28. $\left(\dfrac{5}{7} \div \dfrac{-14}{25}\right) \div \dfrac{15}{21}$

29. $\dfrac{5}{7} \div \left(\dfrac{-14}{25} \div \dfrac{15}{21}\right)$

30. $6 \div 6^{-1}$

31. Complete the following table. Use a ✔ to show that the number listed at the top of a column is a member of the set listed at the side. Use an **X** if the number is not a member of the set.

Sets	5	−7	0	$\frac{4}{5}$	$\frac{-1}{2}$
Natural numbers					
Whole numbers					
Integers					
Rational numbers					

32. Show that the rationals are closed for subtraction.

33. Show that the rationals are closed for nonzero division.

34. Are the integers closed for addition, subtraction, multiplication, and nonzero division?

35. Which properties of addition are satisfied by the rationals? Support your answer.

36. Which properties of multiplication are satisfied by the rationals? Support your answer.

Mind Bogglers

37. A rubber ball is known to rebound half the height it drops. If the ball is dropped from a height of 100 feet, how far will it have traveled by the time it hits the ground for:

  a. The first time?
  b. The second time?
  c. The third time?
  d. The fourth time?
  e. The fifth time?

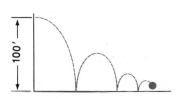

f. Look for a pattern, and try to make a conjecture about the total distance traveled by the ball after it hits the ground for the *n*th time.

g. Is there a maximum distance the ball will travel if we assume that it will bounce indefinitely?

38. *Computer problem.* Write a BASIC program that will print out a table for Problem 37. That is, the output of the program should give the distance the ball travels by the time it hits the ground for the first, second, third, . . . time.

*4.7 Irrational Numbers

𝔥𝔦𝔰𝔱𝔬𝔯𝔦𝔠𝔞𝔩 𝔑𝔬𝔱𝔢

It was considered impious for a member of the Pythagorean society to claim any discovery for himself. Instead, each new idea was attributed to their founder, Pythagoras.

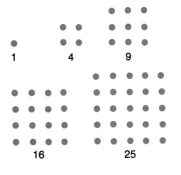

OTHER PERFECT SQUARES:

$$6^2 = 36 \qquad 10^2 = 100$$
$$7^2 = 49 \qquad 11^2 = 121$$
$$8^2 = 64 \qquad 12^2 = 144$$
$$9^2 = 81 \qquad 13^2 = 169$$

FIGURE 4.11   Square numbers

The type of society that we have been considering has used numbers only as they relate to practical problems. However, this need not be the case, since numbers can be appreciated for their beauty and interrelationships. The Pythagoreans, to our knowledge, were among the first to investigate numbers for their own sake. The Pythagoreans were a secret society founded in the 6th century B.C. They had their own philosophy, religion, and way of life. This group investigated music, astronomy, geometry, and number properties. Because of their strict secrecy, much of what we know about them is legend, and it is difficult to tell just what work can be attributed to Pythagoras himself.

Much of the Pythagoreans' life-style was embodied in their beliefs about numbers. They considered the number 1 the essence of reason; the number 2 was identified with opinion; and 4 was associated with justice because it is the first number that is the product of equals (the first perfect squared number other than 1). Of the numbers greater than 1, odd numbers were masculine and even numbers were feminine; thus, 5 represented marriage, since it was the union of the first masculine and feminine numbers   (2 + 3 = 5).

The Pythagoreans were also interested in special types of numbers that had mystical meanings: perfect numbers, friendly numbers, deficient numbers, abundant numbers, prime numbers, triangular numbers, square numbers, and pentagonal numbers. Of these, the perfect square numbers are probably the most important. They are called *perfect squares* because they can be arranged into squares (see Figure 4.11). They are found by squaring the counting numbers.

The Pythagoreans discovered the famous property of square numbers that today bears Pythagoras' name. They found that if they constructed any right triangle and then constructed squares on each of the legs of the triangle, the area of the larger square was equal to the sum of the areas of the smaller squares (see Figure 4.12).

Today we state the Pythagorean theorem algebraically by saying that, if *a* and *b* are the lengths of the legs of a right triangle and *c* is the length of the hypotenuse, then the square of the length of the hypotenuse is equal to the sum of the squares of the length of the other two sides.

*This is an optional section.

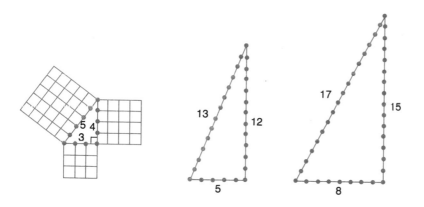

FIGURE 4.12 Relationships of sides of right triangles

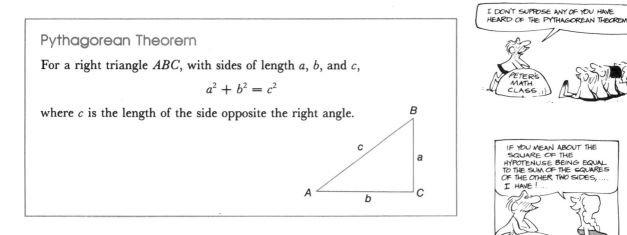

Pythagorean Theorem

For a right triangle ABC, with sides of length a, b, and c,

$$a^2 + b^2 = c^2$$

where c is the length of the side opposite the right angle.

The Pythagoreans were overjoyed with this discovery, but it led to a revolutionary idea in mathematics—one that caused the Pythagoreans many problems.

Legend tells us that one day while the Pythagoreans were at sea, one of their group came up with the following argument. Suppose each leg of a right triangle is 1. Then the hypotenuse must be

$$1^2 + 1^2 = c^2$$
$$2 = c^2$$

If we denote the number whose square is 2 by $\sqrt{2}$, we have $\sqrt{2} = c$. The symbol $\sqrt{2}$ is read "square root of two." This means that $\sqrt{2}$ is that number such that if multiplied by itself, as in

$$\sqrt{2} \times \sqrt{2}$$

the product is 2.

Historical Note

The Egyptians and the Chinese knew of the Pythagorean theorem before the Pythagoreans (but they didn't call it the Pythagorean theorem). An early example is shown below. It is attributed to Chou Pei, who was probably a contemporary of Pythagoras'.

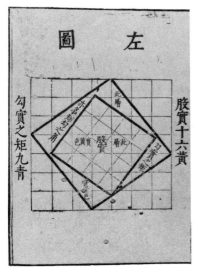

In Examples 1–7, use the definition of square root to find the indicated products.

Example 1: $\sqrt{3} \times \sqrt{3} = 3$ **Example 2:** $\sqrt{4} \times \sqrt{4} = 4$

Example 3: $\sqrt{5} \times \sqrt{5} = 5$ **Example 4:** $\sqrt{15} \times \sqrt{15} = 15$

Example 5: $\sqrt{16} \times \sqrt{16} = 16$ **Example 6:** $\sqrt{144} \times \sqrt{144} = 144$

Example 7: $\sqrt{200} \times \sqrt{200} = 200$

Some square roots are rational. For Example 2,

$$\sqrt{4} \times \sqrt{4} = 4$$

and we know $2 \times 2 = 4$, so it seems reasonable that $\sqrt{4} = 2$. But wait! We also know $(^-2) \times (^-2) = 4$, so isn't it just as reasonable that $\sqrt{4} = {}^-2$? Mathematicians have agreed that the square root symbol be used only to denote positive numbers, so that $\sqrt{4} = 2$ (*not* $^-2$). From Examples 2, 5, and 7 we see:

$$\sqrt{4} = 2, \qquad \sqrt{16} = 4, \qquad \sqrt{144} = 12$$

How about square roots of numbers that are not perfect squares? Is $\sqrt{2}$, for example, a rational number? Remember, if $\sqrt{2} = a/b$, where a/b is some fraction so that

$$\frac{a}{b} \times \frac{a}{b} = 2$$

then it is rational. The Pythagoreans were some of the first to investigate this question. Now remember that, for the Pythagoreans, mathematics and religion were one; they asserted that all natural phenomena could be expressed by whole numbers or ratios of whole numbers. Thus they believed that $\sqrt{2}$ must be some whole number or fraction (ratio of two whole numbers). Suppose we try to find such a rational number:

$$\frac{7}{5} \times \frac{7}{5} = \frac{49}{25} = 1.96$$

We try again:

$$\frac{707}{500} \times \frac{707}{500} = \frac{499,849}{250,000}$$
$$= 1.999396$$

Using a computer, I obtained the following possibility for $\sqrt{2}$:
1.4142135623730950488016887242096907856967187537694. Give a brief argument showing why this could not be $\sqrt{2}$.

This time we might really get down to business and decide to use a calculator, which gives

$$1.414213562$$

If you square this number, do you obtain 2? Notice that the last digit of this multiplication will be 4; what should it be if it were the square root of 2?

There is an interesting argument to show that all such efforts are in vain. However, before we can follow this reasoning, we need to review some of the properties of even and odd numbers.

Recall that an integer is **even** if it is a multiple of 2, that is, if it may be expressed as $2k$, where k stands for an integer. Then the set of even integers is

$$\{\ldots, ^-6, ^-4, ^-2, 0, 2, 4, 6, \ldots\}$$

An integer that is not even is said to be **odd.** Each odd integer may be expressed in the form $2k + 1$, where k stands for an integer. Then the set of odd integers is

$$\{\ldots, ^-7, ^-5, ^-3, ^-1, 1, 3, 5, 7, \ldots\}$$

*Example 8:   Prove that the square of an even integer is an even integer.

Proof: Any even integer may be expressed as $2k$, where k stands for an integer. Then the square of the integer may be expressed as $(2k)^2$.

$$(2k)^2 = 2k \times 2k = 4k^2 = 2 \times 2k^2 = 2(2k^2)$$

which is even.

*Example 9:   Prove that the square of an odd integer is an odd integer.

Proof: Let $2k + 1$ be an odd integer. Show that $(2k + 1)^2$ is odd. The proof is left as a problem.

*Example 10:   Prove that if a is an integer and a^2 is even, then a is even.

Proof: Suppose a is odd. Then a^2 is odd by Example 9. This contradicts the fact that a^2 is even, and, since any integer is either even or odd, it follows that a is even.

Let's return to our argument to show that $\sqrt{2}$ can or cannot be represented as a fraction.* (The following formal argument is generally attributed to Euclid.)

Suppose that $\sqrt{2}$ is a fraction a/b, where a and b are whole numbers. Moreover, to make matters simpler, let us suppose that any factors common to a and b are canceled. (That is, a/b is in reduced form.) We assume, then, that

$$\sqrt{2} = \frac{a}{b} \qquad (1)$$

Square both sides:

$$2 = \frac{a^2}{b^2}$$

Historical Note

Shown below is a Moslem manuscript, written in 1258, which shows the Pythagorean theorem.

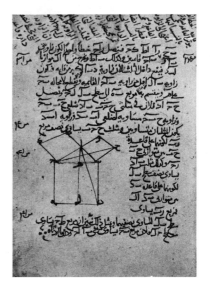

Multiply both sides by b^2:

$$2b^2 = a^2 \qquad (2)$$

The left side of this equation is an even number because it contains a factor of 2. Hence the right side must also be an even number. But if a^2 is even, then, according to Example 10, a must be even. If a is even, it must contain 2 as a factor. That is,

$$a = 2d$$

where d is some integer. If we substitute this value of a into equation (2), we obtain

$$2b^2 = (2d)^2 = 2d \times 2d = 4d^2 \qquad (3)$$

Then, since

$$2b^2 = 4d^2$$

we may divide both sides of this equation by 2 and obtain

$$b^2 = 2d^2 \qquad (4)$$

We now see that b^2 is an even number; so, by again appealing to the result of Example 10, we find that b is an even number.

What we have shown in this argument is that if $\sqrt{2} = a/b$, then a and b must be even numbers. But at the very beginning we had canceled any common factor in a and b; yet we find that a and b still contain 2 as a common factor. This result contradicts the fact that a and b have no common factors.

Historical Note

The word *root* is *radix* in Latin, and it was represented by r. This led to ꞃ, then to √, and finally to √ . Students often say work with radicals is difficult. However, in the 16th century there were four symbols for indicating roots: the letters ℞ and ℓ (both lowercase and capital), √, and the fractional exponent. To complicate matters, these symbols were used not only for roots but for unknown quantities as well. Also, the radical signs for cube and fourth roots had quite different shapes, and by the close of the 16th century there were 25 or more symbols for radicals.

Why do we arrive at a contradiction? Since our reasoning is correct, the only possibility is that the assumption that $\sqrt{2}$ equals a fraction is not correct. In other words, $\sqrt{2}$ cannot be a ratio of two whole numbers (that is, it is not rational).

This result shattered the Pythagorean philosophy. Clearly, they had to change their religion or deny logic. Legend has it that they took the latter course: They set the man who discovered it to sea alone in a small boat, and pledged themselves to strict secrecy.

But secrets will get out (indeed, can there be any mathematical or scientific secrets?), and we see that we are faced with the need for still more numbers. We will invent a new set and call it the set of **irrational numbers.** We have seen that $\sqrt{2}$ is irrational, and it can also be shown that $\sqrt{3}$, $\sqrt{5}$, $\sqrt{7}$, and the square root of any number that is not a perfect square, the cube root of any number that is not a perfect cube, and so on are all irrational numbers. The number π, which is the ratio of the circumference of any circle to its diameter, is also not rational. In everyday work you will use irrational numbers when finding the circumference and area of a circle, which we will consider in Chapter 7. You will also use irrational numbers when applying the Pythagorean theorem.

Since the Pythagorean theorem asserts $a^2 + b^2 = c^2$, and since c is the number whose square is $a^2 + b^2$, then

$$c = \sqrt{a^2 + b^2}$$

Also, if you want to find the length of one of the legs, say a, when you know both b and c, you can use the formula

$$a = \sqrt{c^2 - b^2}$$

as shown by Example 11.

Example 11: If a 13 ft ladder is placed against a building so that the base of the ladder is 5 ft away from the building, how high up does the ladder reach?

Solution: Consider Figure 4.13. Since h is one of the legs, use the formula:

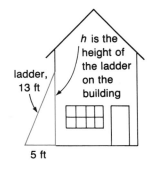

$$h = \sqrt{13^2 - 5^2}$$

    ↑   ↑

Hypotenuse   Subtract the length of the known leg from the known hypotenuse

$$= \sqrt{169 - 25}$$ You must square the numbers *before* you find the square root.

$$= \sqrt{144}$$ Subtract.

$$= 12$$ 144 is a perfect square, so the answer is rational.

FIGURE 4.13

Thus, the ladder reaches 12 ft up the side of the building.

The Pythagorean theorem is valid only when dealing with a right triangle, so if the sides form the relationship $a^2 + b^2 = c^2$, then the triangle must be a right triangle. Carpenters often make use of this property as shown by Example 12.

Example 12: A carpenter wants to make sure that the corner of a room is square (is a right angle). If she measures out sides (legs) of 3 ft and 4 ft, how long should she make the diagonal (hypotenuse)? (See Figure 4.14.)

Solution: The hypotenuse is the unknown so use the formula:

$$c = \sqrt{a^2 + b^2}$$

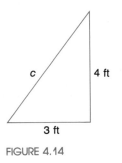

FIGURE 4.14

Thus,

$$c = \sqrt{a^2 + b^2}$$

$$= \sqrt{3^2 + 4^2}$$ The legs are 3 and 4.

$$= \sqrt{9 + 16}$$ Square and simplify *before* finding the square root.

$$= \sqrt{25}$$ Add.

$$= 5$$ 25 is a perfect square, so the answer is rational.

She should make the diagonal 5 ft long.

Your answers to both Examples 11 and 12 were rational. Suppose that the result is irrational. You can either leave the result in **radical form** or approximate your result as shown in Example 13.

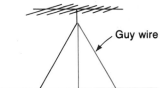

Guy wire

FIGURE 4.15

Example 13:   Suppose you need to attach several guy wires to your TV antenna, as shown in Figure 4.15. If one guy wire is attached 20 ft away from a 30 ft antenna, what is the exact length of one guy wire, and what is the length to the nearest foot?

Solution: The length of the guy wire is the length of the hypotenuse of a right triangle. So

$$c = \sqrt{a^2 + b^2}$$
$$= \sqrt{20^2 + 30^2}$$
$$= \sqrt{400 + 900}$$
$$= \sqrt{1,300}$$

The exact length of the guy wire is $\sqrt{1,300}$ and is irrational since 1,300 is not a perfect square. What is this length as an approximate rational number?

$$30^2 = 900 \qquad \text{1,300 is between 900 and 1,600 so}$$
$$40^2 = 1,600 \qquad \sqrt{1,300} \text{ is between 30 and 40.}$$

For a better approximation, you can use a calculator:

With a square root key: Press: [1300]   [√]
                            Display: 36.05551275

Without a square root key: Make a guess; guess 30 (*any* guess will work). Divide 1,300 (the given number) by 30 (your guess):

[1300] [÷] [30] [=]   43.3333333

Find the average of this answer and the previous guess. (Don't clear your calculator; just continue.)

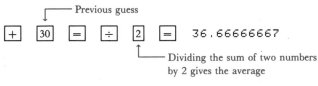

┌─── Previous guess

[+] [30] [=] [÷] [2] [=]   36.66666667

└─── Dividing the sum of two numbers
        by 2 gives the average

Repeat the procedure until you have the degree of accuracy you would like:

[1300] [÷] [36.6666667] [=]   35.45454545
[+] [36.6666667] [=] [÷] [2] [=]   36.06060606

Repeat again:

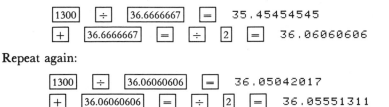

[1300] [÷] [36.06060606] [=]   36.05042017
[+] [36.06060606] [=] [÷] [2] [=]   36.05551311

The guy wire is 36 ft (to the nearest ft). Note, if the application will not allow any length less than $\sqrt{1,300}$, then, instead of rounding, take the *next larger* foot.

Problem Set 4.7

The Pythagorean theorem tells us that the sum of the squares of the lengths of the legs of a right triangle is equal to the square of the length of the hypotenuse. Verify that the theorem is true by tracing the squares in Problem 1 and fitting them onto the pattern of the upper two squares shown in the figure in the margin. Next, cut the squares along the lines and rearrange the pieces so that the pieces all fit into the large square in the figure in the margin.

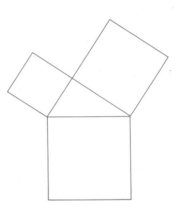

1.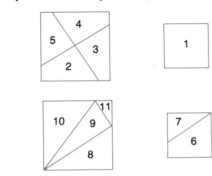

2.

3. With only a straightedge and compass, use a number line and the Pythagorean theorem to construct a segment whose length is $\sqrt{2}$. Measure the segment as accurately as possible, and write your answer in decimal form. *Do not consult any tables.*

Use the definition of square root to find the indicated products in Problems 4–7.

4. a. $\sqrt{6} \times \sqrt{6}$ b. $\sqrt{7} \times \sqrt{7}$ c. $\sqrt{9} \times \sqrt{9}$

5. a. $\sqrt{8} \times \sqrt{8}$ b. $\sqrt{10} \times \sqrt{10}$ c. $\sqrt{25} \times \sqrt{25}$

6. a. $\sqrt{14} \times \sqrt{14}$ b. $\sqrt{30} \times \sqrt{30}$ c. $\sqrt{36} \times \sqrt{36}$

7. a. $\sqrt{807} \times \sqrt{807}$ b. $\sqrt{169} \times \sqrt{169}$ c. $\sqrt{400} \times \sqrt{400}$

Classify each number in Problems 8–19 as rational or irrational. If it is rational, write it without a square root symbol. If it is irrational, approximate it with a rational number correct to the nearest thousandth.

8. $\sqrt{9}$ 9. $\sqrt{25}$ 10. $\sqrt{10}$ 11. $\sqrt{30}$

12. $\sqrt{36}$ 13. $\sqrt{50}$ 14. $\sqrt{169}$ 15. $\sqrt{400}$

16. $\sqrt{500}$ 17. $\sqrt{1,000}$ 18. $\sqrt{1,024}$ 19. $\sqrt{1,936}$

20. How far from the base of a building must a 26 ft ladder be placed so that it reaches 10 ft up the wall?

TWO YOUNG BRAVES and three squaws are sitting proudly side by side. The first squaw sits on a buffalo skin with her 50 pound son. The second squaw is on a deer skin with her 70 pound son. The third squaw, who weighs 120 pounds, is on a hippopotamus skin. Therefore, the squaw on the hippopotamus is equal to the sons of the squaws on the other two hides.

21. How high up on a wall does a 26 ft ladder reach if the bottom of the ladder is placed 10 ft from the building?

22. If a carpenter wants to make sure that the corner of a room is square and measures out 5 ft and 12 ft along the walls, how long should he make the diagonal?

23. If a carpenter wants to be sure that the corner of a building is square and measures out 6 ft and 8 ft along the sides, how long should she make the diagonal?

24. What is the exact length of the hypotenuse if the legs of a right triangle are 2 in. each?

25. What is the exact length of the hypotenuse if the legs of a right triangle are 3 ft each?

26. An empty lot is 400 ft by 300 ft. How many feet would you save by walking diagonally across the lot instead of walking the length and width?

27. A television antenna is to be erected and held by guy wires. If the guy wires are 15 ft from the base of the antenna and the antenna is 10 ft high, what is the exact length of each guy wire? What is the length of each guy wire to the nearest foot? If two guy wires are to be attached, how many feet of wire should be purchased if it can't be bought by a fraction of a foot?

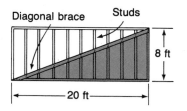

Diagonal brace   Studs

8 ft

20 ft

28. A diagonal brace is to be placed in the wall of a room. The height of the wall is 8 ft and the wall is 20 ft long. What is the exact length of the brace? What is the length of the brace to the nearest foot?

29. A balloon rises at a rate of 4 ft per minute when the wind is blowing horizontally at a rate of 3 ft per minute. After three minutes, how far away from the starting point, in a direct line, is the balloon?

30. In the text we spoke of square numbers. Other kinds of numbers are also of interest.

   a. Continue the sequence of triangular numbers: 1,3,6,10,15, _____, _____, _____.

   1   3   6   10   15

   b. Continue the sequence of tetrahedral numbers: 1,4,10,20, _____, _____, _____.

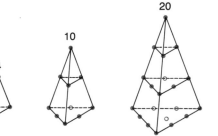

20

10

4

1

c. Answer the question at the bottom of the historical note about Gauss.

31. One of the most famous problems in mathematics is *Fermat's four-square problem.* Fermat (see historical note in Problem Set 4.8) stated that any positive integer can be expressed as the sum of four squares. For example,

$$1 = 1^2 + 0^2 + 0^2 + 0^2 \qquad 6 = 2^2 + 1^2 + 1^2 + 0^2$$
$$2 = 1^2 + 1^2 + 0^2 + 0^2 \qquad 7 = 2^2 + 1^2 + 1^2 + 1^2$$
$$3 = 1^2 + 1^2 + 1^2 + 0^2 \qquad 8 = 2^2 + 2^2 + 0^2 + 0^2$$
$$4 = 2^2 + 0^2 + 0^2 + 0^2 \qquad 9 = 2^2 + 2^2 + 1^2 + 0^2$$
$$5 = 2^2 + 1^2 + 0^2 + 0^2 \qquad 10 = 2^2 + 2^2 + 1^2 + 1^2$$

Write the following numbers using exactly four squares:

a. 11 **b.** 39 **c.** 143 **d.** 1,000 **e.** 1,980

32. **a.** i. Draw a circle.
    ii. Measure the diameter.
    iii. Use a string to approximate the circumference (it is *not* legal to use any formulas here). Measure the string.
    iv. Form the ratio of the diameter into the circumference of the circle. That is, divide the diameter into the circumference. This will serve as an approximation for π.
    b. Repeat these steps for a different circle and compare the results.

Mind Bogglers

33. Prove that the set of even integers is closed for multiplication (that is, prove that the product of two even numbers is even).

34. Prove that the set of odd integers is closed for multiplication.

35. Prove that the square of an odd integer is an odd integer.

36. We have shown that $\sqrt{2}$ is irrational (review this proof). In a similar manner, show that $\frac{1}{2}$ is not an integer.

37. **a.** How are the square numbers embedded in Pascal's triangle?
    b. How are the triangular numbers embedded in Pascal's triangle?
    c. How are the tetrahedral numbers embedded in Pascal's triangle?

Problems for Individual Study

38. *Pythagorean relationship.* What are some unusual proofs of the Pythagorean theorem? What are some of the unusual relationships that exist among Pythagorean numbers? What models can be made to visualize and prove the relationship? How can this relationship be used to measure indirectly?
    Exhibit suggestions: Models of cardboard, marbles, BB shot, or plastic to show the Pythagorean relationship; designs, tiles, cloth, or wallpaper based on this relationship; charts of unusual proofs; historical sidelights; mockups of applications in indirect measurement.

39. *Computer problems.* Write a BASIC program that will output Pythagorean triplets (numbers that satisfy the relation $a^2 + b^2 = c^2$).

4.8 Real Numbers

You are familiar with the rational numbers (fractions, for example) and the irrational numbers (π or square roots of certain numbers, for example), and now we wish to consider the most general set of numbers to be used in elementary mathematics. This set consists of the annexation of the irrational numbers to the set of rational numbers, and is called the set of **real numbers.**

Definition of Real Numbers

The set of *real numbers,* denoted by R, is defined as the union of the set of rationals and the set of irrationals.

𝕳𝖎𝖘𝖙𝖔𝖗𝖎𝖈𝖆𝖑 𝕹𝖔𝖙𝖊

History of the Decimal Point

Author	Time	Notation
Before Simon Stevin		$24\dfrac{375}{1,000}$
Simon Stevin	1585	$24\ 3^{(1)}\ 7^{(2)}\ 5^{(3)}$
Franciscus Vièta	1600	$24|375$
Johannes Kepler	1616	$24(375$ $\ \ |\ ||\ |||$
John Napier	1617	$24{:}3\ 7\ 5$
Henry Briggs	1624	24^{375}
William Oughtred	1631	$24|375$
Balam	1653	$24{:}375$
Ozanam	1691	$\ \ \ \ \ \ \ {}^{(1)\,(2)\,(3)}$ $24\cdot 3\ \ 7\ \ 5$
Modern		24.375

From this definition, we see that the real numbers may be classified as either rational or irrational. The distinguishing characteristic is their decimal representation. Let's now consider some different ways of representing real numbers.

Real numbers that are *rational* can be represented by *terminating decimals* or by *repeating decimals.*

Example 1: $\dfrac{1}{4} = .25$ **Example 2:** $\dfrac{5}{8} = .625$

Example 3: $\dfrac{58}{10} = 5.8$ **Example 4:** $\dfrac{2}{3} = .66\ldots$

Example 5: $\dfrac{1}{6} = .1666\ldots$ **Example 6:** $\dfrac{5}{11} = .4545\ldots$

Recall that the ellipses (...) indicate that the pattern is continuing in the same fashion.

Examples 1–3 are terminating decimals, since the division ends.

$$
\begin{array}{r}
.25 \\
4\overline{)1.00} \\
\underline{8} \\
20 \\
\underline{20} \\
0
\end{array}
\qquad
\begin{array}{r}
.625 \\
8\overline{)5.000} \\
\underline{4\,8} \\
20 \\
\underline{16} \\
40 \\
\underline{40} \\
0
\end{array}
\qquad
\begin{array}{r}
5.8 \\
10\overline{)58.0} \\
\underline{50} \\
8\,0 \\
\underline{8\,0} \\
0
\end{array}
$$

Examples 4–6 are repeating decimals, since the division never ends but repeats some sequence of numerals.

```
    .66...           .4545...          .1666...
 3)2.00          11)5.0000          6)1.0000
   18               44                 6
   20               60                40
   18               55                36
    2...            50                40
                    44                36
                    60                40
                    55                36
                     5...              4...
```

Real numbers that are *irrational* have decimal representations that are *nonterminating* and *nonrepeating*.

$$\sqrt{2} = 1.414214\ldots \qquad \pi = 3.1415926\ldots$$

In each of these examples, the numerals exhibit no repeating pattern and are irrational. Other decimals that do not terminate or repeat are also irrational:

0.12345678910111213... and 0.10110111011110...

We see that the real numbers may be classified in several ways. Any real number is:

1. Positive, negative, or zero
2. A rational number or an irrational number
3. Expressible as a terminating, a repeating, or a nonterminating and nonrepeating decimal:
   a. If it terminates, it is rational.
   b. If it eventually repeats, it is rational.
   c. If it does not terminate or repeat, it is irrational.

In order to change the fractional representation of a rational number to its decimal representation, use the definition of fractions: a/b means a divided by b. (See Table 4.4).

A mnemonic for remembering π:

"Yes, I have a number"

Notice that the number of letters in the words gives an approximation for π (the comma is the decimal point).

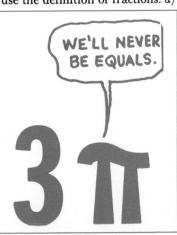

Historical Note

It is not at all obvious that π is an irrational number. In the Bible (I Kings 7:23) π is given as 3. In 1892 *The New York Times* printed an article showing π as 3.2. In 1897, House Bill No. 246 of the Indiana State Legislature defined π as 4; this definition was "offered as a contribution to education to be used only by the State of Indiana free of cost . . . ." Actually, the bill's author had 3.2 in mind but worded the bill so that it came out to be the value 4. The bill was never passed. In a book written in 1934, π was proposed to equal $3\frac{13}{81}$. The symbol itself was first published in 1706 but was not used widely until Euler adopted it in 1737. By 1873 William Shanks had calculated π to 700 decimal places. It took him 15 years to do so; however, computer techniques have shown that his last 100 or so places are incorrect.

Example 7: Change $\frac{4}{5}$ to a decimal.

$$\begin{array}{r} .8 \\ 5\overline{)4.0} \\ \underline{4\,0} \\ 0 \end{array}$$

Thus, $\frac{4}{5} = .8$.

Example 8: Change $\frac{1}{7}$ to a decimal.

$$\begin{array}{l} \quad .142857\ldots \\ 7\overline{)1.000000} \\ \quad \underline{7} \\ \quad 30 \\ \quad \underline{28} \\ \quad\;\; 20 \\ \quad\;\; \underline{14} \\ \quad\;\;\; 60 \\ \quad\;\;\; \underline{56} \\ \quad\;\;\;\; 40 \\ \quad\;\;\;\; \underline{35} \\ \quad\;\;\;\;\; 50 \\ \quad\;\;\;\;\; \underline{49} \\ \quad\;\;\;\;\;\; 1\ldots \end{array}$$

Thus, $\frac{1}{7} = .142857\ldots$. We will also denote a repeating fraction by using a bar over the numerals that are to be repeated; thus, $\frac{1}{7} = .\overline{142857}$.

Example 9: Change $6\frac{2}{9}$ to a decimal. Now $6\frac{2}{9} = 6 + \frac{2}{9}$; thus, we concentrate our attention on $\frac{2}{9}$.

$$\begin{array}{r} .2\ldots \\ 9\overline{)2.0} \\ \underline{1\,8} \\ 2\ldots \end{array}$$

Thus, $6\frac{2}{9} = 6 + .\overline{2} = 6.\overline{2}$.

In order to change from a terminating decimal representation of a rational number to a fractional representation, use expanded notation. Recall that $10^{-1} = \frac{1}{10}$, $10^{-2} = \frac{1}{100}$, $10^{-3} = \frac{1}{1000}, \ldots, 10^{-n} = \frac{1}{10^n}$. Thus, .5 means $5 \times 10^{-1} = 5 \cdot \frac{1}{10} = \frac{5}{10}$, and we have a fractional representation for .5. The remaining step is to reduce the fraction: $.5 = \frac{5}{10} = \frac{1}{2}$.

Example 10: 123 means $(1 \times 10^{-1}) + (2 \times 10^{-2}) + (3 \times 10^{-3})$

$$= \frac{1}{10} + \frac{2}{100} + \frac{3}{1{,}000} = \frac{123}{1{,}000}$$

Thus, $.123 = \dfrac{123}{1{,}000}$.

TABLE 4.4   Decimal Representation of 1/n

n	Decimal Representation of $1/n$
1	1.
2	0.5
3	0.333 . . .
4	0.25
5	0.2
6	0.1666 . . .
7	0.142857142857 . . .
8	0.125
9	0.111 . . .
10	0.1
11	0.0909 . . .
12	0.08333 . . .
13	0.076923076923 . . .
14	0.0714285714285 . . .
15	0.0666 . . .
16	0.0625
17	0.0588235294117647058823529411764 7 . . .
18	0.0555 . . .
19	0.052631578947368421052631578947368421 0 . . .
20	0.05
21	0.047619047619 . . .
22	0.04545 . . .
23	0.0434782608695652173913043478260869565217391 3 . . .
24	0.041666 . . .
25	0.04
26	0.038461538461538 . . .
27	0.037037 . . .
28	0.0357142857142857 . . .
29	0.0344827586206896551724137931034482758620689 6 . . .
30	0.0333 . . .
31	0.032258064516129032258064516129032258064 5 . . .
32	0.03125
33	0.0303 . . .
34	0.02941176470588235294117647058823 5 . . .
35	0.0285714285714285714285 7 . . .

Example 11:   6.28 means $(6 \times 10^0) + (2 \times 10^{-1}) + (8 \times 10^{-2})$

$$= 6 + \frac{2}{10} + \frac{8}{100} = \frac{628}{100} = \frac{157}{25}$$

Thus, $6.28 = \dfrac{157}{25}$.

Example 12: .3479 means $(3 \times 10^{-1}) + (4 \times 10^{-2}) + (7 \times 10^{-3}) + (9 \times 10^{-4})$

$$= \frac{3}{10} + \frac{4}{100} + \frac{7}{1,000} + \frac{9}{10,000} = \frac{3,479}{10,000}$$

Thus, $.3479 = \frac{3,479}{10,000}$.

If we once again consider a number line and associate points on the number line with rational numbers, it appears that the number line is just about "filled up" with the rationals. The reason for this feeling of "fullness" is that the rationals form a **dense** set. That is, between every two rationals we can find another rational (see Figure 4.16).

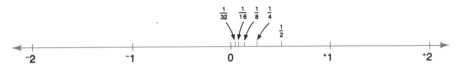

FIGURE 4.16   Real number line

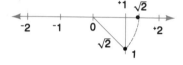

FIGURE 4.17   Construction of a segment of length $\sqrt{2}$

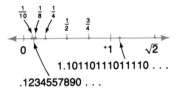

FIGURE 4.18   Real number line

The questions, then, are "Are there any 'holes'?" and "Is there any room left for the irrationals?"

We have shown that $\sqrt{2}$ is irrational. We can show that there is a place on the number line representing this length by using the Pythagorean theorem (see Figure 4.17).

Thus, we have a length equal to the $\sqrt{2}$ on the number line. We could show that other irrationals have their place on the number line (see Figure 4.18).

The relationships among the various sets of numbers we have been discussing are shown in Figure 4.19.

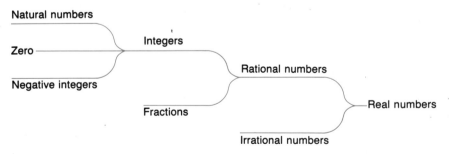

FIGURE 4.19   Sets of numbers

Problem Set 4.8

*1. Draw a Venn diagram showing the natural numbers, integers, rationals, and irrationals where the universe is the set of reals.

*Requires Section 2.1.

2. What is the distinguishing characteristic between the rational and irrational numbers?

3. Tell whether or not each number is an integer, a rational number, an irrational number, and a real number (since these sets are not all disjoint, you may need to list more than one set for each answer).

a. 7	**b.** 4.93	**c.** .656656665 . . .
d. $\sqrt{2}$	**e.** 3.14159 . . .	**f.** $\dfrac{17}{43}$
g. $.0027\overline{27}$	**h.** $\sqrt{9}$	**i.** $\sqrt{4{,}000}$

4. Tell whether or not each number is an integer, a rational number, an irrational number, and a real number (since these sets are not all disjoint, you may need to list more than one set for each answer).

a. 19	**b.** 6.48	**c.** 1.868686 . . .
d. $\sqrt{8}$	**e.** $\sqrt{16}$	**f.** $.00\overline{12}$
g. $\dfrac{22}{7}$	**h.** $\dfrac{19}{41}$	**i.** $\sqrt{1{,}000}$

Express each of the numbers in Problems 5–16 as a decimal.

5.	**a.** $\dfrac{3}{2}$	**b.** $\dfrac{7}{10}$	**6.**	**a.** $\dfrac{5}{2}$	**b.** $\dfrac{9}{10}$
7.	**a.** $\dfrac{2}{5}$	**b.** $\dfrac{47}{100}$	**8.**	**a.** $\dfrac{3}{5}$	**b.** $\dfrac{27}{15}$
9.	**a.** $\dfrac{5}{6}$	**b.** $\dfrac{2}{7}$	**10.**	**a.** $\dfrac{7}{8}$	**b.** $\dfrac{3}{7}$
11.	**a.** $\dfrac{3}{25}$	**b.** $2\tfrac{1}{6}$	**12.**	**a.** $14\tfrac{2}{11}$	**b.** $52\tfrac{4}{9}$
13.	**a.** $\dfrac{2}{3}$	**b.** $2\tfrac{2}{13}$	**14.**	**a.** $\dfrac{15}{3}$	**b.** $\dfrac{12}{11}$
15.	**a.** $\dfrac{^-4}{5}$	**b.** $\dfrac{^-2}{3}$	**16.**	**a.** $\dfrac{^-17}{6}$	**b.** $\dfrac{^-14}{5}$

Express each of the numbers in Problems 17–26 as the quotient of two integers.

17.	**a.** .5	**b.** .8	**18.**	**a.** .25	**b.** .75
19.	**a.** .08	**b.** .12	**20.**	**a.** .65	**b.** 2.3
21.	**a.** .45	**b.** .234	**22.**	**a.** .111	**b.** .52
23.	**a.** 98.7	**b.** .63	**24.**	**a.** .24	**b.** 16.45
25.	**a.** 15.3	**b.** 6.95	**26.**	**a.** .64	**b.** 6.98

27. a. Is it possible to express .999 . . . as $\tfrac{9}{9}$? Are you surprised at this result? Consider the arguments developed in parts b–e.

    b. Show that $\tfrac{1}{3} = .333 . . . .$

c. Multiply both sides of the equation below by 3.

$$\frac{1}{3} = .333\ldots$$

Is your result consistent with part a?

d. Look for a pattern in the following. Verify that:

$$\frac{1}{9} = .111\ldots \qquad \frac{2}{9} = .222\ldots \qquad \frac{3}{9} = .333\ldots$$

$$\frac{4}{9} = .444\ldots \qquad \frac{5}{9} = .555\ldots \qquad \frac{6}{9} = .666\ldots$$

$$\frac{7}{9} = .777\ldots \qquad \frac{8}{9} = .888\ldots$$

What is $\frac{9}{9}$ according to the pattern?

e. Check your answer to this problem by division.

$$1 = .999\ldots$$

$$\frac{9}{9} = .999\ldots$$

and

$$
\begin{array}{r}
.999\ldots \\
9{\overline{\smash{\big)}\,9.000\ldots}} \\
\underline{8\,1} \\
90 \\
\underline{81} \\
90 \\
\underline{81} \\
9 \\
\cdot \\
\cdot \\
\cdot
\end{array}
$$

Is there anything "wrong" with this division? (*Note:* We usually require the remainder to be smaller than the number we're dividing, but this is a matter of definition and is not mandatory.)

28. In the text we stated that every rational number has a decimal representation that is either terminating or repeating. Study Table 4.4 on page 233 and make a conjecture about when a decimal will terminate and when it will repeat. Also, make a conjecture about the number of digits that repeat.

29. *Census taking for the set of rationals.* We've been writing the set of rationals by using the description methods. Devise a scheme that lists the set of rationals by roster. Limit your attention to those rationals between 0 and 1.

30. *Computer problem.* Write a BASIC program to output a roster of the set of rational numbers. A simple version of the program would allow the same rational to appear more than once (for example, $\frac{1}{2}$ also appears as $\frac{2}{4}, \frac{3}{6}, \frac{4}{8}, \ldots$). A more complex version of the program would eliminate such repetitions.

Mind Boggler

31. In 1907, the University of Göttingen offered the Wolfskehl Prize of 100,000 marks to anyone who could prove Fermat's Last Theorem, which seeks any replacements for a, b, and c such that $a^n + b^n = c^n$ (where n is greater than 2 and a, b, and c are counting numbers). In 1937, the mathematician Samuel Krieger announced that 1324, 731, and 1961 solved the equation. He would not reveal n—the power—but said that it was less than 20. That is,

$$1324^n + 731^n = 1961^n$$

However, it is easy to show that this cannot be a solution *for any n*. See if you can explain why by investigating some patterns for powers of numbers ending in 4 and 1.

Problem for Individual Study

32. Find any replacements for a, b, and c such that $a^n + b^n = c^n$, where n is greater than 2 and where a, b, and c are counting numbers. You might wish to do some research for this problem. Look for references to Fermat's Last Theorem. What can you find out about this problem?

Chapter 4 Outline

I. **Early Numeration Systems**
  A. The difference between a number and a numeral
  B. Types of numeration systems
    1. Simple grouping system, such as the Egyptian numeration system
    2. Positional numeration system, such as the Babylonian numeration system
    3. Additive and subtractive properties of a numeration system
II. **Hindu-Arabic Numeration System**
  A. Characteristics
    1. Ten symbols (decimal system)
    2. Positional
    3. Additive
  B. Expanded notation
    1. Used to tell what a number means (counting objects)
    2. Decimal point
    3. Exponents
III. **Large and Small Numbers**
  A. Scientific notation
    1. Changing from decimal notation to scientific notation
    2. Changing from scientific notation to decimal notation
  B. Floating point notation
    1. Calculators
    2. Computers (BASIC)

Historical Note

Pierre de Fermat (1601–1665) was a lawyer by profession, but he was an amateur mathematician in his spare time. He became Europe's finest mathematician, and he wrote well over 3,000 mathematical papers and notes. However, he published only one, because he did them just for fun. Every theorem that Fermat said he proved has subsequently been verified—with one notable exception, *Fermat's Last Theorem*. In the margin of a book he wrote:

> To divide a cube into two cubes, a fourth power, or in general any power whatever above the second, into powers of the same denomination, is impossible, and I have assuredly found an admirable proof of this, but the margin is too narrow to contain it.

Many of the most prominent mathematicians since his time have tried to prove or disprove this conjecture, but the answer is still unresolved. However, in 1983 a 29-year-old German mathematician named Gerd Faltings proved that the number of possible solutions is finite.

   C. Comparing the size of large numbers
      1. Millions; 10^6
      2. Billions; 10^9
      3. Names of large numbers
      4. Googol
IV. Natural Numbers
   A. *Definition:* $N = \{1,2,3,4, \ldots\}$
   B. Operations
      1. Addition
      2. Multiplication: $a \times b$ means $\underbrace{a + a + \cdots + a}_{b \text{ addends}}$

      3. Subtraction: $a - b = x$ means $a = b + x$

      4. Division: $\dfrac{a}{b} = x$ means $a = b \cdot x$

        ($b \neq 0$: never divide by zero)
   C. Properties—Let a, b, and c be natural numbers
      1. Closure for addition: $a + b$ is a natural number
      2. Closure for multiplication: ab is a natural number
      3. Commutative for addition: $a + b = b + a$
      4. Commutative for multiplication: $ab = ba$
      5. Associative for addition: $(a + b) + c = a + (b + c)$
      6. Associative for multiplication: $(ab)c = a(bc)$
      7. Distributive: $a(b + c) = ab + ac$
V. Integers
   A. *Definition:* $Z = \{\ldots, {}^-3, {}^-2, {}^-1, 0, 1, 2, 3, \ldots\}$
   B. Absolute value of x is the undirected distance of x from zero.
   C. Operations
      1. Addition
         a. On a number line
         b. Same signs; add absolute values
         c. Opposite signs; subtract absolute values
      2. Multiplication and division
         a. Same signs; positive
         b. Opposite signs; negative
      3. Subtraction; add the opposite
VI. Rational Numbers
   A. *Definition:* $Q = \left\{ \dfrac{p}{q} \mid p, q \in Z, q \neq 0 \right\}$

   B. Operations—Let a/b and c/d be rational numbers
      1. Addition: $\dfrac{a}{b} + \dfrac{c}{d} = \dfrac{ad + bc}{bd}$

      2. Subtraction: $\dfrac{a}{b} - \dfrac{c}{d} = \dfrac{ad - bc}{bd}$

3. Multiplication: $\dfrac{a}{b} \cdot \dfrac{c}{d} = \dfrac{ac}{bd}$

4. Division: $\dfrac{\dfrac{a}{b}}{\dfrac{c}{d}} = \dfrac{ad}{bc}$ $(c \neq 0)$

*VII. Irrational Numbers
   A. Pythagorean theorem. In a right triangle where c is the length of the hypotenuse and a and b the lengths of the other two sides, $a^2 + b^2 = c^2$.
   B. Square root
   C. Even and odd integers
   D. Irrational: A number whose decimal representation does not terminate and does not repeat.

VIII. Real Numbers
   A. *Definition:*   The set of real numbers, R, is defined by $R = Q \cup Q'$, where $Q = \{$Rationals$\}$ and $Q' = \{$Irrationals$\}$
   B. Fractional and decimal representation of rational numbers
      1. Change from fractional to decimal form.
      2. Change from a terminating decimal to fractional form.
   C. Classifying real numbers
   D. Real number line
      1. Rational numbers are dense.
      2. Every real number corresponds to a point on the line, and every point on the line corresponds to a real number.
   E. Relationships among sets of numbers; see Figure 4.19 on page 234.

Chapter 4 Test

1. a. What do we mean when we say that a numeration system is positional? Give examples.
    b. Is addition easier in a positional system or in a grouping system? Discuss and show examples.
    c. What are some of the characteristics of the Hindu-Arabic numeration system?

2. a. Write 436.20001 in expanded notation.
    b. Write $(4 \times 10^6) + (2 \times 10^4) + (5 \times 10^0) + (6 \times 10^{-1}) + (2 \times 10^{-2})$ in decimal notation.

3. In your own words, explain the closure, commutative, and associative properties.

*Optional section.

4. Write in expanded form.

    a. 5.79×10^{-4} **b.** 4.01×10^{5} **c.** 10^{-1} **d.** 4.321×10^{7}

5. Write in scientific notation.

    a. .0034   **b.** 4,000,300   **c.** 17,400   **d.** 5

6. Let n be any counting number. Define:

    a. b^{n} **b.** b^{-n} **c.** b^{0} $(b \neq 0)$
    d. In the set of rational numbers, why is division by zero excluded?

Perform the indicated operations in each of Problems 7–9.

7. **a.** $8 + {}^{-}15$ **b.** $8 - {}^{-}15$ **c.** ${}^{-}10 + {}^{-}7$ **d.** ${}^{-}10 - {}^{-}7$
   e. ${}^{-}14 + 6$ **f.** ${}^{-}14 - 6$ **g.** $7 - {}^{-}2$ **h.** $({}^{-}5) \times ({}^{-}3)$
   i. ${}^{-}5 + {}^{-}9$ **j.** $\dfrac{52}{{}^{-}13}$

8. **a.** $({}^{-}5) \times (3) - ({}^{-}4)$ **b.** $5 - \dfrac{{}^{-}11}{5}$ **c.** $\dfrac{4}{7} + \dfrac{5}{9}$

   d. $\dfrac{3}{4} - \dfrac{7}{9}$ **e.** $\dfrac{5}{3} \times 7$

9. **a.** $\dfrac{4}{5} \times 3$ **b.** $\dfrac{2}{3} - \dfrac{3}{5}$ **c.** $\dfrac{1}{2} + \dfrac{1}{3} + \dfrac{1}{5}$

   d. $\dfrac{\frac{5}{8}}{\frac{1}{3}}$ **e.** $\dfrac{\frac{7}{8}}{\frac{1}{2}}$

10. **a.** Express $\frac{5}{8}$ as a decimal.
    b. Express .666 as a quotient of two integers.
    c. Simplify $\left(\dfrac{11}{12} + 2\right) + \dfrac{{}^{-}11}{12}$
    d. Simplify $\left(\dfrac{152}{313} + 7\right) + \dfrac{{}^{-}152}{313}$
    e. Simplify $\dfrac{3}{5} \times \dfrac{25}{97} + \dfrac{3}{5} \times \dfrac{72}{97}$

Bonus Questions

*11. A baseball diamond is actually a square, with 90 ft between each base. Compute the distance, to the nearest foot, between the indicated points on the field.

    a. First base to third base
    b. From home plate to second base

12. Define a *happy integer* as follows: Take any number n. Find the sum of the squares of the digits of n. Take this result and find the sum of the squares of the digits of it. Continue this process on each successive result. If 1 is eventually

---

*Requires optional Section 4.7.

reached, then n is a happy integer. For example,

$$13 \to 1^2 + 3^2 = 10 \to 1^2 + 0^2 = 1$$

Thus, 13 is happy.

$$12 \to 1^2 + 2^2 = 5 \to 25 \to 2^2 + 5^2 = 29 \to 85 \to 89$$
$$\to 145 \to 42 \to 20 \to 4 \to 16 \to 37 \to 58 \to 89$$

At this point you can stop because you are stuck in a loop (89 is a repeat). Therefore, 14 is unhappy. Find the happy integers between 1 and 10 (inclusive).

Date	Cultural History
1200	**1206:** Ghengis Khan becomes chief prince of the Mongols **1209:** St. Francis initiates brotherhood **1233:** Start of the Papal inquisition **1240:** Amiens cathedral rebuilt
1250	 **1273:** Thomas Aquinas, *Summa Theologicae* **1275:** Moses de Leon *Zohar,* major source for the cabala **1299:** Florentine bankers are forbidden to use Hindu numerals
1300	**1307–1321:** Dante's *Divine Comedy* **1322:** The pope forbids the use of counterpoint in church music **1347–1351:** Approximately 75 million die of black plague
1350	**1364:** Aztecs build Tenochtitlán **1378:** Beginning of the great schism **1390:** Chaucer's *Canterbury Tales*

St. Francis,
Dürer

Cabalistic emblem

Dante,
Giotto

Aztec sacrifice

Mathematical History

The Nature of Daily Arithmetic

1202: Fibonacci: arithmetic, algebra, geometry, sequences, *Liber Abaci*

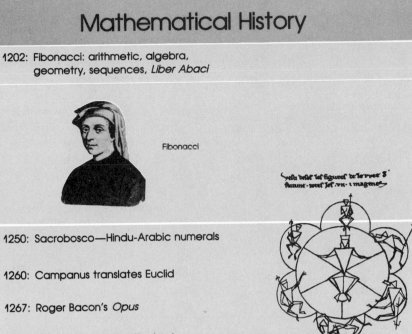

Fibonacci

1250: Sacrobosco—Hindu-Arabic numerals

1260: Campanus translates Euclid

1267: Roger Bacon's *Opus*

1280: Geometry used as the basis of painting

1281: Li Yeh introduces notation for negative numbers

Wheel of Fortune, de Honnecourt

1303: Chu Shï-Kié—algebra, solutions of equations, Pascal's triangle

1325: Thomas Bradwardine—arithmetic, geometry, star polygons

Pascal triangle

1360: Nicole Oresme—coordinates, fractional exponents

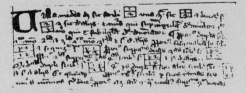

Oresme, *Algorismus proportionum*

There was an old man who said, "Do Tell me how I should add two and two? I think more and more That it makes about four— But I fear that is almost too few."

Anonymous

*5.1 Interest

Although not required, a
calculator will be very helpful for
the material in this chapter.

There are certain arithmetic skills that will enable us to make intelligent decisions about how we spend the money we earn. One of the most fundamental mathematical concepts that consumers, as well as businesspeople, must understand is the notion of interest. Simply stated, **interest** is money paid for the use of money. That is, we receive interest when we let others use our money (when we deposit money in a savings account, for example), and we pay interest when we use the money of others (for example, when we borrow from a bank).

The amount of the deposit or loan is called the **principal,** and the interest rate is stated as a percentage of the principal. The fundamental interest formula is

$$I = P \times r \times t \quad \text{or} \quad Prt$$

PRINCIPAL

The interest rate may be
stated as a percent (8%) or a
decimal (.08). *Percent* means
"hundredth," so

8 percent = 8 hundredths
= .08

In this book, interest rates are
annual interest rates unless
stated otherwise.

where I = Amount of interest
P = Principal
r = Interest rate
t = Time (in years)

Look at the cartoon; suppose Blondie saves 20¢ per day, but only for a year. At the end of the year she will have saved $73. If she then puts the money into a savings account at 8% interest, how much interest does the bank pay her after one year?

$$I = Prt$$
$$= 73(.08)(1)$$

You can do this computation on a calculator by pressing

[73] [×] [.08] [=]   Display: 5.84

to obtain the result $5.84.

Example 1: How much interest does Blondie earn on her initial deposit of $73 in three years?

On a deposit of $73, the total
amount she would have in three
years at simple interest is

$73 + $17.52 = $90.52

Solution $I = Prt$
$= 73(.08)(3)$
$= 17.52$

She earns $17.52 interest.

*This section requires Section 4.3.

This is an example of *simple interest*, but banks pay *compound interest* on savings accounts. Suppose Blondie's bank pays 8% interest *compounded annually*. This means that, at the end of the first year her money is in the bank, she has

$$\text{Principal} + \text{Interest} = 73.00 + 5.84 = 78.84$$

This amount becomes the principal for the second year:

$$I = Prt$$
$$= 78.84 \times (.08) \times (1)$$
$$= 6.3072$$

Notice that, for compound interest, interest is paid on the interest earned during the first period.

or $6.31. For the third year she has $78.84 + $6.31 = $85.15 on which she earns interest:

$$I = Prt$$
$$= 85.15 \times (.08) \times (1)$$
$$= 6.812$$

or $6.81. The total amount she would have in three years compounded annually is $85.15 + $6.81 = $91.96.

Notice that this is $1.44 more than when the amount was compounded at simple interest.

The process described here is fairly easy to understand but rather tedious to apply, particularly if the number of years is very large. However, there is a formula that gives this amount directly. The amount, A, present after the principal, P, is invested for t years at an interest rate of r compounded annually is

$$A = P(1 + r)^t$$

We can verify this formula for the example of Blondie's interest after three years:

$$A = 73(1 + .08)^3$$
$$= 73(1.08)^3$$
$$= 73(1.08) \times (1.08) \times (1.08)$$

On a calculator,

The result is 91.958976 or $91.96.

"The boy that by addition grows
And suffers no subtraction
Who multiplies the thing he knows
And carries every fraction
Who well divides the precious time
The due proportion giving
To sure success aloft will climb
Interest compound receiving."

At this point you might be saying to yourself that this looks like a considerable amount of arithmetic and you don't even own a calculator. Although the use of calculators to work problems such as these is almost universal, it is possible to use interest tables instead.

Table 5.1 shows compound interest for $1 for n periods. This means that, for the previous example, find the column headed 8% and look in the row $n = 3$. The entry is 1.259712. Multiply this number by the principal (73) to obtain 91.958976 or $91.96.

TABLE 5.1   Compound Interest Table:
Compounded Amount of $1.00 for *n* Periods

n	1%	1½%	2%	2½%	3%	3½%	4%	*n*
1	1.010000	1.015000	1.020000	1.025000	1.030000	1.035000	1.040000	1
2	1.020100	1.030225	1.040400	1.050625	1.060900	1.071225	1.081600	2
3	1.030301	1.045678	1.061208	1.076891	1.092727	1.108718	1.124864	3
4	1.040604	1.061364	1.082432	1.103813	1.125509	1.147523	1.169859	4
5	1.051010	1.077284	1.104081	1.131408	1.159274	1.187686	1.216653	5
6	1.061520	1.093443	1.126162	1.159693	1.194052	1.229255	1.265319	6
7	1.072135	1.109845	1.148686	1.188686	1.229874	1.272279	1.315932	7
8	1.082857	1.126493	1.171659	1.218403	1.266770	1.316809	1.368569	8
9	1.093685	1.143390	1.195093	1.248863	1.304773	1.362897	1.423312	9
10	1.104622	1.160541	1.218994	1.280085	1.343916	1.410599	1.480244	10
11	1.115668	1.177949	1.243374	1.312087	1.384234	1.459970	1.539454	11
12	1.126825	1.195618	1.268242	1.344889	1.425761	1.511069	1.601032	12
13	1.138093	1.213552	1.293607	1.378511	1.468534	1.563956	1.665074	13
14	1.149474	1.231756	1.319479	1.412791	1.512590	1.618695	1.731676	14
15	1.160969	1.250232	1.345868	1.448298	1.557967	1.675349	1.800944	15
16	1.172579	1.268986	1.372786	1.484506	1.604706	1.733986	1.872981	16
17	1.184304	1.288020	1.400241	1.521618	1.652848	1.794676	1.947900	17
18	1.196147	1.307341	1.428246	1.559659	1.702433	1.857489	2.025817	18
19	1.208109	1.326951	1.456811	1.598650	1.753506	1.922501	2.106849	19
20	1.220190	1.346855	1.485947	1.638616	1.806111	1.989789	2.191123	20
25	1.282432	1.450945	1.640606	1.853944	2.093778	2.363245	2.665836	25
30	1.347849	1.563080	1.811362	2.097568	2.427262	2.806794	3.243398	30
35	1.416603	1.683881	1.999890	2.373205	2.813862	3.333590	3.946089	35
40	1.488864	1.814018	2.208040	2.685064	3.262038	3.959260	4.801021	40
45	1.564811	1.954213	2.437854	3.037903	3.781596	4.702359	5.841176	45
50	1.644632	2.105242	2.691588	3.437109	4.383906	5.584927	7.106683	50
55	1.728525	2.267944	2.971731	3.888773	5.082149	6.633141	8.646367	55
60	1.816697	2.443220	3.281031	4.399790	5.891603	7.878091	10.519627	60
65	1.909366	2.632042	3.622523	4.977958	6.829983	9.356701	12.798735	65
70	2.006763	2.835456	3.999558	5.632103	7.917822	11.112825	15.571618	70
75	2.109128	3.054592	4.415835	6.372207	9.178926	13.198550	18.945255	75
80	2.216715	3.290663	4.875439	7.209568	10.640891	15.675738	23.049799	80
85	2.329790	3.544978	5.382879	8.156964	12.335709	18.617859	28.043605	85
90	2.448633	3.818949	5.943133	9.228856	14.300467	22.112176	34.119333	90
95	2.573538	4.114092	6.561699	10.441604	16.578161	26.262329	41.511386	95
100	2.704814	4.432046	7.244646	11.813716	19.218632	31.191408	50.504948	100

n	4½%	5%	5½%	6%	6½%	7%	7½%	*n*
1	1.045000	1.050000	1.055000	1.060000	1.065000	1.070000	1.075000	1
2	1.092025	1.102500	1.113025	1.123600	1.134225	1.144900	1.155625	2
3	1.141166	1.157625	1.174241	1.191016	1.207950	1.225043	1.242297	3
4	1.192519	1.215506	1.238825	1.262477	1.286466	1.310796	1.335469	4
5	1.246182	1.276282	1.306960	1.338226	1.370087	1.402552	1.435629	5
6	1.302260	1.340096	1.378843	1.418519	1.459142	1.500730	1.543302	6
7	1.360862	1.407100	1.454679	1.503630	1.553987	1.605781	1.659049	7
8	1.422101	1.477455	1.534687	1.593848	1.654996	1.718186	1.783478	8
9	1.486095	1.551328	1.619094	1.689479	1.762570	1.838459	1.917239	9
10	1.552969	1.628895	1.708144	1.790848	1.877137	1.967151	2.061032	10
11	1.622853	1.710339	1.802092	1.898299	1.999151	2.104852	2.215609	11
12	1.695881	1.795856	1.901207	2.012196	2.129096	2.252192	2.381780	12
13	1.772196	1.885649	2.005774	2.132928	2.267487	2.409845	2.560413	13
14	1.851945	1.979932	2.116091	2.260904	2.414874	2.578534	2.752444	14
15	1.935282	2.078928	2.232476	2.396558	2.571841	2.759032	2.958877	15
16	2.022370	2.182875	2.355263	2.540352	2.739011	2.952164	3.180793	16
17	2.113377	2.292018	2.484802	2.692773	2.917046	3.158815	3.419353	17
18	2.208479	2.406619	2.621466	2.854339	3.106654	3.379932	3.675804	18
19	2.307860	2.526950	2.765647	3.025600	3.308587	3.616528	3.951489	19
20	2.411714	2.653298	2.917757	3.207135	3.523645	3.869684	4.247851	20
25	3.005434	3.386355	3.813392	4.291871	4.827699	5.427433	6.098340	25
30	3.745318	4.321942	4.983951	5.743491	6.614366	7.612255	8.754955	30
35	4.667348	5.516015	6.513825	7.686087	9.062255	10.676581	12.568870	35
40	5.816365	7.039989	8.513309	10.285718	12.416075	14.974458	18.044239	40
45	7.248248	8.985008	11.126554	13.764611	17.011098	21.002452	25.904839	45
50	9.032636	11.467400	14.541961	18.420154	23.306679	29.457025	37.189746	50
55	11.256308	14.635631	19.005762	24.650322	31.932170	41.315001	53.390690	55
60	14.027408	18.679186	24.839770	32.987691	43.749840	57.946427	76.649240	60
65	17.480702	23.839901	32.464587	44.144972	59.941072	81.272861	110.039897	65
70	21.784136	30.426426	42.429916	59.075930	82.124463	113.989392	157.976504	70
75	27.146996	38.832686	55.454204	79.056921	112.517632	159.876019	226.795701	75
80	33.830096	49.561441	72.476426	105.795993	154.158907	224.234388	325.594560	80
85	42.158455	63.254353	94.723791	141.578904	211.211062	314.500328	467.433099	85
90	52.537105	80.730365	123.800206	189.464511	289.377460	441.102980	671.060665	90
95	65.470792	103.034676	161.801918	253.546255	396.472198	618.669748	963.394370	95
100	81.588518	131.501258	211.468636	339.302084	543.201271	867.716326	1383.077210	100

n	8%	9%	10%	11%	12%	13%	14%	n
1	1.080000	1.090000	1.100000	1.110000	1.120000	1.130000	1.140000	1
2	1.166400	1.188100	1.210000	1.232100	1.254400	1.276900	1.299600	2
3	1.259712	1.295029	1.331000	1.367631	1.404928	1.442897	1.481544	3
4	1.360489	1.411582	1.464100	1.518070	1.573519	1.630474	1.688960	4
5	1.469328	1.538624	1.610510	1.685058	1.762342	1.842435	1.925415	5
6	1.586874	1.677100	1.771561	1.870415	1.973823	2.081952	2.194973	6
7	1.713824	1.828039	1.948717	2.076160	2.210681	2.352605	2.502269	7
8	1.850930	1.992563	2.143589	2.304538	2.475963	2.658444	2.852586	8
9	1.999005	2.171893	2.357948	2.558037	2.773079	3.004042	3.251949	9
10	2.158925	2.367364	2.593742	2.839421	3.105848	3.394567	3.707221	10
11	2.331639	2.580426	2.853117	3.151757	3.478550	3.835861	4.226232	11
12	2.518170	2.812665	3.138428	3.498451	3.895976	4.334523	4.817905	12
13	2.719624	3.065805	3.452271	3.883280	4.363493	4.898011	5.492411	13
14	2.937194	3.341727	3.797498	4.310441	4.887112	5.534753	6.261349	14
15	3.172169	3.642482	4.177248	4.784589	5.473566	6.254270	7.137938	15
16	3.425943	3.970306	4.594973	5.310894	6.130394	7.067326	8.137249	16
17	3.700018	4.327633	5.054470	5.895093	6.866041	7.986078	9.276464	17
18	3.996019	4.717120	5.559917	6.543553	7.689966	9.024268	10.575169	18
19	4.315701	5.141661	6.115909	7.263344	8.612762	10.197423	12.055693	19
20	4.660957	5.604411	6.727500	8.062312	9.646293	11.523088	13.743490	20
25	6.848475	8.623081	10.834706	13.585464	17.000064	21.230542	26.461916	25
30	10.062657	13.267678	17.449402	22.892297	29.959922	39.115898	50.950159	30
35	14.785344	20.413968	28.102437	38.574851	52.799620	72.068506	98.100178	35
40	21.724521	31.409420	45.259256	65.000867	93.050970	132.781552	188.883514	40
45	31.920449	48.327286	72.890484	109.530242	163.987604	244.641402	363.679072	45
50	46.901613	74.357520	117.390853	184.564827	289.002190	450.735925	700.232988	50
55	68.913856	114.408262	189.059142	311.002466	509.320606	830.451725	1348.238807	55
60	101.257064	176.031292	304.481640	524.057242	897.596933	1530.053473	2595.918660	60
65	148.779847	270.845963	490.370725	883.066930	1581.872491	2819.024345	4998.219642	65
70	218.606406	416.730086	789.746957	1488.019132	2787.799828	5193.869624	9623.644985	70
75	321.204530	641.190893	1271.895371	2507.398773	4913.055841	9569.368113	18529.506390	75
80	471.954834	986.551668	2048.400215	4225.112750	8658.483100	17630.940454	35676.981807	80
85	693.456489	1517.932029	3298.969030	7119.560696	15259.205681	32483.864937	68692.981028	85
90	1018.915089	2335.526582	5313.022612	11996.873812	26891.934223	59849.415520	132262.467379	90
95	1497.120549	3593.497147	8556.676047	20215.430053	47392.776624	110268.668614	254660.083396	95
100	2199.761256	5529.040792	13780.612340	34064.175270	83522.265727	203162.874228	490326.238126	100

n	15%	16%	17%	18%	19%	20%	n
1	1.150000	1.160000	1.170000	1.180000	1.190000	1.200000	1
2	1.322500	1.345600	1.368900	1.392400	1.416100	1.440000	2
3	1.520875	1.560896	1.601613	1.643032	1.685159	1.728000	3
4	1.749006	1.810639	1.873887	1.938778	2.005339	2.073600	4
5	2.011357	2.100342	2.192448	2.287758	2.386354	2.488320	5
6	2.313061	2.436396	2.565164	2.699554	2.839761	2.985984	6
7	2.660020	2.826220	3.001242	3.185474	3.379315	3.583181	7
8	3.059023	3.278415	3.511453	3.758859	4.021385	4.299817	8
9	3.517876	3.802961	4.108400	4.435454	4.785449	5.159780	9
10	4.045558	4.411435	4.806828	5.233836	5.694684	6.191736	10
11	4.652391	5.117265	5.623989	6.175926	6.776674	7.430084	11
12	5.350250	5.936027	6.580067	7.287593	8.064242	8.916100	12
13	6.152788	6.885791	7.698679	8.599359	9.596448	10.699321	13
14	7.075706	7.987518	9.007454	10.147244	11.419773	12.839185	14
15	8.137062	9.265521	10.538721	11.973748	13.589530	15.407022	15
16	9.357621	10.748004	12.330304	14.129023	16.171540	18.488426	16
17	10.761264	12.467685	14.426456	16.672247	19.244133	22.186111	17
18	12.375454	14.462514	16.878953	19.673251	22.900518	26.623333	18
19	14.231772	16.776517	19.748375	23.214436	27.251616	31.948000	19
20	16.366537	19.460759	23.105599	27.393035	32.429423	38.337600	20
25	32.918953	40.874244	50.657826	62.668627	77.388073	95.396217	25
30	66.211772	85.849877	111.064650	143.370638	184.675312	237.376314	30
35	133.175523	180.314073	243.503474	327.997290	440.700607	590.668229	35
40	267.863546	378.721158	533.868713	750.378345	1051.667507	1469.771568	40
45	538.769269	795.443826	1170.479411	1716.683879	2509.650603	3657.261988	45
50	1083.657442	1670.703804	2566.215284	3927.356860	5988.913902	9100.438150	50
55	2179.622184	3509.048796	5626.293659	8984.841120	14291.666609	22644.802257	55
60	4383.998746	7370.201365	12335.356482	20555.139966	34104.970919	56347.514353	60
65	8817.787387	15479.940952	27044.628088	47025.180900	81386.522174	140210.646915	65
70	17735.720039	32513.164839	59293.941729	107582.222368	194217.025056	348888.956932	70
75	35672.867976	68288.754533	129998.886072	246122.063716	463470.508558	868147.369314	75
80	71750.879401	143429.715890	285015.802412	563067.660386	1106004.544354	2160228.462010	80
85	144316.646994	301251.407222	624882.336142	1288162.407650	2639317.992285	5375339.686589	85
90	290272.325206	632730.879999	1370022.050417	2947003.540121	6298346.150529	13375565.248934	90
95	583841.327636	1328951.025313	3003702.153303	6742030.208228	15030081.387632	33282686.520228	95
100	1174313.450700	2791251.199375	6585460.885837	15424131.905453	35867089.727971	82817974.522015	100

Comparison of Simple and Compound Interest

	Interest, I	Amount, A
Simple interest formula	$I = P \times r \times t$ This is found first for simple interest.	$A = P + I$ This is found after you've used the simple interest formula.
Compound interest	$I = A - P$ This is found after you've used Table 5.1.	Use Table 5.1. This is found first for compound interest. $A = \left(\dfrac{\text{TABLE}}{\text{NUMBER}}\right) \times P$

Example 2: You are considering a six-year, $1,000 certificate of deposit paying 12% compounded annually. How much money will you have at the end of six years?

Solution: By calculator,

$$A = 1,000(1.12)^6$$

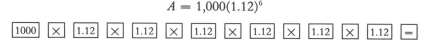

The result 1973.822685 to the nearest cent is $1,973.82.

By Table 5.1, find the column headed 12% and the row headed $n = 6$ to find 1.973823. Multiply this by 1,000 to find $1,973.82.

The disadvantages of working with tables are (1) a different table is needed for each different rate, (2) tables are not as readily available as calculators, and (3) even when using a table, a final multiplication by the principal is necessary.

If you have deposited money or seen savings and loan advertisements lately, you know that most financial institutions will compound interest more frequently than annually. This is to your advantage, because the shorter the compounding period, the sooner you earn interest on your interest. If Blondie's money is deposited in an 8% account that compounds semiannually, then the period is one-half year, and each period pays 4%. The advantage to Blondie is that she begins earning interest on the interest after one-half year instead of after one year. In general, if the money is compounded n times per year, then the rate per period is r/n and the interest formula becomes

$$A = P\left(1 + \frac{r}{n}\right)^{nt}$$

This formula is rather difficult to use (especially if you don't have a calculator with an exponent key). Examples 3 and 4 will show you how to use Table 5.1 for compounding periods of less than one year.

Example 3: Find the number of periods that interest is compounded and give the interest rate per period.

Some calculators have an exponent key $\boxed{y^x}$. If your calculator has this key, you can do Example 2 by pressing

$\boxed{1.12}$ $\boxed{y^x}$ $\boxed{6}$ $\boxed{\times}$ $\boxed{1000}$ $\boxed{=}$

to obtain the result.

Solution

Compounding period	Yearly rate	Time	Number of periods	Rate per period
a. Annual	12%	3 yr	3	12%
b. Semiannual	12%	3 yr	6	6%
c. Quarterly	12%	3 yr	12	3%
d. Monthly	12%	3 yr	36	1%
e. Daily	12%	3 yr	1,080	.03%

To find the number of periods: Multiply the time by

    2 if semiannual
    4 if quarterly
  12 if monthly
360 if daily

For Example 3e, $3 \times 360 = 1,080$.

To find the rate per period: Divide the yearly rate by

    2 if semiannual
    4 if quarterly
  12 if monthly
360 if daily

For Example 3e, $12 \div 360 = .033333\ldots$

Notice that when paying interest, banks use 360 for the number of days in a year; this is called *ordinary interest.* Unless instructed otherwise, use 360 for the number of days in a year. If 365 days are used, it is called *exact interest.*

Example 4: Use Table 5.1 to find the amount you would have if you invested $1,000 for ten years at 8% interest.

a. Compounded annually
$n = 10$, and the rate is 8%:

$$\text{AMOUNT} = \$1,000 \times 2.158925$$
$$= \$2,158.93$$

b. Compounded semiannually
$n = 20$, and the rate per period is 4%:

$$\text{AMOUNT} = \$1,000 \times 2.191123$$
$$= \$2,191.12$$

c. Compounded quarterly
$n = 40$, and the rate per period is 2%:

$$\text{AMOUNT} = \$1,000 \times 2.208040$$
$$= \$2,208.04$$

Table 5.1 is not extensive enough to find the monthly or daily compounding of $1,000 at 8% annual interest.

You have, no doubt, seen advertisements that say

$$8.00\% = 8.33\%^*$$

*Effective annual yield when principal and interest are left in the account.

ALSO NEW. HIGHEST INTEREST EVER ON INSURED SAVINGS.

8.33% annual yield on

8% interest compounded daily

Annual yield based on daily compounding when funds and interest remain on deposit a year. Note: Federal regulations require a substantial interest penalty for early withdrawal of principal from Certificate Accounts.

In order to understand **effective annual yield,** look at the earning of $1 at 8% compounded quarterly:

Annually

$$(1)(1 + .08) = 1.08$$

Quarterly

$$(1)\left(1 + \frac{.08}{4}\right)^4 = (1.02)^4$$
$$= 1.08243216$$

The $1 compounded quarterly at 8% is the same as $1 compounded annually at a rate of 8.243216%. Therefore, 8.24% is called the *effective yield* of 8% compounded quarterly.

Example 5: Verify the claims of the bank in the advertisement in the margin. You will need a calculator with an exponent key for this example.

Solution: The compounding is on a daily basis ($n = 360$), so $1 compounded daily is

$$(1)\left(1 + \frac{.08}{360}\right)^{360} \approx (1.00022222)^{360}$$
This calculation was done on a calculator with an exponent key.

$$\approx 1.083277348$$
≈ means approximately equal to.

This means that if $1 was compounded annually at 8.3277348% the interest would be the same as $1 compounded daily at 8%. The bank is justified in advertising that 8% compounded daily has an effective yield of 8.33%.

The same procedure that we use to calculate interest can also be used to calculate inflation rates. The government releases monthly or annual inflation rates. Example 6 illustrates the procedure.

Example 6: Suppose a person was earning $30,000 a year in 1984. What would her salary need to be in 1990 if we assume a constant 9% inflation rate?

Solution: Use a calculator or Table 5.1.

By calculator: $30,000(1 + .09)^6$ Display: `50313.00333`

By Table 5.1: From the column headed 9% and the row headed $n = 6$, find
1.6771561. Multiply this by $30,000: 50,313

Her salary would need to be about $50,313 to maintain her standard of living.

Problem Set 5.1

*In Problems 1–6, compare the amount of simple interest and the interest if the investment
is compounded annually.*

1. $1,000 at 8% for 5 years

2. $5,000 at 10% for 3 years

3. $2,000 at 12% for 3 years

4. $2,000 at 12% for 5 years

5. $5,000 at 12% for 20 years

6. $1,000 at 14% for 30 years

*In Problems 7–12, compare the amount you would have if the money were invested at
simple interest or invested so that it is compounded annually.*

7. $1,000 at 8% for 5 years

8. $5,000 at 10% for 3 years

9. $2,000 at 12% for 3 years

10. $2,000 at 12% for 5 years

11. $5,000 at 12% for 20 years

12. $1,000 at 14% for 30 years

Fill in the blanks for Problems 13–26.

	Compounding period	Principal	Yearly rate	Time	Period rate	Number of periods	Entry in Table 5.1	Total amount	Total interest
13.	Annual	$1,000	9%	5 yr	——	——	——	——	——
14.	Semiannual	$1,000	9%	5 yr	——	——	——	——	——
15.	Annual	$ 500	8%	3 yr	——	——	——	——	——
16.	Semiannual	$ 500	8%	3 yr	——	——	——	——	——
17.	Quarterly	$ 500	8%	3 yr	——	——	——	——	——
18.	Semiannual	$3,000	18%	3 yr	——	——	——	——	——
19.	Quarterly	$5,000	18%	10 yr	——	——	——	——	——
20.	Quarterly	$ 624	16%	5 yr	——	——	——	——	——
21.	Quarterly	$5,000	20%	10 yr	——	——	——	——	——
22.	Monthly	$ 350	12%	5 yr	——	——	——	——	——
23.	Monthly	$4,000	24%	5 yr	——	——	——	——	——
24.	Quarterly	$ 800	12%	90 days	——	——	——	——	——
25.	Quarterly	$1,250	16%	450 days	——	——	——	——	——
26.	Quarterly	$1,000	12%	900 days	——	——	——	——	——

The prices for certain items in 1984 are given in Problems 27–33. Calculate the expected price of each item in the year 2001, assuming a 6% rate of inflation.

27. Sunday paper, $.75

28. Big Mac, $1.45

29. Monthly rent for a one-bedroom apartment, $285

30. Tuition at a private college, $4,000

31. Small car, $7,000

32. Yearly salary $25,000

33. Average house, $110,000

34. If $5,000 is compounded annually at $5\frac{1}{2}$% for 12 years, what is the total interest received at the end of that time?

35. If $10,000 is compounded annually at 8% for 18 years, what is the total interest received at the end of that time?

36. *Calculator problem.* What is the effective yield of 6% compounded quarterly?

37. *Calculator problem.* What is the effective yield of 6% compounded monthly?

38. *Calculator problem.* What is the effective yield of 8% compounded monthly?

To work Problems 39 and 40 you will need a calculator with an exponent key.

39. *Calculator problem.* Is is better to deposit your money at $5\frac{1}{2}$% compounded annually or $5\frac{1}{4}$% compounded daily?

40. *Calculator problem.* How much money would you have in six years if you purchased a $1,000 six-year savings certificate that paid 8% compounded daily?

41. *Calculator problem.* Blondie's $73 will grow to $78.84 at 8% interest if the interest is compounded annually, but it will grow to $79.08 at the *same* interest if the interest is compounded daily. This doesn't seem like a great deal of difference, but let's take the viewpoint of the savings and loan company paying this interest. One savings and loan advertises "Seven Billion Dollars Strong." If the company has $7 billion on deposit and it pays 8% interest compounded daily, how much more money will they pay out than if they were to pay interest compounded annually?

Mind Bogglers

42. Some savings and loan companies advertise that they pay interest *continuously.* Explain what this means.

43. *Calculator problem.* The problem we solved concerning the Blondie cartoon was simplified for ease of calculation. We assumed that she made a single $73 deposit at the end of a year of saving in a cookie jar. The cartoon, however, implies she will make a $73 deposit *every* year. That is, Blondie will make a series of equal payments over equal time intervals. This is called an **annuity,** A, where P dollars per year are invested at interest rate r compounded continuously over t years. The total amount at the end of t years is approximated by the formula

$$A = \frac{P}{r}(2.718^{rt} - 1)$$

For example, if Blondie deposits $73 every year for ten years at 8%, the amount at the end of ten years would be

$$A = \frac{73}{0.08}(2.718^{0.08(10)} - 1)$$

$$\approx \$1,118.14$$

a. For how many years would Blondie have to deposit $73 a year in order to save $3,650 at 8% interest compounded continuously?

b. How much would she have actually deposited when her annuity reaches $3,650?

c. If inflation was a constant 6% per year during the period Blondie was saving, how much would that same coat cost if the price rose to keep even with inflation?

Problems for Individual Study

44. Conduct a survey of banks, savings and loan companies, and credit unions in your area. Prepare a report on the different types of savings accounts available and the interest rates they pay. Include methods of payment as well as the interest rates.

45. Consult an almanac or government source, and then write a report on the current inflation rate. Project some of these results to the year of your own expected retirement.

5.2 Installment Buying

In the last section we discussed interest in terms of savings accounts. In this section we'll discuss interest in terms of borrowing money. The next two sections will look at those items you are most likely to purchase on credit—an automobile and a home.

There are two types of consumer credit that allow you to make installment purchases. The first, called **closed-end,** is the traditional installment loan. An **installment loan** is an agreement to pay off a loan or a purchase by making equal payments at regular intervals (usually monthly) for some specified period of time. The loan is said to be *amortized* if it is completely paid off by these payments, and the payments are called **installments.** *If the loan is not amortized, then there will be a larger final payment, called a balloon payment.* With an *interest-only* loan, there is a monthly payment equal to the interest with a final payment equal to the amount received when the loan was obtained.

The second type of consumer credit is called an **open-end** loan, or credit card loan (MasterCard®, VISA, Sears, and so on). This type of loan allows for purchases or cash advances up to a specified "line of credit," and it has a flexible repayment schedule.

The most common method for figuring interest for installment purchases is called **add-on interest.**

Historical Note

At the start of the 20th century, a few hotels began to issue credit cards, and as early as 1914 large department stores and gasoline chains were issuing cards. In those days credit cards had three basic functions:

1. They were often prestige items issued only to valued customers.

2. They were more convenient to use than cash.

3. They provided some degree of safety.

During World War II the use of credit cards virtually ceased due to government restraint on consumer spending and credit. After wartime restrictions were lifted, however, many plans were reinstated, and the railroads and airlines began to issue their own travel cards. In 1949 the Diners Club was established and was followed a short time later by American Express and Carte Blanche.

Add-on Interest

$$\text{INTEREST} = \text{PRINCIPAL} \times \text{RATE} \times \text{TIME}$$

$$\text{AMOUNT DUE} = \text{PRINCIPAL} + \text{INTEREST}$$

$$\text{AMOUNT OF EACH PAYMENT} = \frac{\text{AMOUNT DUE}}{\text{NUMBER OF PAYMENTS}}$$

You will notice that add-on interest is nothing more than the simple interest formula discussed in Section 5.1.

Example 1: Suppose you want to purchase a refrigerator for $1,000, and want to pay for it with installments over 3 years, or 36 months. The store tells you the add-on interest rate is 15%. What is the amount of each payment?

Solution:

$$\text{INTEREST} = \$1,000 \times .15 \times 3$$
$$= \$450$$

$$\text{AMOUNT DUE} = \$1,000 + \$450$$
$$= \$1,450$$

$$\text{MONTHLY PAYMENT} = \frac{\$1,450}{36}$$
$$= \$40.28$$

When figuring monthly payments, always *round up* to the next higher penny (not to the nearest penny).

The important thing to notice when working with add-on interest is that you are paying more than the quoted interest rate. The reason for this is that you are not keeping the entire amount for the entire time. In Figure 5.1, we assume

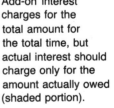

Add-on interest charges for the total amount for the total time, but actual interest should charge only for the amount actually owed (shaded portion).

FIGURE 5.1   Amount owed on a $1,200 loan repaid in one year by making monthly payments

that the amount due is $1,200 with monthly payments of $100, and we compare the amount owed with the amount on which you were charged interest.

In 1969 a *Truth-in-Lending Act* was passed by Congress that requires all lenders to state the true annual interest rate, which is called the **annual percentage rate (APR)**. Regardless of the rate quoted, when you ask the salesperson what the APR is, by law you must be told this rate. This regulation allows you to compare interest rates *before* you sign the contract, which must state the APR even if you have not asked for it.

The actual formula for calculating the APR is too complicated to be useful to most businesspeople, bankers, and consumers. Instead, extensive tables are available from the Federal Reserve System. A portion of one of these tables is shown in Table 5.2.

TABLE 5.2   Annual Percentage Rates

Number of Payments	10%	12%	14%	15%	16%	17%	18%	20%	25%	30%	35%	40%
6	2.94	3.53	4.12	4.42	4.72	5.02	5.32	5.91	7.42	8.93	10.45	11.99
12	5.50	6.62	7.74	8.31	8.88	9.45	10.02	11.16	14.05	16.98	19.96	22.97
18	8.10	9.77	11.45	12.29	13.14	13.99	14.85	16.52	20.95	25.41	29.96	34.59
24	10.75	12.98	15.23	16.37	17.51	18.66	19.82	22.15	28.09	34.19	40.44	46.85
30	13.43	16.24	19.10	20.54	21.99	23.45	24.92	27.89	35.49	43.33	51.41	59.73
36	16.16	19.57	23.04	24.80	26.57	28.35	30.15	33.79	43.14	52.83	62.85	73.20
42	18.93	22.96	27.06	29.15	31.25	33.37	35.51	39.85	51.03	62.66	74.74	87.24
48	21.74	26.40	31.17	33.59	36.03	38.50	41.00	46.07	59.15	72.83	87.06	101.82
60	27.48	33.47	39.61	42.74	45.91	49.12	52.36	58.96	76.11	94.12	112.94	132.51

Procedure for Finding the APR

$$\frac{\text{ACTUAL INTEREST}}{\text{AMOUNT BORROWED}} \times 100 = \text{TABLE 5.2 NUMBER}$$

The procedure for finding the APR using Table 5.2 is illustrated next.

Example 2:   Find the APR for the refrigerator loan described in Example 1.

Solution: The actual interest is $450 and the amount borrowed is $1,000:

$$\frac{\text{ACTUAL INTEREST}}{\text{AMOUNT BORROWED}} \times 100 = \frac{450}{1,000} \times 100$$

$$= 45$$

Now, 45 is the number to be found in Table 5.2. Look across the row of the table corresponding to the number of payments (36 in this example): It is between 43.14 and 52.83, so the APR is between 25% and 30%. Since 45 is a little closer to 43.14, you could estimate the APR at 26%–27%. Remember, if we had room to provide more extensive tables, you could find the APR exactly, but for our purposes an estimate will be sufficient.

The most common types of open-end credit used today are through credit cards such as VISA and MasterCard, and those issued by department stores and oil companies. Because it isn't necessary for you to apply for credit each time you want to charge an item, this type of credit is very convenient.

When comparing the interest rates on any loan, you should always use the APR. Credit card rates, while based on the APR, are often stated as monthly or daily rates rather than yearly rates. Also, note that when dealing with credit cards, a 365-day year is used rather than a 360-day year.

Example 3:   Convert the given credit card rate to the APR.

a. $1\frac{1}{2}\%$ per month   Since there are 12 months per year, multiply by 12 to get
                                       an APR of 18%.
b. Daily rate of .05753%   Multiply by 365:

$$\boxed{.05753} \quad \boxed{\times} \quad \boxed{365} \quad \boxed{=} \quad 20.99845$$

Rounding to the nearest tenth, this is equivalent to 21% APR.

In 1980 credit card rates changed considerably, and many cards introduced a yearly fee. Some are still free, others charge $1 every billing period the card is used by the consumer, but most now charge a $12 yearly fee. This affects the APR differently, depending on the amount the credit card is used during the year and on the balance. If you always pay your credit card bills in full as soon as you receive them, then the card with no yearly fee would obviously be the best for you. On the other hand, if you use your card to stretch out your payments, then the APR is more important than the flat fee. In this book, we won't use this yearly fee in our calculations of APR on credit cards. In addition to fees, the interest rates or APR for credit cards vary greatly. Remember that both VISA and MasterCard credit cards are issued by different banks, so the terms even in one locality can vary greatly.

The finance charges can vary greatly even on credit cards showing the *same* APR, depending on the way the interest is calculated. There are three generally accepted methods for calculating these charges: **adjusted balance, previous balance,** and **average daily balance.**

In Examples 4–6, we will compare the finance charges on a $1,000 credit card purchase using these three different methods. Assume that a bill for $1,000 is received on April 1, and a payment is made. Then another bill is received on May 1, and this bill shows some finance charges. This finance charge is what we are calculating in Examples 4–6.

Adjusted balance   The interest is calculated on the previous month's balance *less* credits and payments. This number is called the *adjusted balance*. Multiply the adjusted balance by the monthly interest rate.

Previous balance   Interest is calculated on the previous month's balance. Multiply the previous balance by the monthly interest rate.

Average daily balance   Add the outstanding balance *each day* in the billing period. Divide by the number of days in the billing period. Multiply by the daily interest rate and then multiply by the number of days in the billing period.

Example 4:   Calculate the interest on a $1,000 credit card bill showing an 18% APR using the adjusted balance method, and assume that $50 is sent and recorded by the due date.

Solution: First, find the adjusted balance:

Previous balance	$1,000
Less credits and payments	$ 50
Adjusted balance	$ 950

$$\text{INTEREST} = \text{ADJUSTED BALANCE} \times \text{MONTHLY RATE}$$
$$= \$950 \times .015$$
$$= \$14.25 \qquad \underline{\qquad} 1\tfrac{1}{2}\% \text{ per month} = 18\% \text{ APR}$$

The interest is $14.25.

Example 5:   Calculate the interest on a $1,000 credit card bill showing an 18% APR using the previous balance method and assuming $50 is sent and recorded by the due date.

Solution: The payment doesn't affect the finance charge for this month unless the bill is paid in full.

$$\text{INTEREST} = \text{PREVIOUS BALANCE} \times \text{MONTHLY RATE}$$
$$= \$1,000 \times .015$$
$$= \$15$$

The interest is $15.

Consumer Reports states that, according to one accounting study, the interest costs for average daily balance and previous balance methods are about 16% higher than those for the adjusted balance method.

Example 6:   Calculate the interest on a $1,000 credit card bill showing an 18% APR using the average daily balance method. Assume that you sent a payment of $50 and that it takes ten days for this payment to be received and recorded.

Solution: For the first ten days the balance is $1,000; then the balance drops to $950. Add the balance for *each day:*

$$10 \text{ days @ } \$1,000 = \$10,000$$
$$20 \text{ days @ } \$950 \ \ = \underline{\$19,000}$$
$$\text{TOTAL} \qquad\qquad \$29,000$$

Divide by the number of days (30 in April):

$$\$29,000 \div 30 = \$966.67$$

This is the average daily balance. Multiply the average daily balance by the daily interest rate and the number of days:

INTEREST = AVERAGE DAILY BALANCE $\times$ DAILY RATE $\times$ NUMBER OF DAYS

$$= \qquad \$966.67 \qquad \times \quad \frac{.18}{365} \quad \times \qquad 30$$

Display: 14.301419

The interest is $14.30.

Problem Set 5.2

Find the amount of interest and the monthly payment for each of the loans described in Problems 1–6. Use add-on interest.

1.   Purchase a living room set for $1,200 at 12% for 3 years

2.   Purchase a wood burning stove for $650 at 11% for 2 years

3.   A $1,500 loan at 11% for 2 years   **4.**   A $1,000 loan at 12% for 2 years

5.   A $2,400 loan at 15% for 2 years   **6.**   A $4,500 loan at 18% for 5 years

Convert each credit card rate in Problems 7–12 to an APR rate. These rates were the listed finance charges on purchases under $500 on a Citibank VISA statement.

7.   Oregon, $1\frac{1}{4}$% per month   **8.**   Arizona, $1\frac{1}{3}$% per month

9.   New York, $1\frac{1}{2}$% per month   **10.**   Tennessee, 0.02740% daily rate

11.   Ohio, .02192% daily rate   **12.**   Nebraska, .03014% daily rate

In Problems 13–18 find the APR for each loan given in Problems 1–6.

13.   See Problem 1.   **14.**   See Problem 2.   **15.**   See Problem 3.

16.   See Problem 4.   **17.**   See Problem 5.   **18.**   See Problem 6.

19. Repeat Example 4 but assume a $500 payment instead of a $50 payment.

20. Repeat Example 5 but assume a $500 payment instead of a $50 payment.

21. Repeat Example 6 but assume a $500 payment instead of a $50 payment.

22. Which method of calculating credit card interest is most advantageous to the consumer? To the bank?

Calculate the monthly finance charge for each credit card transaction in Problems 23–28. Assume that it takes 10 days for a payment to be received and recorded and that the month is 30 days long.

	Balance	Rate	Payment	Method
23.	$ 300	18%	$ 50	Previous balance
24.	$ 300	18%	$ 50	Average daily balance
25.	$ 300	18%	$ 250	Adjusted balance
26.	$3,000	21%	$ 150	Previous balance
27.	$3,000	21%	$ 150	Average daily balance
28.	$3,000	21%	$1,500	Adjusted balance

29. Most credit cards provide for a minimum finance charge of 50¢ per month. Suppose you buy a $30 item and make five payments of $5 and then pay the remaining balance. What is the APR for this purchase?

30. The following appears on a Sears Revolving Charge Card statement. Why do you suppose the limitation on the 50¢ finance charge is for balances under $28.50?

SEARS, ROEBUCK AND CO.
SEARSCHARGE SECURITY AGREEMENT

4. FINANCE CHARGE. If I do not pay the entire New Balance within 30 days (28 days for February statements) of the monthly billing date, a **Finance Charge** will be added to the account for the current monthly billing period. The **FINANCE CHARGE** will be either a minimum of 50¢ if the Average Daily Balance is $28.50 or less, or a periodic rate of 1.75% per month **(ANNUAL PERCENTAGE RATE** of 21%) on the Average Daily Balance.

Mind Bogglers

31. Suppose you wish to purchase a home air conditioner from Sears that costs $800. You have three credit plans from which to choose:

1. Revolving: payments of $80 per month until paid. (APR 18%)
2. Easy Payment: payments of $29.02 per month for 36 months.
3. Modernizing Credit Plan: payments $38.69 per month for 24 months.

What is the APR for each of these choices, and which is the least expensive?

In order to calculate the monthly payment, m, for a loan of P dollars at an interest rate of r for t months, use the formula

$$m = \frac{Pi}{1 - (1 + i)^{-t}}$$

where i = r/12. Use this formula in Problems 32–35 to find the monthly payments for each of the indicated loans.

32. *Calculator problem.* $500 for 12 months at 12%

33. *Calculator problem.* $100 for 18 months at 18%

34. *Calculator problem.* $2,300 for 24 months at 15%

35. *Calculator problem.* $3,000 for 36 months at 17%

Problems for Individual Study

36. Complete the following chart for sources around your own home.

Loan	*Monthly Rate*	*APR*
Credit card		
Credit union—personal loan		
Credit union—car loan		
Bank—personal loan		
Bank—home loan		
Savings and loan—home loan		
Car dealer—car loan		
Life insurance loan		
Finance company		

SELECTING THE CAR
YOU WANT TO BUY

Factors to consider:

1. Crash test results
2. Safety belt comfort
3. Fuel economy
4. Preventive maintenance costs
5. Repair costs
6. Accident repair cost
7. Insurance cost

Classification of cars:
Subcompact, Compact,
Intermediate, and Large. These
factors are used to rate various
compact cars in Table 5.3.

37. Interest rates are sometimes more costly if you pay off the loan early. The "Rule of 78" is a method of charging interest in which the consumer pays for the largest portion of the interest early in the loan. This means that, if it is paid off early, APR rises. Do some research on the "Rule of 78," and prepare a report for the class. As part of your report, apply your findings to Problem 31 for the three types of loans (assume that you decide to make payments for 12 months and then pay off the balance).

5.3 Buying an Automobile

One of the things you are most likely to buy in your lifetime is an automobile. Unless you are paying cash, there are two prices that must be added together to give the true cost of your car—the cost of the car and the cost of the financing. The dealer with the best price for the car may not have the best price for financing, and you should shop for the two items separately.

TABLE 5.3 Each year the U.S. Department of Transportation makes a report of certain cars for sale in this country. Shown here is the *Purchasing Guide for Compact Cars.*

Purchasing Guide

Compact

Car	Crash Test Results[1]	Safety Belt Comfort and Convenience[2]	Fuel Economy[3] mpg	Preventive Maintenance Cost[4]	Repair Cost[5]	Accident Repair Cost[6]	Insurance Cost[7]
BMW 320i		Fair	27	High	High		Surcharge
Buick Skylark	Passed	Poor	22	Medium	High	Low (4 Dr)	Discount
Chevrolet Citation	Passed	Poor	22	Medium	High	Low (4 Dr)	Discount
Datson 200SX	Failed*		26	High	Medium	High	Surcharge
Dodge Aries	Failed		24	Low	Medium		
Dodge Omni	Passed	Fair	25	Low	Low	Low (4 Dr)	Surcharge
Ford Mustang	Passed	Poor	22	Low	Low	High	Surcharge
Maxda RX7			19	Low	High	High	Surcharge
Mazda 626	Failed	Fair	27	Medium	Medium	Medium	Surcharge
Mercury Capri	Passed	Poor	22	Low	Low	High	Surcharge
Oldsmobile Omega	Passed	Poor	23	Medium	High	Low	Discount
Plymouth Horizon	Passed	Fair	25	Low	Low	Medium (4 Dr)	Discount
Plymouth Reliant	Failed		24	Low	Medium		
Pontiac Phoenix	Passed	Poor	22	Medium	High	Low (4 Dr)	Discount
Toyota Celica	Failed	Poor	25	Medium	Medium	High	Surcharge
Toyota Corona			25	Medium	Medium	Medium	

Empty boxes mean data were unavailable at time of printing.

*This car marginally failed the occupant protection test. It may protect its occupants as well as a car of the same class that passed the test.

(The entire book is available free to consumers by writing to: U.S. Department of Transportation, National Highway Traffic Safety Administration, Washington, D.C. 20590)

You can do a great deal of "shopping around," but before you talk to a salesperson about price you should decide on the size, style, and accessories you want. An excellent source for helping you make these decisions is *Consumer Reports* (available in most libraries). Every April they devote an issue to the new cars and car buying. For example, in the April 1983 issue, there were articles on dealing with the dealers, choosing options, car loans, dealer cost versus list price, car ratings, specifications, frequency-of-repair records, and tips for buying used cars.

When you're buying a car from a dealer, it's important to remember that the salesperson is a professional and you are an amateur. The price on the window is not the price that is paid by most people buying the car. It is necessary to negotiate a fair price. Table 5.4 will help you approximate the price the dealer paid for the car. Most dealers will sell the car for 10% over dealer cost, but the higher volume dealers might settle for as little as 5% over their cost. You should be familiar with the following terminology regarding the price of a new automobile.

TABLE 5.4 Dealer's Cost for American Automobiles

Type of Car	Dealer's Cost
Small	.86(.84–.95)
Midsize	.85(.79–.91)
Large	.84(.77–.85)

Pricing an Automobile

The **sticker price** is the price posted on the window of the car. This is the *maximum* price you would pay for the car, and it is the price the dealer would like to obtain for the car. Sometimes the dealer will add a surcharge to the factory price. This surcharge would be included as part of the sticker price.

The **dealer cost** is the price that the dealer actually paid for the car. This is the *minimum* price you would pay for the car. However, since the dealer must cover overhead and profit to stay in business, a certain amount must be added to the cost.

5% price is the dealer cost plus 5% of the dealer's cost. To this amount you must add dealer preparation charges, destination charges, taxes, and license fees. This is the smallest price you should realistically expect to pay for a car.

10% price is the dealer cost plus 10% of the dealer's cost. To this amount you must add dealer preparation charges, destination charges, taxes, and license fees. This is the highest price the dealer should realistically expect you to pay for the car, but remember the dealer is a professional negotiator. Don't give up easily in trying to get this price for the car you want to buy.

Example 1:   Determine the dealer's cost for a car with a sticker price of $8,350 and options of $1,685, destination charges of $200, and dealer preparation charges of $200. Suppose the cost factor is .81 for the car and .78 for the options.

Solution

$$\text{DEALER'S COST FOR THE CAR} = \$8{,}350 \times .81 = \$6{,}730.50$$
$$\text{DEALER'S COST FOR THE OPTIONS} = \$1{,}685 \times .78 = \$1{,}314.30$$
$$\text{DEALER'S COST} = \$6{,}763.50 + \$1{,}314.30 = \$8{,}077.80$$

Example 2:   Make both a 5% offer and a 10% offer on the car described in Example 1.

Solution: From Example 1, the dealer's cost is $8,077.80.

$$5\% \text{ offer:}\quad \$8{,}077.80 \times .05 = 403.89$$
$$\$8{,}077.80 + \$403.89 = \$8{,}481.69$$

10% offer: $8,077.80 × .10 = 807.78

$8,077.80 + $807.78 = $8,885.58

If you wanted the car described in Example 1, you would expect to pay between $8,482 and $8,886 for the car. You would then need to add sales tax, license, destination charges, and dealer preparation charges to this price.

TABLE 5.5   Prices for Selected New 1983 Cars

Automobile	Base Price	Cost Factor	Average Options	Option Factor	Total Price, Base + Options*
Chevrolet Chevette 4-door hatchback	$ 5,616	0.89	$ 868	0.85	$ 6,484
Datsun 200SX 4-door hatchback	$ 8,199	0.86	$ 942	0.86	$ 9,141
Honda Civic 1500 4-door	$ 5,949	0.86	$ 200	0.86	$ 6,149
Chevrolet Celebrity 2-door	$ 8,059	0.86	$1,375	0.85	$ 9,434
Buick Century 2-door custom	$ 8,841	0.86	$1,475	0.85	$10,316
Volkswagen Rabbit 2-door hatchback	$ 7,490	0.88	$1,155	0.88	$ 8,645
Buick LeSabre 4-door limited	$ 9,990	0.85	$ 225	0.85	$10,215
Chevrolet Camaro Z-28	$10,336	0.89	$1,650	0.85	$11,986
Datsun Maxima 4-door wagon	$12,159	0.86	$ 125	0.86	$12,284
Mercedes-Benz 300D	$30,530	0.80	$1,330	0.80	$31,860

*Does not include tax, license, destination, and dealer preparation charges.

The rules for determining the finance rate for a car are the same as the general rules we discussed in the last section. Many people are very careful about negotiating a good price for the car but then take the dealer's first offer on financing. The finance rates vary as much as the price of the car; as you can see from Table 5.6, there are several sources for automobile loans. Even though these rates are subject to frequent change, you can see that they vary considerably. In addition, if you have the dealer arrange the loan through a bank or through a company like the General Motors Acceptance Corporation, the dealer will usually raise the rate to cover the costs of arranging the loan. However, during periods of high interest rates, such loans may be offered at lower rates to stimulate sales. The most economical sources are listed first on Table 5.6.

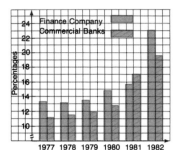

FIGURE 5.2 Car loan interest rates are shown as percentages in this chart covering the period 1977 through 1982.

TABLE 5.6   Sources for New Car Loans*

Passbook Loans—generally 2% above the prevailing rate of passbook account
Life-Insurance Loans—8%
Credit-Union Loans—12%
Commercial-Bank Loans—these can vary considerably depending on the area of the country; 13% to 20%
Dealer Loans—9% to 21%
Finance-Company Loans—15% to 25%

*Keep in mind that interest rates are constantly changing (on an almost daily basis). This table was completed in September 1983 and will not provide the most up-to-date interest rate information. It will, however, show you the comparisons among various types of loans. That is, you would expect all of these interest rates to fluctuate, but the differences shown here should remain about the same.

As I was gathering information for Table 5.6, I commonly obtained quotes of lower rates from the same bank, dealer, or finance company just by asking if they had any other rates. Do not rely on their arithmetic. For example, when I called Beneficial Finance Company and asked for a $4,000 36-month loan for a 1976 Granada, I was told that the payments would be $160 per month and that the interest rate was 18%. When I asked what the total interest for this loan would be, I was told $1,370. Notice that

$$36 \times \$160 = \$5,760$$
$$\text{Less loan} \quad \underline{\quad 4,000}$$
$$\$1,760$$

When I asked about this discrepancy, I was told that the interest rate was 18.05% (it raised $\frac{5}{100}$%), but I still said that the total interest was $1,760. Again I was told that the interest was $1,370. I insisted that I be informed why the additional $390 was included. Then I was told it was for insurance.

I said I didn't want insurance, and then I was given a price of $145 per month. Even though this still did not equal $1,370 interest, I questioned no further and assumed that this "cut rate" was due to my questioning. The reason I relate this story is to give you the warning: BUYER BEWARE AND BE INFORMED—Don't hesitate to question.

$$36 \times \$145 = \$5,220$$
$$\text{Less loan} \quad \underline{\quad 4,000}$$
$$\$1,220$$

If you want insurance for a loan, buy it from an insurance company. A finance company selling insurance is like a car dealer selling financing. Separate these functions if you want the best price. For this example, the finance company was charging $540 for insurance that could be obtained from an

Problem Set 5.3

Use Table 5.5 to determine the dealer's cost (to the nearest dollar) for the list price of each of the cars given in Problems 1–10.

1. Chevrolet Chevette

2. Honda Civic

3. Volkswagen Rabbit

4. Chevrolet Celebrity

5. Datsun 200SX

6. Buick Century

7. Buick LeSabre

8. Chevrolet Camaro

9. Datsun Maxima

10. Mercedes-Benz 300D

Use Table 5.5 to determine the dealer's cost for the options of each of the cars named in Problems 11–20.

11. Chevrolet Chevette

12. Honda Civic

13. Volkswagen Rabbit

14. Chevrolet Celebrity

15. Datsun 200SX

16. Buick Century

17. Buick LeSabre

18. Chevrolet Camaro

19. Datsun Maxima

20. Mercedes-Benz 300D

Make both a 5% offer and a 10% offer for each of the cars listed in Problems 21–30. Use Table 5.5 for prices, cost factors, and options.

21. Chevrolet Chevette

22. Honda Civic

23. Volkswagen Rabbit

24. Chevrolet Celebrity

25. Datsun 200SX

26. Buick Century

27. Buick LeSabre

28. Chevrolet Camaro

29. Datsun Maxima

30. Mercedes-Benz 300D

31. A newspaper advertisement offers a $4,000 car for nothing down and 36 easy monthly payments of $141.62. What is the annual percentage rate?

32. A car dealer will sell you the $6,798 car of your dreams for $798 down and payments of $168.51 per month for 48 months. What is the annual percentage rate?

33. A car dealer carries out the following calculations:

LIST PRICE	$5,368
OPTIONS	$1,625
DESTINATION CGS	$ 200
SUBTOTAL	$7,193
TAX	$ 432
LESS TRADE-IN	$2,932
AMOUNT TO BE FINANCED	$4,693
8% INTEREST FOR 48 MONTHS	$1,501.76
TOTAL	$6,194.76
MONTHLY PAYMENT	$ 129.06

What is the annual percentage rate charged?

insurance company for about $210. If you were interested in life insurance only, a 25-year-old male could buy a *10-year* (not a 3-year) decreasing term insurance with a value of $5,070 for only $2.95 a month. Insurance will be discussed in Section 5.5.

Mind Bogglers

Use the formula $\quad m = \dfrac{Pi}{1 - (1 + i)^{-t}} \quad$ *where* $m =$ *Monthly payment*

$P =$ *Amount of loan*
$r =$ *Interest rate*
$t =$ *Number of months*
$i = \dfrac{r}{12}$

to calculate the monthly payment for the cars indicated in Problems 34–37.

34. *Calculator problem.* Chevrolet Chevette, 10% offer (Problem 21), 36 months at 11.5%

35. *Calculator problem.* Datsun 200SX, full price (Table 5.5), 36 months at 14.5%

36. *Calculator problem.* Honda Civic, 10% offer (Problem 22), 36 months at 12%

37. *Calculator problem.* Chevrolet Camaro, 5% offer (Problem 28), 36 months at 9.7%

Problem for Individual Study

38. Select a car of your choice, find the list price, and calculate a 5% and a 10% offer price. Check out available money sources in your community, and prepare a report showing the different costs for the same car.

5.4 Buying a Home

For many people, buying a home is the single most important financial step in their lives. There are three steps in buying and financing a home. The first is finding a home you would like to buy and then reaching an agreement with the seller on the price and terms. At this step you will need to negotiate a *purchase-and-sale agreement* or *sales contract.* The second step is finding a lender to finance the purchase. You will need to understand interest as you shop around to obtain the best terms. Finally, the third step is payment of certain *closing costs* in a process called *settlement,* or **closing,** where the deal is finalized. The term **closing costs** refers to money exchanged at the settlement, above and beyond the down payment on the property. These costs can include an attorney's fee, the lender's administration fee, taxes to be held in escrow, and *points,* which we will define later in this section.

The first step in buying a home is finding a house you can afford and then coming to an agreement with the seller. A real estate agent can help you find the type of home you want for the money you can afford to pay. A great deal depends on the amount of down payment you can make. A useful rule of thumb to determine the monthly payment you can afford is the following:

1. Subtract any monthly bills (not paid off in the next six months) from your gross monthly income.
2. Multiply by 36%.

Your house payment should not exceed this amount. Another way of determining whether you can afford a house is to multiply gross income by 4; the purchase price should not exceed this amount. Today, the 36% factor is more common and is the one we will use in this book.

What Does It All Mean? Home Buying Terms

To help property purchasers understand the terminology used in real estate transactions, the following definitions are offered in a publication of the U.S. Department of Housing and Urban Development called *Home Buyer's Vocabulary*.

The terms are intentionally general, nontechnical and short. They do not encompass all possible meanings or nuances that a term may acquire in legal use. Also, state laws, as well as custom and use in various states or regions of the country, may modify or completely change the meanings of certain terms defined.

Abstract (of title): A summary of the public records relating to the title to a particular piece of land. An attorney or title insurance company reviews an abstract of title to determine whether there are any title defects which must be cleared before a buyer can purchase clear, marketable, and insurable title.

Acceleration clause: Condition in a mortgage that may require the balance of the loan to become due immediately, if regular mortgage payments are not made or for breach of other conditions of the mortgage.

Agreement of sale: Known by various names, such as "contract of purchase," "purchase agreement," or "sales agreement," according to location or jurisdiction. A contract in which a seller agrees to sell and a buyer agrees to buy, under certain specific terms and conditions spelled out in writing and signed by both parties.

Amortization: A payment plan which enables the borrower to reduce his debt gradually through monthly payments of principal.

Assumption of mortgage: An obligation undertaken by the purchaser of property to be personally liable for payment of an existing mortgage. In an assumption, the purchaser is substituted for the original mortgagor in the mortgage instrument and the original mortgagor is released from further liability under the mortgage.

Closing costs: The numerous expenses which buyers and sellers normally incur to complete a transaction in the transfer of ownership of real estate. These costs are in addition to price of the property and are items prepaid at the closing day.

Conventional mortgage: A mortgage not insured by HUD or guaranteed by the Veterans' Administration.

Deed: A formal written instrument by which the title to real property is transferred from one owner to another.

Deed of trust: Like a mortgage, a security instrument whereby real property is given as security for a debt. However, in a deed of trust there are three parties to the instrument: the borrower, the trustee, and the lender.

Default: Failure to make mortgage payments as agreed to in a commitment based on the terms and at the designated time set forth in the mortgage or deed of trust. Generally, 30 days after the due date if payment is not received, the mortgage is in default.

Down payment: The amount of money to be paid by the purchaser to the seller upon the signing of the agreement of sale. The agreement of sale will refer to the down payment. The down payment is the difference between the sale price and the maximum mortgage amount.

Earnest money: The deposit money given to the seller or his agent by the potential buyer upon the signing of the agreement of sale to show that he is serious about buying the house. If the sale goes through, the earnest money is applied against the down payment.

Equity: The value of a homeowner's unencumbered interest in real estate. When the mortgage and all other debts against the property are paid in full the homeowner has 100% equity in the property.

Escrow: Funds paid by one party to another (the escrow agent) to hold until the occurrence of a specified event.

Mortgage: A lien or claim against real property given by the buyer to the lender as security for money borrowed.

Points: A point is 1% of the amount of the mortgage loan. Points are charged by a lender to raise the yield on the loan at a time when money is tight, interest rates are high, and there is a legal limit to the interest rate that can be charged on a mortgage.

Principal: The basic element of the loan as distinguished from interest and mortgage insurance premium. In other words, principal is the amount upon which interest is paid.

Title: As generally used, the rights of ownership and possession of a particular property.

Trustee: A party who is given legal responsibility to hold property in the best interest or "for the benefit of" another.

Example 1: Suppose your gross monthly income is $2,500, and your current monthly payments on bills are $285. What is the maximum amount you should plan to spend for house payments?

Solution

Step 1. $2,500 − $285 = $2,215
Step 2. $2,215 × .36 = $797.40

The maximum house payment should be under $800 per month.

After you have found a home that you can afford and that you would like to buy, you will be asked to sign a *sales contract* and make a deposit. The deposit is sometimes called *earnest money,* a term that simply means you are serious about buying the house. The next step is to find a lender.

The amount you will need to borrow depends on the price of the home and the amount of your down payment. The down payment can be as large as you wish, but most lenders have some minimum down payment requirements depending on the appraised value of the property. Although it is sometimes possible to buy a home with a 5% down payment, the amount usually runs between 10% and 30% of the appraised value.

Example 2: Determine the down payment for these homes.

a. Purchase price, $76,000

10% down:

DOWN PAYMENT = $76,000 × .10
                          = $7,600

b. Purchase price, $120,000

20% down:

DOWN PAYMENT = $120,000 × .20
                          = $24,000

Next, a lender agrees to provide the money you need to buy a specific home. You, in turn, promise to repay the money based on terms set forth in an agreement, or loan contract, called a **mortgage.** As the borrower, you pledge your home as security. It remains pledged until the loan is paid off. If you fail to meet the terms of the contract, the lender has the right to **foreclose,** which means that the lender may take possession of the property.

There are three types of mortgage loans: (1) conventional loans made between you and a private lender, (2) VA loans made to eligible veterans (these are guaranteed by the Veterans' Administration so they cost less than the other types of loans), and (3) FHA loans made by private lenders and insured by the

Federal Housing Administration. Regardless of the type of loan you obtain, you will pay certain lender costs. By lender costs, we mean all the charges required by the lender: closing costs plus interest.

When you shop around for a loan, certain rates will be quoted:

Interest rate. This is the annual interest rate for the loan; it fluctuates on a daily basis. The APR, as stated on the loan agreement, will generally be just a little higher than the quoted interest rate. This is because the quoted interest rate is usually based on ordinary interest (360-day year) and the APR is based on exact interest (365-day year).

Origination fee. This is a one-time charge to cover the lender's administrative costs in processing the loan. It may be a flat $100–$300 fee, or it may be expressed as a percentage of the loan.

Points. This refers to discount points. This is a one-time charge used to adjust the yield on the loan to what the market conditions demand. It is used to offset constraints placed on the yield by state and federal regulators. Each point is equal to 1% of the amount of the loan.

Example 3:   If you are obtaining a $55,000 loan and the bank charges $5\frac{1}{2}$ points, what is the fee charged?

Solution: FEE = $55,000 × .055 = $3,025

It is desirable to include these three charges—interest rate, origination fee, and points—into one formula so that you can decide which lending institution is giving you the best terms on your loan. This could save you thousands of dollars over the life of your loan. The comparison rate formula shown in the following box can be used to calculate the combined effects of these fees. Even though it is not perfectly accurate, it is usually close enough for meaningful comparisons among lenders.

Comparison Rate for Home Loans

$$\text{COMPARISON RATE} = \text{INTEREST RATE} + .125\left(\text{POINTS} + \frac{\text{ORIGINATION FEE}}{\text{AMOUNT OF LOAN}}\right)$$

Example 4:   Suppose you wish to borrow $60,000 for a new home. Lender A quotes 14.5% + 3 points + $250 and Lender B quotes 14% + 5 points + $150. Which lender is making the better offer?

Solution: Calculate the comparison rate for both lenders (a calculator is helpful for this calculation).

$\frac{1}{8} = .125$; this factor corresponds to a payback period of approximately 15 years. If you intend instead to hold the property for only 5 years and pay off the loan at that time, the factor increases to $\frac{1}{4}$.

Lender A:

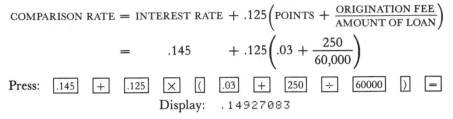

$$\text{COMPARISON RATE} = \text{INTEREST RATE} + .125\left(\text{POINTS} + \frac{\text{ORIGINATION FEE}}{\text{AMOUNT OF LOAN}}\right)$$

$$= \quad .145 \quad + .125\left(.03 + \frac{250}{60,000}\right)$$

Press: $\boxed{.145}$ $\boxed{+}$ $\boxed{.125}$ $\boxed{\times}$ $\boxed{(}$ $\boxed{.03}$ $\boxed{+}$ $\boxed{250}$ $\boxed{\div}$ $\boxed{60000}$ $\boxed{)}$ $\boxed{=}$

Display: .14927083

This is a comparison rate of 14.9%.

Lender B:

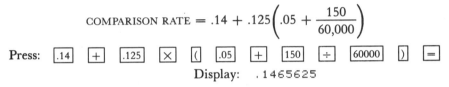

$$\text{COMPARISON RATE} = .14 + .125\left(.05 + \frac{150}{60,000}\right)$$

Press: $\boxed{.14}$ $\boxed{+}$ $\boxed{.125}$ $\boxed{\times}$ $\boxed{(}$ $\boxed{.05}$ $\boxed{+}$ $\boxed{150}$ $\boxed{\div}$ $\boxed{60000}$ $\boxed{)}$ $\boxed{=}$

Display: .1465625

This is a comparison rate of 14.7%.
Lender B is giving the better offer.

Now that you can find the comparable interest rates, the next step is to determine the amount of your mortgage payments. Most lenders assume a 30-year period, but you might be able to afford a 20- or 25-year loan, which would greatly reduce the total finance charge.

The amount of money you borrow, of course, depends on the cost of the home and the size of the down payment. Some financial counselors suggest making as large a down payment as you can afford, and others suggest making as small a down payment as is allowed. There are many factors, such as your tax bracket and your investment potential, that you will have to assess in order to determine how large a down payment you want to make. Table 5.7 may help you with this decision.

TABLE 5.7   Effect of Down Payment on the Cost of a $120,000 Home with Interest at 12%

Down Payment	Percent	Monthly Payment (Principal and Interest)			Total Cost (Interest and Principal)		
		20 Years	25 Years	30 Years	20 Years	25 Years	30 Years
$ 0	0%	$1,321.20	$1,263.60	$1,234.80	$317,160	$379,200	$444,480
6,000	5%	1,255.14	1,200.42	1,173.06	301,000	360,240	422,260
12,000	10%	1,189.08	1,126.44	1,111.32	285,440	341,280	400,030
24,000	20%	1,056.96	1,010.88	987.84	253,730	303,360	355,580
31,000	25%	979.89	937.17	915.81	235,230	281,240	329,660
36,000	30%	924.84	884.52	864.36	222,010	265,440	311,140

Note: Monthly payments are rounded to the nearest cent and total interest to the nearest $10.

Example 5:   The home you select costs $68,500 and you pay 20% down. What is your down payment? How much will be financed?

Solution

$$\text{DOWN PAYMENT} = \$68,500 \times .20 = \$13,700$$

$$\text{AMOUNT TO BE FINANCED} = \text{TOTAL AMOUNT} - \text{DOWN PAYMENT}$$
$$= \$68,500 - \$13,700$$
$$= \$54,800$$

We will use Table 5.8 to calculate the amount of monthly payments for a home loan. You will need to know the amount to be financed, the interest rate, and the length of time the loan is to be financed.

TABLE 5.8   Cost to Finance $1,000 for Selected Years and Rates of Interest

Rate of Interest	Financed for 20 Years		Financed for 25 Years		Financed for 30 Years	
	Monthly Cost	Total Cost	Monthly Cost	Total Cost	Monthly Cost	Total Cost
6 %	$ 7.17	$1,721	$ 6.45	$1,935	$ 6.00	$2,160
7 %	7.76	1,862	7.07	2,121	6.66	2,398
8 %	8.37	2,009	7.72	2,316	7.34	2,642
9 %	9.00	2,160	8.40	2,520	8.05	2,898
10 %	9.66	2,318	9.09	2,727	8.78	3,161
10½%	9.98	2,397	9.44	2,833	9.15	3,294
11 %	10.32	2,477	9.80	2,940	9.52	3,427
11½%	10.66	2,559	10.16	3,049	9.90	3,564
12 %	11.01	2,643	10.53	3,160	10.29	3,704
12½%	11.37	2,729	10.91	3,273	10.68	3,845
13 %	11.52	2,765	11.28	3,384	11.07	3,985
13½%	12.04	2,890	11.66	3,498	11.46	4,126
14 %	12.44	2,986	12.04	3,612	11.85	4,266
14½%	12.80	3,072	12.42	3,726	12.25	4,410
15 %	13.17	3,161	12.81	3,843	12.65	4,554
15½%	13.54	3,250	13.20	3,960	13.05	4,698
16 %	13.92	3,341	13.59	4,077	13.45	4,842
17 %	14.67	3,521	14.38	4,314	14.26	5,134
18 %	15.44	3,706	15.18	4,554	15.07	5,425
19 %	16.21	3,890	15.98	4,794	15.89	5,720
20 %	16.99	4,078	16.79	5,037	16.71	6,016

Example 6:   What is the monthly payment for a home loan of \$54,800 if the interest rate is $14\frac{1}{2}\%$ financed for 30 years?

Solution: Notice that Table 5.8 is in terms of thousands of dollars. This means that you need to divide the amount to be financed by 1,000 (you can do this mentally). For this example it is 54.8. Next, find the entry in Table 5.8 for $14\frac{1}{2}\%$ for 30 years; it shows a monthly cost of \$12.25. This is \$12.25 per thousand, so multiply by 54.8:

$$\$12.25 \times 54.8 = \$671.30$$

The monthly payments for this home are \$671.30.

In 1982, the average price of a new home in the United States was \$110,000 and the interest rate was 15%. There was a considerable amount of press coverage about how this was affecting the economy because most people could not afford a home. The concluding example of this section deals with these figures and asks several questions regarding these average prices and rates.

Example 7:   Suppose you want to purchase a home for \$110,000 with a 30-year mortgage at 15% interest. Suppose also that you can put 20% down.

a. What is the amount of the down payment?
b. What is the amount to be financed?
c. What are the monthly payments?
d. What is the total amount of interest paid on the 30-year loan?

Solution

a.   DOWN PAYMENT $= \$110,000 \times .20 = \$22,000$
b.   AMOUNT TO BE FINANCED $=$ PRICE $-$ DOWN PAYMENT
$$= \$110,000 - \$22,000$$
$$= \$88,000$$

It might be interesting to note that the amount of income necessary to generate this monthly payment is \$3,092 per month. This is an annual salary of more than \$37,000 to buy an average house at an average interest rate!

c.   MONTHLY PAYMENTS $= 88 \times \$12.65 = \$1,113.20$
d.   The total interest can be found from Table 5.8.

$$\text{TOTAL INTEREST} = \text{TOTAL COST} - \text{AMOUNT BORROWED}$$
$$= 88 \times \$4,554 - \$88,000$$
$$= \$400,752 - \$88,000$$
$$= \$312,752$$

Problem Set 5.4

Determine the maximum monthly payment (to the nearest dollar) for a house, given the information in Problems 1–6.

1.   Gross monthly income \$985;   current monthly payments \$147

2.   Gross monthly income \$1,240;   current monthly payments \$215

3. Gross monthly income $1,480; current monthly payments $520

4. Gross monthly income $2,300; current monthly payments $350

5. Gross monthly income $2,800; current monthly payments $540

6. Gross monthly income $3,600; current monthly payments $370

Determine the down payment for each home described in Problems 7–12.

7. $48,500; 5% down

8. $69,900; 20% down

9. $53,200; 10% down

10. $64,350; 10% down

11. $85,000; 20% down

12. $112,000; 30% down

What are the bank charges for the points indicated in Problems 13–18?

13. $48,500; 2 points

14. $69,900; 3 points

15. $53,200; 3 points

16. $64,350; 5 points

17. $85,000; 7 points

18. $112,000; 9 points

For Problems 19–24, determine the comparable interest rate (to two decimal places) for a $50,000 loan, when the quoted information is given.

19. 11.5% + 2 points + $450

20. 14.25% + 3 points + $250

21. 13.7% + 7 points + $250

22. 14.5% + 4 points + $350

23. 15.3% + 5 points + $150

24. 14.8% + 6 points + $200

Use Table 5.8 to estimate the monthly payment for each loan described in Problems 25–30.

25. $48,500; 5% down; 20 years; 14%

26. $48,500; 5% down; 30 years; 14%

27. $69,900; 20% down; 25 years; $14\frac{1}{2}$%

28. $85,000; 20% down; 20 years; 15%

29. $85,000; 20% down; 30 years; 15%

30. $112,000; 30% down; 25 years; 15%

31. Suppose you want to purchase a home for $75,000 with a 30-year mortgage at 15% interest. Suppose that you can put 20% down.

   a. What is the amount of the down payment?
   b. What is the amount to be financed?
   c. What are the monthly payments?
   d. Use Table 5.8 to find the total amount of interest paid on the loan.

32. Suppose you want to purchase a home for $100,000 with a 30-year mortgage at 15% interest, and you can put 20% down. Answer the questions asked in Problem 31.

A sample worksheet for a family purchasing a $35,000 house and obtaining a $30,000 loan is shown below. Line 103 assumes that their total settlement charges are $1,000 and is merely illustrative. The amount may be higher in some areas and for some types of transactions and lower for others.

J.	SUMMARY OF BORROWER'S TRANSACTION	
100.	GROSS AMOUNT DUE FROM BORROWER:	
101.	Contract sales price	35,000.00
102.	Personal property	200.00
103.	Settlement charges to borrower (line 1400)	1,000.00
104.		
105.		
	Adjustments for items paid by seller in advance	
106.	City/town taxes to	
107.	County taxes to	
108.	Assessments 6/30 to 7/31 (owner's assn).	20.00
109.	Fuel oil 25 to gal. @ 50/gal	12.50
110.		
111.		
112.		
120.	GROSS AMOUNT DUE FROM BORROWER	36,232.50
200.	AMOUNTS PAID BY OR IN BEHALF OF BORROWER:	
201.	Deposit or earnest money	1,000.00
202.	Principal amount of new loan(s)	30,000.00
203.	Existing loan(s) taken subject to	
204.		
205.		
206.		
207.		
208.		
209.		
	Adjustments for items unpaid by seller	
210.	City/town taxes to	
211.	County taxes 1/1 to 6/30 @$600/yr	300.00
212.	Assessments 1/1 to 6/30 @$100/yr	50.00
213.		
214.		
215.		
216.		
217.		
218.		
219.		
220.	TOTAL PAID BY/FOR BORROWER	31,350.00
300.	CASH AT SETTLEMENT FROM/TO BORROWER	
301.	Gross amount due from borrower (line 120)	36,232.50
302.	Less amounts paid by/for borrower (line 220)	31,350.00
303.	CASH (⊠ FROM) (☐ TO) BORROWER	4,882.50

33. Suppose you want to purchase a home for $150,000 with a 30-year mortgage at 15% interest, and you can put 25% down. Answer the questions asked in Problem 31.

Problem for Individual Study

34. Do some research and write a report about buying a home in your community. Interview a real estate agent, various lenders, and some escrow agents. Be sure to include specific information about the local customs and interest rates.

*5.5 Buying Insurance

In the last two sections we have discussed buying an automobile and buying a home. After you buy them you will have to insure them. In addition, you may want to consider buying some life insurance, which would pay off the loans if you were to die before they were repaid.

Automobile Insurance

Buying insurance is like buying any other consumer item, and, if you are willing to do some preliminary work, you can generally find the kind of auto insurance you need at a substantial savings over that recommended by the car dealer. The first step is to decide what coverage you want. The main types of automobile coverage are liability, medical, collision, comprehensive, uninsured motorist, and wage loss.

Liability insurance pays others for damage that was your fault. It is usually quoted using a numerical shorthand such as 10/20/5 or 100/300/25. These numbers represent amounts in thousands of dollars, and the first two numbers state the maximum amount that can be paid for *bodily injury* that you cause. The first represents the maximum amount the company will pay for any one person who is injured or killed; the second number represents the maximum amount the company will pay for any one accident to all the victims or their survivors. The third number represents the maximum amount the company will pay for *property damage.*

Example 1:   If you have 10/20/5 coverage and there is a judgment against you for $15,000 for bodily injury to another person and $4,000 for damage to that person's car, how much would your insurance pay and how much would you have to pay?

---

*This is an optional section.

Solution: Your insurance will pay $10,000 for bodily injury, so you would be liable for $5,000. Since your coverage will pay up to $5,000 for property damage, you would not need to pay anything for property damage.

Collision coverage pays for the damage you do to your own car; it requires that you pay something before you can collect from the company. The amount you have to pay is called the *deductible*. The higher the deductible, the lower the cost of the insurance. For example, the following rates are based on a basic coverage of $100 deductible.

$ 150 deductible;   80% of basic coverage

$ 250 deductible;   60% of basic coverage

$ 500 deductible;   45% of basic coverage

$1,000 deductible;   25% of basic coverage

Example 2:   If your collision insurance costs $145 for $100 deductible, how much would you expect to pay for $250 deductible?

Solution: 60% of $145 = .60 × $145

= $87

The last major type of automobile coverage, called *comprehensive,* pays for loss or damage to your car due to something other than collision, such as damage done during a storm.

After you decide what kinds of coverage you want, check with three or four companies and make your selection. You can use Table 5.9 to help you with this comparison.

For an excellent discussion of automobile insurance in general and no-fault insurance in particular, see *Consumer Reports,* June, July, and August 1977, for a three-part series on automobile insurance.

TABLE 5.9   Comparison of Automobile Insurance

	Coverage Desired*	Company A Annual Rate	Company B Annual Rate	Company C Annual Rate
Liability	_____	_____	_____	_____
Medical	_____	_____	_____	_____
Collision	_____	_____	_____	_____
	Deductible			
Comprehensive	_____	_____	_____	_____
	Deductible			
Uninsured motorist	_____	_____	_____	_____
Towing	_____	_____	_____	_____
Wage loss	_____	_____	_____	_____
Other fees	_____	_____	_____	_____
TOTAL	_____	_____	_____	_____

*Fill in the "Coverage Desired" column first; then obtain quotations from three companies. Be sure to compare annual rates since some companies make quotations for less than one year.

Home Insurance

Protection for your home and personal property is available in various kinds of insurance policies. To help you decide on the kind and amount of coverage you need for your home and its contents, discuss your situation with a qualified insurance agent in your area. Be sure to ask about the cost of different policies. Then you can select the policy that best suits your needs at the least cost to you.

A standard *fire insurance* policy protects you against losses from damage caused by fire and lightning. Fire insurance can be extended to protect your house against losses caused by hailstorm, tornado, wind, explosion, riot, aircraft damage, vehicle damage, and smoke damage. This is known as *extended coverage*. The extra protection of the extended coverage is well worth the small additional cost. Earthquake and flood insurance usually require special policies and are not covered even by extended coverage policies.

A *personal liability insurance* policy may be desirable, in addition to the fire and extended coverage insurance. With a liability policy, you are protected in the event that someone is injured on your property. Injuries or damage resulting from activities of members of your family are also covered. Some policies have special provisions to pay medical costs within certain limits, regardless of your liability.

A *theft insurance* policy protects your personal property against robbery, burglary, and larceny. Property-theft policies vary widely, so it's important to select a policy carefully. You can get a policy in a broad form that protects your possessions both in your home and away from home, but it costs more than the more limited form.

A *homeowners'* policy provides insurance on the house, garage, and other buildings, insurance on personal property, and coverage for personal liability. Policies for renters are also available. You can buy a homeowners' policy, with its combined coverage, for less money than if you bought these coverages separately. Insurance companies usually have more than one homeowners' policy from which to choose. The policies differ in the extent of coverage, and some package policies may include more protection than you need.

THE BORN LOSER

by Art Sansom

Life Insurance

There are two basic types of life insurance, *term* and *whole life*. With term insurance you are buying life insurance only, and there is no cash-value buildup. The premium stays the same for the period of the term. You can continue your insurance without the need for a new medical exam if you have a *renewable* term policy. However, the premium goes up after each term, as you can see by looking at Table 5.10. A term policy is said to be *convertible* if you can convert it to whole life insurance.

Instead of increasing the payments to maintain constant coverage, you can keep the same payments while the amount of the insurance decreases. Such a policy is called *decreasing term insurance*. It can be used to protect the mortgage on your home, car, or other installment purchase. As we noted in Section 5.3, it is usually cheaper to arrange your own insurance than to pay for it as part of the item you are buying.

For younger people, term insurance may be the best insurance buy. But as you can see from Table 5.10, the premiums for term insurance become very expensive as you grow older. At some point you will probably want to purchase a *whole life insurance policy*. Be aware of the fact that insurance agents make a larger commission on whole life policies than on other types of insurance.

An example of decreasing term insurance is a policy that costs about $12 per month and provides the following coverage:

Age	Amount of Insurance
Under 30	$40,000
30–34	35,000
35–39	30,000
40–44	25,000
45–49	18,000
50–54	11,000
55–59	7,000
60–66	4,000

TABLE 5.10 Comparison of Annual Premium Rates of Life Insurance for Term, Whole Life, and Endowment Policies per $1,000 of Insurance

Age	5-Year Term	10-Year Term	Whole Life	20-Year Endowment 20 Pay Life at Age 65*
20	$ 3.39	$ 4.90	$ 9.68	$26.07
25	3.49	5.36	11.44	28.63
30	3.71	6.09	13.69	31.60
35	4.26	7.22	16.64	35.13
40	5.55	9.27	20.49	39.34
45	7.56	12.61	25.51	44.41
50	12.06	17.88	31.71	50.70
55	17.95	26.01	40.05	58.85
60	26.95	n.a.	51.36	69.85

The ages given are for males. For females, add three years to these ages. For example, rates for a 23-year-old female would be the same as the 20-year-old male rates from this table. Women pay lower premiums than men of the same age because they live longer. The figures given in this table are for a participating policy and may vary among companies. This table is given for illustrative purposes only, and you will need to consult your local agent for a quotation.
*This means that you pay the premiums for 20 years and then receive the endowment at age 65.

A *whole life* policy is characterized by a constant payment throughout the life of the policy, until your death or retirement. The premium is determined by your age when you take out the policy, but it does not become greater as you grow older. It also builds up a *cash value,* which is a form of forced savings or investment program.

The extreme step in a forced savings plan with insurance is the *endowment policy.* This type of policy guarantees payment at the end of some period if you live, as well as payment if you die during the interim. These policies are far

The cash value is the amount you would receive if you cancelled the policy.

Glossary of Life Insurance Terms

Beneficiary: The person named in the policy to receive the insurance proceeds at the death of the insured.

Face amount: The amount stated on the face of the policy that will be paid in case of death or at the maturity of the contract. It does not include dividend additions, or additional amounts payable under accidental death or other special provisions.

Insured: The person on whose life an insurance policy is issued.

Mutual life insurance company: A life insurance company without stockholders whose management is directed by a Board elected by policyowners. Mutual companies, in general, issue participating insurance.

Nonparticipating insurance: Insurance on which the premium is calculated to cover as closely as possible the anticipated cost of the insurance protection and on which no dividends are payable.

Participating insurance: Insurance on which the policyholder is entitled to receive policy dividends reflecting the difference between the premium charged and actual experience. The premium is calculated to provide some margin over the anticipated cost of the insurance protection.

Policy/Plan: The printed document stating the terms of the insurance contract that is issued to the policyholder by the company.

Premium: The payment, or one of the periodical payments, a policyholder agrees to make for an insurance policy.

Stock life insurance company: A life insurance company owned by stockholders who elect a Board to direct the company's management. Stock companies, in general, issue nonparticipating insurance, but may also issue participating insurance.

As we've said before in this chapter, it is generally more expensive to buy from a company that is not in that particular business. Just as a finance company charges more for insurance than an insurance company does, an insurance company doesn't usually pay you as well for your investment as a bank or other type of savings program.

more expensive than the other types of insurance, and you would probably be better off buying a term or whole life policy while investing the difference in premium rates directly into the bank or other investment program. A new type of policy, called *adjustable life,* is also available, which provides a combination of term insurance and whole life insurance. This type of insurance may be the best type to consider, depending on your own situation.

Example 3: If a $10,000 10-year term policy is purchased at age 45 and renewed at age 55, for a total of 20 years, what is the total cost?

Solution: 10 years at $126.10 per year (from Table 5.10), and
10 years at $260.10 per year = $10(126.10) + 10(260.10)$
$= \$3,862.00$

Example 4: If a $10,000 20-year endowment is purchased at age 45, what is the total cost?

Solution: 20 years at $444.10 per year = $20(444.10)$ See Table 5.10
$= \$8,882$ for these figures.

At age 65 the amount returned is $10,000, so the net result is a gain of $1,118. Remember that all during that 20 years you had $10,000 insurance coverage.

The savings that result from buying term insurance rather than an endowment policy are:

First ten years:
444.10 − 126.10 = 318 per year

Second ten years:
444.10 − 260.10 = 184 per year

From Examples 3 and 4, it might appear that a 20-year endowment is a better buy, but let's look at it from another viewpoint. Suppose you are 45 years old and want a 20-year endowment of $10,000. You also want $10,000 of life insurance protection. However, instead of purchasing an endowment policy, you decide to purchase term insurance and put the difference in the premium prices into your own savings account. At the end of 20 years you would have $11,927 in the bank (as shown by the calculation in the margin) instead of only $10,000 from the endowment policy. By buying term insurance and saving the difference, you will have $1,927 more than if you had bought an endowment policy.

All life insurance policies are either *participating* or *nonparticipating*. **Participating** policies pay dividends to the policyholder, while nonparticipating policies do not.

Example 5: Compare the costs for a female purchasing a $10,000 10-year term policy at age 53 or two consecutive 5-year term policies at ages 53 and 58.

Solution: 10-year term: 10(178.80) = 1,788

> Use the figure for 50-year-old males from Table 5.10.

First 5-year term: 5(120.60) = 603.00
Second 5-year term: 5(179.50) = 897.50
TOTAL = 1,500.50

This person would save $287.50 ($1,788 − $1,500.50) if she purchased two consecutive 5-year term policies.

In Problem 43 of Problem Set 5.1, the idea of an annuity was discussed. An annuity for ten years at 6% with an annual deposit of $318 would grow to $4,357. If this amount were then deposited in a 10-year time certificate paying 8% it would become $9,406. The second ten years would be a 10-year annuity at 6% with an annual deposit of $184, which would grow to $2,521. Therefore, the grand total saved is $11,927.

Problem Set 5.5

In your own words, explain each of the terms given in Problems 1–8.

1. Deductible
2. Convertible life
3. Cash value
4. Collision insurance
5. Comprehensive
6. Bodily injury insurance
7. Extended coverage
8. Property damage insurance

9. In your own words, compare and contrast term, whole life, and endowment policies.

10. If you have 20/30/10 automobile coverage and there is a judgment against you for $120,000 for bodily injury to another person and $6,000 damage to that person's car, how much would your insurance pay and how much would you have to pay?

11. Explain what each of the following sets of numbers means with regard to your automobile insurance.

    a. 40/80/10
    b. 100/200/20

In Problems 12–15 state the premium cost for a collision insurance policy with a $100 deductible. For each given deductible, estimate the new cost of the policy.

12. $195; $250 deductible
13. $155; $500 deductible
14. $175; $150 deductible
15. $175; $1,000 deductible

Use Table 5.10 to determine the annual cost for each of the policies indicated in Problems 16–21.

16. $50,000 10-year term for a 35-year-old male

17. $20,000 20-year endowment for a 45-year-old male

18. $10,000 whole life for a 28-year-old female

19. $15,000 5-year term for a 33-year-old female

20. $60,000 whole life for a 28-year-old female

21. $100,000 5-year term for a 40-year-old male

22. If a man purchases a $50,000 10-year term at age 35, renews it at age 45, and renews it again at age 55, for a total of 30 years, what is the total cost?

23. If a $50,000 20-year endowment is purchased by a male at age 35, paid for 20 years, what is the total cost?

24. If a $50,000 whole life is purchased by a male at age 35 and kept for 30 years, what is the total cost?

25. If a $20,000 10-year term is purchased by a male at age 45 and renewed at age 55, for a total of 20 years, what is the total cost?

26. If a $20,000 20-year endowment is purchased by a male at age 45 and paid for 20 years, what is the total cost?

27. If a $20,000 whole life is purchased by a male at age 45 and kept for 20 years, what is the total cost?

28. If a $10,000 10-year term is purchased by a male at age 25 and renewed at ages 35, 45, and 55, for a total of 40 years, what is the total cost?

29. If a $10,000 20-year endowment is purchased by a male at age 25 and paid for 20 years, what is the total cost?

30. If a $10,000 whole life is purchased by a male at age 25 and kept for 40 years, what is the total cost?

Mind Boggler

31. Compare the costs of the policies shown in Problems 22–24 by investing the difference in premiums in an 8% savings account and then cashing in at age 65.

The most extensive source of information about life insurance companies in the United States is *Best's Insurance Reports—Life/Health* published annually by the A. M. Best Company. This hefty volume will probably be available in the reference room of your public library. If not, a local insurance agent may have a copy.

For Problem 31, see Problem 43 in Problem Set 5.1 about annuities.

Chapter 5 Outline

I. Interest
  A. Simple interest
    1. Formula; $I = Prt$
    2. Amount is the principal plus interest

Chapter 5 Test

1. **a.** What is the simple interest for $1,000 at 8% for 20 years?
   b. Use a calculator or Table 5.1 on page 247 to find the compound interest for $1,000 at 8% for 20 years, compounded annually.
   c. What is the effective yield of 8% compounded quarterly?

*Optional section.

2. **a.** Explain the difference between closed-end and open-end credit.
    b. What is APR, and why is it useful?

3. **a.** If the simple add-on interest rate for a $1,000 loan is quoted at 9% for 3 years, use Table 5.2 on page 255 to find the APR for monthly payments.
    b. If a $4,195 automobile is advertised at $121.42 per month for 48 months with no down payment, what is the total amount of interest paid for the car?
    c. *Calculator problem.* Use Table 5.2 to determine the APR for the automobile described in part b.

4. A certain car has a list price of $8,920 with a cost factor of .87, and options totaling $1,450 with a cost factor on the options of .90.

    a. What is the dealer's cost?
    b. Make a 5% offer.
    c. Make a 10% offer.
    d. What is the sticker price for this car including options?

5. Suppose you purchase a $550 TV set on a VISA credit card and receive the bill. You then make a $50 payment; the APR is 18%. Show the total interest charged on your next statement if that credit card uses the following methods of calculating interest.

    a. Average daily balance (allow 10 days for payment and assume a 30-day month)
    b. Adjusted balance
    c. Previous balance

For Problems 6–9 assume that you select a home that costs $85,000 and you will finance the home for 30 years.

6. Determine the following down payments.

    a. 10% down   **b.** 20% down   **c.** 25% down

7. Use Table 5.8 on page 271 and Problem 6 to estimate the monthly payments for each 30-year loan described.

    a. 10% down, 14% interest rate
    b. 20% down, 13% interest rate
    c. 25% down, 12% interest rate

8. Suppose you decide to pay 30% down and can obtain a 12% loan. Use Table 5.8 on page 271 to estimate the total interest paid if you pay off the loan in:

    a. 20 years   **b.** 25 years   **c.** 30 years

9. Suppose you decide to pay 30% down. Lender A quotes 12% + 2 pts + $200, and Lender B quotes $12\frac{1}{4}$% + 1 pt + $300. Which lender is giving you the better offer, and by how much?

*10. What is the total cost of a $50,000 10-year term policy purchased by a male at age 45 and renewed at age 55 for a total of 20 years? Use Table 5.10 on page 277.

---

*Requires optional Section 5.5.

Bonus Question

11. If your gross monthly income is $2,200 with monthly payments of $335, estimate the maximum monthly house payment you could afford. What down payment would be necessary for you to make on an $85,000 home if you can obtain a 12% 30-year loan?

Date	Cultural History
1400	1400: Beginning of early Renaissance 1408: End great schism 
1425	1429: Joan of Arc raises siege of Orleans 1435: Rogier Van der Weyden 1436: Fra Angelico begins frescoes at San Marco 
1450	1450: Florence is center of Renaissance 1454: Gutenberg prints Bible 1465: First printed music 1470: First illustrated books 
1475	 1484: Botticelli's *Birth of Venus* 1492: Columbus discovers America 1497: Vasco de Gama rounds Cape of Good Hope

Fresco, Fra Angelico

St. George, Donatello

Woodcut from *Ars Morendi*

Mathematical History

1400: In Florence, commercial activity results in several books on mercantile arithmetic

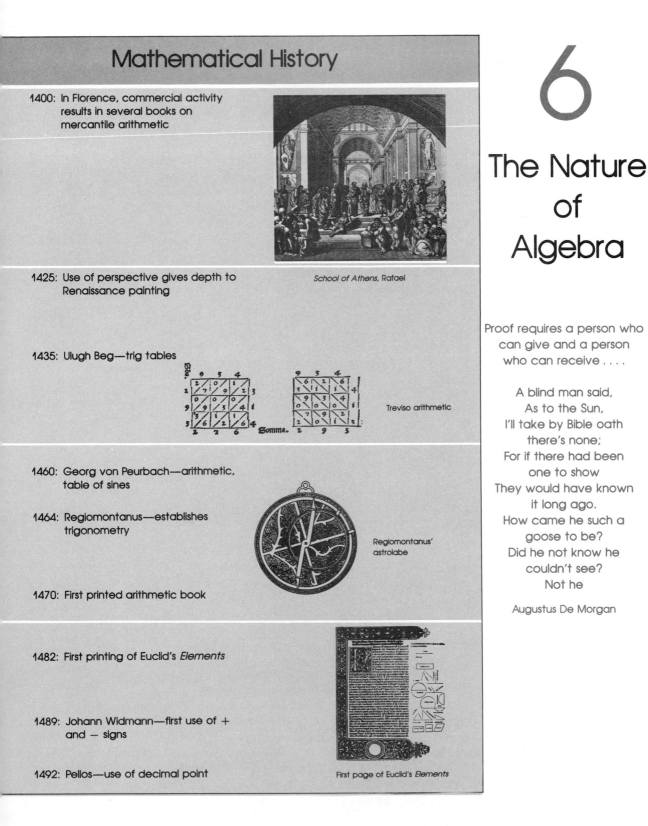

School of Athens, Rafael

1425: Use of perspective gives depth to Renaissance painting

1435: Ulugh Beg—trig tables

Treviso arithmetic

1460: Georg von Peurbach—arithmetic, table of sines

1464: Regiomontanus—establishes trigonometry

Regiomontanus' astrolabe

1470: First printed arithmetic book

1482: First printing of Euclid's *Elements*

1489: Johann Widmann—first use of + and − signs

1492: Pellos—use of decimal point

First page of Euclid's Elements

Proof requires a person who can give and a person who can receive . . . .

A blind man said,
As to the Sun,
I'll take by Bible oath
there's none;
For if there had been
one to show
They would have known
it long ago.
How came he such a
goose to be?
Did he not know he
couldn't see?
Not he

Augustus De Morgan

6.1 Mathematical Systems

Many people think of algebra as simply a high school mathematics course in which variables are manipulated. It is not the intent of this chapter to duplicate the material of such a course but rather to give you some insight into the nature of algebra. In mathematics, the word *algebra* refers to a structure, or a set of axioms, that form the basis for what is accepted and what is not when manipulating the symbols of that system. There are many different algebras, most of them too advanced to discuss in this book, but they are all studied and defined according to some basic properties. In Chapter 4 we saw some of these properties: closure, commutative properties, associative properties, and a distributive property.

The basic understanding of an algebra is related to what is called a mathematical system.

Historical Note

Algebra and algebraic ideas date back 4,000 years to the Babylonians and the Egyptians; the Hindus and the Greeks also solved algebraic problems. The title of Arab mathematician al-Khowârizmî's text (about 825 A.D.), *Hisâb al-jabr w'almugâbalah,* is the origin of the word *algebra*. The book was widely known in Europe through Latin translations. The word *al-jabr* or al-ge-bra became synonymous with equation solving. Interestingly enough, the Arabic word *al-jabr* was also used in connection with medieval barbers. The barber, who also set bones and let blood in those times, usually called himself an *algebrista*. In the 19th century there was a significant change in the prevalent attitude toward mathematics. Up to that time, mathematics was expected to have immediate and direct applications, but now mathematicians became concerned with the structure of their subject and with the *validity,* rather than the practicality, of their conclusions. Thus there was a move toward *pure mathematics* and away from *applied mathematics.*

> **A mathematical system** is a set with at least one defined operation and some developed properties.

An operation is the process of carrying out rules of procedure, such as addition, subtraction, multiplication, or division. However, the word *operation* is not limited to these four fundamental operations. For example, taking square roots is an operation; other abstract operations were discussed in Chapter 4.

As an example of a mathematical system, let's consider the set of real numbers with the operations of addition and multiplication. As we saw in Chapter 4, certain properties are satisfied in the set of real numbers.

	Addition	*Multiplication*
Closure:	$(a + b)$ is a real number	ab is a real number
Associative:	$(a + b) + c = a + (b + c)$	$(ab)c = a(bc)$
Commutative:	$a + b = b + a$	$ab = ba$

Distributive for multiplication over addition:
$$a(b + c) = ab + ac$$

There are two additional properties that are important in the set of real numbers: the identity and the inverse properties.

In Chapter 4 we saw that the development of the concept of zero and a numeral for it did not take place at the same time as the development of the counting numbers. The Greeks were using the letter "oh" (omicron) for zero as early as 150 A.D., but the predominant system in Europe was the Roman numeration system, which did not include zero. It was not until the 15th century, when the Hindu-Arabic numeration system finally replaced the Roman system, that the zero symbol came into common usage.

The number 0 (zero) has a special property for addition that allows it to be added to any real number without changing the value of that number. This property is called the **identity property for addition** of real numbers.

Identity for Addition

There exists a real number 0, called zero, so that

$$0 + a = a + 0 = a$$

for any real number a.

Remember, when you are studying algebra, you're studying ideas and not just rules about specific numbers. A mathematician would attempt to isolate the *concept* of an identity. First, does an identity property apply to other operations?

Historical Note

The Egyptians did not have a zero symbol, but their numeration system did not require such a symbol. On the other hand, the Babylonians, with their positional system, had a need for a zero symbol but did not really use one until around 150 A.D. The Mayan Indians' numeration system was one of the first to use a zero symbol, not only as a place-holder but as a number zero. The exact time of its use is not known, but the symbol was noted by the early 16th century Spanish expeditions into Yucatan. Evidently, the Mayans were using the zero long before Columbus arrived in America.

Multiplication	*Subtraction*	*Division*
$\square \times a = a \times \square = a$	$\triangle - a = a - \triangle = a$	$\bigcirc \div a = a \div \bigcirc = a$
Same number	Same number	Same number

Is there a real number that will satisfy any of the blanks for multiplication, subtraction, or division?

Identity for Multiplication

There exists a real number 1, called one, so that

$$1 \times a = a \times 1 = a$$

for any real number a.

Notice that there are no real numbers satisfying either the subtraction or division identity properties. There may be identities for other operations or for sets other than the set of real numbers.

*Example 1:   Is there an identity set for the operation of union?

Solution: We wish to find a single set $\square$ so that

$$\square \cup X = X \cup \square = X$$

holds for every set X. We see that the empty set satisfies this condition; that is,

$$\varnothing \cup X = X \cup \varnothing = X$$

Thus, $\varnothing$ is the identity set for the operation of union.

*This example requires Section 2.1.

In Chapter 4 we spoke of opposites when we were adding and subtracting integers. Recall the property of opposites:

$$5 + (^-5) = 0$$
$$^-128 + 128 = 0$$
$$a + (^-a) = 0$$

When opposites are added, the result is zero, the identity for addition. This idea, which can be generalized, is called the **inverse property.**

Inverse Property for Addition

For each real number a, there is a unique real number ^-a, called the opposite of a, so that

$$a + (^-a) = {}^-a + a = 0$$

For multiplication, we replace the identity element for addition (0) with the identity element for multiplication (1) and check:

$$5 \times \square = \square \times 5 = 1 \qquad\qquad ^-128 \times \triangle = \triangle \times (^-128) = 1$$

$$5 \times \frac{1}{5} = \frac{1}{5} \times 5 = 1 \qquad\qquad ^-128 \times \frac{1}{^-128} = \frac{1}{^-128} \times (^-128) = 1$$

and $\frac{1}{5}$ is a real number   and $\frac{1}{^-128}$ is a real number

To show the inverse property for multiplication, we seek to find a replacement for the box for each and every real number a:

$$a \times \square = \square \times a = 1$$

Does the inverse property for multiplication hold? No, because if $a = 0$, then

$$0 \times \square = \square \times 0 = 1$$

does not have a replacement for the box in the set of real numbers. However, the inverse property for multiplication holds for all *nonzero* replacements of a.

Inverse Property for Multiplication

Notice that every *nonzero* real number has a reciprocal.

For each real number a, $\boldsymbol{a \neq 0}$, there exists a number, $1/a$, called the *reciprocal* of a, so that

$$a \times \frac{1}{a} = \frac{1}{a} \times a = 1$$

Example 2: Given the set $A = \{^-1,0,1\}$, the operation of multiplication has the identity 1 and satisfies the inverse property:

$$1 \times 1 = 1, \quad \text{so the inverse of 1 is 1}$$
$$^-1 \times {}^-1 = 1, \quad \text{so the inverse of } {}^-1 \text{ is } {}^-1$$

The only element in the set that does not have an inverse is 0, so we say that the inverse property is satisfied.

We have now introduced 11 properties satisfied by the set of real numbers. *Any* set satisfying these properties is called a *field*.

Definition

A **field** is a set, F, with two operations, $+$ and $\times$, satisfying the following properties for any a, b, and c elements of F.

1. *Closure for addition.* $(a + b)$ is an element of F
2. *Closure for multiplication.* $a \times b$ is an element of F
3. *Commutative for addition.* $a + b = b + a$
4. *Commutative for multiplication.* $a \times b = b \times a$
5. *Associative for addition.* $(a + b) + c = a + (b + c)$
6. *Associative for multiplication.* $(a \times b) \times c = a \times (b \times c)$
7. *Identity for addition.* There exists a real number 0 so that
   $0 + a = a + 0 = a$ for any element a in F.
8. *Identity for multiplication.* There exists a real number 1 such that
   $1 \times a = a \times 1 = a$ for any element a in F.
9. *Inverse for addition.* For each number a in F there is a unique number ^-a in F so that

$$a + (^-a) = (^-a) + a = 0$$

10. *Inverse for multiplication.* For each nonzero a in F there is a unique number $1/a$ in F so that

$$a \times \frac{1}{a} = \frac{1}{a} \times a = 1$$

11. *Distributive for multiplication over addition.*

$$a \times (b + c) = a \times b + a \times c$$

The set of real numbers is a field, but there are other fields. In our definition of a field we used the operations of addition and multiplication for the sake of understanding, but, for the mathematician, a field is defined as a set with *any two* operations satisfying the 11 stated properties.

Problem Set 6.1

In Problems 1–6, explain, in your own words, the meaning of each property, and justify your explanations by giving examples.

1. Closure
2. Associative
3. Identity
4. Inverse
5. Commutative
6. Distributive

TABLE 6.1

*	1	4	7	9
1	9	7	1	4
4	7	9	4	1
7	1	4	7	9
9	4	1	9	7

7. Give the identity (if any) for the set $\{1,4,7,9\}$ with operation * defined by Table 6.1.

8. Does the set $\{1,4,7,9\}$ with operation * defined by Table 6.1 satisfy the inverse property?

Identify each of the properties in Problems 9–18.

9. $5 + 7 = 7 + 5$

10. $3 \times (4 + 8) = 3 \times 4 + 3 \times 8$

11. Mustard + Catsup = Catsup + Mustard

12. $5 \cdot 1 = 5$

13. $5 \cdot \dfrac{1}{5} = 1$

14. $a + (10 + b) = (a + 10) + b$

15. $a + (10 + b) = (10 + b) + a$

16. (Red + Blue) + Yellow = Red + (Blue + Yellow)

17. $15 + (a + {}^-a) = 15 + 0$

*18. $A \cap (B \cup C) = (A \cap B) \cup (A \cap C)$

In Problems 19–23, you are asked to check various field properties for the sets N (natural numbers), W (whole numbers), Z (integers), Q (rationals), and R (reals). Fill in the table with yes or no answers, but be prepared to justify your answers. For these exercises, division means nonzero division.

	19.				20.				21.				22.				23.			
Set	N				W				Z				Q				R			
Operation	+	×	−	÷	+	×	−	÷	+	×	−	÷	+	×	−	÷	+	×	−	÷
Closure																				
Associative																				
Identity																				
Inverse																				
Commutative																				
Distributive																				

*This problem requires Section 2.1.

24. a. Given the set $\{1,2,3,4\}$ and the operation $\#$ defined by

$$a \mathbin{\#} b = ab$$

construct a table for $\#$ showing all possible answers for the numbers in the set. That is, complete the table in the margin.

b. Given the set $\{1,2,3,4\}$ and the operation $*$ defined by

$$a * b = 2a$$

construct a table for $*$ showing all possible answers for the numbers in the set. That is, complete the table in the margin.

c. Verify the field properties for the operations of $\#$ and $*$.

25. *Symmetries of a square.* Cut out a small square and label it as shown in the figure in the margin. Be sure that 1 is in front of $1'$, 2 is in front of $2'$, 3 is in front of $3'$, and 4 is in front of $4'$. We will study certain *symmetries* of this square—that is, the results that are obtained when the square is moved around according to certain rules that we will establish.

Hold the square with the front facing you and the 1 in the top right-hand corner as shown. This is called the *basic position.*

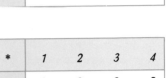

$\#$	1	2	3	4
1	1	2	3	4
2	2	4		
3			.	
4				.

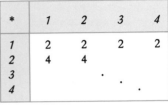

$*$	1	2	3	4
1	2	2	2	2
2	4	4		
3			.	
4				.

Front   Back

Square construction

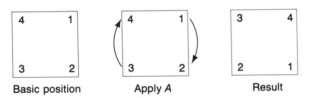

Basic position   Apply A   Result

Now rotate the square 90° clockwise so that 1 moves into the position formerly held by 2 and so that 4 and 3 end up on top. We use the letter A to denote this rotation of the square. That is, A indicates a clockwise rotation of the square through 90°.

Other symmetries can be obtained similarly according to the table on the next page.

You should be able to tell how each of the results in the table was found. Do this before continuing with the problem.

We now have part of a mathematical system: a set of elements $\{A,B,C,D,E,F,G,H\}$. We must define an operation that combines a pair of these symmetries. Define an operation $\star$ which means the following:

$A \star B$ is defined as "A followed by B"; that is, begin with the square in basic position, apply A, and then apply B to get the result as shown:

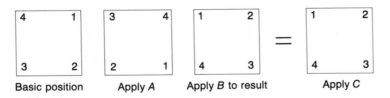

Basic position   Apply A   Apply B to result   Apply C

Thus we say $A \star B = C$, meaning "A followed by B is equivalent to the single element C."

Complete a table for the operation $\star$ and the set $\{A,B,C,D,E,F,G,H\}$.

Symmetries of the Square

Letter	Description	Result
A	90° clockwise rotation	3 4 / 2 1
B	180° clockwise rotation	2 3 / 1 4
C	270° clockwise rotation	1 2 / 4 3
D	360° clockwise rotation	4 1 / 3 2
E	Flip about a horizontal line through the middle of the square	3' 2' / 4' 1'
F	Flip about a vertical line through the middle of the square	1' 4' / 2' 3'
G	Flip along a line drawn from upper left to lower right	4' 3' / 1' 2'
H	Flip along a line drawn from lower left to upper right	2' 1' / 3' 4'

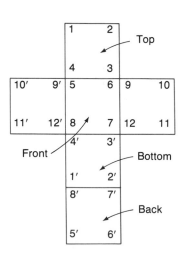

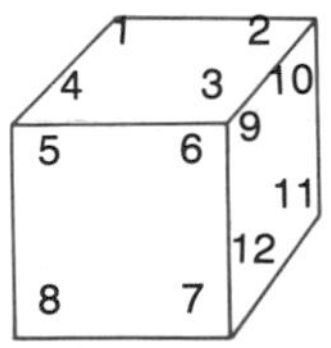

A cube

26. Is the set $\{A,B,C,D,E,F,G,H\}$ for the operation "followed by" of Problem 25:

   a. Closed?   **b.** Associative?   **c.** Commutative?

27. Does the set $\{A,B,C,D,E,F,G,H\}$ for the operation "followed by" of Problem 25 satisfy the:

   a. Identity property?   **b.** Inverse property?

Mind Bogglers

28. *Symmetries of a cube.* Consider a cube labeled as shown in the figure. List all the possible symmetries of this cube (patterning your work after that done for the symmetries of a square). Determine which of the following properties are satisfied: closure, associative, identity, inverse, and commutative.

29. Suppose we use a place-holder ∘ to stand for an operation (addition, subtraction, multiplication, or division). See if you can determine what the operation ∘ stands for in each part.

a. $5 \circ 3 = 8$
    $7 \circ 2 = 9$
    $9 \circ 1 = 10$
    $8 \circ 2 = 10$

b. $5 \circ 3 = 15$
    $7 \circ 2 = 14$
    $9 \circ 1 = 9$
    $8 \circ 2 = 16$

c. $5 \circ 3 = 2$
    $7 \circ 2 = 5$
    $9 \circ 1 = 8$
    $8 \circ 2 = 6$

30. Using the idea presented in Problem 29, we can let ∘ represent operations other than the ordinary ones, as shown by the examples in the margin. Describe the operation ∘ for each part. Also, fill in the blanks by using the same operational rule.

a. $1 \circ 9 = 11$
    $2 \circ 7 = 10$
    $9 \circ 0 = 10$
    $9 \circ 8 = 18$
    $4 \circ 4 = \underline{\hphantom{xx}}$
    $4 \circ 8 = \underline{\hphantom{xx}}$
    $8 \circ 6 = \underline{\hphantom{xx}}$

b. $8 \circ 0 = 1$
    $5 \circ 4 = 21$
    $1 \circ 0 = 1$
    $5 \circ 6 = 31$
    $6 \circ 2 = \underline{\hphantom{xx}}$
    $2 \circ 3 = \underline{\hphantom{xx}}$
    $4 \circ 7 = \underline{\hphantom{xx}}$

c. $4 \circ 6 = 20$
    $8 \circ 2 = 20$
    $7 \circ 9 = 32$
    $6 \circ 8 = 28$
    $5 \circ 2 = \underline{\hphantom{xx}}$
    $4 \circ 3 = \underline{\hphantom{xx}}$
    $8 \circ 3 = \underline{\hphantom{xx}}$

EXAMPLES:

$5 \circ 3 = 5$ $5 \circ 3 = 125$ $5 \circ 3 = 6$
$7 \circ 2 = 7$ $7 \circ 2 = 49$ $7 \circ 2 = 4$
$9 \circ 1 = 9$ $9 \circ 1 = 9$ $9 \circ 1 = 2$
$8 \circ 2 = 8$ $8 \circ 2 = 64$ $8 \circ 2 = 4$

Solutions:

$a \circ b = a$ $a \circ b = a^b$ $a \circ b = 2b$

31. Describe the operation ∘ and fill in the blanks. (See Problems 29 and 30.)

a. $4 \circ 7 = 1$
    $4 \circ 5 = 3$
    $7 \circ 3 = 11$
    $12 \circ 9 = 15$
    $7 \circ 7 = \underline{\hphantom{xx}}$
    $9 \circ 5 = \underline{\hphantom{xx}}$
    $6 \circ 9 = \underline{\hphantom{xx}}$

b. $4 \circ 7 = 17$
    $5 \circ 6 = 26$
    $6 \circ 4 = 37$
    $2 \circ 8 = 5$
    $0 \circ 3 = \underline{\hphantom{xx}}$
    $2 \circ 7 = \underline{\hphantom{xx}}$
    $7 \circ 9 = \underline{\hphantom{xx}}$

c. $5 \circ 9 = 7$
    $6 \circ 9 = 18$
    $6 \circ 7 = 22$
    $3 \circ 0 = 9$
    $6 \circ 2 = \underline{\hphantom{xx}}$
    $8 \circ 6 = \underline{\hphantom{xx}}$
    $4 \circ 5 = \underline{\hphantom{xx}}$

32. *Pretty Kitty Kelly.* By looking for a pattern, formulate a rule that tells what Pretty Kitty Kelly likes. After you find the rule, don't divulge it—simply add more examples to the list. You'll know when you have it!

Pretty Kitty Kelly likes school but doesn't like homework;
          she likes green but doesn't like blue;
          she likes apples but doesn't like oranges;
          she likes happy but doesn't like sad;
          she likes good but doesn't like bad.
Pretty Kitty Kelly likes the bedroom but doesn't like the kitchen;
          she likes trees but doesn't like leaves;
          she likes puppies but doesn't like dogs;
          she likes three but doesn't like four;
          she likes little but doesn't like more.

*6.2 Linear Equations

The field properties of the last section cannot be proved and must be assumed as postulates of the set of real numbers. If they are all assumed, then the resulting

*This section requires Section 4.5.

By "ordinary algebra" we mean the algebra that is taught in high school.

A **variable** is a symbol that is used to represent an unspecified member of some set. A variable is a "place-holder" for the name of some member of a set.

Some values for the variable make the equation true (as with $x = 13$), and some values make the equation false (as with $x = 10$).

mathematical system leads to ordinary algebra. Even though there are many aspects of ordinary algebra that are important to the scientist and mathematician, the ability to solve simple equations is important to the layperson and can be used in a variety of everyday applications.

An **equation** is a statement of equality that may be true or false or may depend on the values of the variable. If it is always true, as in

$$2 + 3 = 5$$

then it is called an *identity*. If it is always false, as in

$$2 + 3 = 15$$

then it is called a *contradiction*. If it depends on the values of the variable, as in

$$2 + x = 15$$

then it is called a *conditional equation*. The values that make the conditional equation true are said to *satisfy* the equation and are called the **solutions** or **roots** of the equation. Generally, when we speak of equations we mean conditional equations. Our concern when solving equations is to find the numbers that satisfy a given equation, so we look for things to do to equations to make the solutions or roots more obvious. Two equations with the same solutions are called **equivalent equations.** An equivalent equation may be easier to solve than the original equation, so we try to get successively simpler equivalent equations until the solution is obvious. There are certain procedures you can use to create an equivalent equation. The following properties specify the operations that are permissible.

MYSTERY THRILLER

An equation is like a mystery
  thriller,
  It grips you once you've
  begun it.
You are the sleuth who stalks
  the killer,
  X represents "whodunit";
The scene of the crime must
  first be cleared,
  The suspects called into
  session;
You look for clues to prove
  your case,
  Till you wring from X a
  confession.

Tom Sampson
Blakelack High School
Oakville, Ontario

Equation Properties

1. *Addition property.* Adding the same number to both sides of a given equation results in an equivalent equation.
2. *Subtraction property.* Subtracting the same number from both sides of a given equation results in an equivalent equation.
3. *Multiplication property.* Multiplying both sides of a given equation by the same nonzero number results in an equivalent equation.
4. *Division property.* Dividing both sides of a given equation by the same nonzero number results in an equivalent equation.

When these properties are used to obtain equivalent equations, *the goal is to isolate the variable on one side of the equation,* as illustrated in Examples 1–4. You can always check the solution to see if it is correct; substituting the solution into the original equation will verify that it satisfies the equation. Notice how the field properties are used when solving these equations.

Example 1: $x + 15 = 25$

$x + 15 - 15 = 25 - 15$ Subtract 15 from both sides.

$x = 10$

The goal is to isolate the variable on one side of the equals sign. Perform the appropriate opposite operations to find a simpler equivalent equation.

The root of the simpler equivalent equation is 10. We often display the answer in the form of an equation, $x = 10$, whose root is obvious.

Example 2: $x - 36 = 42$

$x - 36 + 36 = 42 + 36$ Add 36 to both sides.

$x = 78$

The root is 78; you may leave your answer in the form $x = 78$.

Example 3: $52 = 14 + x$

$52 - 14 = 14 + x - 14$ Actually do the work here. Don't just put down the answer you expect to get.

$38 = x$

Example 4: $15 - x = 0$

$15 - x + x = 0 + x$

$15 = x$

$15 = x$ is the same as $x = 15$. This is a general property of equality called the *symmetric property of equality*: if $a = b$, then $b = a$.

In algebra, it is easy to confuse the variable x with the times sign ($\times$), so the general agreement is to avoid the use of the times sign. Instead, a dot, parentheses, or juxtaposition are often used as follows:

$3 \cdot 5$, $3(5)$, $(3)5$, $(3)(5)$ all mean 3 times 5

$3 \cdot x$, $3x$, $3(x)$, $(3)(x)$ all mean 3 times x

ab, $a(b)$, $a \cdot b$, $(a)(b)$ all mean a times b

It is also sometimes necessary, in algebra, to *combine similar terms* when solving equations. This is done by using the distributive property. For example,

$$5x + 3x = (5 + 3)x \quad \text{By the distributive property}$$

$$= 8x$$

In Examples 5 and 6, simplify the given expressions by combining similar terms.

Example 5: $3a + 7a = (3 + 7)a$

$= 10a$

Example 6: $5a + 4b + 3a = (5a + 3a) + 4b$ Associative and commutative properties

$= 8a + 4b$ $8a + 4b$ cannot be simplified further.

We now use the idea of combining similar terms to solve more complicated equations. In Examples 7–9, find the root of the given equations.

Example 7:
$$5x + 3 - 4x + 5 = 6 + 9$$
$$(5x - 4x) + (3 + 5) = 15$$
$$x + 8 = 15 \qquad \text{Combine similar terms.}$$
$$x + 8 - 8 = 15 - 8 \qquad \text{Subtract 8 from both sides.}$$
$$x = 7$$

Example 8:
$$6x + 3 - 5x + 7 = 11 + 2$$
$$x + 10 = 13 \qquad \text{Combine similar terms first.}$$
$$x = 3 \qquad \textbf{Mentally} \text{ subtract 10 from both sides.}$$

Remember that the goal is to isolate the variable on one side. Use the addition property for solving equations to obtain an equivalent equation that has all the variables on one side.

Example 9:
$$5x + 2 = 4x - 7$$
$$5x + 2 - 4x = 4x - 7 - 4x$$
$$x + 2 = {}^-7$$
$$x + 2 - 2 = {}^-7 - 2$$
$$x = {}^-9$$

Once we have a root, we might want to check our answer. Remember that a root is a value that satisfies the equation. If we substitute the root into the original equation, we should obtain an equality. For example, we can check the solution of Example 9 above:

Since we are checking the equality, we write $\stackrel{?}{=}$ until we verify the equality by doing the arithmetic.

$$5x + 2 = 4x - 7 \qquad \text{Check for } x = {}^-9.$$
$$5({}^-9) + 2 \stackrel{?}{=} 4({}^-9) - 7$$
$${}^-45 + 2 \stackrel{?}{=} {}^-36 - 7$$
$${}^-43 = {}^-43$$

Solve the equations in Examples 10–12.

Example 10:
$$4x + x = 20$$
$$5x = 20 \qquad \text{Combine similar terms.}$$
$$\frac{5x}{5} = \frac{20}{5} \qquad \text{Divide both sides by 5.}$$
$$x = 4 \qquad \text{Simplify.}$$

Example 11:
$$3(m + 4) + 5 = 5(m - 1) - 2$$
$$3m + 12 + 5 = 5m - 5 - 2 \qquad \text{Eliminate parentheses.}$$
$$3m + 17 = 5m - 7 \qquad \text{Combine similar terms.}$$
$$24 = 2m \qquad \text{Add 7 to both sides; subtract } 3m$$
$$\qquad\qquad \text{from both sides.}$$
$$12 = m \qquad \text{Divide both sides by 2.}$$

Example 12:
$$5(y + 2) + 6 = 3(2y) + 5(3y)$$
$$5y + 10 + 6 = 6y + 15y$$
$$5y + 16 = 21y$$
$$16 = 16y$$
$$1 = y$$

I'M SICK OF BEING
AN UNKNOWN!

Problem Set 6.2

Solve the equations in Problems 1–45.

1. $x - 5 = 10$
2. $x - 8 = 14$
3. $6 = x - 2$
4. $^-13 = x - 49$
5. $x + 2 = 7$
6. $x + 8 = 2$
7. $8 + x = 4$
8. $12 + x = 15$
9. $18 + x = 10$
10. $\dfrac{x}{4} = 8$
11. $\dfrac{x}{2} = 18$
12. $\dfrac{x}{^-4} = 11$
13. $7 = \dfrac{x}{^-8}$
14. $^-4 = \dfrac{x}{^-10}$
15. $4x = 12$
16. $4x = ^-48$
17. $^-8x = ^-96$
18. $^-x = 5$
19. $^-x = 14$
20. $13x = 0$
21. $^-19x = 0$
22. $A + 13 = 18$
23. $5 = 3 + B$
24. $6 = C - 4$
25. $2D + 2 = 10$
26. $15E - 5 = 0$
27. $16F - 5 = 11$
28. $6 = 5G - 24$
29. $4(H + 1) = 4$
30. $5(I - 7) = 0$
31. $\dfrac{J}{5} = 3$
32. $\dfrac{2K}{3} = 6$
33. $\dfrac{3L}{4} = 5$
34. $\dfrac{2M}{3} + 7 = 1$
35. $\dfrac{2N}{3} + 11 = 7$
36. $^-5 = \dfrac{2P + 1}{3}$
37. $\dfrac{2 - 5Q}{3} = 4$
38. $\dfrac{5R - 1}{2} = 5$
39. $4(6S - 81) = ^-3(4 + 5S)$
40. $5T + 3(T + 2) + (T + 4) + 17 = 0$
41. $3(U - 3) - 2(U - 12) = 18$
42. $6(V - 2) - 4(V + 3) = 10 - 42$
43. $5(W + 3) - 6(W + 5) = 0$
44. $6(Y + 2) = 4 + 5(Y - 4)$
45. $5(Z - 2) - 3(Z + 3) = 9$

Our modern approach to algebra can be traced to the work of Niels Abel (1802–1829). Abel, now considered to be Norway's greatest mathematician, was, for the most part, ignored during his lifetime. He investigated fifth-degree equations, infinite series, functions, and integral calculus, as well as solutions of the general class of equations now called "Abelian." Although he died at the age of 26 "of poverty, disappointment, malnutrition, and chest complications," he left mathematicians enough material to keep them busy for several hundreds of years.

Mind Bogglers

46. This problem was designed to help you check your work in Problems 22–45. Look at the puzzle form shown at the top of the next page and notice that each small box has a number in the top left-hand corner. Fill in the capital letters from Problems 22–45 to correspond with their numerical values as shown by the number in the box. (The letter *O* has been filled in for you.) Some letters may not appear in the boxes. When you finish putting letters in the boxes, darken all the blank spaces to separate the words in the message. (One of these has been done for you.)

12	−3	0	7	8	−7	−8	$\frac{11}{5}$	O	2	$\frac{20}{3}$	$\frac{1}{3}$	−9	$\frac{1}{5}$	■	11
−15	7	$\frac{20}{3}$	$\frac{20}{3}$	−1	3	−6	4	O	3	2	−3	$\frac{1}{3}$	4	$\frac{20}{3}$	−28
−1	5	8	8	7	8	−3	−5	−28	O	3	$\frac{1}{6}$	$\frac{1}{6}$	7	−6	12
1	7	−6	4	7	−6	6	$\frac{1}{4}$	$\frac{1}{3}$	$\frac{11}{5}$	$\frac{11}{5}$	O	$\frac{11}{5}$	8	/	/

47. A hippopotamus and a little bird want to play on a teeter-totter. The bird says that it's impossible, but the hippopotamus assures the little bird that it will work out and that she will prove it, since she has had a little algebra. She presents the following argument:

Let H = Weight of the hippopotamus
b = Weight of the bird

Now there must be some weight w (probably very large) so that

$$H = b + w$$

Multiply both sides by $H - b$:

$$H(H - b) = (b + w)(H - b)$$

Using the distributive property:

$$H^2 - Hb = bH + wH - b^2 - wb$$

Subtract wH from both sides:

$$H^2 - Hb - wH = bH - b^2 - wb$$

Use the distributive property to factor both sides:

$$H(H - b - w) = b(H - b - w)$$

Divide both sides by $H - b - w$:

$$H = b$$

Thus, the weight of the hippopotamus is the same as the weight of the bird. "Now," says the hippopotamus, "since our weights are the same, we'll have no problem on the teeter-totter."

"Wait!" hollers the bird. "Obviously this is false." But where is the error in the reasoning?

48. What is wrong with the following "proof"?

i.   12 **eggs** = 1 **dozen**   Multiply both sides of step i by 2.
ii.   24 **eggs** = 2 **dozen**   Divide both sides of step i by 2.
iii.   6 **eggs** = $\frac{1}{2}$ **dozen**   Multiply step iii times step ii
iv. 144 **eggs** = 1 **dozen**   (equals times equals are equal).
v.   12 **dozen** = 1 **dozen**   Substitute, since 144 eggs = 12 dozen.

*6.3 Prime Numbers

A set of numbers that is important, not only in algebra but in all of mathematics, is the set of prime numbers. In order to understand prime numbers, you must first understand the idea of divisibility, along with some new terminology and notation.

The natural number 10 is divisible by 2, since there is a counting number 5 so that $10 = 2 \cdot 5$; it is not divisible by 3, since there is no counting number k such that $10 = 3 \cdot k$. This leads us to the following definition.

Definition

If m and d are natural numbers, and if there is a natural number k so that $m = d \cdot k$, we say that:

$$d \text{ is a } divisor \text{ of } m$$
$$d \text{ is a } factor \text{ of } m$$
$$d \text{ } divides \text{ } m$$
$$m \text{ is a } multiple \text{ of } d$$

Let's denote this relationship as $d|m$. That is, $5|30$ means that 5 divides 30 (because there exists some counting number k—namely 6—so that $30 = 5 \cdot k$). Do not confuse this notation with the notation sometimes used for fractions: 5/30—which is $\frac{1}{6}$. In Examples 1–4, which are true?

Example 1: $7|63$

Solution: $7|63$ says "7 divides 63," which is true, since we can find a natural number k—namely 9—such that $63 = 7 \cdot k$.

*This section requires Section 4.3.

Historical Note

Evariste Galois (1811–1832)

Galois is another important name in the history of algebra. Like Abel, he also was not recognized during his lifetime. He had trouble with his teachers in school because of his ability to do a great amount of work in his head; his teachers always wanted him to "show all of his work." At 16 he wrote up several of his discoveries and sent them to a famous mathematician of the time, Augustin-Louis Cauchy, who promised to present Galois' results. However, Cauchy lost

Galois' work. Galois tried again by sending his work to another mathematician, Joseph Fourier. This time Galois' work was lost because Fourier died before he could read it. When Galois was 20, he had an affair with a flirtatious young girl and subsequently became involved in a duel over "her honor." Knowing that he would die in the encounter, Galois arranged for his best friend to marry his own fiancee. He then went home and spent the entire night before the duel writing out as many of his discoveries as he could. On May 20, 1832, at the age of 20, Galois was shot, and he died the next day. Two days after his death a letter was received stating that he was appointed to a professorship at the University of Berlin—the fulfillment of his lifelong dream.

Example 2: $8|104$

Solution: $8|104$ is true, since $104 = 8 \cdot 13$.

Example 3: $14|2$

Solution: $14|2$ is false. We write $14\nmid2$ to say that 14 does not divide 2.

Example 4: $6|15$

Solution: $6|15$ is false. That is, we can find no natural number k so that $15 = 6 \cdot k$.

It is easy to see that 1 divides every counting number m, since

$$m = 1 \cdot m$$

Also, we see that every counting number m divides itself, since

$$m = m \cdot 1$$

We have proved the following theorem.

Theorem

Every counting number greater than 1 has at least two distinct divisors, itself and 1.

Since every counting number greater than 1 has at least two divisors, can any numbers have more than two?

Checking: 2 has exactly two divisors: 1, 2

3 has exactly two divisors: 1, 3

4 has more than two divisors: 1, 2, and 4

Thus some numbers (such as 2 and 3) have exactly two divisors, and some (such as 4 and 6) have more than two divisors. Do any counting numbers have fewer than two divisors?

We now state a definition that classifies the counting numbers according to the number of divisors of each.

Definition

A *prime number* is a counting number that has exactly two divisors. The counting numbers that have more than two divisors are called *composite numbers*.

Thus 2 is prime, 3 is prime, 4 is composite (since it is divisible by three counting numbers), 5 is prime, 6 is composite (since it is divisible by 1, 2, 3, and 6). Note that every counting number greater than 1 is either prime or composite. The number 1 is neither prime nor composite.

One method for finding primes smaller than some given number was first used by a Greek mathematician named Eratosthenes more than 2,000 years ago. The technique is known as the *Sieve of Eratosthenes*. Suppose we wish to find the primes less than 100. First, we prepare a table of the counting numbers through 100, as is shown in Table 6.2.

TABLE 6.2   Finding Primes Using the Sieve of Eratosthenes

1	2	3	4	5	6	7	8	9	10
11	12	13	14	15	16	17	18	19	20
21	22	23	24	25	26	27	28	29	30
31	32	33	34	35	36	37	38	39	40
41	42	43	44	45	46	47	48	49	50
51	52	53	54	55	56	57	58	59	60
61	62	63	64	65	66	67	68	69	70
71	72	73	74	75	76	77	78	79	80
81	82	83	84	85	86	87	88	89	90
91	92	93	94	95	96	97	98	99	100

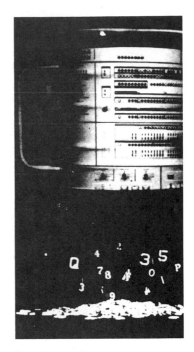

Sieve for numbers

Cross out 1, since it is not classified as a prime number.

Draw a circle around 2, the smallest prime number. Then cross out every following multiple of 2, since each one is divisible by 2 and thus is not prime.

Draw a circle around 3, the next prime number. Then cross out each succeeding multiple of 3. Some of these numbers, such as 6 and 12, will already have been crossed out because they are also multiples of 2.

Circle the next open number, 5, and cross out all subsequent multiples of 5.

The next open number is 7; circle 7 and cross out multiples of 7. Since 7 is the largest prime less than $\sqrt{100} = 10$, we will now show that all of the remaining numbers are prime.

The process is a simple one, since you do not have to cross out the multiples of 3 (for example) by checking for divisibility by 3 but can simply cross out every third number. Thus anyone who can count can find primes by this method. Also, notice that, in finding the primes under 100, we had crossed out all the composite numbers by the time we crossed out the multiples of 7. That is, to find all primes less than 100: (1) Find the largest prime smaller than or equal to

$\sqrt{100} = 10$ (7 in this case); (2) cross out multiples of primes up to and including 7; and (3) all the remaining numbers in the chart are primes.

This result generalizes. If we wish to find all primes smaller than n:

1. Find the *largest* prime smaller than or equal to $\sqrt{n}$.
2. Cross out the multiples of primes less than or equal to $\sqrt{n}$.
3. All the remaining numbers in the chart are primes.

Phrasing this another way, if n is composite, then one of its factors must be less than or equal to $\sqrt{n}$. That is, if $n = ab$, then it can't be true that *both a* and b are greater than $\sqrt{n}$ (otherwise $ab > \sqrt{n}\sqrt{n} = n = ab$, so $ab > ab$ is a contradiction). Thus one of the factors must be less than or equal to $\sqrt{n}$.

We use this property to tell us when we have excluded all composite numbers from a set. In the set of numbers $\{1,2,3,\ldots,100\}$, we have considered the primes 2, 3, 5, and 7. The next prime is 11. Since $11^2 = 121$, and since we have 100 numbers in our table, we know we are finished and all remaining numbers are primes. Hence the last prime we need to consider is 7.

This method of Eratosthenes' is not very satisfactory to use if we wish to determine whether or not a given number n is a prime. For centuries, mathematicians have tried to find a formula that would yield *every* prime—or even a formula that would yield only primes. Let's try to find a formula that results in giving only primes. A possible candidate is

$$n^2 - n + 41$$

If we try this formula for $n = 1$, we obtain $1^2 - 1 + 41 = 41$. For $n = 2$: $2^2 - 2 + 41 = 43$.   For $n = 3$: $3^2 - 3 + 41 = 47$.

So far, so good. That is, we are obtaining only primes. Continuing, we have Table 6.3.

Notice that all the numbers in the right columns are primes. If we reason inductively and conclude from the pattern that we have a formula for primes, our conjecture would be shattered by the next case, $n = 41$.

$$41^2 - 41 + 41$$
$$1{,}681 - 41 + 41 = 1{,}681$$

Now, 1,681 is divisible by 1, 1,681, and 41, and therefore is not a prime.

A more serious attempt to find a prime number formula was made by Fermat, who tried

$$2^{2^n} + 1$$

For $n = 1$, $2^{2^1} + 1 = 5$, and 5 is a prime.
For $n = 2$, $2^{2^2} + 1 = 2^4 + 1 = 17$, and 17 is a prime.
For $n = 3$, $2^{2^3} + 1 = 2^8 + 1 = 257$, and 257 is a prime.

Now the numbers get large rather quickly:

For $n = 4$, $2^{2^4} + 1 = 2^{16} + 1 = 65{,}537$

TABLE 6.3 Evaluation of the Formula $n^2 - n + 41$

Value of n	Value of $n^2 - n + 41$	Value of n	Value of $n^2 - n + 41$
4	53	23	547
5	61	24	593
6	71	25	641
7	83	26	691
8	97	27	743
9	113	28	797
10	131	29	853
11	151	30	911
12	173	31	971
13	197	32	1033
14	223	33	1097
15	251	34	1163
16	281	35	1231
17	313	36	1301
18	347	37	1373
19	383	38	1447
20	421	39	1523
21	461	40	1601
22	503		

Checking this number with a computer, we find that it, too, is a prime number. But, alas, this formula fails for $n = 5$:

$$2^{2^5} + 1 = 2^{32} + 1 = 4{,}294{,}967{,}297$$

It turns out that this number is not prime! It is divisible by 641. Whether this formula generates any other primes is still an unsolved problem.

In 1970 a young Russian named Matyasievich discovered several explicit polynomials of this sort that generate only prime numbers, but all of those he discovered are too complicated to reproduce here. The largest prime number known at that time was

$$2^{11,213} - 1$$

and was discovered at the University of Illinois through the use of number-theoretic techniques and computers. The mathematicians were so proud of this discovery that the following ad was used on the university's postage meter:

Historical Note

Fermat was not able to show that 4,294,967,297 was not a prime. Euler later discovered that it is divisible by 641. In the 1800s there was a young American named Colburn who had a great capacity to do things in his head. He was shown this number and asked if it was a prime. He said, "No, because 641 divides it." When asked how he knew that, he replied "Oh, I just felt it." He was unable to explain this great gift—he just had it.

𝔥istorical 𝔑ote

Laura Nickel and Curt Noll, two 18-year-old students at Cal State Hayward, spent three years finding the then largest known prime number:

$$2^{21,701} - 1$$

At the time of this writing they are at work trying to find an even larger prime. As you can see from Figure 6.1, they used a computer to help them find this number, and they spent over 1,900 hours on the problem. The number shown in Figure 6.1 was discovered at 9 P.M. on October 30, 1978.

MATHEMATICIANS SOLVE A 3,000-YEAR-OLD PUZZLE

Los Angeles Times

European mathematicians, building on recent work by theorists in the United States, have solved a 3,000-year-old problem in mathematics, an achievement that raises questions about the security of recently developed secret codes.

The problem, one of the oldest in mathematics, is how to determine whether a number is prime, that is, whether it can be divided by any number other than itself and 1. The smallest prime numbers are 2, 3, 5, 7, and 11. The largest known contains 18,000 digits.

Using the new method, a test on a 97-digit number, which previously could have taken as long as a hundred years with the fastest computer, was done in 77 seconds.

It's possible that the method may help determine the divisors of a large number that is not prime.

Another large prime is

$$2^{19,937} - 1$$

which was found by Bryant Tuckerman on March 4, 1971, at the **IBM Research Center**. It is more than

$$1.6 \times 10^{2617}$$

times as great as the University of Illinois' number.

A larger prime number wasn't found until 1978 when it was discovered by students Laura Nickel and Curt Noll. They began their work in 1975 as high school students and three years later found the prime

$$2^{21,701} - 1$$

by using the computer at Cal State Hayward. This prime number is shown in Figure 6.1. A few months later Curt Noll found a larger prime

$$2^{23,209} - 1$$

and in May 1979 Harry Nelson of the Livermore Lab found a prime almost twice as large:

$$2^{44,497} - 1$$

At the date of this printing, this is the largest known prime.

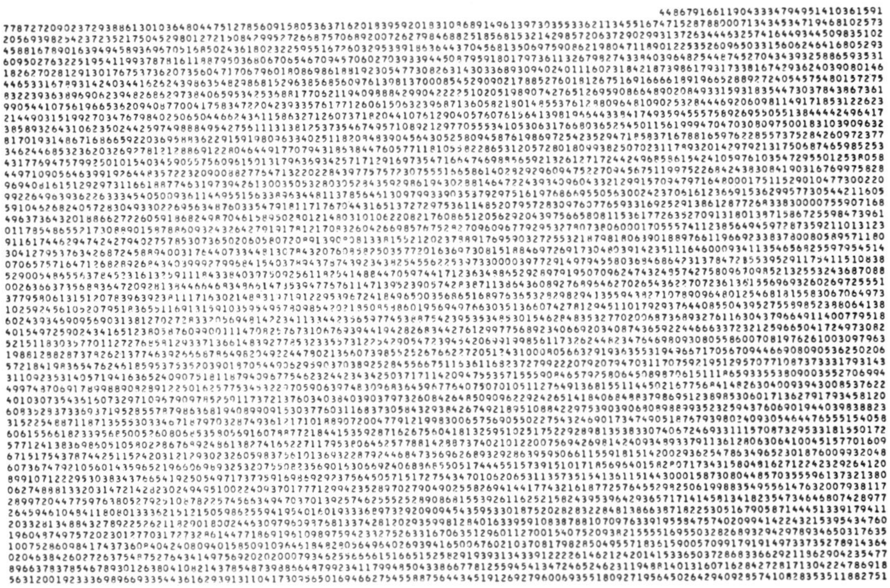

FIGURE 6.1 The largest known prime in 1978. This *single* number has no divisors other than itself and 1.

Problem Set 6.3

1. What is a prime number?

2. Which of the following numbers are primes?

   a. 59   **b.** 57   **c.** 1   **d.** 97   **e.** 79

3. Which of the following numbers are primes?

   a. 63   **b.** 73   **c.** 37   **d.** 43   **e.** 87

4. **a.** Do any counting numbers have fewer than two distinct divisors?
   b. Are all counting numbers either prime or composite?

5. Are the following true or false?

   a. $8 | 48$ **b.** $6 \nmid 39$ **c.** $15 | 5$ **d.** $2 | 628,174$ **e.** $10 | 148,729,320$

6. Are the following true or false?

   a. $7 | 65$ **b.** $12 | 156$ **c.** $16 \nmid 576$ **d.** $5 | 817,543$
   e. $10 \nmid 518,728,641$

7. **a.** Is every odd number a prime number?
   b. Is every prime number greater than 2 an odd number?

8. Find a pair of prime numbers that differ by 1, and show that there is only one such pair possible.

9. Find all the prime numbers less than or equal to 300.

10. What is the largest prime you need to consider to be sure that you have excluded, in the Sieve of Eratosthenes, all primes less than or equal to:

    a. 200?   **b.** 500?   **c.** 1,000?

11. We used the Sieve of Eratosthenes in Table 6.2 by arranging the first 100 numbers into 10 rows and 10 columns. Repeat the sieve process for the first 100 numbers by arranging the numbers in the following patterns.

    a. By 6:

1	2	3	4	5	6
7	8	9	10	11	12
13	14	15			

    $\vdots$

    b. By 7:

1	2	3	4	5	6	7
8	9	10	11	12	13	14
15						

    $\vdots$

    c. By 21:

    As you are using these sieves, look for patterns. Name some of the patterns you notice. Do you think that these sieves are better than the one shown in Table 6.2? Why or why not?

12. Using the sieves of Problem 11, make a conjecture about primes and multiples of 6.

One way to tell if a number is prime is to try to divide it by other numbers. If the number is very large, trying to divide it by other numbers is hopeless.

In 1640, Pierre de Fermat invented a test for primes. In the last decade, mathematicians decided that they could run Fermat's test on a suspected prime number many times, and if it consistently passed, there was a strong probability that the number was prime.

There still were many steps involved and the probabilistic nature of the test left mathematicians unsatisfied.

To overcome the problem, Pomerance, Rumely, and Adleman bombard the suspected prime with tests using not ordinary numbers but closely related structures called algebraic number fields. That reduces the steps required and yields a definitive answer.

1	2̸	3	4̸	5̸	6̸
7	8̸	9	10̸	11̸	12̸
13	14̸	15	16̸	17̸	18̸
19̶	20̸	21	22̸	23̸	24̸
25	26̸	27̸	28̸	29̸	30̸
31	32̸	33	34̸	35̸	36̸
37	38̸	39̶	40̸	41̸	42̸
43	44̸	45	46̸	47̸	48̸

. . .

𝔥istorical 𝔑ote

The Russian mathematician L. Schnirelmann (1905–1938) proved that every positive integer can be represented as the sum of not more than 300,000 primes (a step in the right direction?). Later another Russian mathematician, I. M. Vinogradoff, proved that there exists a number N such that all numbers larger than N can be written as the sum of, at most, four primes.

13. *Lucky numbers.* Set up a sieve similar to the one illustrated in the margin.
    (1) Start counting with 1 *each time.*
    (2) Cross out every second number (shown as /).
    (3) The next uncrossed number is 3; cross out every third number that remains (shown as ✕).
    (4) The next uncrossed number is 7; cross out every seventh number that remains (shown as —).
    (5) Continue in the same fashion.

    What are the lucky numbers less than 100?

14. Pairs of consecutive odd numbers that are primes are called *prime twins.* For example, 3 and 5, 11 and 13, and 41 and 43 are prime twins. Can you find any others?

15. For what values of n is $11 \cdot 14^n + 1$ a prime? [*Hint:* Consider n even and then consider n odd.]

16. Suppose three new states were added to the United States. Then the number of states would be 53, a prime. Could we still arrange the stars into rows and columns so that the design would be symmetric?

17. Three consecutive odd numbers that are primes are called *prime triplets.* It is easy to show that 3, 5, and 7 are the only prime triplets. Can you explain why this is true?

18. *Goldbach's conjecture.* In a letter to Leonhard Euler in 1742, Christian Goldbach observed that every even integer except 2 seemed representable as the sum of two primes. Euler was unable to prove or disprove this conjecture, and it remains unsolved to this day. Write the following numbers as the sum of two primes (the first three are worked for you):

$4 = 2 + 2$	$6 = 3 + 3$
$8 = 5 + 3$	$10 =$
$12 =$	$14 =$
$16 =$	$18 =$
$20 =$	$40 =$
$80 =$	$100 =$

19. Construct a 3-by-3 magic square consisting solely of primes or the number 1. [*Hint:* The entries of the magic square include 1, 31, 37, 43, 61, 67, and 73. The others are less than 11.]

Mind Bogglers

20. A man goes to a well with two cans whose capacities are 3 gallons and 5 gallons. Explain how he can obtain exactly 4 gallons of water from the well.

21. Some primes are 1 more than a square. For example, $5 = 2^2 + 1$. Can you find any other primes p so that $p = n^2 + 1$?

22. Some primes are 1 less than a square. For example, $3 = 2^2 - 1$. Can you find any other primes p so that $p = n^2 - 1$?

23. *Calculator problem.* In the text we tried some formulas that might have generated only primes, but, alas, they failed. Below are some other formulas. Show that these, too, do not generate only primes.

a. $n^2 + n + 41$ b. $n^2 - 79n + 1{,}601$
c. $2n^2 + 29$ d. $9n^2 - 489n + 6{,}683$
e. $n^2 + 1$, n an even integer

Problems for Individual Study

24. *Infinitude of primes.* A proof that there are infinitely many primes using $p! + 1$ is attributed to Euclid. Consult some elementary number-theory textbook, and then write a convincing argument showing that there are infinitely many prime numbers.

25. *Primes.* Investigate some of the properties of primes not discussed in the text. Why do primes hold such fascination for mathematicians? Why are primes important in mathematics? What are some of the important theorems concerning primes?

    Reference: Martin Gardner, "The Remarkable Lore of the Prime Number," *Scientific American,* March 1964 (vol. 210, no. 3).

6.4 Prime Factorization

Figure 6.2 shows the frequency of primes among the first 10,000 or the first 65,000 integers. But is there a largest prime? In this section we will answer this question conclusively after showing that every counting number greater than 1 is either a prime or a product of primes.

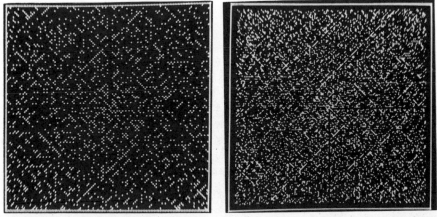

FIGURE 6.2 Photographs of a computer grid showing primes as a spiral of integers from 1 to about 10,000 (left) and from 1 to about 65,000 (right). For a description, see the Historical Note on page 308.

The operation of factoring is the reverse of the operation of multiplying. Multiplying 3 by 6 yields

$$3 \cdot 6 = 18$$

Historical Note

In 1964, Stanislaw Ulam of the
Los Alamos Scientific Laboratory
was passing the time by arranging
the counting numbers into a grid
in a spiral pattern as shown:

17	16	15	14	13	
	5	4	3	12	
	6	1	2	11	
	7	8	9	10	

Next, he crossed out the primes
and noticed that the primes
seemed to congregate into straight
lines:

100	99	98	97	96	95	94	93	92	91
65	64	63	62	61	60	59	58	57	90
66	37	36	35	34	33	32	31	56	89
67	38	17	16	15	14	13	30	55	88
68	39	18	5	4	3	12	29	54	87
69	40	19	6	1	2	11	28	53	86
70	41	20	7	8	9	10	27	52	85
71	42	21	22	23	24	25	26	51	84
72	43	44	45	46	47	48	49	50	83
73	74	75	76	77	78	79	80	81	82

How would the grid look, Ulam
wondered, if this process were
extended to thousands of primes?
The result is shown in Figure 6.2.

and the answer is unique (only one answer is possible). The reverse process is
called **factoring.** Suppose we are given the number 18 and are asked for numbers that can be multiplied together to give 18. Then we can list the different
possible factorization of 18:

$$18 = 1 \cdot 18$$
$$= 18 \cdot 1$$
$$= 2 \cdot 9$$
$$= 1 \cdot 1 \cdot 2 \cdot 9$$
$$= 3 \cdot 6$$
$$= 2 \cdot 3 \cdot 3 = 2 \cdot 3^2$$
$$\vdots$$

In this problem we see that the answer is certainly not unique; in fact, there
are infinitely many possibilities. Some agreements are in order.

1. We will not consider the order in which the factors are listed as important.
   That is, $2 \cdot 9$ and $9 \cdot 2$ will be considered the same factorization.
2. We will not consider 1 as a factor when writing out any factorizations. That
   is, prime numbers will not have factorizations.
3. Recall that we are working in the set of counting numbers; thus,
   $18 = 36 \cdot \frac{1}{2}$ and $18 = (^-2)(^-9)$ are not considered factorizations of 18.

With these agreements, we have greatly reduced the possibilities.

$$18 = 2 \cdot 9$$
$$= 3 \cdot 6$$
$$= 2 \cdot 3^2$$

are the only possible factorizations. Notice that the last factorization contains
only prime factors; thus it is called the **prime factorization** of 18.
   It should be clear that, if a number is composite, it can be factored into two
counting numbers greater than 1. These two numbers themselves will be prime
or composite. If they are prime, then we have a prime factorization. If one or
more is composite, we repeat the process. This continues until we have written
the original number as a product of primes. It is also true that this representation is unique. This is one of the most important results in arithmetic, and we
state it formally:

Fundamental Theorem of Arithmetic

Every counting number greater than 1 is either a prime or a product of
primes, and the factorization is unique (except for the order in which the
factors appear).

Example 1:   Find the prime factorization of 385.

Solution: One of the easiest ways of finding the prime factors of a number is to try division by each of the prime numbers in order:

$$2, 3, 5, 7, 11, 13, 17, \text{ and } 19$$

If none of these primes divides 385, then 385 must be prime. We see by inspection that 385 is not divisible by 2 or 3. It is divisible by 5, so

$$385 = 5 \cdot 77$$

Since 77 is a composite number that is divisible by 7 and 11, we write

$$385 = 5 \cdot 7 \cdot 11$$

Some people prefer to write these steps using a "factor tree."

Recall that we need only check primes less than or equal to $\sqrt{385}$, since the next prime is 23 and $23^2 = 529$.

A number is divisible by
  2 if the last digit is even;
  3 if the sum of the digits is
    divisible by 3;
  5 if the last digit is 0 or 5.

```
      385
      /\
    5 · 77
    /  /\
  5 · 7 · 11
```

Example 2:   Find the prime factorization of 1,400.

Solution: Using the factor tree, we may find *any* factors of 1,400:

Now find any factors of 10 and 140:

Continue finding factors of nonprimes:

Write the factorization using exponents:

```
       1,400
       /\
     10 · 140
    /\     /\
  2 · 5 · 10 · 14
  /  /  /\    /\
2 · 5 · 2 · 5 · 2 · 7
```

$$2^3 \cdot 5^2 \cdot 7$$

This answer is written in what is called canonical form. The **canonical representation** of a number is the representation of that number as a product of primes using exponential notation with the factors arranged in order of increasing magnitude.

Example 3:   Find the prime factorization of 3,465 using a factor tree.

Solution:

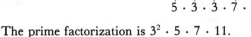

```
           3,465
           /\
         5 · 693
        /   /\
      5 · 3 · 231
     /   /   /\
    5 · 3 · 3 · 77
   /   /   /   /\
  5 · 3 · 3 · 7 · 11
```

The prime factorization is $3^2 \cdot 5 \cdot 7 \cdot 11$.

TABLE 6.4

n	% of primes $\leq n$
10	40
100	25
500	19
1,000	16.8
10,000	12.37
1,000,000	6 (approx.)

This is a sequence of two consecutive nonprimes.

This is a sequence of 18 consecutive nonprimes.

By the Fundamental Theorem of Arithmetic this is the only possible prime factorization of 3,465. Regardless of how we begin the factor tree, the end result will always be the same.

You might wonder how many primes exist. If we count the primes less than or equal to 10, we have 2, 3, 5, and 7 (or 40% of the numbers). If we do this for other numbers, as shown by Table 6.4, we find that the primes are "thinning out" and, indeed, may run out. In fact, we can create a sequence of consecutive nonprime numbers that is as large as we please. Consider

$$3 \cdot 2 \cdot 1 + 1$$

The next two consecutive numbers are

$$3 \cdot 2 \cdot 1 + 2 \qquad \text{Divisible by 2}$$
$$3 \cdot 2 \cdot 1 + 3 \qquad \text{Divisible by 3}$$

and these are not prime. Consider $19 \cdot 18 \cdot 17 \cdots \cdot 3 \cdot 2 \cdot 1 + 1$ along with the next 18 consecutive numbers:

$$19 \cdot 18 \cdot 17 \cdots \cdot 3 \cdot 2 \cdot 1 + 2 \qquad \text{Divisible by 2}$$
$$19 \cdot 18 \cdot 17 \cdots \cdot 3 \cdot 2 \cdot 1 + 3 \qquad \text{Divisible by 3}$$
$$19 \cdot 18 \cdot 17 \cdots \cdot 3 \cdot 2 \cdot 1 + 4 \qquad \text{Divisible by 4}$$
$$\vdots \qquad\qquad \vdots$$
$$19 \cdot 18 \cdot 17 \cdots \cdot 3 \cdot 2 \cdot 1 + 18 \qquad \text{Divisible by 18}$$
$$19 \cdot 18 \cdot 17 \cdots \cdot 3 \cdot 2 \cdot 1 + 19 \qquad \text{Divisible by 19}$$

Notice that $1,001 \cdot 1,000 \cdot 999 \cdots \cdot 3 \cdot 2 \cdot 1 + 1$ is followed by 1,000 consecutive nonprimes. You can thus show that there exist a million consecutive nonprimes, or a billion, or any number you wish.

Indeed, the primes do seem to be thinning out. However, using the Fundamental Theorem, we can present an argument that will lead us to conclude that there must be infinitely many prime numbers. Let's begin by considering a proof that 19 is not the largest prime. There are two ways to proceed. The obvious way is not much fun. That is, we could simply find a larger prime—say, 23—and be finished.

Let's proceed differently. We will prove that 19 is not the largest prime by contradiction. That is, assume that there are only finitely many primes, ending with 19. *We are assuming that 19 is the largest prime.*

Consider the number $2 \cdot 3 \cdot 5 \cdot 7 \cdot 11 \cdot 13 \cdot 17 \cdot 19$. Certainly this number is larger than 19, but it is not a prime (why?). So we have no contradiction yet.

Consider the number

$$M = (2 \cdot 3 \cdot 5 \cdot 7 \cdot 11 \cdot 13 \cdot 17 \cdot 19) + 1$$

This number is larger than 19. Is it a prime? According to our assumption, it must be composite, since it is larger than 19. But if it is composite, it has a prime divisor (Fundamental Theorem). Check all primes: 2 does not divide M, since 2 divides $2 \cdot 3 \cdot 5 \cdot 7 \cdot 11 \cdot 13 \cdot 17$, and it cannot divide that number plus 1; 3 does not divide M for the same reason; but then 5, 7, 11, 13, 17, and 19 do not divide M. Thus, if M is not divisible by any prime, then either it must be prime or there is a prime divisor larger than 19. In either case, we have found a prime larger than 19.

Now if *anyone* claims to be in possession of the largest prime, we need only carry out an argument like the above to find a larger prime. Thus we are saying that there are infinitely many primes, since it is impossible to have a largest prime.

Problem Set 6.4

Write the prime factorization for each of the numbers in Problems 1–10.

1. **a.** 24   **b.** 30   2. **a.** 300   **b.** 144

3. **a.** 28   **b.** 76   4. **a.** 215   **b.** 125

5. **a.** 108   **b.** 740   6. **a.** 699   **b.** 123

7. **a.** 120   **b.** 90   8. **a.** 75   **b.** 975

9. **a.** 490   **b.** 4,752   10. **a.** 143   **b.** 51

Indicate which of the numbers in Problems 11–22 are prime. If the number is not prime, write its prime factorization.

11. 83   12. 97   13. 127   14. 113

15. 377   16. 151   17. 105   18. 187

19. 67   20. 229   21. 315   22. 111

23.   In your own words, state the Fundamental Theorem of Arithmetic.

24.   *Calculator problem.* Find the prime factorization of 101,101.

25.   *Calculator problem.* Find the prime factorization of 514,080.

26.   Use an argument similar to the one in the text on page 310 to show that 23 is not the largest prime.

Find the prime factorization for each of the numbers in Problems 27–30.

27. 567   28. 568   29. 2,869   30. 793

31.   In the text, we showed that 19 was not the largest prime by considering

$$M = (2 \cdot 3 \cdot 5 \cdot 7 \cdot 13 \cdot 17 \cdot 19) + 1$$

Now, M is either prime or composite. If it is prime, then it is larger than 19, and we have a prime larger than 19. If it is composite, it has a prime divisor larger than 19. In either case, we find a prime larger than 19. Show that this number M does not always generate primes. That is,

Historical Note

A statement equivalent to the Fundamental Theorem of Arithmetic is found in Book IX of Euclid's *Elements* (Proposition IX 14). Very little is known about Euclid except through these books. This work is not only the earliest known major Greek mathematical book, but it is also the most influential textbook of all time. It was composed around 300 B.C. and first printed in 1482. Except for the Bible, no other book has been through so many editions and printings. Even though we do not know the year of Euclid's birth or death, we do know he lived around 300 B.C. because he was the first professor of mathematics at the Museum of Alexandria, which opened in 300 B.C. He was also the author of at least ten other books, five of which have been lost to us. However, Euclid's 13-volume *Elements* is a masterpiece of mathematical thinking. It was described by Augustus De Morgan (1806–1871) with the statement: "The thirteen books of Euclid must have been a tremendous advance, probably even greater than that contained in the *Principia* of Newton."

$$2 + 1 = 3, \quad \text{a prime}$$
$$2 \cdot 3 + 1 = 7, \quad \text{a prime}$$
$$2 \cdot 3 \cdot 5 + 1 = 31, \quad \text{a prime}$$

Find an example in which the product of consecutive primes plus 1 does not yield a prime.

32. We showed that 19 was not the largest prime. Generalize the argument in the text to show that there is *no* largest prime.

33. *Calculator problem.* Find the one composite number in the following set:

31
331
3331   *Hint:* The composite number has
33331   a prime factor under 19.
333331
3333331
33333331
333333331

Mind Bogglers

34. There are only seven prime number years in the period 1950–2000. What are the prime number years, if any, from 1980 to 1989?

35. The Pythagoreans studied numbers in order to find certain mystical properties in them. We mentioned that they had masculine and feminine numbers, amicable numbers, deficient numbers, and so on.

A *perfect number* is a counting number that is equal to the sum of all its divisors that are less than itself. These divisors less than the number itself are called *proper divisors.* The proper divisors of 6 are {1,2,3} and $1 + 2 + 3 = 6$, so 6 is the smallest perfect number.

On the other hand, 24 is not perfect, since its proper divisors, {1,2,3,4,6,8,12}, have the sum

$$1 + 2 + 3 + 4 + 6 + 8 + 12 = 36$$

All even perfect numbers are of the form $2^{p-1}(2^p - 1)$, where p and $(2^p - 1)$ are prime numbers.

Find the perfect number furnished by the formula

$$2^{p-1}(2^p - 1) \quad \text{when} \quad p = 5$$

Show that it is perfect.

36. Two numbers are called *amicable,* or *friendly,* if each is the sum of the proper divisors of the other (a proper divisor includes the number 1 but not the number itself). For example, the numbers 220 and 284 are friendly. The proper divisors of 220 are {1,2,4,5,10,11,20,22,44,55,110}, and

$$1 + 2 + 4 + 5 + 10 + 11 + 20 + 22 + 44 + 55 + 110 = 284$$

The proper divisors of 284 are {1,2,4,71,142}, and

$$1 + 2 + 4 + 71 + 142 = 220$$

Show that 1,184 and 1,210 are friendly.

Problems for Individual Study

37. *Computer problem.* Do some research to find additional perfect numbers. Then use a computer to help show that each of them is a perfect number. [*Hint:* There are no odd perfect numbers less than or equal to 10^{200}, according to Bryant Tuckerman, IBM Research Paper RC-1925.]

Reference: Robert Prielipp, "Perfect Numbers, Abundant Numbers, and Deficient Numbers," *Mathematics Teacher,* vol. LXIII, no. 8 (December 1970), pp. 692–696.

38. Write out a proof of the Fundamental Theorem of Arithmetic. [*Hint:* You must show uniqueness.]

*6.5 Applications of Prime Factorization

We can make use of unique prime factorization in many situations. Recall that if $m = d \cdot k$, then m is a multiple of d and k, and d and k are factors of m. The first situation we'll consider is useful when simplifying fractions. Begin by finding the set of factors of two numbers, say 24 and 30.

$$A = \text{Set of factors of } 24 = \{1,2,3,4,6,8,12,24\}$$
$$B = \text{Set of factors of } 30 = \{1,2,3,5,6,10,15,30\}$$

The intersection of these two sets is the set of *common factors* of 24 and 30:

$$A \cap B = \{1,2,3,6\}$$

The number 1 is the least element in this set, and 6 is the greatest. Now, 1 will always be a common factor for any pair of counting numbers, but the greatest of the common factors depends on the numbers chosen. Thus we will be concerned with finding the **greatest common factor (g.c.f.)**, and we will say that 6 is the g.c.f. of 24 and 30. That is, 6 is the *largest* number to divide *both* 24 and 30.

The method just used is lengthy. We would like to find an algorithm that will give us the g.c.f. directly.

We find the g.c.f. by first finding the prime factorization of each number:

$$24 = 2^3 \cdot 3$$
$$30 = 2 \ \cdot 3 \cdot 5$$

We write these factorizations in *canonical form* as follows:

$$24 = 2^3 \cdot 3^1 \cdot 5^0 \qquad \text{Recall that } b^0 = 1.$$
$$30 = 2^1 \cdot 3^1 \cdot 5^1$$

*This section requires Sections 4.3 and 4.6.

Historical Note

The Pythagoreans discovered that 220 and 284 are friendly. The next pair of friendly numbers was found by Pierre de Fermat: 17,296 and 18,416. In 1638 the French mathematician René Descartes found a third pair, and the Swiss mathematician Leonhard Euler found more than 60 pairs. Another strange development was the discovery in 1866 of another pair of friendly numbers: 1184 and 1210. The discoverer, Nicolo Pagonini, was a 16-year-old Italian schoolboy. Evidently the great mathematicians had overlooked this simple pair. Today there are more than 600 known pairs of friendly numbers.

Writing the canonical form of a number means writing its prime factorization using exponential notation with the factors arranged in order of increasing magnitude.

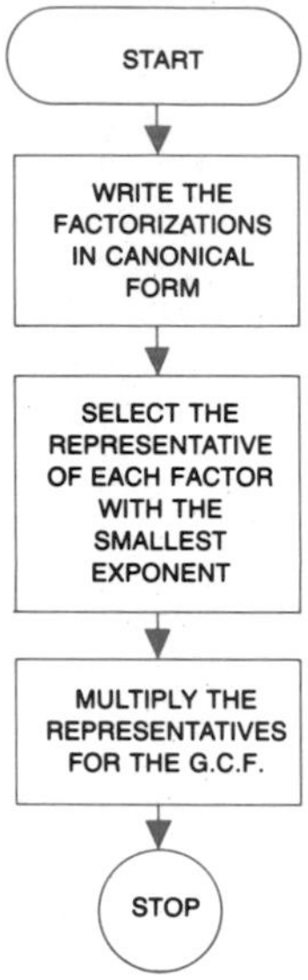

START

↓

WRITE THE
FACTORIZATIONS
IN CANONICAL
FORM

↓

SELECT THE
REPRESENTATIVE
OF EACH FACTOR
WITH THE
SMALLEST
EXPONENT

↓

MULTIPLY THE
REPRESENTATIVES
FOR THE G.C.F.

↓

STOP

FIGURE 6.3 Flowchart for finding
the greatest common factor

Now we select one representative from each of the columns in the factorization. The representative we select when finding the g.c.f. will be the one with the smallest exponent:

$$24 = 2^3 \cdot 3^1 \cdot 5^0$$
$$30 = 2^1 \cdot 3^1 \cdot 5^1$$
$$\text{g.c.f.} = 2^1 \cdot 3^1 \cdot 5^0$$

Finally, we take the product of these representatives to find the g.c.f. (See also Figure 6.3.)

$$\text{g.c.f} = 2^1 \cdot 3^1 \cdot 5^0 = 2 \cdot 3 \cdot 1 = 6$$

Example 1:   Find the g.c.f. of 300, 144, and 108.

Solution: $$300 = 2^2 \cdot 3^1 \cdot 5^2$$
$$144 = 2^4 \cdot 3^2 \cdot 5^0$$
$$108 = 2^2 \cdot 3^3 \cdot 5^0$$
$$\text{g.c.f.} = 2^2 \cdot 3^1 \cdot 5^0 = 4 \cdot 3 \cdot 1 = 12$$

Example 2:   Find the g.c.f. of 15 and 28.

Solution: $$15 = 2^0 \cdot 3^1 \cdot 5^1 \cdot 7^0$$
$$28 = 2^2 \cdot 3^0 \cdot 5^0 \cdot 7^1$$
$$\text{g.c.f.} = 2^0 \cdot 3^0 \cdot 5^0 \cdot 7^0 = 1$$

If the g.c.f. of two numbers is 1, we say that the numbers are **relatively prime.** (Notice that 15 and 28 are relatively prime numbers but are not themselves prime. It is possible for relatively prime numbers to be themselves composites.)

Example 3:   15 and 33 are not relatively prime, since their g.c.f. is not 1 (it is 3).

Example 4:   15 and 28 are relatively prime, since their g.c.f. is 1. Notice that these are not prime numbers—they are simply *prime to each other.*

The second application of prime factorizations is also useful when dealing with fractions. Instead of considering the factors of 24 and 30, this time we'll consider the set of multiples of 24 and 30:

C = Set of multiples of 24: $\{24,48,72,96,120,144,168,192,216,240, \ldots\}$
D = Set of multiples of 30: $\{30,60,90,120,150,180,210,240,270,300, \ldots\}$

The set of common multiples of 24 and 30 is infinite:

$$C \cap D = \{120,240,360, \ldots\}$$

There is no largest member in this set, so we focus attention on the smallest member, 120, which is called the **least common multiple** (l.c.m.). In general, the l.c.m. of two or more counting numbers is the smallest number into which the counting numbers divide. Notice that the l.c.m. is never smaller than the given numbers.

Again we use an algorithm to find the l.c.m. We proceed as we did with the g.c.f. to find the canonical representations of the numbers involved.

$$24 = 2^3 \cdot 3^1 \cdot 5^0$$
$$30 = 2^1 \cdot 3^1 \cdot 5^1$$

Now, for the l.c.m., we choose the representative of each factor with the largest exponent. The l.c.m. is the product of these representatives. (See also Figure 6.4.)

$$\text{l.c.m.} = 2^3 \cdot 3^1 \cdot 5^1 = 120$$

Example 5: Find the l.c.m. of 300, 144, and 108.

Solution:
$$300 = 2^2 \cdot 3^1 \cdot 5^2$$
$$144 = 2^4 \cdot 3^2 \cdot 5^0$$
$$108 = 2^2 \cdot 3^3 \cdot 5^0$$
$$\text{l.c.m.} = 2^4 \cdot 3^3 \cdot 5^2 = 16 \cdot 27 \cdot 25 = 10{,}800$$

This means that the *smallest* number that 300, 144, and 108 *all* divide into is $2^4 \cdot 3^3 \cdot 5^2$ or 10,800.

These procedures are summarized in the following box:

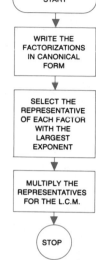

FIGURE 6.4 Flowchart for finding the least common multiple

To find the greatest common factor:	*To find the least common multiple:*
1. Find the prime factorization.	**1.** Find the prime factorization.
2. Write in canonical form.	**2.** Write in canonical form.
3. Choose the representative of each factor with the *smallest* exponent.	**3.** Choose the representative of each factor with the *largest* exponent.
4. Take the product of the representatives.	**4.** Take the product of the representatives.

The greatest common factor will be less than or equal to the numbers involved. The least common multiple will be greater than or equal to the numbers involved.

We may use the ideas of greatest common factor and least common multiple when working with rational numbers. If the greatest common factor of the numerator and denominator is 1, then we say the fraction is in lowest terms or *reduced.* That is, a fraction is reduced if the numerator and denominator are relatively prime.

Example 6:   Reduce $\frac{24}{30}$.

Solution: Note: $24 = 2^3 \cdot 3^1$
$30 = 2^1 \cdot 3^1 \cdot 5^1$ The g.c.f. is 6.

Thus,

$$\frac{24}{30} = \frac{\overbrace{(2 \cdot 3)}^{\text{g.c.f.}} \cdot 2^2}{\underbrace{(2 \cdot 3)} \cdot 5} = \frac{2^2}{5} = \frac{4}{5}$$

Notice that $\frac{24}{30}$ reduces to $\frac{4}{5}$.

After finding the g.c.f. (in parentheses), we wish to simplify the fraction. But how can we justify the step shown with the arrow? We will need one of the most fundamental ideas of arithmetic, and that is the changing of a rational number from one form to another. It is well known that one element in the set of rational numbers has several representations. For example,

$$\frac{1}{2} = \frac{2}{4} = \frac{19}{38} = \frac{17{,}475}{34{,}950} = \cdots$$

The reason why we know these are all equal is summarized with the following property, which follows directly from the identity property of multiplication discussed in Section 6.1.

Fundamental Property of Fractions

If a/b is any rational number and x is any nonzero integer, then

$$\frac{a \cdot x}{b \cdot x} = \frac{x \cdot a}{x \cdot b} = \frac{a}{b}$$

That is, given some fraction to simplify:

1. Find the g.c.f. of the numerator and denominator (this is the x of the Fundamental Property).
2. Use the Fundamental Property of Fractions to simplify the fraction. Notice that this property works only for *factors* and not for terms.

Example 7:   Reduce $\frac{300}{144}$.

Solution: $\dfrac{300}{144} = \dfrac{(2^2 \cdot 3^1) \cdot 5^2}{(2^2 \cdot 3^1) \cdot (2^2 \cdot 3^1)}$

"Nonsense, Miss Murgle! Their eager, little minds are ready for fractions. There must be something wrong with you as a teacher."

FRACTIONS MADE EASY

Since the g.c.f. is 12, or $(2^2 \cdot 3)$, we write:

$$\frac{300}{144} = \frac{12 \cdot 5^2}{12 \cdot 2^2 \cdot 3}$$

$$= \frac{25}{12}$$

We see that $\frac{25}{12}$ is reduced, since the g.c.f. of 25 and 12 is 1.

As we pointed out in Section 4.6, the definitions for addition and subtraction do not require that we first obtain a least common denominator. However, it is sometimes convenient to find one before adding. We can use the least common multiple of the denominators of the two fractions when we add or subtract fractions. For example, the least common multiple of 24 and 30 is 120. Thus,

$$\frac{5}{24} = \frac{5}{24} \cdot \frac{5}{5} = \frac{25}{120}$$

$$+ \frac{7}{30} = \frac{7}{30} \cdot \frac{4}{4} = \frac{28}{120}$$

$$\frac{53}{120}$$

We are simply multiplying each fraction by the identity 1, since $\frac{5}{5}$ and $\frac{4}{4}$ are both equal to 1.

The answer is in reduced form, since 53 and 120 are relatively prime.

Example 8:   Add: $\dfrac{19}{300} + \dfrac{55}{144} + \dfrac{25}{108}$

Solution: In Example 5 we found the l.c.m. of 300, 144, and 108 to be $2^4 \cdot 3^3 \cdot 5^2$. Thus we write

$$\frac{19}{300} = \frac{19}{300} \cdot \frac{2^2 \cdot 3^2}{2^2 \cdot 3^2} = \frac{19 \cdot 2^2 \cdot 3^2}{2^4 \cdot 3^3 \cdot 5^2}$$

$$\frac{55}{144} = \frac{5 \cdot 11}{144} \cdot \frac{3 \cdot 5^2}{3 \cdot 5^2} = \frac{3 \cdot 5^3 \cdot 11}{2^4 \cdot 3^3 \cdot 5^2}$$

$$+ \frac{25}{108} = \frac{5^2}{108} \cdot \frac{2^2 \cdot 5^2}{2^2 \cdot 5^2} = \frac{2^2 \cdot 5^4}{2^4 \cdot 3^3 \cdot 5^2}$$

This is the step during which we multiply each fraction by 1.

Adding:

$$\frac{19 \cdot 2^2 \cdot 3^2}{2^4 \cdot 3^3 \cdot 5^2} + \frac{3 \cdot 5^3 \cdot 11}{2^4 \cdot 3^3 \cdot 5^2} + \frac{2^2 \cdot 5^4}{2^4 \cdot 3^3 \cdot 5^2}$$

$$= \frac{684}{2^4 \cdot 3^3 \cdot 5^2} + \frac{4,125}{2^4 \cdot 3^3 \cdot 5^2} + \frac{2,500}{2^4 \cdot 3^3 \cdot 5^2} = \frac{7,309}{2^4 \cdot 3^3 \cdot 5^2}$$

Now, 7,309 is not divisible by 2, 3, or 5; thus, 7,309 and $2^4 \cdot 3^3 \cdot 5^2$ are relatively prime, and the solution is complete:

$$\frac{7,309}{2^4 \cdot 3^3 \cdot 5^2} \quad \text{or} \quad \frac{7,309}{10,800}$$

Problem Set 6.5

1. What do we mean when we say that numbers are relatively prime?

2. Find the prime factorization of each of the following, if possible:

    a. 60   **b.** 72   **c.** 95   **d.** 1425

3. Find the prime factorization for each of the following, if possible:

    a. 12   **b.** 54   **c.** 171   **d.** 53

Find the g.c.f. of the sets of numbers in Problems 4–12.

4. $\{60,72\}$ 5. $\{95,1425\}$ 6. $\{12,54,171\}$

7. $\{11,13,23\}$ 8. $\{12,20,36\}$ 9. $\{6,8,10\}$

10. $\{9,12,14\}$ 11. $\{3,6,15,54\}$ 12. $\{90,75,120\}$

Find the l.c.m. of the sets of numbers in Problems 13–21.

13. $\{60,72\}$ 14. $\{95,1425\}$ 15. $\{12,54,171\}$

16. $\{11,13,23\}$ 17. $\{12,20,36\}$ 18. $\{6,8,10\}$

19. $\{9,12,14\}$ 20. $\{3,6,15,54\}$ 21. $\{90,75,120\}$

22. **a.** Find the g.c.f. of 6 and 15.
    b. Find the l.c.m. of 6 and 15.
    c. Find the product of your answers to parts a and b.
    d. Compare you answer to part c with the product of 6 and 15.
    e. Repeat parts a–d for 8 and 12.
    f. Repeat parts a–d for 5 and 10.
    g. Repeat parts a–d for 5 and 12.

23. Make a conjecture based on your findings in Problem 22.

For Problems 24–27 consider the set $\{1,2,3,5,6,10,15,30\}$. Let us define an operation $\cancel{D}$, meaning the greatest common factor. For example,

$$5 \cancel{D} 10 = 5$$
$$5 \cancel{D} 6 = 1$$
$$10 \cancel{D} 30 = 10$$

Let us also define a second operation $\cancel{M}$, meaning the least common multiple. For example,

$$5 \cancel{M} 10 = 10$$
$$5 \cancel{M} 6 = 30$$
$$10 \cancel{M} 30 = 30$$

24. Is the set closed for the operations of $\cancel{D}$ and $\cancel{M}$?

25. Are the operations associative?

26. The identity for $\not{D}$ is 30 and for $\not{M}$ is 1. Does the set have the inverse property for $\not{D}$ and $\not{M}$?

27. Does the set form a field for the operations of $\not{D}$ and $\not{M}$?

28. Find the g.c.f. and the l.c.m. of $\{120,75,975\}$.

29. Reduce:

a. $\dfrac{78}{455}$ **b.** $\dfrac{75}{500}$ **c.** $\dfrac{240}{672}$ **d.** $\dfrac{5,670}{12,150}$

Perform the indicated operations in Problems 30–53.

30. $\dfrac{\dfrac{1}{2} + \dfrac{^-2}{3}}{\dfrac{5}{6} - \dfrac{^-3}{5}}$

31. $\dfrac{2^{-1} + 3^{-2}}{2^{-1} + 3^{-1}}$

32. $\dfrac{^-3}{4} \cdot \dfrac{119}{200} + \dfrac{^-3}{4} \cdot \dfrac{81}{200}$

33. $\dfrac{^-3}{4}\left(\dfrac{119}{200} + \dfrac{81}{200}\right)$

34. $\dfrac{2}{3} - \dfrac{7}{12}$

35. $\dfrac{^-7}{24} + \dfrac{^-13}{16}$

36. $\dfrac{28}{9} - \dfrac{4}{27}$

37. $\dfrac{1}{^-8} + \dfrac{1}{^-7}$

38. $^-.3 + .16 + .476$

39. $5^{-1} + 5^{-2} + 5$

40. $^-5 + (^-5)^2$

41. $^-3 - \dfrac{5}{29}$

42. $\dfrac{^-2}{15} + \dfrac{3}{5} + \dfrac{7}{12}$

43. $^-2\frac{3}{5} + 4\frac{1}{8} - 7\frac{1}{10}$

44. $\dfrac{1}{10} \cdot \dfrac{^-2}{5}$

45. $\left(\dfrac{3}{7} \cdot \dfrac{3}{5}\right) \div \dfrac{1}{2}$

46. $\dfrac{2}{^-3} \cdot \dfrac{^-2}{15}$

47. $\dfrac{4}{5}\left(\dfrac{17}{95}\right) + \dfrac{4}{5}\left(\dfrac{78}{95}\right)$

48. $\dfrac{^-7}{8} \cdot \dfrac{131}{147} + \dfrac{^-7}{8} \cdot \dfrac{16}{147}$

49. $\dfrac{\left(\dfrac{3}{5} \cdot \dfrac{1}{7}\right) + \dfrac{1}{3}}{2}$

50. $\dfrac{\dfrac{1}{3} - \dfrac{^-1}{4}}{\dfrac{7}{8} - \dfrac{3}{16}}$

51. $\dfrac{^-3}{8} + \dfrac{3}{4} - \dfrac{9}{10}$

52. $\dfrac{\dfrac{15}{16}}{\dfrac{3}{8}}$

53. $\dfrac{\dfrac{4}{5} + \dfrac{^-2}{3}}{\dfrac{1}{5}} + 2$

54. *Euclidean Algorithm.* The g.c.f. of two numbers can be found using another method, which is attributed to Euclid. It is called the Euclidean Algorithm and is based on repeated division. For example, find the g.c.f. of 108 and 300.

Procedure:
(1) Divide 300 by 108. List the quotient and remainder.

$$108{\overline{\smash{\big)}\,300}} \qquad 300 = 2 \cdot 108 + 84$$
$$\underline{216}$$
$$84$$

(2) Divide 108 by remainder of step 1.

$$84{\overline{\smash{\big)}\,108}} \qquad 108 = 1 \cdot 84 + 24$$
$$\underline{84}$$
$$24$$

EXAMPLE: {357,629}

$$357{\overline{\smash{\big)}\,629}} \qquad\qquad 272{\overline{\smash{\big)}\,357}}$$
$$\underline{357} \qquad\qquad\qquad \underline{272}$$
$$272 \quad r = 272 \qquad 85 \quad r = 85$$

(3) Repeat the process.

$$24{\overline{\smash{\big)}\,84}} \qquad 84 = 3 \cdot 24 + 12$$
$$\underline{72}$$
$$12$$

$$85{\overline{\smash{\big)}\,272}} \qquad\qquad 17{\overline{\smash{\big)}\,85}}$$
$$\underline{255} \qquad\qquad\qquad \underline{85}$$
$$17 \quad r = 17 \qquad 0 \quad r = 0$$
Thus, we stop.

(4) Repeat until the remainder is 0.

$$12{\overline{\smash{\big)}\,24}} \qquad \text{The remainder is 0, so we stop.}$$
$$\underline{24}$$
$$0$$

The last divisor is the g.c.f. Therefore the g.c.f. of 357 and 629 is 17.

(5) The last divisor (in this case, 12) is the g.c.f.

Find the g.c.f. using the Euclidean Algorithm:

a. {91,107}   **b.** {126,180}   **c.** {51,1995}

Mind Boggler

55. The sum and difference of two squares may be primes. For example,

$$9 - 4 = 5 \qquad \text{and} \qquad 9 + 4 = 13$$

Can the sum and difference of two primes be square? Can you find more than one example?

Problems for Individual Study

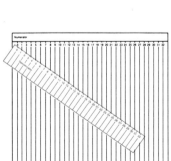

A fraction machine

56. *A fraction machine.* Take two sheets of lined paper. Label one *numerator* and one *denominator,* and number the lines consecutively from 0 as shown in the figure. Fasten the pages together at 0 so that the top page pivots at this point (or else always be sure that the 0 lines coincide).

   Suppose you wish to find a fraction equal to $\frac{5}{6}$ using a denominator of 18. Slide the top page down until the lines 5 (numerator) and 6 (denominator) coincide. Now you can read other equivalent fractions from the fraction machine by noticing those places where the lines coincide. From the figure we see that

$$\frac{5}{6} = \frac{15}{18}$$

Also notice from the figure that $\frac{5}{6} = \frac{10}{12}$ and $\frac{5}{6} = \frac{20}{24}$. Build a fraction machine and fill in the following blanks.

a. $\dfrac{5}{7} = \dfrac{?}{21}$ **b.** $\dfrac{4}{9} = \dfrac{?}{18}$ **c.** $\dfrac{5}{8} = \dfrac{?}{4}$ **d.** $\dfrac{3}{4} = \dfrac{1}{?}$

Demonstrate a fraction machine to the class. Can you think of any other uses for a fraction machine?

Reference: M. Wassmansdorf, "Reducing Fractions Can Be Easy, Maybe Even Fun," *Arithmetic Teacher*, February 1974, pp. 99–102.

*6.6 Ratio, Proportion, and Problem Solving

Powerful problem-solving tools in algebra are ratios and proportions. Ratios are a way of comparing two numbers or quantities—for example, the compression ratio of a car, the gear ratio of a transmission, the pitch of a roof, the steepness of a road, or a player's batting average. The ratio of the number a to the number b is written as

$$a \text{ to } b, \qquad a{:}b, \qquad \frac{a}{b}$$

and in this book we will usually use the last form.

$\dfrac{a}{b}$ is called the **ratio** of a to b. The two parts a and b are called its **terms.**

Historical Note

The symbol : for a ratio was first used by the astronomer Vincent Wing in 1651. Prior to this, a ratio was represented by a single dot, as in 4·5, to mean $\frac{4}{5}$. In a book published in 1633, *Johnson's Arithmetik*, the colon (:) is used to represent a fraction.

In Examples 1–3, write the given ratios as reduced fractions.

Example 1: The ratio of men to women is 5 to 4.

Answer: $\dfrac{5}{4}$ When writing ratios, do not write improper fractions as mixed numbers.

Example 2: The ratio of people to cars is 10 to 2.

Answer: $\dfrac{10}{2} = 5$ Write this as $\dfrac{5}{1}$ because a ratio compares two numbers.

*This section requires Section 4.6.

Example 3:   The ratio of 3 ft to 8 in.

Answer: When comparing quantities that are measured in different units, first change one of the units to the other, if possible. Since 3 ft is 36 in., write

$$\frac{36}{8} = \frac{9}{2}$$

A **proportion** is a statement of equality between two ratios. In symbols,

$$\frac{a}{b} = \frac{c}{d}$$

which is read

$$\frac{a}{b} \quad = \quad \frac{c}{d}$$

$$\uparrow \qquad \uparrow \qquad \uparrow$$

"*a* is to *b*" "as" "*c* is to *d*"

The notation used in some books is

$$a:b::c:d$$

Even though we won't use this notation, we will use words associated with this notation to name the terms:

Extremes
Means
↓ ↓
a:b::c:d

Thus, in fractional notation,

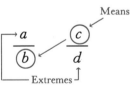

To check and see if two ratios form a proportion, use the following *cross-product method.*

Cross-Product Method for Finding a Proportion

If the product of the means equals the product of the extremes, then the ratios form a proportion.

In Examples 4 and 5, tell whether each pair of ratios forms a proportion.

Example 4: $\dfrac{3}{4}$; $\dfrac{36}{48}$

Means: Extremes:
4×36 3×48
144 $=$ 144

Thus, $\dfrac{3}{4} = \dfrac{36}{48}$ so the fractions form a proportion.

Example 5: $\dfrac{5}{16}$; $\dfrac{7}{22}$

Means: Extremes:
16×7 5×22
112 $\neq$ 110

Thus, $\dfrac{5}{16} \neq \dfrac{7}{22}$ so the fractions do not form a proprotion.

François Viète (1540–1603)

Viète was one of the greatest French mathematicians of the 16th century. He wrote a number of books on trigonometry, algebra, and geometry, most of which were printed at his own expense. He was among the first to use letters to represent numbers. Viète was a lawyer at the court of Henry IV of France, and the story is told of an ambassador from another country bragging to Henry IV that France had no mathematician able to solve a certain 45th degree equation. Viète was able to quickly give 2 solutions, and then later 21 more. In return, Viète challenged the originator of the problem, Adrianus Romanus (1561–1615), to a type of "mathematical duel." Romanus could not solve the challenge problem, but traveled to France to meet Viète with the result that a lifelong friendship developed.

The usefulness of proportions is not apparent when you are verifying that two ratios form a proportion; instead you are usually given three of the terms of a proportion and need to find out the value of the missing term. This process is called **solving a proportion.** It is always possible to find the missing term if you know that the ratios form a proportion. To do this, use the following result (which is the converse of the cross-product method).

Method for Solving a Proportion

If the ratios form a proportion, then the product of the means equals the product of the extremes. In symbols,

$$\frac{a}{b} = \frac{c}{d}$$

$$\underbrace{b \cdot c}_{} = \underbrace{a \cdot d}_{}$$

PRODUCT OF MEANS $=$ PRODUCT OF EXTREMES

In Examples 6–10, find the missing term of each proportion.

Example 6: $\dfrac{3}{1} = \dfrac{x}{5}$

$$\overbrace{\text{PRODUCT OF MEANS}}\qquad\overbrace{\text{PRODUCT OF EXTREMES}}$$

$$1 \cdot x \qquad = \qquad 3 \cdot 5$$

$$x = 15$$

Thus, the missing term is 15.

Example 7: $\dfrac{5}{2} = \dfrac{y}{4}$ $\overbrace{\text{PRODUCT OF MEANS}}\qquad\overbrace{\text{PRODUCT OF EXTREMES}}$

$$2y \qquad = \qquad 20$$

$$y = 10$$

Example 8: $\dfrac{3}{4} = \dfrac{9}{D}$ **Example 9:** $\dfrac{G}{9} = \dfrac{7}{3}$

$$36 = 3D \qquad\qquad\qquad 63 = 3G$$

$$12 = D \qquad\qquad\qquad 21 = G$$

Example 10: $\dfrac{16}{K} = \dfrac{4}{9}$ *Alternate method:*

$$4K = 144 \qquad\qquad 4K = 9 \cdot 16 \qquad \text{Do not multiply yet;}$$
$$K = 36 \qquad\qquad\qquad\qquad\qquad \text{leave it in factored form.}$$

$$K = \dfrac{9 \cdot \overset{4}{\cancel{16}}}{\underset{1}{\cancel{4}}}$$

$$K = 36$$

Example 10 illustrates that you might wish to simplify the fraction before multiplying. You will want to do this whenever the numbers are very large.

In Examples 11–13, solve the proportion for the number represented by the letter.

Example 11: $\dfrac{5}{6} = \dfrac{55}{y}$

$$5y = 6 \cdot 55 \qquad \text{This is the same as } 6 \cdot 55 = 5y.$$

$$y = \dfrac{6 \cdot 55}{5} \qquad \begin{array}{l}\leftarrow \text{Product of means} \\ \leftarrow \text{Number opposite the unknown.}\end{array}$$

$$= \dfrac{6 \cdot \overset{11}{\cancel{55}}}{\underset{1}{\cancel{5}}} \qquad \text{You can cancel to simplify many of these problems.}$$

$$= 66$$

Example 12: $\dfrac{5}{x} = \dfrac{10}{14}$

$10x = 5 \cdot 14$

$$x = \dfrac{\overset{1}{\cancel{5}} \cdot \overset{7}{\cancel{14}}}{\underset{\underset{1}{\cancel{2}}}{\cancel{10}}} \quad \begin{matrix}\leftarrow\text{Product of extremes}\\[2pt]\leftarrow\text{Number opposite the unknown}\end{matrix}$$

$= 7$

Example 13: $\dfrac{8,000}{32} = \dfrac{17,000}{x}$

$8,000x = 32 \cdot 17,000$

$$x = \dfrac{\overset{4}{\cancel{32}} \cdot \overset{17}{\cancel{17,000}}}{\underset{\underset{1}{\cancel{8}}}{\cancel{8,000}}} \quad \begin{matrix}\leftarrow\text{Product of means}\\[2pt]\leftarrow\text{Number opposite the unknown}\end{matrix}$$

$x = 68$

Problems like Example 13 might make you ask whether you can use a calculator efficiently to solve proportions. By studying Examples 11–13, we can formulate the following statement, which is particularly useful when using a calculator.

To Solve a Proportion for an Unknown Term:

1. Find the product of the means or the product of the extremes—whichever does not contain the unknown term.
2. Divide the product from step 1 by the number that is opposite the unknown.

In Examples 14–16, show the calculator steps for solving the proportions.

Example 14: $\dfrac{5}{6} = \dfrac{55}{y}$

y is the value displayed after pressing the following sequence of buttons:

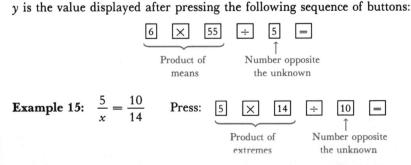

Example 15: $\dfrac{5}{x} = \dfrac{10}{14}$ Press:

Example 16: $\dfrac{8{,}000}{32} = \dfrac{17{,}000}{x}$ Press: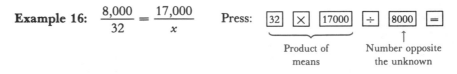

Many applied problems can be solved by using a proportion. When setting up a proportion with units, be sure that like units occupy corresponding positions, as illustrated by Examples 17–19. Answer the questions by setting up a proportion.

Example 17: If 4 cans of cola sell for 89¢, how much will 12 cans cost?

Correct arrangements: *Incorrect arrangements:*

$$\dfrac{\text{Cans}}{\text{Cents}} = \dfrac{\text{Cans}}{\text{Cents}} \qquad\qquad \dfrac{\text{Cans}}{\text{Cents}} = \dfrac{\text{Cents}}{\text{Cans}}$$

$$\dfrac{\text{Cents}}{\text{Cans}} = \dfrac{\text{Cents}}{\text{Cans}} \qquad\qquad \dfrac{\text{Cans}}{\text{Cents}} = \dfrac{\text{Cans}}{\text{Dollars}}$$

$$\dfrac{\text{Cents}}{\text{Cents}} = \dfrac{\text{Cans}}{\text{Cans}} \qquad\qquad \dfrac{\text{Dollars}}{\text{Cans}} = \dfrac{\text{Cents}}{\text{Cans}}$$

$$\dfrac{\text{Cans}}{\text{Dollars}} = \dfrac{\text{Cans}}{\text{Dollars}}$$

$$\dfrac{4 \text{ cans}}{89 \text{ cents}} = \dfrac{12 \text{ cans}}{x \text{ cents}}$$

Knowing units can help you set up a correct proportion. After you have a correct proportion, you can drop the unit values when solving the proportion.

$$\dfrac{89 \times \overset{3}{\cancel{12}}}{\underset{1}{\cancel{4}}} = x$$

$$267 = x$$

So, 12 cans cost 267¢ or \$2.67.

Example 18: If a 120 mile trip took $8\frac{1}{2}$ gallons of gas, then how much gas is needed for a 240 mile trip?

Solution: We'll use

$$\dfrac{\text{Miles}}{\text{Gallons}} = \dfrac{\text{Miles}}{\text{Gallons}}$$

$$\dfrac{120 \text{ miles}}{8\frac{1}{2} \text{ gallons}} = \dfrac{240 \text{ miles}}{x \text{ gallons}}$$

$$\dfrac{8\frac{1}{2} \times 240}{120} = x$$

$$\dfrac{\frac{17}{2} \times 240}{120} = x$$

← If you have a calculator you can find the solution directly from this step.

Press: $\boxed{8.5}$ $\boxed{\times}$ $\boxed{240}$ $\boxed{\div}$ $\boxed{120}$ $\boxed{=}$

The display will show the value for x.

$$\frac{17 \times \overset{1}{\cancel{120}}}{\underset{1}{\cancel{120}}} = x$$

$$17 = x$$

The trip will require 17 gallons of gas.

Example 19: If the property tax on a $65,000 home is $416, what is the tax on an $85,000 home?

Solution:

$$\frac{\text{Value}}{\text{Tax}} = \frac{\text{Value}}{\text{Tax}}$$

$$\frac{\$65,000}{\$416} = \frac{\$85,000}{\$x}$$

← On a calculator,

Press: ⎡416⎤ ⎡×⎤ ⎡85000⎤ ⎡÷⎤ ⎡65000⎤ ⎡=⎤

The display will show the value for x.

$$\frac{\overset{32}{\cancel{416}} \times \overset{17}{\cancel{85,000}}}{\underset{\underset{1}{13}}{\cancel{65,000}}} = x$$

$$544 = x$$

The tax is $544.

Problem Set 6.6

Write the ratios given in Problems 1–6 as reduced fractions.

1. The ratio of yes to no answers is 2 to 1.

2. The ratio of cars to people is 1 to 3.

3. The ratio of cats to dogs is 4 to 7.

4. The ratio of dogs to cats is 7 to 4.

5. The ratio of gallons to miles is 4 to 60.

6. The ratio of dollars to people is 92 to 4.

Tell whether each pair of ratios in Problems 7–15 form a proportion.

7. $\dfrac{7}{1}$; $\dfrac{21}{3}$ 8. $\dfrac{6}{8}$; $\dfrac{9}{12}$ 9. $\dfrac{3}{6}$; $\dfrac{5}{10}$

10. $\dfrac{7}{8}$; $\dfrac{6}{7}$ 11. $\dfrac{9}{2}$; $\dfrac{10}{3}$ 12. $\dfrac{6}{5}$; $\dfrac{42}{35}$

13. $\dfrac{20}{70}$; $\dfrac{4}{14}$ 14. $\dfrac{3}{4}$; $\dfrac{75}{100}$ 15. $\dfrac{2}{3}$; $\dfrac{67}{100}$

Solve each proportion in Problems 16–36 for the number represented by a letter.

16. $\dfrac{5}{1} = \dfrac{A}{3}$ **17.** $\dfrac{1}{9} = \dfrac{4}{B}$ **18.** $\dfrac{C}{2} = \dfrac{11}{1}$

19. $\dfrac{7}{D} = \dfrac{1}{8}$ **20.** $\dfrac{12}{18} = \dfrac{E}{12}$ **21.** $\dfrac{12}{15} = \dfrac{20}{F}$

22. $\dfrac{G}{24} = \dfrac{14}{16}$ **23.** $\dfrac{4}{H} = \dfrac{3}{15}$ **24.** $\dfrac{2}{3} = \dfrac{I}{7}$

25. $\dfrac{4}{5} = \dfrac{3}{J}$ **26.** $\dfrac{3}{K} = \dfrac{2}{5}$ **27.** $\dfrac{L}{2} = \dfrac{5}{6}$

28. $\dfrac{4}{9} = \dfrac{M}{6}$ **29.** $\dfrac{12}{3} = \dfrac{8}{N}$ **30.** $\dfrac{P}{5} = \dfrac{21}{35}$

31. $\dfrac{5}{R} = \dfrac{15}{21}$ **32.** $\dfrac{4}{1} = \dfrac{1}{T}$ **33.** $\dfrac{4,000}{U} = \dfrac{12,000}{3}$

34. $\dfrac{1}{V} = \dfrac{5}{2}$ **35.** $\dfrac{X}{150} = \dfrac{14}{6}$ **36.** $\dfrac{50}{30} = \dfrac{40}{Y}$

37. If you drive 119 miles with $8\frac{1}{2}$ gallons of gas, what is the simplified ratio of miles to gallons?

38. If you drive 180 miles with $7\frac{1}{2}$ gallons of gas, what is the simplified ratio of miles to gallons?

39. The **pitch** of a roof is the ratio of the rise to the span. If a roof has a rise of 8 feet and a span of 24 feet, what is the pitch?

40. What is the pitch of a roof (see Problem 39) with a 3 foot rise and a span of 12 feet?

41. If Jello is on sale for four boxes for $1, how much would seven boxes cost?

42. If 10 pounds of potatoes cost $1.19, how much (to the nearest cent) would 25 pounds cost?

43. If tomato sauce is on sale for six cans for $1, how much would four cans cost (to the nearest cent)?

44. If 2 pounds of coffee cost $4.30, how much should be charged for 3 pounds?

45. If a 184 mile trip took $11\frac{1}{2}$ gallons of gas, then how much gas is needed for a 160 mile trip?

46. If a 121 mile trip took $5\frac{1}{2}$ gallons of gas, how many miles can be driven with a full tank of 13 gallons?

47. If a family uses $3\frac{1}{2}$ gallons of milk per week, how much milk will this family need for 4 days?

48. Last year 1,200 people at a 4-H chicken-que ate 450 pounds of chicken. This year they plan on 1,800 people. How many pounds of chicken should be ordered?

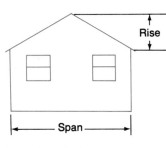

Pitch of a roof

Mind Bogglers

49. This problem has been devised to help you check your work in Problems 16–36. Look at the puzzle form shown here; each small box has a number in the top left-hand corner. Using your solutions to Problems 16–36 place a letter in each box to make up a secret message. For example, if

$$\frac{W}{7} = \frac{10}{14}$$

then

$$\frac{\overset{1}{\cancel{7}} \times \overset{5}{\cancel{10}}}{\underset{\underset{1}{\cancel{2}}}{\cancel{14}}} = W$$

$$5 = W$$

Now find the box or boxes with the number 5 in the corner and fill in the letter *W*. This has already been done for you. The letters *O* and *S* have also been filled in as a bonus. Now, do the same thing for each letter in Problems 16–36. Some letters may not appear in the boxes. When you are finished, darken in all the blank spaces to separate the words in the secret message. (Notice that one of the blank spaces has also been filled in to help you.)

15	25	1/4	8	7	■	22 (O)	8/3	3	5/3	8	1/4	14/3	2	21	30
1/4	20	14/3 (S)	4	■ (S)	8	22	1/4	14/3	(O)	2	9	24	(O)	1	27
5 (W)	14/3	5/3	5/3	10	36	8	11 (O)	2	1/7	24	(O)	1	7	16	16
5 (W)	15	24	18	1/4 (O)	10	8	350	22	8	5/3	5/3	8	2	22	8

50. Answer the question in the following *Peanuts* cartoon strip.

"A BANANA PEEL WEIGHS 1/8 THE TOTAL WEIGHT OF A BANANA"

"IF AN UNPEELED BANANA BALANCES A PEELED BANANA OF THE SAME WEIGHT PLUS 7/8 OF AN OUNCE..."

"..HOW MUCH DOES THE BANANA WEIGH WITH PEEL?"

ABANDON SHIP!!

*6.7 A Finite Algebra

At the beginning of this chapter we mentioned that there are many types of algebra. We then looked at the algebraic system taught in high school and solved some equations with that algebra. In this section we'll look at an algebra based on a finite set rather than on the infinite set of real numbers.

*This is an optional section.

FIGURE 6.5   A 12-hour clock

We now consider a mathematical system based on the 12-hour clock. We'll need to define some operations for this set of numbers, and we'll use the way we tell time as a guide to our definitions. For example, if you have an appointment at 4:00 P.M. and you are two hours late, you arrive at 6:00 P.M. On the other hand, if your appointment is at 11:00 A.M. and you are two hours late, you arrive at 1:00 P.M. That is,

$$4 + 2 = 6 \quad \text{and} \quad 11 + 2 = 1 \quad \text{on a 12-hour clock}$$

Using the clock as a guide, we define an operation called "clock addition" (which is different from ordinary addition) according to Table 6.5.

TABLE 6.5   Addition on a 12-Hour Clock

+	1	2	3	4	5	6	7	8	9	10	11	12
1	2	3	4	5	6	7	8	9	10	11	12	1
2	3	4	5	6	7	8	9	10	11	12	1	2
3	4	5	6	7	8	9	10	11	12	1	2	3
4	5	6	7	8	9	10	11	12	1	2	3	4
5	6	7	8	9	10	11	12	1	2	3	4	5
6	7	8	9	10	11	12	1	2	3	4	5	6
7	8	9	10	11	12	1	2	3	4	5	6	7
8	9	10	11	12	1	2	3	4	5	6	7	8
9	10	11	12	1	2	3	4	5	6	7	8	9
10	11	12	1	2	3	4	5	6	7	8	9	10
11	12	1	2	3	4	5	6	7	8	9	10	11
12	1	2	3	4	5	6	7	8	9	10	11	12

Example 1: $7 + 5 = 12$ **Example 2:** $9 + 5 = 2$

Example 3: $4 + 11 = 3$

Subtraction, multiplication, and division may be defined as they are in ordinary arithmetic:

Subtraction is defined as the inverse or opposite operation of addition:

$$a - b = x \quad \text{means} \quad a = b + x$$

Multiplication is defined as repeated addition ($b \neq 0$):

$$a \times b \quad \text{means} \quad \underbrace{a + a + a + \cdots + a}_{b \text{ addends}}$$

If $b = 0$,   then $a \cdot 0 = 0$.

Division is defined as the inverse or opposite of multiplication ($b \neq 0$):

$$\frac{a}{b} = x \quad \text{means} \quad a = b \cdot x$$

Example 4: Find $4 - 9$. This means find the number that, when added to 9, produces 4. Thus, solve $4 = 9 + t$. From the table, $t = 7$.

Example 5: Find 4×9. This means find $9 + 9 + 9 + 9 = 12$.

Example 6: Find $4 \div 7$ or $\frac{4}{7} = t$. This means find the number that, when multiplied by 7, produces 4. Thus, solve $4 = 7 \times t$. Since $7 \times 4 = 4$, we see that $\frac{4}{7} = 4$.

Example 7: Find $\frac{4}{9} = t$. This means find the number that, when multiplied by 9, produces 4. There is no such number.

Since the addition and multiplication tables for the 12-hour clock are rather big, we shorten our arithmetic system by considering a clock with fewer than 12 hours.

Consider a mathematical system based on a 5-hour clock, numbered 0, 1, 2, 3, and 4, as shown in Figure 6.6. Clock addition on the 5-hour clock is the same as on an ordinary clock, except that the only numerals are 0, 1, 2, 3, and 4 (notice that 5 has been replaced by 0, and we agree that 0 and 5 have the same meaning in this system).

We define the operations of multiplication, subtraction, and division as before, and thus we can complete Table 6.6 (since subtraction and division are the opposite operations of addition and multiplication, we need only two tables).

Instead of speaking about "arithmetic on the 5-hour clock," mathematicians usually speak of "modulo 5 arithmetic." The set $\{0,1,2,3,4\}$, together with the operations defined in Table 6.6, is called a *modulo 5* or *mod 5* system.

Notice that in a mod 5 system (or on a 5-hour clock):

$$4 + 9 = 3 \quad \text{and} \quad 2 + 1 = 3$$

Thus, $4 + 9 = 2 + 1$.

Since we do not wish to confuse this with ordinary arithmetic, in which

$$4 + 9 \neq 2 + 1$$

we use the following notation:

$$4 + 9 \equiv 2 + 1 \ (\text{mod } 5)$$

where $\equiv$ is read

"$4 + 9$ is congruent to $2 + 1$ in modulo 5."

We define congruence as follows.

FIGURE 6.6 A 5-hour clock

TABLE 6.6
Addition on a 5-Hour Clock

+	0	1	2	3	4
0	0	1	2	3	4
1	1	2	3	4	0
2	2	3	4	0	1
3	3	4	0	1	2
4	4	0	1	2	3

Multiplication on a 5-Hour Clock

×	0	1	2	3	4
0	0	0	0	0	0
1	0	1	2	3	4
2	0	2	4	1	3
3	0	3	1	4	2
4	0	4	3	2	1

Definition

a and *b* are *congruent modulo m*, written $a \equiv b \pmod{m}$, if *a* and *b* differ by a multiple of *m*.

Example 8: $3 \equiv 8 \pmod 5$, since $8 - 3 = 5$, and 5 is a multiple of 5

Example 9: $3 \equiv 53 \pmod 5$, since $53 - 3 = 50$, and 50 is a multiple of 5

Example 10: $3 \not\equiv 19 \pmod 5$, since $19 - 3 = 16$, and 16 is not a multiple of 5

Another way of determining two congruent numbers mod m is to divide each by m and check the remainders. If the remainders are the same, then the numbers are congruent mod m. For example, $3 \div 5$ gives a remainder 3, and $53 \div 5$ gives a remainder 3; thus $3 \equiv 53 \pmod 5$.

Example 11: $4 + 9 \equiv 2 + 1 \pmod 5$,
                   since $13 \equiv 3 \pmod 5$

Example 12: $15 + 92 \equiv x \pmod 5$
                    $107 \equiv x \pmod 5$
                      $2 \equiv x \pmod 5$

Notice that every whole number is congruent modulo 5 to exactly one element in the set $I = \{0,1,2,3,4\}$. This set I contains all possible remainders when dividing by 5. Since the number of elements in I is rather small, we can easily solve equations by trying the numbers 0, 1, 2, 3, and 4.

Example 13: $2 + 4 \equiv x \pmod 5$

Method 1: $2 + 4 = 6$; so

$6 \equiv x \pmod 5$ We found this number by putting down the remainder

$1 \equiv x \pmod 5$ obtained when we divided 6 by the modulus 5.

Method 2: Check all possibilities for x, and determine whether any make the sentence true.

$$x \equiv 0? \quad 2 + 4 \equiv 0 \quad \text{No}$$
$$x \equiv 1? \quad 2 + 4 \equiv 1 \quad \textit{Yes}$$
$$x \equiv 2? \quad 2 + 4 \equiv 2 \quad \text{No}$$
$$x \equiv 3? \quad 2 + 4 \equiv 3 \quad \text{No}$$
$$x \equiv 4? \quad 2 + 4 \equiv 4 \quad \text{No}$$

You must try all possibilities, since, for mod m, there may be more than one solution. Be careful not to assume that you can "solve" these equations using ordinary rules of algebra. This is *not* ordinary algebra—it is modular arithmetic. For example, in mod 6 you cannot divide both sides of the equivalence relation by 3.

Example 14: $2 - 4 \equiv x \pmod 5$
Now, $2 - 4 \equiv x$ means $2 \equiv 4 + x$. Directly, or by substitution, we find that $x \equiv 3 \pmod 5$.

Example 15: $2 \times 4 \equiv x \pmod 5$
$$8 \equiv x \pmod 5$$
$$3 \equiv x \pmod 5$$

Example 16: The manager of a TV station hired a college student, U. R. Stuck, as an election-eve runner. The request for reimbursement turned in is shown. The station manager refused to pay the bill, since the mileage was not honestly recorded. How did he know? (*Note:* It has nothing to do with the illegible digits in the mileage report.)

REQUEST FOR REIMBURSEMENT FOR MILEAGE
NAME: _U. R. Stuck_
ADDRESS: _1234 Fifth st._
S.S. NO.: _576-38-4459_
ENDING MILEAGE: _14,128_
BEGINNING MILEAGE: _14,613_
NO. OF TRIPS _8_

Solution: Let x = Number of miles in one round trip. From the reimbursement request we see there are 8 trips, so the total mileage is $8x$, and since the last digits are legible we see that the difference is 5 (mod 10). Thus,

$$8x \equiv 5 \pmod{10}$$

Solving:

$$x = 0: 8(0) \equiv 0 \pmod{10}$$
$$x = 1: 8(1) \equiv 8 \pmod{10}$$
$$x = 2: 8(2) \equiv 6 \pmod{10}$$
$$x = 3: 8(3) \equiv 4 \pmod{10}$$
$$x = 4: 8(4) \equiv 2 \pmod{10}$$
$$x = 5: 8(5) \equiv 0 \pmod{10}$$
$$x = 6: 8(6) \equiv 8 \pmod{10}$$
$$x = 7: 8(7) \equiv 6 \pmod{10}$$

These are the only possible values for x (mod 10), and none gives a mileage reading of 5 (mod 10) in the units digit. Therefore, Stuck did not report the mileage honestly.

Example 17: Suppose you are planning to buy paper to cover some shelves. You need to cover between 30 and 50 11-inch shelves and 1 shelf 100 inches long. The paper can be purchased in multiples of 36 inches. How much paper should you buy to eliminate as much waste as possible?

Solution: Since the paper comes in multiples of 36 inches, you must buy $36x$ inches of paper, where x represents the number of multiples you buy. Also, you will cover k shelves at 11 inches and one shelf at 100 inches for a total of $11k + 100$ inches. We can write this as a congruence:

$$36x \equiv 100 \pmod{11}$$

or

$$3x \equiv 1 \pmod{11}$$

This congruence has a solution of $x \equiv 4 \pmod{11}$, which means that we can buy 4, 15, 26, ... multiples of the 36-inch paper. If $x = 15$, then we buy

$$36 \cdot 15 = 540$$

inches of paper. This amount allows us to cover the 100-inch board and 40 11-inch boards.

Notice that $36 \equiv 3$ and $100 \equiv 1$ (mod 11). We found this solution by considering:

$x = 0$: $3(0) \equiv 0 \pmod{11}$
$x = 1$: $3(1) \equiv 3 \pmod{11}$
$x = 2$: $3(2) \equiv 6 \pmod{11}$
$x = 3$: $3(3) \equiv 9 \pmod{11}$
$x = 4$: $3(4) \equiv 1 \pmod{11}$
$x = 5$: $3(5) \equiv 4 \pmod{11}$
$x = 6$: $3(6) \equiv 7 \pmod{11}$

The next example requires the idea of finding the greatest integer part of a number. Let $[x]$ denote the greatest integer less than or equal to x. This means, for example: $[2.5] = 2$; $[6.99] = 6$; $[7.01] = 7$; $[2\frac{1}{2}] = 2$.

Example 18: Find the dates of Friday the 13th in any given year by using Table 6.7, the greatest integer part of a number, and modular arithmetic.

If there were no leap years, the first day of each year would move up by one day, but because of leap years the problem becomes more complicated. Let $y =$ Year in question.

The Gregorian calendar has a leap year every four years, except century years. In addition, every fourth century is a leap year.

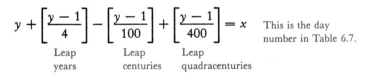

$$y + \left[\frac{y-1}{4}\right] - \left[\frac{y-1}{100}\right] + \left[\frac{y-1}{400}\right] = x$$

$\quad\quad$ Leap $\quad\quad$ Leap $\quad\quad$ Leap
$\quad\quad$ years $\quad$ centuries $\quad$ quadracenturies

This is the day number in Table 6.7.

Find x mod 7 and use Table 6.7 to find the months with a Friday the 13th. For example, in 1985, to find the day number:

$$1985 + \left[\frac{1984}{4}\right] - \left[\frac{1984}{100}\right] + \left[\frac{1984}{400}\right]$$

$$= 1985 + \quad 496 \quad - \quad 19 \quad + \quad 4$$

$$= 2466$$

$$\equiv 2 \pmod 7$$

The calculation $\left[\frac{1984}{400}\right]$ is easily done on a calculator:

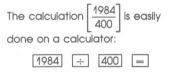

The result is 4.96; disregard the decimal part for this calculation since you are considering the greatest integer part.

Thus, from Table 6.7, the first day of 1985 is a Tuesday and it is a non-leap year, so there will be a Friday the 13th in September and December. Notice that if you happen to know the day of the week of January 1st, you need no calculation, but can go directly to Table 6.7.

TABLE 6.7   Months with Friday the 13th*

Day #	January 1	Non-Leap Year	Leap Year
0	Sunday	January, October	January, April, July
1	Monday	April, July	September, December
2	Tuesday	September, December	June
3	Wednesday	June	March, November
4	Thursday	February, March, November	February, August
5	Friday	August	May
6	Saturday	May	October

*This table, along with the justification for the formula, is found in "An Aid to the Superstitious," by G. L. Ritter, S. R. Lowry, H. B. Woodruff, and T. L. Isenhour, *The Mathematics Teacher,* May 1977, pp. 456–457.

Example 19:   Solve: $\frac{2}{4} \equiv x \pmod 5$. This means $2 \equiv 4 \cdot x \pmod 5$. Checking all possibilities:

$$4 \cdot 0 = 0 \not\equiv 2 \pmod 5$$
$$4 \cdot 1 = 4 \not\equiv 2 \pmod 5$$
$$4 \cdot 2 = 8 \not\equiv 2 \pmod 5$$
$$4 \cdot 3 = 12 \equiv 2 \pmod 5$$
$$4 \cdot 4 = 16 \not\equiv 2 \pmod 5$$

Thus, there is one solution, $x \equiv 3 \pmod 5$.

Can you say $\frac{2}{4} \equiv \frac{1}{2} \pmod 5$? You must be careful! In this case, we see that

$$\frac{2}{4} \equiv 3 \pmod 5 \qquad \text{and} \qquad \frac{1}{2} \equiv 3 \pmod 5$$

Thus, $\frac{2}{4} \equiv \frac{1}{2} \pmod 5$. Consider the following example in modulo 6.

Example 20:   Solve: $\frac{2}{4} \equiv x \pmod 6$. This means $2 \equiv 4 \cdot x \pmod 6$.

$$4 \cdot 0 = 0 \not\equiv 2 \pmod 6$$
$$4 \cdot 1 = 4 \not\equiv 2 \pmod 6$$
$$4 \cdot 2 = 8 \equiv 2 \pmod 6$$
$$4 \cdot 3 = 12 \not\equiv 2 \pmod 6$$
$$4 \cdot 4 = 16 \not\equiv 2 \pmod 6$$
$$4 \cdot 5 = 20 \equiv 2 \pmod 6$$

We see that it has the solutions $x \equiv 2$ and $x \equiv 5 \pmod 6$. Also, $\frac{1}{2} \equiv x \pmod 6$ has no solution (check all possibilities). Thus, $\frac{2}{4} \not\equiv \frac{1}{2} \pmod 6$.

Historical Note

Sir Isaac Newton (1642–1727) was one of the greatest mathematicians of all time. He was a genius of the highest order but was often absent-minded. One story about Newton is that, when he was a boy, he was sent to cut a hole in the bottom of the bar door for the cats to go in and out. He cut two holes—a large one for the cat and a small one for the kittens. Newton's influence on mathematics was so great that the field is sometimes divided into pre-Newtonian mathematics and post-Newtonian mathematics. Post-Newtonian mathematics is characterized by the changing and the infinite rather than by the static and the finite. As an example of his tremendous ability, when Newton was 74 years old the mathematician Leibniz (see Historical Note on page 63) posed a challenge problem to all the mathematicians in Europe. Newton received the problem after a day's work at the mint (remember he was 74!) and solved the problem that evening. His intellect was monumental.

It appears that some mod systems "behave" like ordinary algebra and others do not. The distinction is found by determining which systems form a field.

Let's explore the field properties for the set $I = \{0,1,2,3,4\}$ and the operations of addition and multiplication as defined by Table 6.6. The equivalence relation is congruence, $\equiv$.

1. *Closure for addition.* The set I of elements in arithmetic modulo 5 is closed with respect to addition. That is, for any pair of elements, there is a unique element that represents their sum and that is also a member of the original set.
2. *Closure for multiplication.* The set is closed for multiplication as you can see from Table 6.6.
3. *Commutative for addition.* Addition in arithmetic modulo 5 satisfies the commutative property. That is,

$$a + b \equiv b + a$$

where a and b are any elements of the set I. Specifically, we see that the entries in Table 6.6 are symmetric with respect to the principal diagonal.

4. *Commutative for multiplication.* The set is commutative for multiplication as you can see from Table 6.6.
5. *Associative for addition.* Addition of elements in arithmetic modulo 5 satisfies the associative property. That is,

$$(a + b) + c \equiv a + (b + c)$$

for all elements a, b, and c of the set I.

As a specific example, we evaluate $2 + 3 + 4$ in two ways:

$$(2 + 3) + 4 \quad \text{and} \quad 2 + (3 + 4) \quad (\text{mod } 5)$$
$$0 \quad + 4 \quad \equiv \quad 2 + \quad 2 \quad (\text{mod } 5)$$

6. *Associative for multiplication.* This property is satisfied, and the details are left for you.
7. *Identity for addition.* The set I of elements modulo 5 includes an identity element for addition. That is, the set contains an element zero such that the sum of any given element and zero is the given element. That is,

$$0 + 0 \equiv 0, \quad 1 + 0 \equiv 1, \quad 2 + 0 \equiv 2, \quad 3 + 0 \equiv 3, \quad 4 + 0 \equiv 4$$

8. *Identity for multiplication.* The identity element for multiplication is 1, since

$$0 \cdot 1 \equiv 0, \quad 1 \cdot 1 \equiv 1, \quad 2 \cdot 1 \equiv 2, \quad 3 \cdot 1 \equiv 3, \quad 4 \cdot 1 \equiv 4$$

9. *Inverse for addition.* Each element in arithmetic modulo 5 has an inverse with respect to addition. That is, for each element a of set I, there exists a unique element a' of I such that $a + a' = 0$, the identity element. The element a' is said to be the inverse of a. Specifically, we have the following:

The inverse of 0 is 0; $0 + 0 \equiv 0$ (mod 5)

The inverse of 1 is 4; $1 + 4 \equiv 0$ (mod 5)

The inverse of 2 is 3; $2 + 3 \equiv 0$ (mod 5)

The inverse of 3 is 2; $3 + 2 \equiv 0$ (mod 5)

The inverse of 4 is 1; $4 + 1 \equiv 0$ (mod 5)

$\uparrow$ $\uparrow$ $\uparrow$

Elements | Identity

Inverses

10. *Inverse for multiplication.* The inverse can be checked by finding the following:

There is no inverse for 0; $0 \cdot ? \equiv 1$ (mod 5)

The inverse of 1 is 1; $1 \cdot 1 \equiv 1$ (mod 5)

The inverse of 2 is 3; $2 \cdot 3 \equiv 1$ (mod 5)

The inverse of 3 is 2; $3 \cdot 2 \equiv 1$ (mod 5)

The inverse of 4 is 4; $4 \cdot 4 \equiv 1$ (mod 5)

$\uparrow$ $\uparrow$ $\uparrow$

Elements | Identity

Inverses

Since every nonzero element has an inverse for multiplication, we say the inverse property for multiplication is satisfied.

11. *Distributive for multiplication over addition.* We need to check

$$a(b + c) \equiv ab + ac$$

For example,

$2(3 + 4) \equiv 2(2)$ (mod 5) $2 \cdot 3 + 2 \cdot 4 \equiv 1 + 3$ (mod 5)

and

$\equiv 4$ (mod 5) $\equiv 4$ (mod 5)

Thus, for this example, the distributive property holds. It would be possible (but very tedious) to check all possibilities, since it is a finite system, and find that this property is satisfied.

Therefore, modulo 5 arithmetic with addition and multiplication forms a field. You should expect that the solving of equations in mod 5 is consistent with ordinary algebra.

On the other hand, modulo 6 arithmetic does not form a field for addition and multiplication, since it does not satisfy the inverse property for multiplication (you are asked to show this in Problem Set 6.7). You should not expect that the solving of equations in mod 6 is consistent with ordinary algebra. For example, consider

$$4x \equiv 2 \qquad (\text{mod } 6)$$

In ordinary algebra, you would write

$$4x = 2$$
$$x = \frac{2}{4}$$
$$x = \frac{1}{2} \qquad \text{Remember, } \tfrac{1}{2} \text{ means 1 divided by 2.}$$

In mod 5, you could write

$$4x \equiv 2$$
$$x \equiv \frac{2}{4} \qquad (\text{mod } 5)$$
$$\equiv \frac{1}{2} \qquad (\text{mod } 5)$$
$$\equiv 3 \qquad (\text{mod } 5)$$

But in mod 6, which is not a field, you would need to check each possibility:

$$4x \equiv 2 \qquad \text{This problem was solved in Example 19.}$$
$$x \equiv 2 \quad \text{and} \quad x \equiv 5$$

Notice that in ordinary algebra a first-degree equation has one root. In mod 5, which is a field, there is one root. However, in mod 6, there may be more than one solution to a first-degree equation.

Problem Set 6.7

There was a young fellow named Ben
Who could only count modulo ten
He said when I go
Past my last little toe
I shall have to start over again.

Perform the indicated operations in Problems 1–5 using arithmetic for a 12-hour clock.

1. **a.** $9 + 6$ **b.** $5 - 7$ **c.** 5×3 **d.** $2 \cdot 7$

2. **a.** $7 + 10$ **b.** $4 - 8$ **c.** 6×7 **d.** $\dfrac{1}{5}$

3. **a.** $5 + 7$ **b.** $7 - 9$ **c.** $2 - 6$ **d.** $9 \cdot 3$

4. **a.** $4 \cdot 8$ **b.** $2 \cdot 3$ **c.** $3 \cdot 5 - 7$ **d.** $5 \cdot 2 - 11$

5. **a.** $\dfrac{1}{12}$ **b.** $5 \cdot 8 + 5 \cdot 4$

6. **a.** Define precisely the concept of "congruence modulo m."
   b. Discuss the meaning of this definition in your own words.

7. Which of the following are true?

   a. $5 + 8 \equiv 1 \pmod 6$ **b.** $4 + 5 \equiv 1 \pmod 7$
   c. $5 \equiv 53 \pmod 8$ **d.** $102 \equiv 1 \pmod 2$

8. Which of the following are true?

   a. $47 \equiv 2 \pmod 5$ **b.** $108 \equiv 12 \pmod 8$

   c. $5{,}570 \equiv 270 \pmod{365}$ **d.** $2{,}001 \equiv 39 \pmod{73}$

9. Which of the following are true?

   a. $1{,}975 \equiv 0 \pmod{1975}$ **b.** $246 \equiv 150 \pmod 6$

   c. $126 \equiv 1 \pmod 7$ **d.** $144 \equiv 12 \pmod{144}$

Perform the indicated operations in Problems 10–13.

10. **a.** $9 + 6 \pmod 5$ **b.** $7 - 11 \pmod{12}$

    c. $4 \cdot 3 \pmod 5$ **d.** $\frac{1}{2} \pmod 5$

11. **a.** $5 + 2 \pmod 4$ **b.** $2 - 4 \pmod 5$

    c. $6 \cdot 6 \pmod 8$ **d.** $5 \div 7 \pmod 9$

12. **a.** $4 + 3 \pmod 5$ **b.** $6 - 12 \pmod 8$

    c. $121 \cdot 47 \pmod{121}$ **d.** $\frac{2}{3} \pmod 7$

13. **a.** $7 + 41 \pmod 5$ **b.** $5 \cdot 4 \pmod{11}$

    c. $62 \cdot 4 \pmod 2$ **d.** $\frac{7}{12} \pmod{13}$

Solve for x in Problems 14–18.

14. **a.** $x + 3 \equiv 0 \pmod 7$ **b.** $4x \equiv 1 \pmod 5$

    c. $x \cdot x \equiv 1 \pmod 4$ **d.** $\frac{x}{4} \equiv 5 \pmod 9$

15. **a.** $x + 5 \equiv 2 \pmod 9$ **b.** $4x \equiv 1 \pmod 6$

    c. $x^2 \equiv 1 \pmod 5$ **d.** $\frac{4}{6} \equiv x \pmod{13}$

16. **a.** $5x \equiv 2 \pmod 7$ **b.** $7x + 1 \equiv 3 \pmod{11}$

    c. $\frac{x}{3} \equiv 4 \pmod 8$ **d.** $4k + 2 \equiv x \pmod 4$, where k is any counting number

17. **a.** $x - 2 \equiv 3 \pmod 6$ **b.** $3x \equiv 2 \pmod 7$

    c. $4k \equiv x \pmod 4$, where k is any counting number   **d.** $4k + 3 \equiv x \pmod 4$, where k is any counting number

18. **a.** $2x^2 - 1 \equiv 3 \pmod 7$ **b.** $5x^3 - 3x^2 + 70 \equiv 0 \pmod 2$

19. Your doctor tells you to take a certain medication every 8 hours. If you begin at 8:00 A.M., show that you will not have to take the medication between midnight and 7 A.M.

20. **a.** Suppose you make six round trips to visit a sick aunt and wish to record your mileage. You forget the original odometer reading, but you do remember that the units digit has increased by 8 miles. What are the possible distances between your house and your aunt's house?

b. If you know that your aunt lives somewhere between 10 and 15 miles from your house, how far exactly is her house, given the information in part a?

21. Suppose you are planning to purchase some rope. You need between 15 and 20 pieces that are 7 inches long and one piece that is 80 inches long. The rope can be purchased in multiples of 12 inches. How much rope should you buy to eliminate as much waste as possible?

Find the months containing a Friday the 13th for each of the years indicated in Problems 22–24.

22. 1984   **23.** 2001   **24.** 1943

25. Consider the set of numbers used in 12-hour clock arithmetic:

a. Is the set closed with respect to addition?
b. Does the set of numbers have an identity for addition?
c. Does the set of numbers on the clock contain inverse elements with respect to addition?

26. Investigate the following addition properties of a modulo 6 system.

a. Closure   **b.** Associative
c. Identity   **d.** Inverse
e. Commutative

27. Investigate the following multiplication properties of a modulo 6 system.

a. Closure   **b.** Associative
c. Identity   **d.** Inverse
e. Commutative

28. Make a table for the addition and multiplication facts, mod 11.

29. Using the addition and multiplication tables for mod 11 found in Problem 28, answer the following questions:

a. Is the set closed for the operations of addition and multiplication?
b. Is the set commutative for these operations?
c. Is the set associative for these operations?

30. Using the addition and multiplication tables for mod 11 found in Problem 28, answer the following questions:

a. Does the set have an identity for addition? For multiplication?
b. What is the additive inverse of each element?
c. What is the multiplicative inverse of each element?

31. *Computer problem.* We can easily write a BASIC program that will allow us to input a positive integer and a mod m and have the computer convert the integer into a mod m numeral.

Notice that we have used a new command, INT(X), called the greatest integer, denoted by $[x]$ in this section. It gives the greatest number not exceeding the one in the brackets, or parentheses in INT(X). For example,

INT(46.2) = 46   INT(19.9) = 19
INT(108) = 108   INT(π) = 3

Some sample runs of the program given in Problem 31 are shown below:

```
RUN
WHAT IS THE INTEGER? 23
WHAT IS THE MOD? 5
THE INTEGER IN MOD 5
IS 3

READY

RUN
WHAT IS THE INTEGER?
108
WHAT IS THE MOD? 13
THE INTEGER IN MOD 13
IS 4

READY

RUN
WHAT IS THE INTEGER?
23453
WHAT IS THE MOD? 61
THE INTEGER IN MOD 61
IS 29

READY
```

```
10  PRINT "WHAT IS THE INTEGER";
20  INPUT I
30  PRINT "WHAT IS THE MOD";
```

```
40  INPUT M
50  LET A = I - INT(I/M)*M
60  PRINT "THE INTEGER IN MOD";
70  PRINT M;
80  PRINT " IS ";
90  PRINT A
100 END
```

Use this program to find x:

a. $5,872 \equiv x$ (mod 119)   **b.** $4,872,879 \equiv x$ (mod 365)
c. $5,827,300 \equiv x$ (mod 365)   **d.** $5,710,000 \equiv x$ (mod 4,380)

Mind Bogglers

32. If it is now 2 P.M., what time will it be 99,999,999,999 hours from now?

33. Write a schedule for 12 teams so that each team will play every other team once and no team will be idle. [*Hint:* Consider mod 11 arithmetic.]

34. Write a BASIC program to determine the months in a given year that have a Friday the 13th.

35. A man buys 100 birds for $100. A rooster is worth $10, a hen is worth $3, and chicks are worth $1 a pair. How many roosters, hens, and chicks did he buy if he bought at least one of each type? (This is the Chinese problem of "One Hundred Fowls.")

36. *Computer problem.* A band of 17 pirates decided to divide their doubloons into equal portions. When they found that they had 3 coins remaining, they agreed to give them to their Chinese cook, Wun Tu. But 6 of the pirates were killed in a fight. Now when the treasure was divided equally among them, there were 4 coins left that they considered giving to Wun Tu. Before they could divide the coins, there was a shipwreck and only 6 pirates, the coins, and the cook were saved. This time equal division left a remainder of 5 coins for the cook. Now Wun Tu took advantage of his culinary position to concoct a poison mushroom stew so that the entire fortune in doubloons became his own. What is the smallest number of coins that the cook would have finally received? [*Hint:* See the flowchart in the margin.]

37. What is the next smallest number of coins that would satisfy the conditions of Problem 36?

Problems for Individual Study

38. What is a Diophantine equation?

   Reference: Warren J. Himmelberger, "Puzzle Problems and Diophantine Equations," *Mathematics Teacher*, February 1973, pp. 136–138.

   For a more complete reference, see any number theory textbook.

39. *Modular art.* When constructing addition and multiplication tables for the various modular systems, you probably noticed some patterns. Consider, for example, the

Hint for Problem 36:
Study the flowchart below.

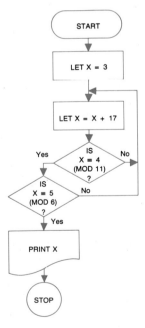

multiplication tables for mod 19 and mod 21 (see Tables 6.8 and 6.9). The patterns from these tables can be translated into interesting figures.

TABLE 6.8 Mod 19

×	0	1	2	3	4	5	6	7	8	9	10	11	12	13	14	15	16	17	18
0	0	0	0	0	0	0	0	0	0	0	0	0	0	0	0	0	0	0	0
1	0	1	2	3	4	5	6	7	8	9	10	11	12	13	14	15	16	17	18
2	0	2	4	6	8	10	12	14	16	18	1	3	5	7	9	11	13	15	17
3	0	3	6	9	12	15	18	2	5	8	11	14	17	1	4	7	10	13	16
4	0	4	8	12	16	1	5	9	13	17	2	6	10	14	18	3	7	11	15
5	0	5	10	15	1	6	11	16	2	7	12	17	3	8	13	18	4	9	14
6	0	6	12	18	5	11	17	4	10	16	3	9	15	2	8	14	1	7	13
7	0	7	14	2	9	16	4	11	18	6	13	1	8	15	3	10	17	5	12
8	0	8	16	5	13	2	10	18	7	15	4	12	1	9	17	6	14	3	11
9	0	9	18	8	17	7	16	6	15	5	14	4	13	3	12	2	11	1	10
10	0	10	1	11	2	12	3	13	4	14	5	15	6	16	7	17	8	18	9
11	0	11	3	14	6	17	9	1	12	4	15	7	18	10	2	13	5	16	8
12	0	12	5	17	10	3	15	8	1	13	6	18	11	4	16	9	2	14	7
13	0	13	7	1	14	8	2	15	9	3	16	10	4	17	11	5	18	12	6
14	0	14	9	4	18	13	8	3	17	12	7	2	16	11	6	1	15	10	5
15	0	15	11	7	3	18	14	10	6	2	17	13	9	5	1	16	12	8	4
16	0	16	13	10	7	4	1	17	14	11	8	5	2	18	15	12	9	6	3
17	0	17	15	13	11	9	7	5	3	1	18	16	14	12	10	8	6	4	2
18	0	18	17	16	15	14	13	12	11	10	9	8	7	6	5	4	3	2	1

TABLE 6.9 Mod 21

×	0	1	2	3	4	5	6	7	8	9	10	11	12	13	14	15	16	17	18	19	20
0	0	0	0	0	0	0	0	0	0	0	0	0	0	0	0	0	0	0	0	0	0
1	0	1	2	3	4	5	6	7	8	9	10	11	12	13	14	15	16	17	18	19	20
2	0	2	4	6	8	10	12	14	16	18	20	1	3	5	7	9	11	13	15	17	19
3	0	3	6	9	12	15	18	0	3	6	9	12	15	18	0	3	6	9	12	15	18
4	0	4	8	12	16	20	3	7	11	15	19	2	6	10	14	18	1	5	9	13	17
5	0	5	10	15	20	4	9	14	19	3	8	13	18	2	7	12	17	1	6	11	16
6	0	6	12	18	3	9	15	0	6	12	18	3	9	15	0	6	12	18	3	9	15
7	0	7	14	0	7	14	0	7	14	0	7	14	0	7	14	0	7	14	0	7	14
8	0	8	16	3	11	19	6	14	1	9	17	4	12	20	7	15	2	10	18	5	13
9	0	9	18	6	15	3	12	0	9	18	6	15	3	12	0	9	18	6	15	3	12
10	0	10	20	9	19	8	18	7	17	6	16	5	15	4	14	3	13	2	12	1	11
11	0	11	1	12	2	13	3	14	4	15	5	16	6	17	7	18	8	19	9	20	10
12	0	12	3	15	6	18	9	0	12	3	15	6	18	9	0	12	3	15	6	18	9
13	0	13	5	18	10	2	15	7	20	12	4	17	9	1	14	6	19	11	3	16	8
14	0	14	7	0	14	7	0	14	7	0	14	7	0	14	7	0	14	7	0	14	7
15	0	15	9	3	18	12	6	0	15	9	3	18	12	6	0	15	9	3	18	12	6
16	0	16	11	6	1	17	12	7	2	18	13	8	3	19	14	9	4	20	15	10	5
17	0	17	13	9	5	1	18	14	10	6	2	19	15	11	7	3	20	16	12	8	4
18	0	18	15	12	9	6	3	0	18	15	12	9	6	3	0	18	15	12	9	6	3
19	0	19	17	15	13	11	9	7	5	3	1	20	18	16	14	12	10	8	6	4	2
20	0	20	19	18	17	16	15	14	13	12	11	10	9	8	7	6	5	4	3	2	1

Divide a circle into $n - 1$ parts, where n is the given modulus. For example, with mod 19 we divide the circle into 18 parts (for our purposes, we disregard the 0 element). Next, select some nonzero number in the table—say, 9. That is, we consider:

	1	2	3	4	5	6	7	8	...
9	9	18	8	17	7	16	6	15	...

We now connect the points 1 and 9, 2 and 18, 3 and 8, 4 and 17, and so on. The result is shown below.

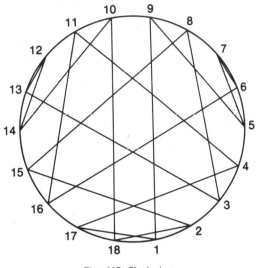

The (19, 9) design

The (19, 2) design

We call this the (19, 9) design. By shading in alternate regions, we obtain the pattern shown below. Create some of your own modular art patterns, and make a presentation to the class.*

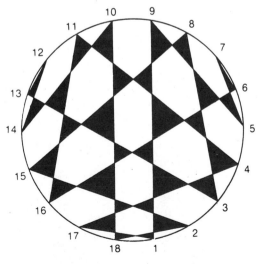

The (19, 9) design with shading

The (19, 18) design

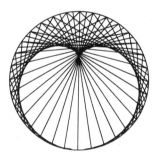

The (65, 2) design

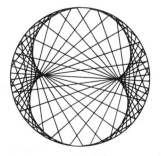

The (65, 3) design

*My thanks to Phil Locke for this problem.

40. *Modular arithmetic.* Many interesting designs such as those shown here can be created using patterns based on modular arithmetic. Prepare a report for class presentation based on the article "Using Mathematical Structures to Generate Artistic Designs" by Sonia Forseth and Andria Price Troutman, *Mathematics Teacher,* May 1974, pp. 393–398. Another source is "Mod Art: The Art of Mathematics," by Susan Morris, *Technology Review,* March/April 1979.

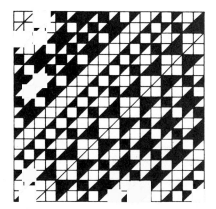

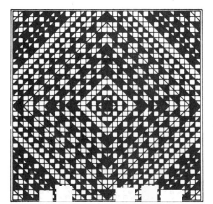

Chapter 6 Outline

I. Mathematical Systems
    A. Definition
    B. Identity properties
    C. Inverse properties
    D. Field

II. Linear Equations
    A. Equation
    B. Equation properties
    C. Solving equations

III. Prime Numbers
    A. Divisibility
    B. Definition
    C. Finding primes
        1. Sieve of Eratosthenes
        2. Other possible methods

IV. Prime Factorization
    A. Definitions
        1. Set of factors
        2. Factorization
        3. Prime factorization
        4. Canonical representation
    B. Fundamental Theorem of Arithmetic
    C. Factor tree
    D. There is no largest prime

V. Applications of Prime Factorization
    A. Greatest common factor (g.c.f.)
        1. Definition
        2. Relatively prime numbers
        3. Procedure for finding the g.c.f.
    B. Least common multiple (l.c.m.)
        1. Definition
        2. Procedure for finding the l.c.m.
    C. Working with rational numbers
        1. Fundamental property of fractions
        2. Adding and subtracting fractions
        3. Reducing fractions

VI. Ratio and Proportion
    A. Ratios as reduced fractions
    B. Determining if two ratios form a proportion
    C. Solving proportions

*VII. A Finite Algebra
    A. Operations
        1. Addition: Defined by table according to the "addition on a clock."
        2. Subtraction: $a - b = x$ means $a = b + x$.
        3. Multiplication: $a \times b$ means $\underbrace{a + a + \cdots + a + a}_{b \text{ addends}}$.

        4. Division: $\dfrac{a}{b} = x$ means $a = b \cdot x$.
    B. Clock arithmetic
    C. Modular arithmetic
        1. Definition
        2. Operations
        3. Equation solving
        4. Field properties

Chapter 6 Test

1.   **a.** Discuss what we mean when we write $a \mid b$.

      Which of the following are true? If true, use the definition to prove it is true.

      b. $6 \mid 714$ **c.** $8 \mid 735{,}812$ **d.** $11 \mid 6{,}825$ **e.** $9 \mid 598{,}364$

2.   **a.** Consider arithmetic on a 4-hour clock. Construct the addition and multiplication tables.
      b. Is the set $\{0,1,2,3\}$ a field for the operations of addition and multiplication (mod 4)? Why or why not? Discuss each of the properties involved.

3. Find the prime factorization of each of the given numbers. If it is a prime, so state.

      a. 89   **b.** 101   **c.** 349   **d.** 1,001   **e.** 6,825

4. Find the greatest common factor and least common multiple for each of the following sets of numbers.

      a. $\{30,42,99\}$ **b.** $\{49,1001,2401\}$

5. Reduce: $\dfrac{3{,}536}{3{,}952}$

6. Perform the indicated operations.

      a. $\dfrac{3^{-1} + 4^{-1}}{6}$ **b.** $\dfrac{^{-}7}{9} \cdot \dfrac{99}{174} + \dfrac{^{-}7}{9} \cdot \dfrac{75}{174}$

      c. $\dfrac{5}{49} - \dfrac{1}{1{,}001}$ **d.** $\dfrac{7}{30} + \dfrac{5}{42} + \dfrac{5}{99}$

*Optional section.

7. Solve for x.

   a. $x + 11 = 14$ **b.** $2x + 5 = 13$

   c. $3x + 2 = x + 6$ **d.** $3x + 1 = 7x$

   e. $\dfrac{2x}{3} = 6$

8. Solve for x.

   a. $2x - 7 = 5x$ **b.** $14 = 5x - 1$

   c. $3(2x + 3) = 12$ **d.** $\dfrac{2x - 1}{2} = 5$

   e. $\dfrac{3(x - 1)}{2} = 7$

*9. Which of the following are true?

   a. $4 + 9 \equiv 1 \ (\text{mod } 12)$ **b.** $2 \equiv 2{,}001 \ (\text{mod } 2)$

   c. $7 \cdot 9 \equiv 22 \ (\text{mod } 8)$ **d.** $4 - 10 \equiv 6 \ (\text{mod } 12)$

   e. $5 - 7 \equiv 24 \ (\text{mod } 11)$

*10. Solve for x.

   a. $\dfrac{x}{5} \equiv 2 \ (\text{mod } 8)$ **b.** $2x \equiv 3 \ (\text{mod } 7)$

   c. $\dfrac{x}{3} \equiv 5 \ (\text{mod } 6)$ **d.** $5x \equiv 6 \ (\text{mod } 7)$

   e. $2x^2 + 7x + 1 \equiv 0 \ (\text{mod } 2)$

Bonus Question

*11. John Davidson was born on December 13, 1941. Was he born on a Friday the thirteenth?

---

*Requires optional Section 6.7.

Date	Cultural History
1500	 **1500:** Michaelangelo's *Pieta* **1507:** da Vinci begins *Leda and the Swan* **1513:** Machiavelli's *The Prince*  **1520:** Magellan discovers the straits Luther excommunicated
1525	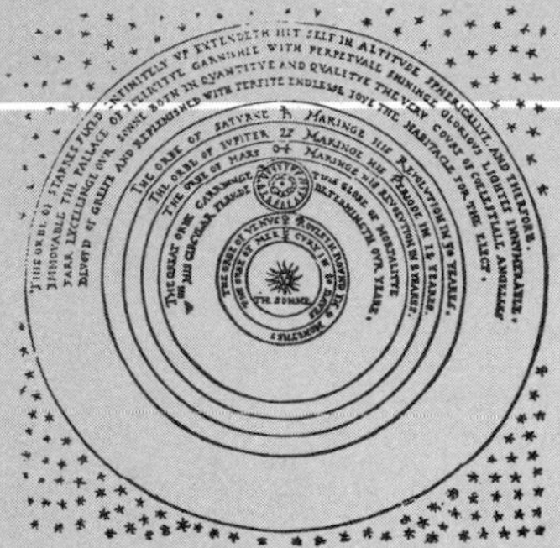 **1534:** Henry VIII becomes head of the Church of England **1543:** Publication of Copernicus' work
1550	 **1558:** Elizabeth crowned Queen of England  **1567:** Bothwell abducts Mary, Queen of Scots **1569:** Tycho Brahe begins construction of a 19-foot quadrant **1573:** Francis Drake sees Pacific Ocean
1575	**1582:** Pope Gregory XIII creates new calendar **1588:** England defeats Spanish Armada 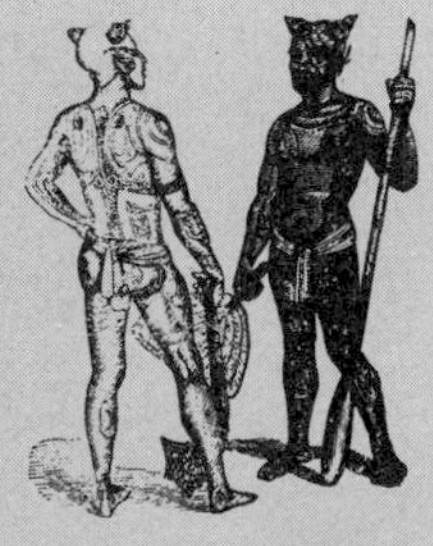 **1596:** Discovery of the Marquesas

da Vinci polyhedron

Sketch for *Leda and the Swan*

Copernican system

Queen Elizabeth I

Tycho Brahe

Marquesan chieftains

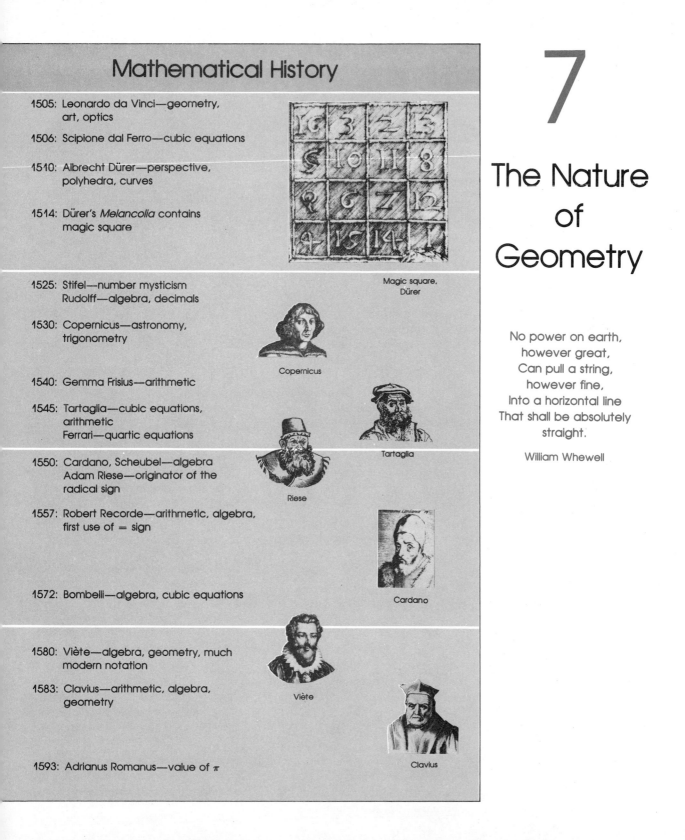

Mathematical History

1505: Leonardo da Vinci—geometry, art, optics

1506: Scipione dal Ferro—cubic equations

1510: Albrecht Dürer—perspective, polyhedra, curves

1514: Dürer's *Melancolia* contains magic square

Magic square, Dürer

1525: Stifel—number mysticism
Rudolff—algebra, decimals

1530: Copernicus—astronomy, trigonometry

Copernicus

1540: Gemma Frisius—arithmetic

1545: Tartaglia—cubic equations, arithmetic
Ferrari—quartic equations

Tartaglia

1550: Cardano, Scheubel—algebra
Adam Riese—originator of the radical sign

Riese

1557: Robert Recorde—arithmetic, algebra, first use of = sign

1572: Bombelli—algebra, cubic equations

Cardano

1580: Viète—algebra, geometry, much modern notation

1583: Clavius—arithmetic, algebra, geometry

Viète

1593: Adrianus Romanus—value of π

Clavius

No power on earth,
however great,
Can pull a string,
however fine,
Into a horizontal line
That shall be absolutely
straight.

William Whewell

7.1 Geometry

Historical Note

Early civilizations observed from nature certain simple shapes such as triangles, rectangles, and circles. The study of geometry began with the need to measure and understand the properties of these simple shapes.

Geometry, or "earth measure," was one of the first branches of mathematics. Both the Egyptians and the Babylonians needed geometry for construction, for land measurement, and for commerce. As we've seen, they both discovered the Pythagorean theorem, although it was not proved until the Greeks developed geometry formally. This formal development utilizes the logic in Chapter 2 by beginning with certain undefined terms. For example, what is a line? You might say "I know what a line is!" But try to define a line. Is it a set of points? Any set of points? What is a point?

1. A point is something that has no length, width, or thickness.
2. A point is a location in space.

Certainly these are not satisfactory definitions, because they involve other terms that are not defined. We take, therefore, the terms *point, line,* and *plane* as **undefined.** Next, certain properties associated with these terms are assumed as **axioms.** By accepting different sets of axioms, we can develop different geometries. The axioms you accepted in high school seemed to correspond to the world around us and led to a geometry that we call *Euclidean geometry.* Around 300 B.C. a mathematician named Euclid was a professor of mathematics at the University of Alexandria. He wrote a textbook called the *Elements,* which was an introductory textbook covering all elementary mathematics: arithmetic (number theory), geometry (synthetic), and algebra (not symbolic, but geometrical). The *Elements* did not contain any new discoveries, and it was not Euclid's only book, but it is the most successful mathematics textbook in history (only the Bible has had more printings). Based on the undefined terms and assumed axioms, logic is used to deduce other results called **theorems.** By accepting axioms different from those accepted in high school geometry, other geometries are developed. We will take a look at some of these toward the end of this chapter.

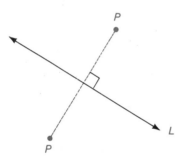

FIGURE 7.1   A reflection

One modern approach to geometry is called *transformational.* It takes the viewpoint of transforming one set (a geometric figure, for example) into another. There are several ways of doing this. For example, given a line L and a point P as shown in Figure 7.1, we call the point P' the *reflection* of P about the line L if line PP' is perpendicular to L and is also bisected by L. Each point in the plane has exactly one reflection point corresponding to a given line L. This means that there is a one-to-one correspondence between the set of points of a plane and their image points with respect to a given line L. *Any operation,* such as a reflection, *in which there is a one-to-one correspondence between the points of the plane and their image points is called a* **transformation,** and the line of reflection is called the **line of symmetry.** (See Figure 7.2 for examples of symmetry in nature and in design.)

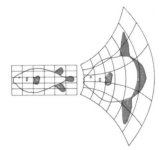

We find an excellent example of a grid transformation in this scheme devised by the Scottish biologist D'Arcy Wentworth Thompson. The fish in the grid at the left is transformed with the grid at the right. This scheme is reproduced from Thompson's 1917 work *On Growth and Form.* His method is useful in understanding the forces that shape sheets of cells into the tissue layers of the embryo.

Other transformations include *translations, rotations, dilations, contractions,* and *inversions,* which are illustrated in Figure 7.3. The investigation of these ideas is beyond the scope of this book, but transformational geometry is an important part of mathematics.

FIGURE 7.2 Symmetry in nature is illustrated by snowflakes. Symmetry in design is illustrated by the *Apollo* space capsule.

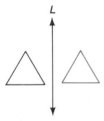  

Reflection transformation · Rotation transformation · Translation transformation

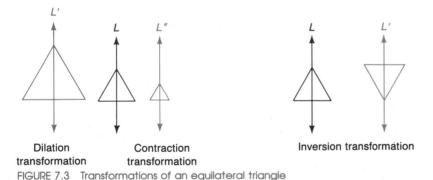

Dilation transformation · Contraction transformation · Inversion transformation

FIGURE 7.3 Transformations of an equilateral triangle

Geometry is also concerned with the study of the relationships between geometric figures. A primary relationship is that of **congruence.** Two figures are said to be congruent if one can be made to coincide with the other. A second relationship is called **similarity.** Two figures are said to be *similar* if they have the same shape, although not necessarily the same size. These ideas are usually studied in a high school geometry course.

In geometry we often draw physical models or pictures to represent the figures or concepts; however, you must be careful not to base your conjectures or conclusions on pictures. A figure or picture may contain hidden assumptions or ambiguities. For example, consider Figure 7.4a.

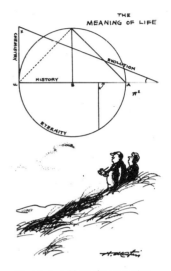

"I wish I could see the expression on the faces of my students who said there was no value in studying geometry."

If we were to develop the course formally, we would have to be very careful about the statement of our postulates. The notion of a mathematical proof requires a very precise formulation of all postulates and theorems. For example, Euclid based many proofs about congruence on a postulate that said "Things that coincide with one another are equal to one another" (Common Notions, Book I, *Elements*). This statement is not formulated precisely enough to be used as a justification for some of the things Euclid did with congruence. Even the explanation we gave for congruence is not precise enough to be used in a formal course. However, since we are not formally developing geometry, we simply accept the general drift of the statement. Remember, though, that you cannot base a mathematical proof on general drift.

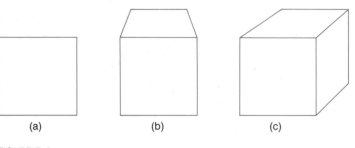

FIGURE 7.4

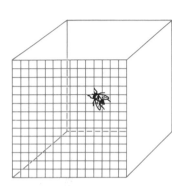

FIGURE 7.5   Fly on a cube

What do you see? A square? Perhaps. But what if we take a slightly different view of the same object, as shown in Figure 7.4b. What do you see now? But what if we have in mind a cube, as shown in Figure 7.4c? Even if you view this object as a cube, do you see the same cube as everyone else? Is the fly in Figure 7.5 inside or outside the cube? Or is it perhaps on an edge of the cube? Thus, although we may use a figure to help us understand a problem, we cannot prove results by this technique.

Problem Set 7.1

1. What is the difference between an axiom and a theorem?

2. Is the woman in the figure a young woman or an old woman?

3. Which of the following pictures illustrate(s) line symmetry? That is, in each picture can a line be drawn so that the picture will be symmetric with respect to this line?

a.   *Allegorical Garden of Geometry.*
      Engraving by Francesco Curti

b.   *Winged Lion.* China—Early T'ang
      Dynasty, 7th century A.D.

A young woman or an old
woman?

c. *Portrait of St. John.* Book of Kells

d. *Kaiser porcelain vase.* Created to commemorate Queen Elizabeth's 25th year on the throne

Study the patterns shown in the margin. When folded they will form cubes spelling CUBE. Letter each pattern in Problems 4–9 so that when folded they too will form a cube spelling CUBE. You can check your answers by cutting out the patterns and folding them into cubes.

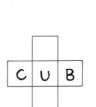

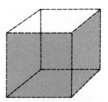

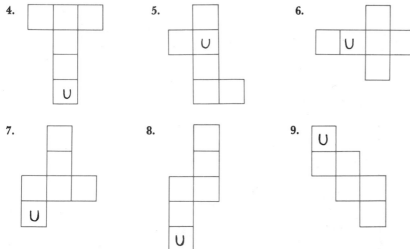

4.

5. U
U

6. U

7. U

8. U

9. U

10. The figure in the margin shows a cube with the top cut off. Use solid lines and shading to depict seven other different views of a cube with one side cut off.

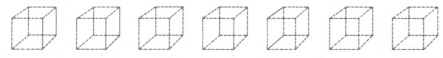

11. Following are two sets of three intersecting planes. Use solid lines and shading to illustrate different ways of viewing the planes. Can you find more than two ways?

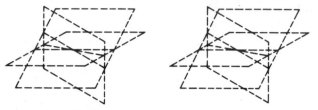

Problems 12–14 all build on the previous one of the sequence. Compare the results of these problems with the results obtained in Section 1.5. (However, Section 1.5 is not required for these problems.)

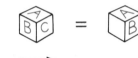

To the right

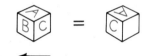

To the left

Upward

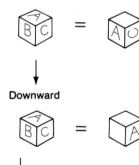

Downward

Downward, then to the right

12. **a.** Draw a square *ABCD*.
　　b. Divide the square into two parts by completing the following steps:
　　　　i. Call the midpoint of *BC* point *E*.
　　　　ii. Call the midpoint of *AD* point *F*.
　　　　iii. Draw line segment *EF*.
　　c. Draw a circle with center at *F* and radius *FC*.
　　d. Extend the base line *AD* to intersect the circle at point *G*.
　　e. This line forms the base of a rectangle *ABHG* where *H* is the point on the extension of *BC* such that *HG* forms a right angle with *AG*.
　　f. Measure the sides of rectangle *ABHG*.
　　g. Divide the width by the length.

13. **a.** Remove the square *ABCD* from the rectangle *ABHG* of Problem 12. The result is a rectangle *DCHG*.
　　b. Measure the sides of rectangle *DCHG*.
　　c. Divide the width by the length.
　　d. Compare your answers to Problems 12g and 13c.

14. **a.** Draw a rectangle that is aesthetically pleasing to you.
　　b. Measure the sides.
　　c. Divide the width by the length.
　　d. Compare your answer to part c with your answer to Problems 12g and 13c.

The arrows in Problems 15–19 indicate the direction(s) that the given cube has been rotated 90°. Study the examples shown in the margin. If no arrows are shown, you must deduce the direction of rotation. In Problems 15–19, pick the cube illustrating the appropriate rotation.

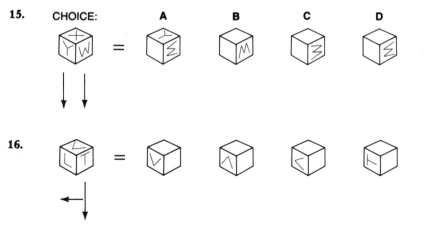

17.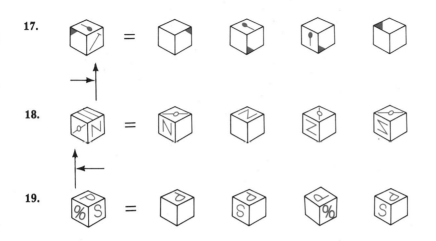

18.

19.

Mind Boggler

20. The "Mirror Image" illustration shows a problem condensed from *Games* magazine. Answer the questions asked in that problem.

Mirror Image

by Diane Dawson

In the Fabulous Kingdom, anything can happen—and usually does.
When we held a looking glass up to this page, we noticed a few incongruities
between it and its "reflection" on the facing page. Can you spot 24
substantial differences here, with or without the help of a mirror?
(Since both illustrations were hand-drawn there are bound
to be hairline differences—these should be ignored.)

Mirror Image

by Diane Dawson

In the Fabulous Kingdom, anything can happen—and usually does.
When we held a looking glass up to this page, we noticed a few incongruities
between it and its "reflection" on the facing page. Can you spot 24
substantial differences here, with or without the help of a mirror?
(Since both illustrations were hand-drawn there are bound
to be hairline differences—these should be ignored.)

Problem for Individual Study

21. *Optical illusions.* What are optical illusions? Why do these patterns cause illusions? Are these illusions still apparent with three-dimensional representations? What tricks depend on these illusions? How are these illusions used in advertising, dress designing, and architecture?

Exhibit suggestions: charts, models, ads, pictures, or illusions.

References: Martin Gardner, "Mathematical Games," *Scientific American,* May 1970.

Richard Gregory, "Visual Illusions," *Scientific American,* November 1968.

Lionel Penrose, "Impossible Objects: A Special Type of Visual Illusion," *The British Journal of Psychology,* February 1958.

7.2 Measuring Length and Area

In the everyday world, numbers arise naturally in either of two ways—by counting or by measuring. In this chapter we'll focus on measuring.

> To measure a length is to assign a number to its size. The number is called its **measure** or **length.**

In order to measure length, you need a measuring device appropriate to the length you are measuring. For example, you would not use the same ruler to measure the length of this page, the length of a football field, and the distance between Los Angeles and San Francisco. The measuring device used is dependent on the system of measuring. In the United States we use a system of measuring originally based on parts of the human body (see Historical Note). However, the most common measurement system used throughout the world today is the *metric system.* Many people in the United States resist changing to the metric system. There is a common misbelief that changing to the metric system will require complex multiplying and dividing and the use of confusing decimal points. For example, in a recent popular article, James Collier states:

> For instance, if someone tells me it's 250 miles up to Lake George, or 400 out to Cleveland, I can pretty well figure out how long it's going to take and plan accordingly. Translating all of this into kilometers is going to be an awful headache. A kilometer is about 0.62 miles, so to convert miles into kilometers you divide by six and multiply by ten, and even that isn't accurate. Who can do that kind of thing when somebody is asking me are we almost there, the dog is beginning to drool and somebody else is telling you you're driving too fast?
>
> Of course, that won't matter, because you won't know how fast you're going anyway. I remember once driving in a rented car on a superhighway in France, and everytime I looked down at the speedometer we were going 120. That kind of thing can give you the creeps. What's it going to be like when your wife keeps

Historical Note

The system of measurement used in the United States, called the British system, goes back to the Babylonians and Egyptians. Measurements were made in terms of the human body (digit, palm, span, and foot). Eventually, measurements were standardized in terms of the physical measurements of certain monarchs. (King Henry I, for example, decreed that one yard was the distance from the tip of his nose to the end of his thumb.)

ONE YARD

In 1790, the French Academy of Science was asked by the government to develop a logical system. The original metric

shouting, "Slow down, you're going almost 130"?
    But if you think kilometers will be hard to calculate . . . .*

The author of this article has missed the whole point. Why are kilometers hard to calculate? How does he know that it's 400 miles to Cleveland? He knows because the odometer on his car or a road sign told him. Won't it be just as easy to read an odometer calibrated to kilometers or a metric road sign telling him how far it is to Cleveland?

system came into being, and by 1900 it was adopted by over 35 major countries. In 1906 there was a major effort to make the conversion to the metric system in the United States mandatory, but it was opposed by big business and the attempt failed. In 1960 the metric system was revised and simplified into what is now known as the *SI system*. In 1972 and 1973, when the United States was the only major country not using the metric system, further attempts to make it mandatory failed. However, in 1975, Congress declared conversion to the metric system to be "national policy." Now, with the backing of big business, adoption of the system in the United States seems inevitable.

Measuring is never exact, and you will need to decide how precise the measurement should be. You will also need to decide on some standard unit of measurement. Since you are probably familiar with the U.S. system of measurement, we will focus on the metric system in this chapter. The real advantage of using the metric system is the ease of conversion from one unit of measurement to another.

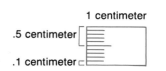

The U.S. system has four standard units of length:	The metric system has one standard unit of length:
inch (in.) foot (ft) yard (yd) mile (mi)	meter (m)

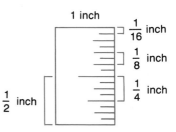

FIGURE 7.6   Actual size comparison of an inch and a centimeter.

*James Lincoln Collier, "Fourth Down, 50 Centimeters to Go," *Kansas City Star Magazine*, January 8, 1978.

In order to work with different lengths, you must memorize some conversion factors:

U.S. Conversion Factors for Length	Metric Conversion Factors for Length

12 in. = 1 ft
3 ft = 1 yd
1,760 yd = 1 mi
5,280 ft = 1 mi
	centi means $\frac{1}{100}$;
	   centimeter is abbreviated cm
	kilo means 1,000;
	   kilometer is abbreviated km

One application of both measurement and geometry is finding the distance around a polygon. A **polygon** is made up of three or more points (called *vertices*) and the segments connecting those points (called *sides*). The sides also have no common point except their end points (the vertices). The segments should all lie on a flat surface, or **plane,** so that the starting point and the ending point are the same.

Polygons are classified according to the number of sides, as shown in Figure 7.7. A **regular polygon** is a polygon with all sides the same length and all angles the same size.

> So full of shapes is fancy,
> That it alone is high
>   fantastical.
>
> *Shakespeare*
> Twelfth Night

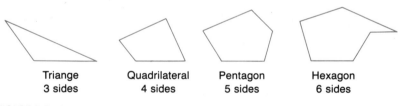

Triange	Quadrilateral	Pentagon	Hexagon
3 sides	4 sides	5 sides	6 sides

FIGURE 7.7   Definition and examples of polygons. If we keep adding sides, we get a heptagon with 7 sides, octagon with 8 sides, nonagon with 9 sides, decagon with 10 sides, dodecagon with 12 sides, etc.

The distance around a polygon is called the *perimeter*.

> The **perimeter** of a polygon is the sum of the lengths of the sides of that polygon.

Example 1:   Find the perimeter of the polygon shown on the next page by measuring each side to the nearest $\frac{1}{10}$ centimeter. Use a metric ruler like that shown in Figure 7.8.

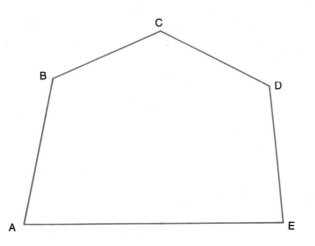

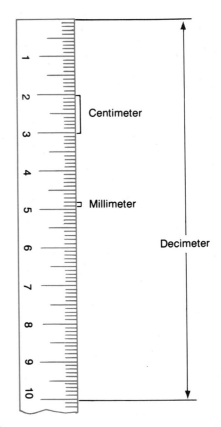

FIGURE 7.8 Actual size metric ruler. Metric rulers are usually marked in centimeters and divided into millimeters.

Solution:

Side	Length
AB	3.9 cm
BC	3.1 cm
CD	3.2 cm
DE	3.6 cm
EA	6.8 cm
TOTAL	20.6 cm perimeter

Here are some formulas for finding the perimeters of the most common polygons.

An **equilateral triangle** is a triangle with sides that are equal.

$$\text{Perimeter} = 3(\text{Side})$$
$$P = 3s$$

A **rectangle** is a quadrilateral with angles that are all right angles.

$$\text{Perimeter} = 2(\text{Length}) + 2(\text{Width})$$
$$P = 2L + 2W$$

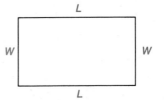

A **square** is a rectangle with sides that are equal.

$$\text{Perimeter} = 4(\text{Side})$$
$$P = 4s$$

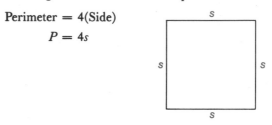

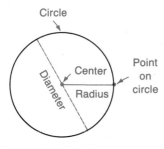

Circle

FIGURE 7.9   Circle

A **circle** is not a polygon, but you may find it useful to find the distance around a circle or a part of a circle. Figure 7.9 illustrates some of the terminology associated with a circle. The distance around the circle is called the **circumference**. In Section 4.7, an irrational number π was defined as the ratio of the circumference to the diameter of a circle. This definition leads to the following formula for finding the circumference of a circle:

$$C = d\pi \quad \text{or} \quad C = 2\pi r$$

where C = Circumference
d = Diameter
r = Radius
$\pi \approx 3.14 \text{ or } \frac{22}{7}$ The symbol $\approx$ means approximately equal to. It needs to be used here because π is an irrational number and does not have a terminating or repeating decimal representation.

In Examples 2–6 find the perimeter or circumference of the given figures.

Example 2:

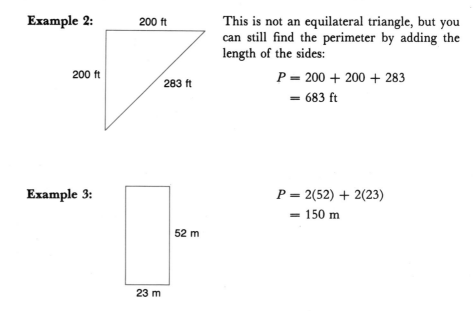

This is not an equilateral triangle, but you can still find the perimeter by adding the length of the sides:

$$P = 200 + 200 + 283$$
$$= 683 \text{ ft}$$

Example 3:

$$P = 2(52) + 2(23)$$
$$= 150 \text{ m}$$

52 m

23 m

Example 4:

5 in.

$$C = \pi(5) \quad \text{or} \quad 5\pi \text{ in.}$$

This is about 15.7 in.

Example 5:

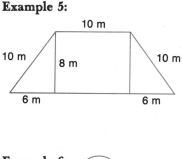

10 m

10 m   8 m   10 m

6 m   6 m

The perimeter of an irregularly shaped figure can be found if you know the lengths of each side. For this example,

$$P = 10 + 10 + 10 + 6 + 10 + 6$$
$$= 52 \text{ m}$$

Example 6:

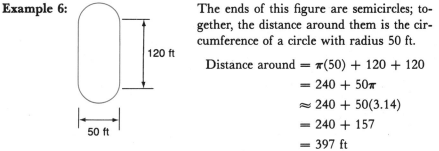

120 ft

50 ft

The ends of this figure are semicircles; together, the distance around them is the circumference of a circle with radius 50 ft.

$$\text{Distance around} = \pi(50) + 120 + 120$$
$$= 240 + 50\pi$$
$$\approx 240 + 50(3.14)$$
$$= 240 + 157$$
$$= 397 \text{ ft}$$

The important thing to remember about the conversion to the metric system is simply that you will use different measuring devices. Instead of a *yardstick* you will use a *meterstick*. *You should not convert from one system to the other.* Simply remeasure the length using a different standard unit, as we did in the previous examples. A comparison between the yardstick and the meterstick is shown in Figure 7.10.

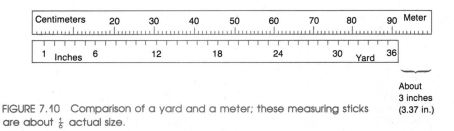

Centimeters   20   30   40   50   60   70   80   90 Meter

1 Inches 6   12   18   24   30   Yard 36

About
3 inches
(3.37 in.)

FIGURE 7.10 Comparison of a yard and a meter; these measuring sticks are about $\frac{1}{6}$ actual size.

We don't usually measure miles or kilometers directly. We rely on road signs, maps, and the odometers on our cars to tell us the distance between cities. When the road signs, maps, and odometers give distances in kilometers, we will be able to estimate large metric distances as easily as we can estimate miles. However, during the changeover period, there may be times when you need to convert from one system to the other.

The exact conversion is more complicated, but for all practical purposes you need only use these estimates. It is interesting to compare this conversion constant for changing kilometers to miles,

.621 . . .

with the golden ratio of Section 1.5:

.618 . . .

> *To change from kilometers to miles:*
>   1 kilometer is about .6 mile, so multiply kilometers by .6
>
> *To change from miles to kilometers:*
>   1 mile is about 1.6 kilometers, so multiply miles by 1.6

The exact conversion is more complicated, but for all practical purposes you can use these estimates. Since these are just estimates and are not exact, we use the approximately equal symbol, $\approx$.

In Examples 7 and 8 change from kilometers to miles.

Example 7:   600 km;   multiply by .6: $600 \times .6 \approx 360$ mi

Example 8:   200 km; $200 \times .6 \approx 120$ mi

In Examples 9 and 10 change from miles to kilometers.

Example 9:   500 miles;   multiply by 1.6: $500 \times 1.6 \approx 800$ km

Example 10:   340 miles; $340 \times 1.6 \approx 544$ km

Suppose you want to carpet your living room. The price of carpet is quoted in terms of a price per *square yard*. A square yard is a measurement of **area.** To measure the area of a plane figure, we fill it with **square units** (see Figure 7.11).

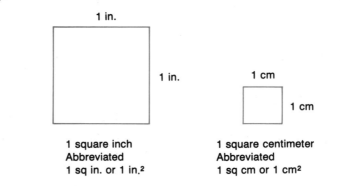

FIGURE 7.11   Actual size comparison of a square inch and a square centimeter.

The basic units of U.S. and metric units of square measure are given in the following box.

U.S. Measurement	Metric Measurement
square inch (in.^{2})	square centimeter (cm^2)
square foot (ft^2)	square meter (m^2)
square yard (yd^2)	are (a, pronounced "air")
square mile (mi^2)	hectare (ha)
acre (A)	

Example 11:   What is the area of the given rectangle?

Area is 12 cm^2

1	2	3	4
8	7	6	5
9	10	11	12

3 cm

4 cm

When we say "area is 12 cm^2" we mean that 12 squares of size 1 cm by 1 cm will fit inside the region.

Example 11 illustrates the meaning of area, but in practice you don't actually fit squares into the larger figure. Instead, you find the area according to one of the following **area formulas:**

Square:
Area = Side × Side
    = (Side)^2
$A = s^2$

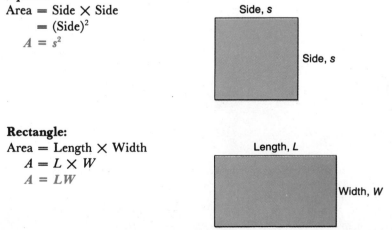

Side, *s*

Side, *s*

Rectangle:
Area = Length × Width
$A = L \times W$
$A = LW$

Length, *L*

Width, *W*

Parallelogram:
Area = Base × Height
$$A = b \times h$$
$$A = bh$$

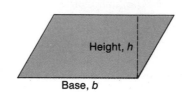

Height, h

Base, b

Triangle:

Area = $\dfrac{1}{2}$ × Base × Height

$$A = \dfrac{1}{2} \times b \times h$$

$$A = \dfrac{1}{2} bh$$

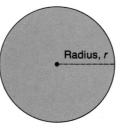

Height, h

Base, b

The height of a triangle is the perpendicular
distance from a vertex to the base.
It is not necessarily a side of a triangle.

Circle:
Area = π × Radius × Radius
$$A = \pi r^2$$

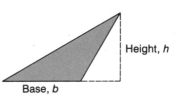

Radius, r

In Examples 12–17 find the area of each shaded region.

Example 12:

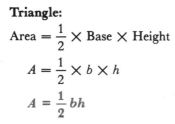

6.7 m

6 m

3 m

The height is 6 m and the base
is 3 m, so

$$A = 3 \text{ m} \times 6 \text{ m}$$
$$= 18 \text{ m}^2$$

Example 13:

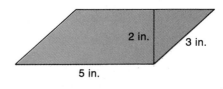

2 in.

3 in.

5 in.

The height is 2 in. and the base
is 5 in., so

$$A = 2 \text{ in.} \times 5 \text{ in.}$$
$$= 10 \text{ in.}^2$$

Example 14:

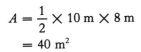

$$A = \frac{1}{2} \times 10 \text{ m} \times 8 \text{ m}$$
$$= 40 \text{ m}^2$$

Example 15:

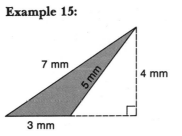

$$A = \frac{1}{2} \times 4 \text{ mm} \times 3 \text{ mm}$$
$$= 6 \text{ mm}^2$$

Example 16:

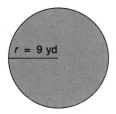

$$A = \pi \times (9 \text{ yd})^2$$
$$\approx 3.14 \times 81 \text{ yd}^2$$
$$= 254.34 \text{ yd}^2$$

On a calculator:

254.469005

or

$$\boxed{3.1416} \times 9 \times 9 =$$
254.4696

Round you answer from the calculator. Notice that the rounded calculator answer (254.5) is not the same as the rounded answer found using $\pi \approx 3.14$.

Example 17:

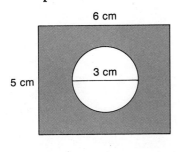

The area of the shaded portion is the area of the rectangle minus the area of the circle.

$$A = 5(6) - \pi(3)^2$$
$$= 30 - 9\pi$$
$$\approx 1.73$$ (1.74 if you use the approximation 3.14)

Problem Set 7.2

As we said in the text, it is important to be able to think in metric. Answer the multiple-choice questions in Problems 1–10 without converting to the U.S. system.

1. An adult's height is most likely to be:

    A. 6 m   **B.** 50 cm   **C.** 170 cm

2. The width of this book is about:

    A. 2.1 cm   **B.** 21 cm   **C.** 2.1 m

3. The width of a dollar bill is about

    A. .65 cm   **B.** 6.5 cm   **C.** .65 m

4. The distance from your home to the nearest grocery store is most likely to be:

    A. 1 cm   **B.** 1 m   **C.** 1 km

5. The prefix *centi* means:

    A. One thousand   **B.** One thousandth   **C.** One hundredth

6. The prefix *kilo* means:

    A. One thousand   **B.** One thousandth   **C.** One hundredth

7. If the distance from San Francisco to New York is about 3,000 mi, then the distance in kilometers is:

    A. Less than 3,000 km   **B.** More than 3,000 km
    C. About 3,000 km

8. The length of a Volkswagen bug is about:

    A. 1 m   **B.** 4 m   **C.** 10 m

9. The distance from floor to ceiling in a typical home is about:

    A. 2.5 m   **B.** .5 m   **C.** 4.5 m

10. In 1964 Bob Hayes ran the 100 m dash in 10 seconds flat. At this same rate, could he run a 100 yd dash in:

    A. Less than 10 seconds?   **B.** More than 10 seconds?
    C. 10 seconds?

From memory, and without any measuring devices, estimate the requested lengths in Problems 11–16 and draw line segments with those lengths.

11. 1 in.   **12.** 2 in.   **13.** 3 in.

14. 1 cm   **15.** 5 cm   **16.** 10 cm

Find the perimeters of the polygons given in Problems 17 and 18 to the nearest $\frac{1}{10}$ cm.

17.

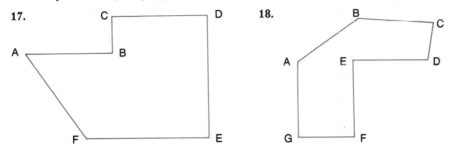

18.

Change the distances given in Problems 19–24 from kilometers to miles. Use the approximation given in this section.

19. Boston to Chicago, 1,580 km

20. Dallas to Memphis, 750 km

21. Louisville to Birmingham, 600 km

22. Washington, D.C., to Moscow, 7,800 km

23. London to Sydney, 17,000 km

24. New York to Cape Town, 12,500 km

Change the distances given in Problems 25–30 from miles to kilometers. Use the approximation given in this section.

25. San Francisco to Los Angeles, 400 miles

26. New York to Washington, D.C., 230 miles

27. Miami to New Orleans, 890 miles

28. Honolulu to New York, 4,960 miles

29. San Francisco to Paris, 5,560 miles

30. Tokyo to Buenos Aires, 11,400 miles

Find the perimeter, circumference, or distance around each shaded figure in Problems 31–47. Also find the area of each shaded region. Use 3.14 as an approximation for π.

31.

32.

33.

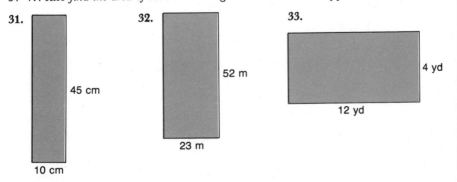

45 cm

10 cm

52 m

23 m

4 yd

12 yd

MY WIFE AND I STOPPED for lunch in a Nebraska town on our way to California, and I asked the waitress how much snow the area usually got. "About as deep as a meter," she replied. Impressed by her use of the metric system, I asked where she had learned it. She was momentarily baffled, then said, "That's the one I mean," and pointed out the window to the parking meter in front of the restaurant.

N. A. Norris

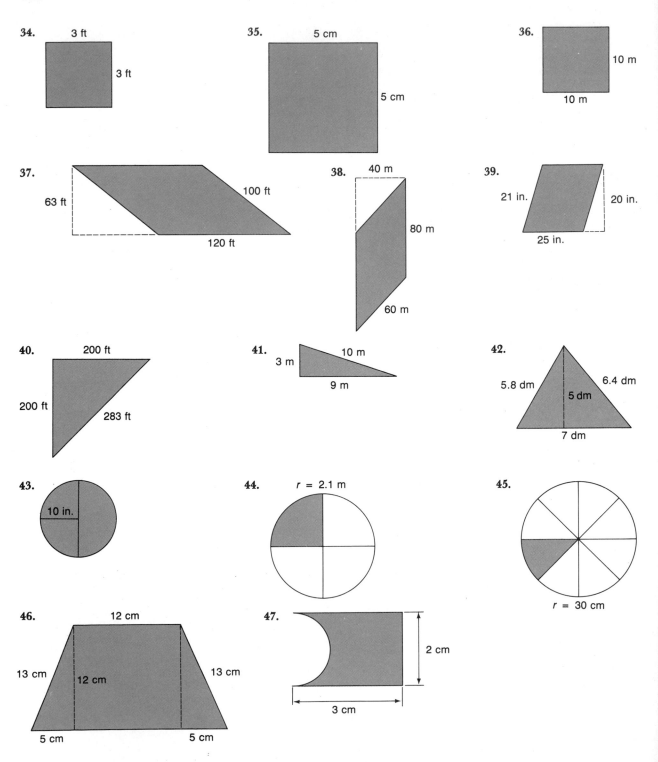

34. 3 ft   3 ft

35. 5 cm   5 cm

36. 10 m   10 m

37. 63 ft   100 ft   120 ft

38. 40 m   80 m   60 m

39. 21 in.   20 in.   25 in.

40. 200 ft   200 ft   283 ft

41. 3 m   10 m   9 m

42. 5.8 dm   6.4 dm   5 dm   7 dm

43. 10 in.

44. *r* = 2.1 m

45. *r* = 30 cm

46. 12 cm   13 cm   12 cm   13 cm   5 cm   5 cm

47. 2 cm   3 cm

THE WIZARD OF ID

48. Find your own metric measurements.

Men

a. Height
b. Chest
c. Waist
d. Seat
e. Neck
f. Length of shoe

Women

a. Height
b. Bust
c. Waist
d. Hips
e. Distance from waist
    to hemline
f. Fist (Measure around largest part
    of hand over knuckles while making
    a fist, excluding thumb.)

8″ × 8″ = 64 sq in.

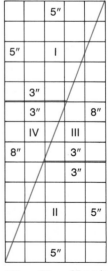

13″ × 5″ = 65 sq in.

Mind Bogglers

49. *The extra square inch.* Here is a strange and interesting relationship. The 8″ × 8″ square in the margin has an area of 64 square units. When it is rearranged as indicated by the 13″ × 15″ rectangle, it appears to have an area of 65 square units. Where did the "extra" square unit come from?

Construct your own square, 8 inches on a side, and then cut it into the four pieces as illustrated. Place the four pieces together as shown. Be sure to do your measuring and cutting very carefully. Satisfy yourself that this "extra" square inch has appeared. Can you explain this?

Where did the extra square inch come from?

50. How many squares are in the following figure?

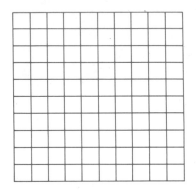

51. *Santa Rosa street problem.* On Saturday evenings a favorite pastime of the high school students in Santa Rosa, California, is to cruise certain streets. The selected routes are shown on the map in the figure below. Is it possible to choose a route so that all of the permitted streets are traveled exactly once?

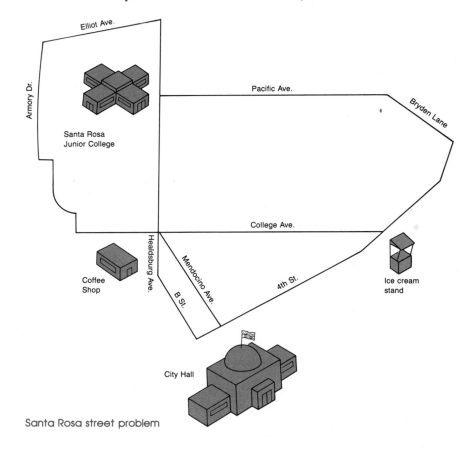

Santa Rosa street problem

52. Is it possible to take an entire trip through either house A or B, whose floor plans are shown here, and pass through each door once and only once?

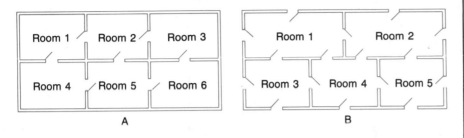

A

B

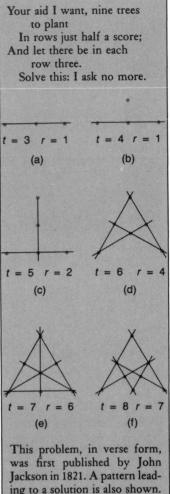

Your aid I want, nine trees
to plant
In rows just half a score;
And let there be in each
row three.
Solve this: I ask no more.

$t = 3$ $r = 1$ $t = 4$ $r = 1$

(a)   (b)

$t = 5$ $r = 2$ $t = 6$ $r = 4$

(c)   (d)

$t = 7$ $r = 6$ $t = 8$ $r = 7$

(e)   (f)

This problem, in verse form, was first published by John Jackson in 1821. A pattern leading to a solution is also shown. Can you answer the question asked in the verse?

7.3 Angles

In Section 7.1, we referred to Euclidean geometry. When it is studied in high school as an entire course, it is usually presented in a *formal* manner using definitions, axioms, and theorems. The development in this chapter is *informal*, which means that we base our results on observations and intuition. We begin by assuming that you are familiar with the ideas of *point, line,* and *plane.*

If you consider a point on a line, that point separates the line into three parts: two **half-lines** and the point itself. A **ray** is the union of a half-line and its end point and is denoted by naming the end point as well as any point on the ray, as shown in Figure 7.12.

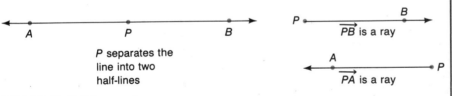

A P B

P separates the
line into two
half-lines

P ————————— B

$\overrightarrow{PB}$ is a ray

A

$\overleftarrow{}$ ———— P

$\overrightarrow{PA}$ is a ray

FIGURE 7.12 Half-lines and rays

One of the most important geometric figures is an angle, with which you, no doubt, have some familiarity.

> An **angle** is the union of two rays that have the same end point.

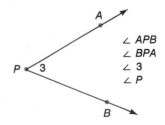

A

$\angle APB$
$\angle BPA$
$\angle 3$
$\angle P$

P 3

B

In Figure 7.13, the rays $\overrightarrow{PA}$ and $\overrightarrow{PB}$ are called the **sides** of the angle, and P (the common end point of the rays) is called the **vertex.** The symbol $\angle$ is used to denote angles, as shown in Figure 7.13.

FIGURE 7.13 Angle and angle notation

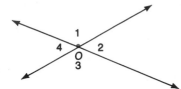

FIGURE 7.14

Consider any two distinct (different) intersecting lines in a plane, and let O be the point of intersection, as shown in Figure 7.14. These lines must form four angles. Angles with a common ray are called **adjacent angles;** nonadjacent angles are called **vertical angles.**

Example 1:   Name the adjacent angles in Figure 7.14.

Answer: $\angle 1$ and $\angle 2$; $\angle 2$ and $\angle 3$; $\angle 3$ and $\angle 4$; $\angle 4$ and $\angle 1$

Example 2:   Name the vertical angles in Figure 7.14.

Answer: $\angle 1$ and $\angle 3$; $\angle 2$ and $\angle 4$

Angles are often measured using a unit called a **degree,** which was invented by the Babylonians almost 4,000 years ago. To see the size of a degree, use a device called a *protractor* as shown in Figure 7.15. As you can see, most protractors show from 1 to 180 degrees, which are written as $1°$ and $180°$.

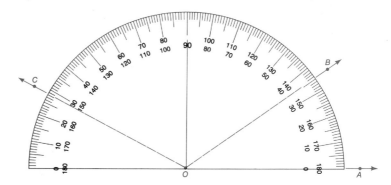

FIGURE 7.15   Protractor measuring angles

Figure 7.15 shows how to measure an angle. The measure of angle A is denoted by $m\angle AOB$ so that $m\angle AOB = 35°$. Also notice $m\angle AOC = 152°$. Angles are often classified by their measures, as shown in Table 7.1.

TABLE 7.1   Classification of Angles

Angle Measure	Name
90°	Right angle
180°	Straight angle
Between 0° and 90°	Acute angle
Between 90° and 180°	Obtuse angle

Two angles with the same measure are said to be **congruent.** If the sum of the measures of two angles is $90°$, they are called **complementary angles** and if the sum is $180°$ they are called **supplementary angles.**

Example 3:   Classify the requested angles shown in Figure 7.16.

a. ∠*AOB* is an acute angle.
b. ∠*BOC* is an obtuse angle.
c. ∠*AOB* and ∠*BOC* are supplementary angles.
d. ∠*BOC* and ∠*OPE* are congruent angles.
e. ∠*DAO* is a right angle.
f. ∠*AOB* and ∠*COP* are vertical angles.

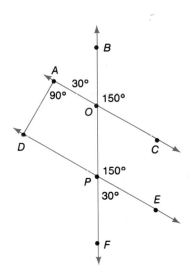

FIGURE 7.16

Consider three lines arranged similarly to those shown in Figure 7.16. Suppose that two of the lines, say ℓ_1 and ℓ_2, are parallel (they are in the same plane and never intersect), and also that the third line ℓ_3 intersects the parallel lines at points P and Q, as shown in Figure 7.17.

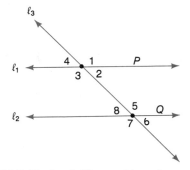

FIGURE 7.17   Parallel lines cut by a transversal

The line ℓ_3 is called a **transversal.** Notice that there are several angles formed:

Vertical angles are congruent.

Alternate interior angles are pairs of angles whose interiors lie between the parallel lines, but on opposite sides of the transversal. Alternate interior angles are congruent.

Alternate exterior angles are pairs of angles that lie outside the parallel lines, but on opposite sides of the transversal. Alternate exterior angles are congruent.

Corresponding angles are two nonadjacent angles whose interiors lie on the same side of the transversal such that one angle lies between the parallel lines and the other side lies on the outside. Corresponding angles are congruent.

Example 4:   Name the vertical angles in Figure 7.17.

Answer: ∠1 and ∠3;   ∠2 and ∠4;   ∠5 and ∠7;   ∠6 and ∠8

Example 5:   Name the alternate interior angles in Figure 7.17.

Answer: ∠2 and ∠8;   ∠3 and ∠5

Example 6: Name the alternate exterior angles in Figure 7.17.

Answer: ∠1 and ∠7;   ∠4 and ∠6

Example 7: Name the corresponding angles in Figure 7.17.

Answer: ∠1 and ∠5;   ∠2 and ∠6;   ∠3 and ∠7;   ∠4 and ∠8

Summarizing the results from Examples 4–7, notice that the following angles are congruent (written ≃):

$$∠1 ≃ ∠3 ≃ ∠5 ≃ ∠7 \quad \text{and also} \quad ∠2 ≃ ∠4 ≃ ∠6 ≃ ∠8$$

Also, the pairs of adjacent angles are all supplementary.

Example 8: Find the measures of all the angles in Figure 7.18.

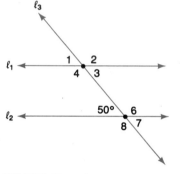

FIGURE 7.18

Solution:

$m∠1 = 50°$	Corresponding angles
$m∠2 = 180° - 50°$ $= 130°$	Angles 1 and 2 are supplementary (in this example)
$m∠3 = 50°$	Vertical angles (∠3 and ∠1 are vertical angles, and vertical angles are congruent)
$m∠4 = 130°$	Vertical angles
$m∠6 = 130°$	Supplementary angles
$m∠7 = 50°$	Vertical angles
$m∠8 = 130°$	Supplementary angles

Problem Set 7.3

1. Name the angle shown in the margin four different ways.

2. Name the vertex of the angle shown in the margin.

Use the protractor shown below to find the measure of the angles in Problems 3–8.

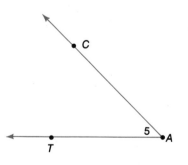

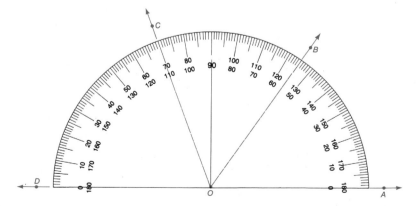

3. ∠AOB

4. ∠AOC

5. ∠AOD

6. ∠DOC

7. ∠DOB

8. ∠COB

Classify the requested angles shown in the margin, where ℓ₁ and ℓ₂ are parallel.

9. ∠3 if ∠1 is 30°

10. ∠6 if ∠5 is 90°

11. ∠18 if ∠17 is 105°

12. ∠19 if ∠11 is 70°

13. ∠10 if ∠11 is 90°

14. ∠16 if ∠15 is 30°

15. ∠6 if ∠16 is 120°

16. ∠9 if ∠19 is 110°

17. ∠2 if ∠4 is 90°

In Problems 18–26, classify the pairs of angles, as shown in the margin.

18. ∠2 and ∠4

19. ∠13 and ∠14

20. ∠9 and ∠12

21. ∠9 and ∠17

22. ∠9 and ∠11

23. ∠12 and ∠18

24. ∠7 and ∠13

25. ∠10 and ∠20

26. ∠5 and ∠15

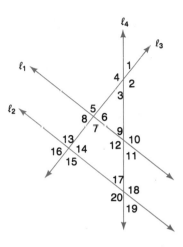

27. Find the measures of all the angles in Figure 7.17 if m∠7 = 110°.

28. Find the measures of all the angles in Figure 7.17 if m∠2 = 65°.

29. Find the measures of all the angles in Figure 7.17 if m∠6 = 19°.

30. Find the measures of all the angles in Figure 7.17 if m∠1 = 153°.

Mind Bogglers

31. Without the poetry, the 1821 puzzle given at the beginning of this section could be stated: Arrange nine trees so that they occur in ten rows of three trees each. If *t* represents the number of trees, and *r* the number of rows, figures *a–f* below the problem create a pattern leading to the solution. Solve this problem.

32. a. Draw any triangle. Measure each angle using a protractor. What is the sum of the measures of these angles?

  b. Repeat part a for a triangle that has one obtuse angle.

  c. Repeat part a for a triangle that has one right angle.

  d. Form a conjecture based on your answers to parts a–c.

7.4 The Metric System—
Volume, Mass, and Temperature

In the last two sections we studied plane figures. However, we live in a three-dimensional world and are concerned with measurements of three-dimensional objects, such as the box shown in Figure 7.19. A box is made up of *faces*, which meet at *edges;* the "corners" of the box are called *vertices* (plural of *vertex*). The box in Figure 7.19 is one kind of three-dimensional object that is part of a more general category called **polyhedrons.** Polyhedrons have faces that are made up

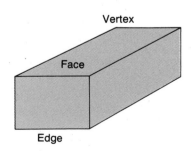

FIGURE 7.19 A box is a rectangular solid.

of only polygons. Figure 7.20 shows the five **regular polyhedrons,** which are polyhedrons whose faces are regular polygons.

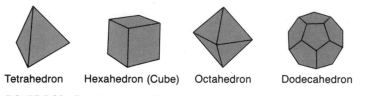

| Tetrahedron | Hexahedron (Cube) | Octahedron | Dodecahedron | Icosahedron |

FIGURE 7.20   Regular polyhedrons

Figure 7.20 shows all five types of regular polyhedrons; there are no others. There are, however, other types of (nonregular) polyhedrons which are familiar. Some of these are shown in Figure 7.21.

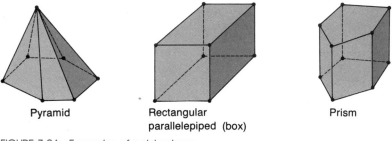

Pyramid   Rectangular   Prism
                           parallelepiped (box)

FIGURE 7.21   Examples of polyhedrons

In this section we'll limit our study of volumes to cubes and rectangular parallelepipeds (boxes). To find the volume of a solid object, we fill it with cubes. Our basic unit is a cubic inch and a cubic centimeter, as shown in Figure 7.22.

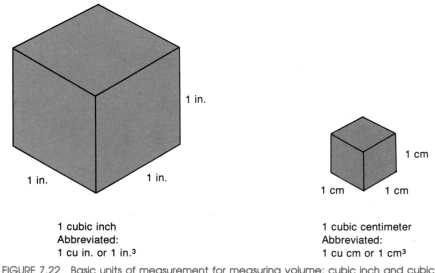

1 in.

1 in.   1 in.

1 cm

1 cm   1 cm

1 cubic inch
Abbreviated:
1 cu in. or 1 in.³

1 cubic centimeter
Abbreviated:
1 cu cm or 1 cm³

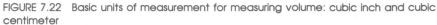

FIGURE 7.22   Basic units of measurement for measuring volume: cubic inch and cubic centimeter

We could fill various boxes to deduce the following volume formulas.

Cube:

The volume of a cube with edge s is found by

Volume = Edge $\times$ Edge $\times$ Edge
$$V = s^3$$

Rectangular parallelepiped (box):

The volume of a box with edges L, W, and H is

Volume = Length $\times$ Width $\times$ Height
$$V = L \times W \times H$$

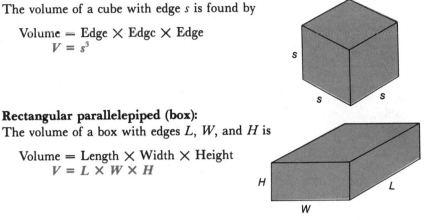

In Examples 1–4 find the volume of each box.

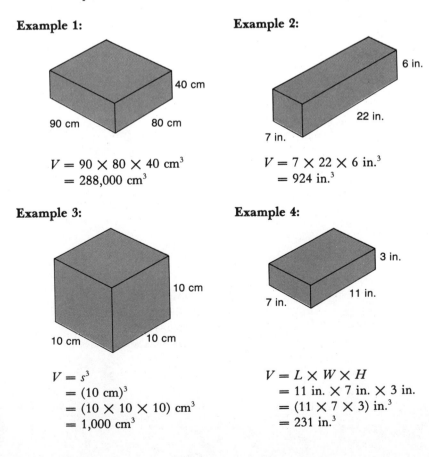

Example 1:

40 cm

90 cm   80 cm

$$V = 90 \times 80 \times 40 \text{ cm}^3$$
$$= 288{,}000 \text{ cm}^3$$

Example 2:

6 in.

22 in.

7 in.

$$V = 7 \times 22 \times 6 \text{ in.}^3$$
$$= 924 \text{ in.}^3$$

Example 3:

10 cm

10 cm   10 cm

$$V = s^3$$
$$= (10 \text{ cm})^3$$
$$= (10 \times 10 \times 10) \text{ cm}^3$$
$$= 1{,}000 \text{ cm}^3$$

Example 4:

3 in.

11 in.

7 in.

$$V = L \times W \times H$$
$$= 11 \text{ in.} \times 7 \text{ in.} \times 3 \text{ in.}$$
$$= (11 \times 7 \times 3) \text{ in.}^3$$
$$= 231 \text{ in.}^3$$

The capacity of a box can be measured in both the U.S. and the metric systems.

The U.S. system has seven common units of capacity for liquid measure:	The metric system has one standard unit of capacity:
teaspoon (tsp) tablespoon (tbsp) ounce (oz) cup (c) pint (pt) quart (qt) gallon (gal)	liter (ℓ)

Capacities of Common Grocery Items as Shown on Labels

Item	U.S. Capacity	Metric Capacity
Milk	$\frac{1}{2}$ gal	1.89 ℓ
Milk	1.06 qt	1 ℓ
Budweiser	12 oz	355 mℓ
Coke	67.6 oz	2 ℓ
Hawaiian Punch	1 qt	.95 ℓ
Del Monte Pickles	1 pt 6 oz	651 mℓ

In order to work with different capacities, you must memorize some conversion factors.

U.S. Conversion Factors for Capacity	Metric Conversion Factors for Capacity

3 tsp = 1 tbsp
1 oz = 2 tbsp
16 tbsp = 1 c
8 oz = 1 c
16 oz = 1 pt
2 c = 1 pt
2 pt = 1 qt
4 qt = 1 gal
	milli means $\frac{1}{1,000}$;
	    milliliter is abbreviated mℓ
	kilo means 1,000;
	    kiloliter is abbreviated kℓ

A liter can be defined as the capacity of a box that is 10 cm on a side. The box in Example 3 has a capacity of 1 ℓ.

**VOLUME 1,000 CUBIC CENTIMETERS
CAPACITY 1 LITER**

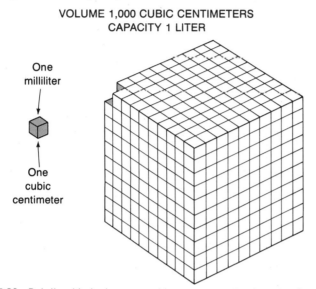

FIGURE 7.23   Relationship between metric measures of volume and capacity

One cubic centimeter is one-thousandth of a liter. Notice that this is the same as a milliliter. For this reason, you will sometimes see cc used to mean cm^3 or $m\ell$.

In the U.S. measurement system, the relationship between volume and capacity is a little more difficult to remember. The box in Example 4 has a volume of 231 in.^{3}. A box this size will hold exactly one gallon of water.

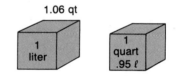

FIGURE 7.24   Comparison of a quart and a liter

Volume/Capacity

$$1 \text{ liter} = 1,000 \text{ cm}^3$$
$$1 \text{ gallon} = 231 \text{ in.}^3$$

In Examples 5 and 6, how much water would each container hold?

Example 5:

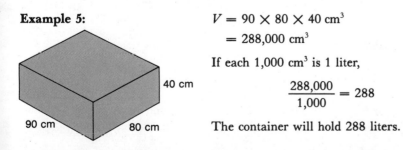

$$V = 90 \times 80 \times 40 \text{ cm}^3$$
$$= 288,000 \text{ cm}^3$$

If each 1,000 cm^3 is 1 liter,

$$\frac{288,000}{1,000} = 288$$

The container will hold 288 liters.

Example 6:

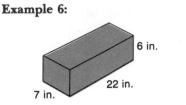

6 in.

22 in.

7 in.

$V = 7 \times 22 \times 6$ in.^{3}

$\quad = 924$ in.^{3}

If each 231 in.^{3} is 1 gallon,

$$\frac{924}{231} = 4$$

The container will hold 4 gallons.

In the U.S. system it is not always convenient to measure in cubic inches as was done in Example 6. You will also find it helpful to remember the following relationship:

---

1 ft^3 is about $7\frac{1}{2}$ gallons

Example 7:   An ecology swimming pool is advertised as being 20 ft $\times$ 25 ft $\times$ 5 ft. How many gallons will it hold?

Solution: $V = 20$ ft $\times$ 25 ft $\times$ 5 ft

$\quad\quad\quad = 2,500$ ft^3

Each cubic foot is about 7.5 gallons, so multiply:

$$2,500 \times 7.5 = 18,750$$

The pool will hold about 18,750 gallons.

The correct usage is mass, but in everyday usage you will often hear it referred to as weight.

The basic metric unit for measuring the mass of an object is the **gram** (g). It is also related to the other measurements we've considered. Suppose a cubic centimeter is filled with one milliliter of water. The weight of one milliliter of water is a gram. It is helpful to remember that a gram is about the weight of one paper clip. Since the gram is small, you will probably use the kilogram, which is about 2.2 pounds, for most weight measurements.

Paper clip

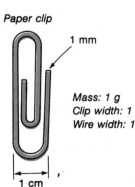

1 mm

Mass: 1 g
Clip width: 1 cm
Wire width: 1 mm

1 cm

The U.S. System Has Three Common Units of Mass:	The Metric System Has One Standard Unit of Mass:
ounce (oz) pound (lb) ton (T)	gram (g)

In order to work with items of different mass, you must memorize some conversion factors.

U.S. Conversion Factors for Mass	Metric Conversion Factors for Mass

16 oz = 1 lb
2,000 lb = 1 T
	milli means $\frac{1}{1,000}$;
	  milligram is abbreviated mg
	kilo means 1,000;
	  kilogram is abbreviated kg

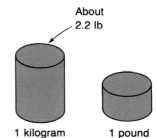

About
2.2 lb

1 kilogram   1 pound

FIGURE 7.25 Comparison of a pound and a kilogram

Table 7.2 shows some equivalencies for selected masses. Remember, if you say you weigh 155 lb, you probably mean that you stepped onto a scale and it read 155.

As we have mentioned, one of the advantages of the metric system is the ease with which you can change from one metric unit to another. Also, conversions for length, capacity, and mass are all done the same way—by simply moving the decimal point. Recall, the basic units of metric measurement:

BASIC UNIT OF MEASUREMENT

Length:	meter	(m)
Capacity:	liter	(ℓ)
Mass:	gram	(g)

TABLE 7.2 Equivalencies (Approximate) between Pounds and Kilograms

Pounds	Kilograms	Pounds	Kilograms
50	23	140	64
55	25	145	66
60	27	150	68
65	29	155	70
70	32	160	73
75	34	165	75
80	36	170	77
85	39	175	80
90	41	180	82
95	43	185	84
100	45	190	86
105	48	195	89
110	50	200	91
115	52	205	93
120	55	210	95
125	57	215	98
130	59	220	100
135	61	225	102

We have seen the prefixes centi, milli, and kilo. There are others with which you should be familiar; these are shown in Table 7.3.

TABLE 7.3   Common Metric Prefixes

| Larger Unit Prefixes: | | | Smaller Unit Prefixes: | | |
Prefix	Symbol	Meaning	Prefix	Symbol	Meaning
kilo	k	Thousand	deci	d	One-tenth
hecto	h	Hundred	**centi**	c	One-hundredth
deka	dk	Ten	**milli**	m	One-thousandth

To change from one unit to another, notice the following pattern.

1 meter equals		*1 liter equals*		*1 gram equals*		
.001	km	.001	kℓ	.001	kg	⎤ Larger unit prefixes cause a
.01	hm	.01	hℓ	.01	hg	⎟ smaller number of units.
.1	dkm	.1	dkℓ	.1	dkg	⎦
1	m	1	ℓ	1	g	} Basic unit
10	dm	10	dℓ	10	dg	⎤
100	cm	100	cℓ	100	cg	⎟ Smaller unit prefixes cause a
1,000	mm	1,000	mℓ	1,000	mg	⎦ larger number of units.

Therefore, if the height of this book is .225 m, then it is also

Larger prefixes
$$\begin{bmatrix} .000225 & \text{km} & \leftarrow \text{Three decimal places} \\ .00225 & \text{hm} & \leftarrow \text{Two decimal places} \\ .0225 & \text{dkm} & \leftarrow \text{One decimal place} \end{bmatrix}$$

Smaller prefixes
$$\begin{bmatrix} 2.25 & \text{dm} & \leftarrow \text{One decimal place} \\ 22.5 & \text{cm} & \leftarrow \text{Two decimal places} \\ 225 & \text{mm} & \leftarrow \text{Three decimal places} \end{bmatrix}$$

Remember, smaller unit prefixes cause a larger number of units, and larger unit prefixes cause a smaller number of units.

Example 8:   43 km equals

All prefixes are smaller so units become larger
$$\begin{bmatrix} 430 & \text{hm} & \leftarrow \text{One decimal place} \\ 4,300 & \text{dkm} & \leftarrow \text{Two decimal places} \\ 43,000 & \text{m} & \leftarrow \text{Three decimal places} \\ 430,000 & \text{dm} & \leftarrow \text{Four decimal places} \\ 4,300,000 & \text{cm} & \leftarrow \text{Five decimal places} \end{bmatrix}$$

Example 9:   60 ℓ equals

Larger prefixes, smaller numbers
$$\begin{bmatrix} 0.06 & \text{k}\ell & \leftarrow \text{Three decimal places} \\ 0.6 & \text{h}\ell & \leftarrow \text{Two decimal places} \\ 6 & \text{dk}\ell & \leftarrow \text{One decimal place} \end{bmatrix}$$

Smaller prefixes, larger numbers
$$\begin{bmatrix} 600 & \text{d}\ell & \leftarrow \text{One decimal place} \\ 6,000 & \text{c}\ell & \leftarrow \text{Two decimal places} \\ 60,000 & \text{m}\ell & \leftarrow \text{Three decimal places} \end{bmatrix}$$

METRICATION

Mike Keedy, Purdue University, says
Go Metric—Be a Liter Bug!

Example 10:   14.1 cg equals

Smaller prefix	{141	mg	←One decimal place
Larger prefixes	.141	g	←Two decimal places
mean smaller units	.0141	dkg	←Three decimal places
(count the number	.00141	hg	←Four decimal places
of decimal places)	.000141	kg	←Five decimal places

To make a single conversion, use the pattern illustrated by Examples 8–10 and simply count the number of decimal places.

Example 11:   287 cm = ____ km   Fill in the blank.

— Larger unit implies smaller number

— Decimal point moves five places

Answer: .00287

Example 12:   1.5 kℓ = ____ ℓ   Fill in the blank.

— Smaller prefix implies larger number

— Decimal point moves three places

Answer: 1,500

Example 13:   4.8 kg = ____ g   Fill in the blank.

— Smaller unit implies larger number

— Decimal point moves three places

Answer: 4,800

The final quantity of measure that we'll consider in this chapter is the measurement of temperature.

Remember, to find the number of places to move the decimal point, count the lines in

> kilo
> hecto
> deka
> BASIC UNIT
> deci
> centi
> milli

The U.S. System Has One Common Temperature Unit	The Metric System Has One Common Temperature Unit
Fahrenheit (°F)	Celsius (°C)

In order to work with temperatures, it is necessary to have some reference points.

U.S. Temperature		Metric Temperature	
Water freezes:	32°F	Water freezes:	0°C
Water boils:	212°F	Water boils:	100°C

We are usually interested in measuring temperature in three areas. The first is atmospheric temperature (usually given in weather reports on the radio or television); the second is body temperature (used to determine illness); and the third is oven temperature (used in cooking). The same scales are used, of course, for measuring all of these temperatures. But notice the difference in the range of temperatures we're considering. The comparisons for Fahrenheit and Celsius are shown in Figure 7.26.

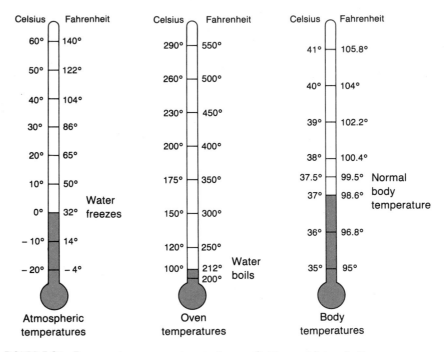

FIGURE 7.26   Temperature conversions between Celsius and Fahrenheit

Problem Set 7.4

Without measuring, pick the best choice in each of Problems 1–20 by estimating.

1. A hamburger patty would weigh about

     A.   170 g   **B.**   240 mg   **C.**   2 kg

2. A can of carrots at the grocery store most likely weighs about

   A. 40 kg   **B.** 4 kg   **C.** .4 kg

3. A newborn baby would weigh about

   A. 490 mg   **B.** 4 kg   **C.** 140 kg

4. John tells you he weighs 150 kg. If John is an adult, do you think he is

   A. underweight   **B.** about average   **C.** overweight

5. You have invited 15 people over for Thanksgiving dinner. You should buy a turkey that weighs about

   A. 795 mg   **B.** 4 kg   **C.** 12 kg

6. Water boils at

   A. 0°C   **B.** 100°C   **C.** 212°C

7. If it is 32°C outside, you would most likely find people

   A. ice skating   **B.** water skiing

8. If the doctor said that your child's temperature was 37°C, would you think your child's temperature was

   A. low   **B.** normal   **C.** high

9. You would most likely broil steaks at

   A. 120°C   **B.** 500°C   **C.** 290°C

10. A kilogram is __?__ a pound.

    A. more than   **B.** less than   **C.** about the same as

11. The prefix used to mean 1,000 is

    A. centi   **B.** milli   **C.** kilo

12. The prefix used to mean $\frac{1}{1,000}$ is

    A. centi   **B.** milli   **C.** kilo

13. The prefix used to mean $\frac{1}{100}$ is

    A. centi   **B.** milli   **C.** kilo

14. 15 kg is a measure of

    A. length   **B.** capacity
    C. mass   **D.** temperature

15. 28.2 m is a measure of

    A. length   **B.** capacity
    C. mass   **D.** temperature

16. 6 ℓ is a measure of

    A. length   **B.** capacity
    C. mass   **D.** temperature

There once was a teacher named Streeter
Who taught stuff like meter and liter
The kids moaned and groaned,
Until they were shown,
That meter and liter were neater!

17. 38°C is a measure of

    A. length   **B.** capacity
    C. mass   **D.** temperature

18. 7 mℓ is a measure of

    A. length   **B.** capacity
    C. mass   **D.** temperature

19. 68 km is a measure of

    A. length   **B.** capacity
    C. mass   **D.** temperature

20. 14.3 cm is a measure of

    A. length   **B.** capacity
    C. mass   **D.** temperature

Find the volumes in Problems 21–24.

21.
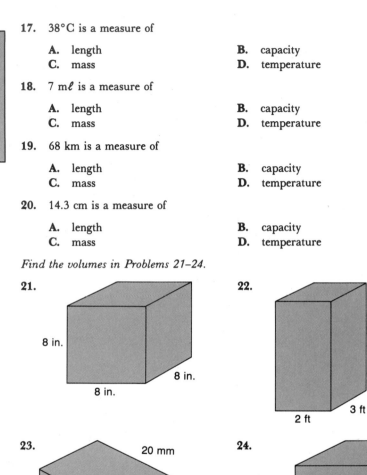
8 in.
8 in.
8 in.

22.
4 ft
2 ft
3 ft

23.
20 mm
2 mm
10 mm

24.
1 yd
1 yd
1 yd

How much water would each of the containers in Problems 25–30 hold? (Give your answer to the nearest tenth of a gallon or tenth of a liter.)

25.
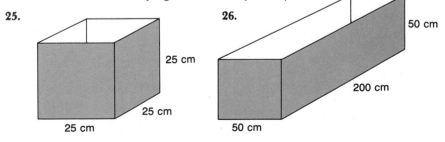
25 cm
25 cm
25 cm

26.
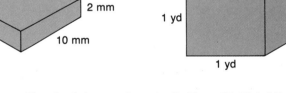
50 cm
200 cm
50 cm

27.

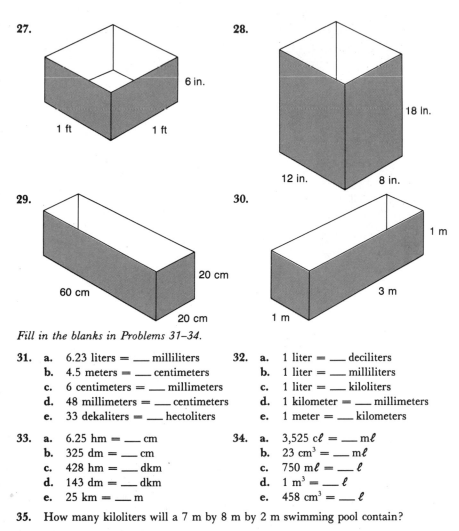

6 in.

1 ft   1 ft

28.

18 In.

12 in.   8 in.

29.

20 cm

60 cm

20 cm

30.

1 m

3 m

1 m

Fill in the blanks in Problems 31–34.

31.
a. 6.23 liters = ___ milliliters
b. 4.5 meters = ___ centimeters
c. 6 centimeters = ___ millimeters
d. 48 millimeters = ___ centimeters
e. 33 dekaliters = ___ hectoliters

32.
a. 1 liter = ___ deciliters
b. 1 liter = ___ milliliters
c. 1 liter = ___ kiloliters
d. 1 kilometer = ___ millimeters
e. 1 meter = ___ kilometers

33.
a. 6.25 hm = ___ cm
b. 325 dm = ___ cm
c. 428 hm = ___ dkm
d. 143 dm = ___ dkm
e. 25 km = ___ m

34.
a. $3{,}525 \, c\ell$ = ___ $m\ell$
b. $23 \, \text{cm}^3$ = ___ $m\ell$
c. $750 \, m\ell$ = ___ ℓ
d. $1 \, \text{m}^3$ = ___ ℓ
e. $458 \, \text{cm}^3$ = ___ ℓ

35. How many kiloliters will a 7 m by 8 m by 2 m swimming pool contain?

36. How many gallons will a 21 ft by 24 ft by 4 ft swimming pool contain?

Mind Bogglers

37. A rather surprising relationship, called the **Euler-Descartes Formula,** exists among the vertices, edges, and sides of polyhedra. See if you can discover it by looking for patterns among the figures below and filling in the blanks in the following table.

a. Triangular pyramid   **b.** Quadrilateral pyramid   **c.** Pentagonal pyramid   **d.** Regular tetrahedron

e. Regular hexa–hedron (cube) **f.** Regular octahedron **g.** Regular dodecahedron **h.** Regular icosahedron

Cube

Octahedron

Icosahedron

Dodecahedron

Tetrahedron

Examples of the five regular solids as they can be found in nature. These are skeletons of marine animals called *radiolaria*.

Figure	*Sides*	*Vertices*	*Edges*
a.	4	4	6
b.	5	——	8
c.	——	6	10
d.	4	4	——
e.	——	——	12
f.	——	6	——
g.	——	——	30
h.	——	——	30

38. *Spider and fly problem.* A room is 80 feet long, 40 feet wide, and 20 feet high.

  a. A fly wishes to get from one lower corner of the room to the opposite upper corner by traveling the shortest possible distance. How far does the fly fly?

  b. A spider also wishes to get to the upper corner from the opposite lower corner (to eat the fly, no doubt). What is the shortest distance that the spider must travel?

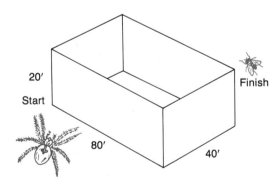

Problems for Individual Study

39. Construct models for the regular polyhedra.

  References: H. S. M. Coxeter, *Introduction to Geometry* (New York: Wiley, 1961).

  Max Sobel and Evan Maletsky, *Teaching Mathematics: A Source-book* (Englewood Cliffs, N.J.: Prentice-Hall, 1975), pp. 173–184.

  Charles W. Trigg, "Collapsible Models of the Regular Octrahedron," *The Mathematics Teacher,* October 1972, pp. 530–533.

40. *Mathematical forms in nature.* What forms in nature represent mathematical patterns? What shapes in nature are geometric and/or symmetric? What activities in nature illustrate mathematical functions?

Exhibit suggestions: Samples of spiral shells, crystals, ellipsoidal stones or eggs, spiraling sunflower seed pods, branch distributions illustrating Fibonacci series, snowflake patterns, body bones acting as levers, ratio of food consumed to body size.

References: David Bergamini, *Mathematics* (New York: Time, Inc., Life Science Library, 1963), chapter 4.

Judithlynne Carson, "Fibonacci Numbers and Pineapple Phyllotaxy," *Two Year College Mathematics Journal,* June 1978, pp. 132–136.

James Newman, *The World of Mathematics* (New York: Simon and Schuster, 1956). "Crystals and the Future of Physics," pp. 871–881. "On Being the Right Size," pp. 952–957. "On Magnitude," pp. 1001–1046. "The Soap Bubble," pp. 891–900.

See also the *Fibonacci Quarterly.*

*7.5 Coordinate Geometry

Have you ever drawn a picture by connecting the dots or found a city or street on a map? If you have, then you've used the ideas we'll be discussing in this section. Suppose we wish to find Fisherman's Wharf in San Francisco. We look in the index for the map in Figure 7.27 and find Fisherman's Wharf listed as (6, D). Can you locate section (6, D) in Figure 7.27?

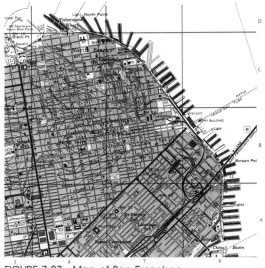

FIGURE 7.27   Map of San Francisco

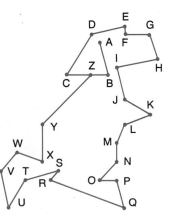

*This section requires Section 6.2.

Have you ever been given the coordinates for a street and still not been able to find it easily? For example, suppose we want to find Main Street. Without the index the task is next to impossible, but even with the index listing Main at (7, B), the task of finding the street is not an easy one. How could we improve the map?

Let's create a smaller grid, as shown in Figure 7.28.

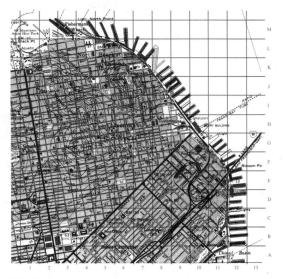

FIGURE 7.28   Map of San Francisco with refined grid

Main Street is located at (9, G) and is easier to find. However, a smaller grid means that a lot of letters are needed for the vertical scale, and we might need more letters than we have. So let's use a notation that will allow us to represent points on the map with pairs of numbers. A pair of numbers written as

$$(2, 3)$$

is called an **ordered pair** to remind you that the order in which the numbers 2 and 3 are listed is important. In this example, 2 is the **first component** and 3 is the **second component.**

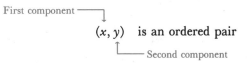

For our map, suppose we relabel the vertical scale with numbers and change both scales so that we label the lines instead of the spaces, as shown in Figure 7.29. (By the way, most technical maps number lines instead of spaces.)

Now we can fix the location of any street on the map quite precisely. Notice that if we use an ordered pair of numbers (instead of a number and a letter), it is important to know which component of the ordered pair represents the hori-

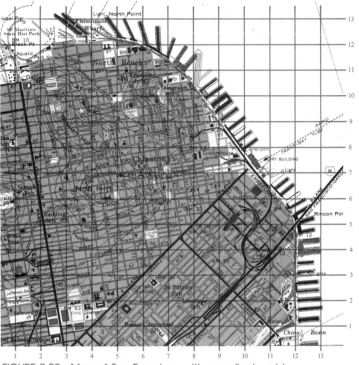

FIGURE 7.29 Map of San Francisco with coordinate grid

zontal distance and which component represents the vertical distance. Now, if you use Figure 7.29, you'll see that the coordinates of Main Street are about (9.5, 6.3), or we can even say that Main Street runs from about (9.5, 6.3) to (11.2, 4.5). Notice that, by using ordered pairs and numbering the lines instead of the spaces, we have refined our grid. We've refined it even more with decimal components. When using pairs of numbers instead of numbers and letters, we must remember that the first component is on the horizontal axis and the second component is on the vertical axis.

Example 1: Find the major landmarks located in Figure 7.29 by the following coordinates:

Answers:

a. (3.5, 1.5)   Civic Center
b. (3.5, 12.5) Fisherman's Wharf
c. (5, 11)   North Beach
d. (4, 6)   Nob Hill
e. (6, 7.3)   Chinatown

There are many ways to use ordered pairs to find particular locations. For example, a teacher may make out a seating chart like the one shown in Figure 7.30 on the next page.

	1	2	3	4	5
5	Otis Morehouse	Wayne Savick	Harvey Dunker	Richard Giles	Shannon Smith
4	Vicki Switzer	Harold Peterson	Bob Anderson	Josephine Lee	Milt Hoehn
3	Jeff Atz	Carol Olmstead	Todd Humann	Jim Kintzi	Sharon Boschen
2	Terry Shell	Ralph Earnest	Steve Switzer	Clint Stevenson	
1	Rosamond Foley	Amy Olmstead	Missy Smith	Niels Sovndal	Eva Mikalson
	1	2	3 Front	4	5

FIGURE 7.30   Seating chart

In the grade book, the teacher records

Anderson (3, 4)   Remember, first component is horizontal direction,
Atz (1, 3)   second component is vertical direction.

     .
     .
     .

Anderson's seat is represented in column 3, row 4. Can you think of some other ways in which ordered pairs could be used to locate a position?

The idea of using an ordered pair to locate a certain position is associated with particular terminology. **Axes** are two perpendicular real number lines, such as those in Figure 7.31a. The point of intersection of the axes is called the **origin.** We associate the direction to the right or up with positive numbers. The arrows in Figure 7.31a are pointing in the positive direction. These perpendicular lines are usually drawn so that one is horizontal and the other is vertical. The horizontal axis is called the **x-axis,** and the vertical axis is called the **y-axis.** Notice that these axes divide the plane into four parts, which are called **quadrants** and are arbitrarily labeled as shown in Figure 7.31b.

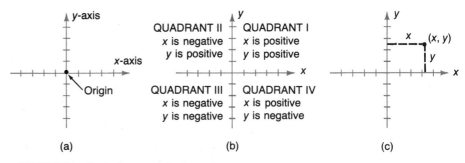

FIGURE 7.31   Cartesian coordinate system

We can now label points in the plane by using ordered pairs. The first component of the pair gives the horizontal distance and the second component gives the vertical distance, as shown in Figure 7.31c. If x and y are components of the point, then this representation of the point (x, y) is called the **rectangular** or **Cartesian coordinates** of the point.

Example 2: Plot (graph) the given points.

a. (5, 2)   **b.** (3, 5)   **c.** ($^-$2, 1)
d. (0, 5)   **e.** ($^-$6, $^-$4)   **f.** (3, $^-$2)
g. ($\frac{1}{2}$, 0)   **h.** (0, 0)   **i.** (5, 0)

Solution

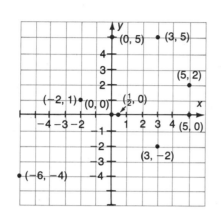

Historical Note

René Descartes (1596–1650)

The Cartesian coordinate system is named in honor of the French mathematician René Descartes. Because of his frail health he was accustomed to staying in bed as long as he wished. Legend tells us that he thought of this coordinate system while he was lying in bed watching a fly crawl around on the ceiling of his bedroom. He noticed that the path of the fly could be described if he knew the relation connecting the fly's distances from each of the walls. This idea of a coordinate system ties together the two great branches of mathematics, algebra and geometry.

Many situations involve two variables or unknowns related in some specific fashion.

Example 3: Linda is on a business trip and needs to rent a car for the day. The car rental agency has the following options on the car she wants to rent:

Option A: $40 plus 50¢ per mile

Option B: Flat $60 with unlimited mileage

Which car should she rent?

Solution: The answer, of course, depends on the number of miles Linda intends to drive. Let's relate the cost of the car rental, c, to the number of miles driven in the following way:

Option A:

COST = BASIC CHARGE + MILEAGE CHARGE

COST =   40   + .5(NUMBER OF MILES DRIVEN)

$c = 40 + .5m$

Option B: COST = FLAT FEE

c =   60

Suppose we represent these relationships on a **graph** as described by a Cartesian coordinate system. We will find ordered pairs (m, c) that make these equations true.

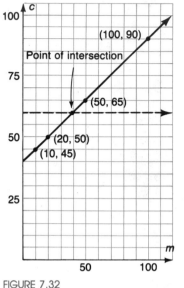

FIGURE 7.32

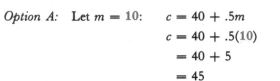

Option A: Let $m = 10$: $c = 40 + .5m$
$$c = 40 + .5(10)$$
$$= 40 + 5$$
$$= 45$$

If Linda drives 10 miles $(m = 10)$,   the cost of the rental is \$45. Plot the point (10, 45) as shown in Figure 7.32.

Let $m = 20$: $c = 40 + .5m$
$$c = 40 + .5(20)$$
$$= 40 + 10$$
$$= 50$$ If she drives 20 miles, the cost of the rental is \$50. Plot (20, 50).

Let $m = 50$: $c = 40 + .5m$
$$c = 40 + .5(50)$$
$$= 40 + 25$$
$$= 65$$ Plot (50, 65).

Let $m = 100$: $c = 40 + .5m$
$$c = 40 + .5(100)$$
$$= 40 + 50$$
$$= 90$$ Plot (100, 90).

From the graph, we can see that these points all lie on the same line. Draw a line through these points, as shown in Figure 7.32.

Option B: $c = 60$

The second component of the ordered pair (m, c) is always 60 in this case, regardless of the number of miles, m. This is drawn as a dashed line in Figure 7.32. We can now use Figure 7.32 to estimate the mileage for which both rates are the same. It is the point of intersection—it looks like (40, 60). This means that if Linda expects to drive more than 40 miles, she should take the fixed rate.

In order to graph a line, find two ordered pairs that lie on the line. Then find a third point as a check. Draw the line passing through these three points. If the three points don't lie on a straight line, then you have made an error.

Example 4:   Graph: $y = 2x + 2$

Solution: It is generally easier to pick x and find y.

If $x = 0$: $y = 2 \times 0 + 2$
$$= 0 + 2$$
$$= 2$$ Plot (0, 2) as shown in Figure 7.33.

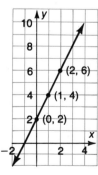

FIGURE 7.33 Graph of $y = 2x + 2$

If $x = 1$: $y = 2 \times 1 + 2$
      $= 2 + 2$
      $= 4$ Plot $(1, 4)$.

If $x = 2$: $y = 2 \times 2 + 2$
      $= 4 + 2$
      $= 6$ Plot $(2, 6)$. This is the checkpoint. Two points
                determine a line—this point checks your work.

Draw the line passing through the three plotted points.

Problem Set 7.5

Find the landmarks given by the coordinates in Problems 1–9. Use the map in Figure 7.29.

1. $(7, 10)$ **2.** $(1, 2)$ **3.** $(11, 6)$

4. $(12, 0.5)$ **5.** $(11, 7.3)$ **6.** $(1.5, 11.3)$

7. $(6.5, 2.5)$ **8.** $(6.5, 4.5)$ **9.** $(6.9, 9.8)$

Using the seating chart in Figure 7.30, name the occupant of the seat given by the coordinates in Problems 10–15.

10. $(5, 1)$ **11.** $(4, 1)$ **12.** $(1, 2)$

13. $(4, 2)$ **14.** $(2, 3)$ **15.** $(3, 1)$

Find the points given in Problems 16–20. Use the indicated scale.

16. Scale: 1 square on your paper $= 1$ unit

    a. $(1, 2)$ **b.** $(3, ^-3)$ **c.** $(^-4, 3)$
    d. $(^-6, ^-5)$ **e.** $(0, 3)$

17. Scale: 1 square on your paper $= 1$ unit

    a. $(^-1, 4)$ **b.** $(6, 2)$ **c.** $(1, ^-5)$
    d. $(3, 0)$ **e.** $(^-1, ^-2)$

18. Scale: 1 square on your paper $= 10$ units

    a. $(10, 25)$ **b.** $(^-5, 15)$ **c.** $(0, 0)$
    d. $(^-50, ^-35)$ **e.** $(^-30, ^-40)$

19. Scale: 1 square on your paper $= 50$ units

    a. $(100, ^-225)$ **b.** $(^-50, ^-75)$ **c.** $(0, 175)$
    d. $(^-200, 125)$ **e.** $(50, 300)$

20. Scale, x-axis: 1 square on your paper $= 1$ unit
    Scale, y-axis: 1 square on your paper $= 10$ units

    a. $(4, 40)$ **b.** $(^-3, 0)$ **c.** $(2, ^-80)$
    d. $(^-5, ^-100)$ **e.** $(0, 0)$

21. Plot the following coordinates on graph paper, and connect each point with the preceding one: $(^-2, 2)$, $(^-2, 9)$, $(^-7, 2)$, $(^-2, 2)$. Start again: $(^-6, ^-6)$, $(^-6, ^-8)$, $(10, ^-8)$, $(10, ^-6)$, $(6, ^-6)$, $(6, 0)$, $(10, 0)$, $(10, 2)$, $(6, 2)$, $(6, 16)$, $(^-2, 14)$, $(^-11, 2)$, $(^-9, ^-1)$, $(^-2, ^-1)$, $(^-2, ^-6)$, $(^-6, ^-6)$.

22. Plot the following coordinates on graph paper, and connect each point with the preceding one: $(9, 0)$, $(7, ^-\frac{1}{2})$, $(6, ^-2)$, $(8, ^-5)$, $(5, ^-8)$, $(^-7, ^-10)$, $(^-1, ^-5)$, $(0, ^-2)$, $(^-2, ^-1)$, $(^-6, ^-5)$, $(^-5, ^-3)$, $(^-6, ^-2)$, $(^-5, ^-1)$, $(^-6, 0)$, $(^-5, 1)$, $(^-6, 2)$, $(^-5, 3)$, $(^-6, 4)$, $(^-5, 5)$, $(^-6, 7)$, $(^-2, 3)$, $(0, 4)$, $(^-1, 10)$, $(^-7, 15)$, $(3, 10)$, $(6, 5)$, $(4, 3)$, $(5, 2)$, $(7, 2)$, $(9, 0)$. Finally, plot a point at $(7, 1)$.

23. Suppose we place coordinate axes on the connect-the-dots figure shown in the margin. Let A be $(2, 9)$ and Z be $(1, 5)$. Write directions similar to those of Problems 21 and 22 about how to sketch this figure.

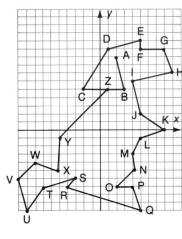

Graph the lines in Problems 24–44.

24. $y = x + 3$
25. $y = x + 1$
26. $y = 2x + 1$
27. $y = 3x - 2$
28. $y = x - 1$
29. $y = 2x + 3$
30. $x + y = 0$
31. $x - y = 0$
32. $2x + y = 3$
33. $3x + y = 10$
34. $3x + 4y = 8$
35. $x + 2y = 4$
36. $x = 5$
37. $x = ^-4$
38. $x = ^-1$
39. $y = ^-2$
40. $y = 6$
41. $y = 4$
42. $x + y + 100 = 0$
43. $5x - 3y = 45$
44. $y = 50x$

45. Suppose a car rental agency gave you the following choices:

    Option A: $30 per day plus 40¢ per mile

    Option B: Flat $50 per day

    a. Write the equations for Options *A* and *B*.
    b. Graph the equations from part a.
    c. Use the graph to estimate the mileage for which both rates are the same.

46. A paint sprayer rents for $4 per hour or $24 per day. Estimate the number of hours for which both rates are the same.

Mind Bogglers

47. Draw a picture similar to the ones in Problems 21 and 22, and then describe the picture using ordered pairs.

48. Internal Revenue Service (IRS) regulations allow taxpayers a $1,000 deduction for each dependent and a 15% standard deduction on the amount of income remaining. For a person with income *I* and four dependents (including herself), the amount to be taxed can be found as follows:

$$\text{INCOME} - \left(\begin{array}{c}\text{DEPENDENT} \\ \text{DEDUCTION}\end{array}\right) - \left(\begin{array}{c}\text{STANDARD} \\ \text{DEDUCTION}\end{array}\right)$$

$$= \quad I \quad - \quad 4(1{,}000) \quad - \quad .15I$$

$$= .85I - 4{,}000$$

If the tax is 25% of the final amount, then the relationship between income and tax can be expressed

$$T = .25(.85I - 4{,}000)$$

$$T = .2125I - 1{,}000$$

Graph this equation. Choose $I = 10{,}000$, $I = 12{,}000$, and $I = 14{,}000$ as your choices for the first component.

7.6 Königsberg Bridge Problem

In the 18th century, in the German town of Königsberg (now the Russian city of Kaliningrad), a popular pastime was to walk along the bank of the Pregel River and cross over some of the seven bridges that connected two islands, as is shown in Figure 7.34.

Compare the experience of the people of Königsberg with your findings in the Santa Rosa street problem, page 370.

FIGURE 7.34 Königsberg bridges

One day a native asked a neighbor this question: "How can you take a walk so that you cross each of our seven bridges once and only once?" The problem intrigued the neighbor and soon caught the interest of many other people of Königsberg as well. Whenever people tried it, they ended up either not crossing a bridge at all or else crossing one bridge twice.

This problem was brought to the attention of Swiss mathematician Leonhard Euler, who was serving at the court of the Russian empress Catherine the Great in St. Petersburg. His study of the problem laid the groundwork for a branch of modern mathematics called **topology,** which is discussed in the next section.

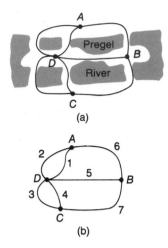

FIGURE 7.35 Networks for the Königsberg bridge problem.

Euler began by drawing a diagram of the problem. With this diagram, the land became points and the bridges became lines connecting these points, as shown in Figure 7.35a. Now we can work the problem as Euler did. We begin by trying to trace a diagram like the one in Figure 7.35b. In order to solve the bridge problem, we need to draw the figure without lifting our pencil from the paper. Figures similar to the one in Figure 7.35b are called **networks.**

In a network, the points where the lines cross are called *vertices,* and the lines representing bridges are called *arcs.* A network is *traversed* by passing through all the arcs exactly once. You may pass through the vertices any number of times. In the network of the Königsberg bridges, the vertices are A, B, C, and D. The number of arcs to vertex A is 3, so the vertex at A is called an *odd vertex.* In the same way, D is an odd vertex, since 5 arcs go to it. Euler discovered that there must be a certain number of odd vertices in any network if you are to travel it in one journey without retracing any arcs. You may start at any vertex and end at any other vertex, as long as you travel the entire network. Also, the network must connect each point (this is called a *closed network*).

Let's examine the network more carefully and look for a pattern.

TABLE 7.4

Number of Arcs Connecting a Vertex	Possibilities	
1	One departure; starting point one arrival; ending point	
2	One arrival/one departure	
3	One arrival/two departures; starting point two arrivals/one departure; ending point	
4	Two arrivals/two departures	
5	Two arrivals/three departures; starting point three arrivals/two departures; ending point	
	⋮	

We see that, if the vertex is odd, then it must be a starting point or an ending point. What is the largest number of starting and ending points in any network? Note the following results:

1. Odd vertices always occur in pairs, so there will be an even number of odd vertices if it is traversible.
2. If a network has more than two odd vertices, it is not traversible.
3. If a network has two odd vertices, it is traversible; one odd vertex must be a starting point and the other odd vertex must be an ending point.
4. If all vertices are even, then it is traversible.

𝔥istorical Note

Example 1.:   Are the following networks traversible?

a.   **b.**

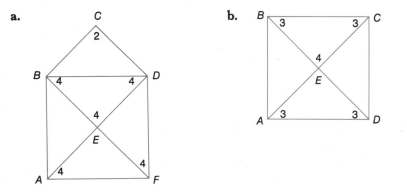

Solutions: **a.**   You may remember seeing this network before. It has four even vertices (*B*, *C*, *D*, and *E*) and two odd vertices (*A* and *F*) and is therefore traversible. In order to traverse it, you must start at *A* or *F* (at an odd vertex).

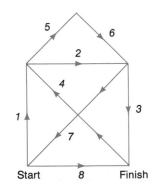

In 1783 Czarina Catherine II appointed Princess Daschkoff to the directorship of the Imperial Academy of Sciences in St. Petersburg. Since women were seldom so highly honored in those days, the appointment received wide publicity. The Princess decided to commence her directorship with a short address to the assembled members of the Academy. She invited Leonhard Euler (see the Historical Note on page 190) who at the time was elderly and blind. Since Euler was, at the time, the most respected scientist in Russia, he was to have the special seat of honor. However, as the Princess sat down, a local professor named Schtelinn maneuvered into Euler's seat. When Princess Daschkoff saw Schtelinn settling himself next to her, she turned to Euler and said, "Please be seated anywhere, and the chair you choose will naturally be the seat of honor." This act charmed Euler and all present—except the arrogant Professor Schtelinn.

b.   This network has one even vertex and four odd vertices, and it is therefore not traversible.

We can now solve the Königsberg bridge problem. Referring to Figure 7.35a, we see there are four odd vertices and the network is not traversible.

Another example of a network problem was given by Problem 52 on page 371. We will call this the *floor-plan problem;* it involves taking a trip through all the rooms and passing through each door only once. Let's label the rooms in Figure 7.36a A, B, C, D, E, and F. Rooms A, C, E, and F have two doors; rooms B and D have three doors. Make a conjecture about which room we should begin in if we wish to solve this problem.

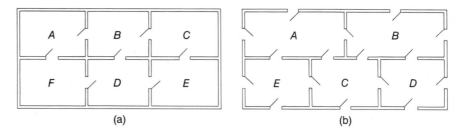

(a)   (b)

FIGURE 7.36   Floor-plan problem

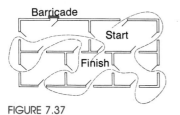

FIGURE 7.37

How about the floor plan in Figure 7.36b? Rooms A, B, and C have an odd number of doors; rooms E and D have an even number of doors. It is not possible to choose a route to traverse the floor plan of this house. Why not? If we barricade one of the doors in room A, the problem can then be solved as shown in Figure 7.37.

Problem Set 7.6

Which of the networks in Problems 1–10 are traversable? If a network can be traversed, show how.

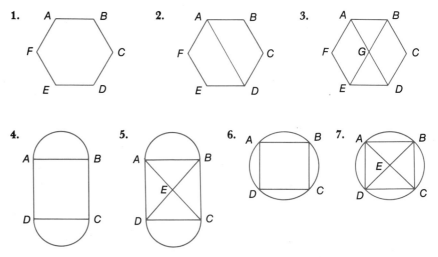

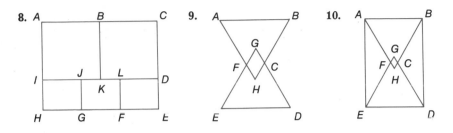

11. Present a complete solution to the Santa Rosa street problem, originally posed in Problem 51, page 370.

12. The edges of a cube form a three-dimensional network. Are the edges of a cube traversible?

13. After Euler solved the Königsberg bridge problem, an eighth bridge was built as shown in the figure. Is this network traversible? If so, show how.

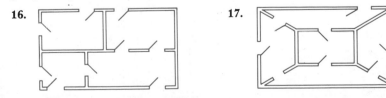

Königsberg with eight bridges

14. A portion of London's Underground transit system is shown in the figure. Is it possible to travel the entire system and visit each station while taking each route exactly once?

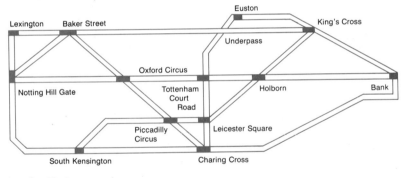

London Underground

15. A simplified map of New York City, showing the subway connections between Manhattan and The Bronx, Queens, and Brooklyn, is shown in the margin. Is it possible to travel on the New York Subway system and use each subway exactly once? You can visit each borough (The Bronx, Queens, Brooklyn, or Manhattan) as many times as you wish.

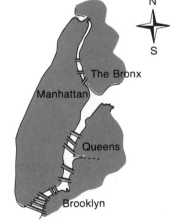

New York City subways

For which of the floor plans in Problems 16–19 can you pass through all the rooms while going through each door only once?

16.

17.

18. 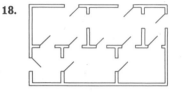   **19.**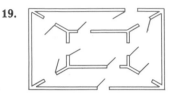

Escher print: *Möbius Strip II*

20. Make a conjecture about when the floor plans of Problems 16–19 can be solved.

21. Make a conjecture about which room to begin with to solve a floor-plan problem.

22. Count the number of vertices, arcs, and regions for each of Problems 1–10.
Let V = Number of vertices
A = Number of arcs
R = Number of regions
Compare $V + R$ with A. Make a conjecture relating V, R, and A. This finding is called *Euler's Formula for Networks*.

23. About a century ago, August Möbius made the discovery that, if you take a strip of paper and give it a single half-twist and paste the ends together, you will have a piece of paper with only one side! Construct a Möbius strip, and verify that it has only one side. How many edges does it have?

24. Cut your Möbius strip from Problem 23 in half down the center. What is the result?

25. Cut your resulting band from Problem 24 in half again. What is the result?

26. Construct a Möbius strip, and cut along a path that is about one-third the distance from the edge. What is the result?

27. Construct a Möbius strip, and mark a point A on it. Draw an arc from A around the strip until you return to the point A. Do you think you could connect *any* two points on the sheet of paper without lifting your pencil?

28. Take a strip of paper $11'' \times 1''$, and give it three half-twists; join the ends together. How many edges and sides does this band have? What happens if you cut down the center of this piece?

Mind Bogglers

29. *The paper-boot puzzle.* The boot puzzle consists of three pieces, all of which are cut out of stiff paper. One piece is shaped like a pair of boots joined together at the top, as in figure a. The remaining pieces are shaped as shown in figures b and c.

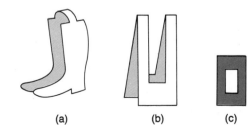

(a)   (b)   (c)

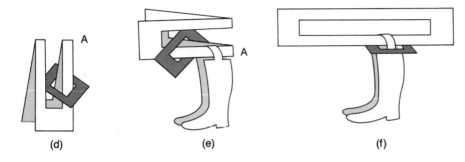

(d)   (e)   (f)

To assemble the puzzle, fold the large rectangular piece as in figure b and slip the smaller piece over one of its arms, as in figure d. Then hang the boots over part of the same arm, as in figure e. Pull the small piece to the right and over the end of the arm at *A*. The unfold the large piece, and the puzzle will be assembled as in figure f.

The problem is to remove the boots without tearing the paper. Or better yet, give the assembled piece shown in figure f to someone else and ask them to remove the boots.

30. In the figure there are eight square rooms making up a maze. Each square room has two walls that are mirrors and two walls that are open spaces. Identify the mirrored walls.

Mirror maze

31. This is a network problem that depends on the correct answer to Problem 30. Solve the maze shown in the figure. That is, find the sequence of open walls that allows you to pass through all eight rooms consecutively without going through the same room twice. If that is not possible, tell why.

Problem for Individual Study

32. What is a Klein bottle?

Hint: This limerick should give you some clues for Problem 32:

A mathematician named Klein
Thought the Möbius strip
    was divine.
        Said he, "If you glue
        The edges of two,
You'll get a weird bottle
    like mine."

—*Anonymous*

Klein bottle

Can you build or construct a physical model?

*7.7 Euclid Does Not Stand Alone

> Mighty is geometry; joined with art, resistless.
>
> —*Euripides*

As Europe passed out of the Middle Ages and into the Renaissance, artists were at the forefront of the intellectual revolution. No longer satisfied with flat-looking scenes, they wanted to portray people and objects as they looked in real life. The artists' problem was one of dimension. How could a flat surface be made to look three-dimensional? Some of these artists tried to solve the problem by studying mathematics; however, the Euclidean geometry of the day did not provide all the answers they needed.

One of the first (but rather unsuccessful) attempts at portraying depth in a painting is shown in Figure 7.38, Duccio's *Last Supper*. Notice that the figures are in a boxed-in room. This technique is characteristic of the period and made the perspective easier to define.

Artists finally solved the problem of perspective by considering the surface of the picture to be a window through which to view the object to be painted. This technique was pioneered by Paolo Uccello (1397–1475), Piero della Francesca (1416–1492), Leonardo da Vinci (1452–1519), and Albrecht Dürer (1471–1528). As the lines of vision from the object converge at the eye, the picture captures a cross section of them, as shown in Figure 7.39.

*This is an optional section.

FIGURE 7.38 Duccio's *Last Supper*. Notice the receding wall and ceiling lines. However, the perspective is not complete. The table and the room are not seen from the same vantage point. The table seems to slant toward the front.

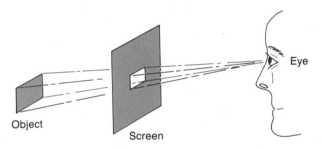

FIGURE 7.39 An attempt to understand perspective

The next step was the execution of this idea. Figure 7.40 shows an artist painting a picture using a grid system and a fixed viewpoint. As he changes his line of vision, the points projected onto the screen change.

FIGURE 7.40 Albrecht Dürer's *Designer of the Lying Woman* shows how the problem of perspective can be overcome. The point in front of the artist's eye fixes the point of viewing the painting. The grid on the window corresponds to the grid on the artist's canvas.

However, the method illustrated in this figure certainly has its limitations. Thus artists continued to seek mathematical solutions to their problems.

In response to the needs of these artists, a geometry quite different from Euclidean geometry developed. This new geometry is called *projective geometry* and is studied in more advanced courses. It is a generalization of the Euclidean geometry you learned in high school.

When Euclid formalized geometry, he used axioms and postulates that appeared to conform to the world in which we live. The so-called fifth postulate caused problems from the time it was stated. It somehow doesn't seem like the other postulates but, rather, like a theorem that should be proved. In fact, this postulate even bothered Euclid himself, since he didn't use it until he had proved his 29th theorem. Many mathematicians tried to find a proof for this postulate.

Historical Note

There are many ways of stating Euclid's fifth postulate, and several historical ones are repeated here.

1. *Poseidonius* (about 135–51 B.C.): Two parallel lines are equidistant.

2. *Proclus* (410–485 A.D.): If a line intersects one of two parallel lines, then it also intersects the other.

3. *Saccheri* (1667–1733): The sum of the interior angles of a triangle is two right angles.

4. *Legendre* (1752–1833): A line through a point in the interior of an angle other than a straight angle intersects at least one of the arms of the angle.

5. *Bolyai* (1802–1860): There is a circle through every set of three noncollinear points.

6. *John Playfair* (1748–1819): Given a straight line and any point not on this line, there is one and only one line through that point that is parallel to the given line.

Playfair's statement is the one most often used in high school geometry textbooks. Remember that all of these statements are *equivalent,* and, if you accept any one of them, you must accept them all.

Euclid's Postulates

1. A straight line can be drawn from any point to any other point in only one way.
2. A finite straight line can be drawn continuously in a straight line.
3. A circle can be described with any point as center and with a radius equal to any finite straight line drawn from the center.
4. All right angles are equal to each other.
5. If a straight line falling on two straight lines makes the interior angles on the same side less than two right angles, the two straight lines, if produced infinitely, meet on that side on which the angles are less than the two right angles.

One of the first serious attempts to prove Euclid's fifth postulate was made by Girolamo Saccheri (1667–1733), an Italian Jesuit. He constructed a quadrilateral with base angles A and B right angles and with sides AD and BC the same length (see Figure 7.41).

FIGURE 7.41   A Saccheri quadrilateral

As you probably know from high school geometry, the summit angles C and D are also right angles. However, this result used the fifth postulate. Now, it is also true that, *if* the summit angles are right angles, *then* Euclid's fifth postulate holds. The problem, then, was to establish the fact that angles C and D were right angles. First, Saccheri assumed the angles to be obtuse, and he didn't have too much difficulty in arriving at a contradiction. Next, he assumed the angles to be acute. No contradiction arose, and finally he gave up the quest because it "led to results that were repugnant to the nature of a straight line."

Saccheri never realized the significance of what he had started, and his work was forgotten until 1889. However, in the meantime, Johann Lambert (1728–1777) and Adrien-Marie Legendre (1752–1833) similarly investigated the possibility of eliminating Euclid's fifth postulate by proving it from the other postulates.

By the early years of the 19th century, three accomplished mathematicians began to suspect that the parallel postulate was independent and could not be eliminated by deducting it from the others. The great mathematician Karl Gauss, whom we've mentioned before, was the first to reach this conclusion, but, since he didn't publish this finding of his, the credit goes to two others. In 1811, an 18-year-old Russian named Nikolai Lobachevski pondered the possibility of a "non-Euclidean" geometry—that is, a geometry that did not assume Euclid's fifth postulate. In 1829, he published his ideas in an article entitled "Geometrical Researches on the Theory of Parallels." The postulate he used was subsequently named after him:

Saccheri's plan was a simple one:

1. Assume that the angles are obtuse and deduce a contradiction.
2. Assume that the angles are acute and deduce a contradiction.
3. Therefore, by the first two steps, the angles must be right angles.
4. From step 3, Euclid's fifth postulate can be deduced.

The Lobachevskian Postulate

The summit angles of a Saccheri quadrilateral are acute.

Historical Note

Another mathematician, Janos Bolyai (1802–1860), was working on this same idea at about the same time as Lobachevski. (Clearly the world was nearly ready to accept the idea that a geometry need not be based on our physical experience.) Because of his work with this geometry, it is sometimes called *Bolyai-Lobachevski* geometry.

This axiom, in place of Euclid's fifth postulate, leads to a geometry that we call *hyperbolic geometry*.

If we use the plane as a model for Euclidean geometry, what model will serve for hyperbolic geometry? A rough model for this geometry can be seen by placing two trumpet bells together. It is called a *pseudosphere* and is generated by a curve called a *tractrix* (as shown in Figure 7.42 on page 408). The tractrix is rotated about the line AB. It has the property that, through a point not on a line, there are many lines parallel to the given one.

Georg Bernhard Riemann (1826–1866), who also worked in this area, pointed out that, although a straight line may be extended indefinitely, it need

𝕳istorical 𝕹ote

The set theory of Cantor (see the Historical Note on page 49) provided a basis for topology, which was presented for the first time by Henri Poincaré (1854–1912) in *Analysis Situs*. A second branch of topology was added in 1914 by Felix Hausdorff (1868–1942) in *Basic Features of Set Theory*. Earlier mathematicians, such as Euler, Möbius, and Klein, had touched on some of the ideas we study in topology, but the field was given its major impetus by L. E. J. Brouwer (1882–1966). Today much research is being done in topology, which has practical applications in astronomy, chemistry, economics, and electrical circuitry.

Elliptic geometry is sometimes called Riemannian geometry in honor of Riemann.

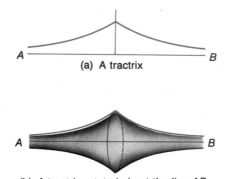

(a) A tractrix

(b) A tractrix rotated about the line *AB*

FIGURE 7.42

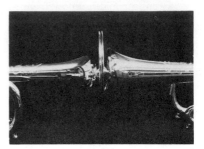

(c) Two trumpets serve as a model of a tractix rotated about a line.

not have infinite length. It could instead be similar to the arc of a circle, which eventually begins to retrace itself. Such a line is called *re-entrant*. An example of a re-entrant line is found by considering a great circle on a sphere. A *great circle* is a circle on the sphere with a diameter equal to the diameter of the sphere. Taking this as a model, a Saccheri quadrilateral is constructed on a sphere with the summit angles obtuse. The resulting geometry is called **elliptic geometry.** The shortest path between any two points on a sphere is an arc of the great circle through those points; thus these arcs correspond to line segments in Euclidean geometry. In 1854 Riemann showed that, with some other slight adjustments in the remaining postulates, another consistent non-Euclidean geometry can be developed. Notice that the fifth, or parallel, postulate fails to hold because any two great circles on a sphere must intersect at two points (see Figure 7.43).

𝕳istorical 𝕹ote

For centuries, we've assumed that the physical space in our universe is Euclidean; the notion that it wasn't was repugnant to Saccheri. However, Einstein's theory of relativity is based on a non-Euclidean geometry. This fact leads us to believe that, at least on a cosmic scale, Euclidean geometry is not sufficient to describe our universe.

FIGURE 7.43   A sphere showing the intersection of two great circles

We have not, by any means, discussed all possible geometries. We have merely shown that the Euclidean geometry that is taught in high school is not the only model. A comparison of some of the properties of these geometries is shown in Table 7.5.

TABLE 7.5 Comparison of Major Two-Dimensional Geometries

Euclidean Geometry	Hyperbolic Geometry	Elliptic Geometry
Euclid (about 3000 B.C.)	*Gauss, Bolyai, Lobachevski* (about 1830)	*Riemann* (about 1850)
Given a point not on a line, there is one and only one line through the point and parallel to the given line.	Given a point not on a line, there are an infinite number of lines through the point that do not intersect the given line.	There are no parallels.
Geometry on a plane D = 90°	Geometry on a pseudosphere D < 90°	Geometry on a sphere D > 90°
The sum of the angles of a triangle is 180°.	The sum of the angles of a triangle is less than 180°.	The sum of the angles of a triangle is more than 180°.
Lines are infinitely long.	Lines are infinitely long.	Lines are finite in length.

In the 17th century, space came to be conceptualized as a set of points, and, with the non-Euclidean geometries of the 19th century, mathematicians gave up the notion that geometry had to describe the physical universe. By the end of the 19th century, the idea that a geometry could be described as an abstract set of axioms along with a body of deduced theorems was generally accepted. Space was thought of simply as a set of points together with an abstract set of relations in which these points are involved. The time was right for geometry to be considered as the theory of such a space, and in 1895 Henri Poincaré published a book using this notion of space and geometry in a systematic development. This book was called *Vorstudien zur Topologie* (*Introductory Studies in Topology*). However, topology was not the invention of any one person, and the names of Cantor, Euler, Fréchet, Hausdorff, Möbius, and Riemann are associated with the origins of **topology.** Today it is a broad and fundamental branch of mathematics.

The children and their distorted images are topologically equivalent.

To obtain an idea of the nature of topology, consider a *geoboard*. You will recall that geoboards are used to illustrate a wide variety of geometric ideas, especially in elementary schools. Suppose we stretch one rubber band over the nails to form a square and another to form a triangle, as shown in Figure 7.44.

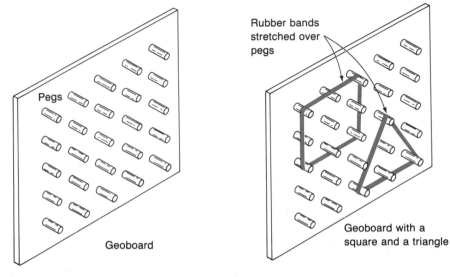

FIGURE 7.44

In the geometries we've been considering, in which we've been concerned with congruence, similarity, angles, number of sides, and so on, these figures are seen as different. However, in topology, we are concerned with the *mathematics of distortion*. The general idea is one called *elastic motion,* which includes bending, stretching, shrinking, or distorting the figure in any way that allows the points to remain distinct. It does not include cutting a figure unless we "sew up" the cut *exactly* as it was before.

Definition

Two geometric figures are said to be **topologically equivalent** if one figure can be elastically twisted and stretched into the same shape as the other.

Rubber bands can be stretched into a wide variety of shapes. All the forms in Figure 7.45 are topologically equivalent.

These curves are *simple closed curves*. A curve is *closed* if it divides the plane into three disjoint subsets: the set of points on the curve itself, the set of points *interior* to the curve, and the set of points *exterior* to the curve. It is *simple*

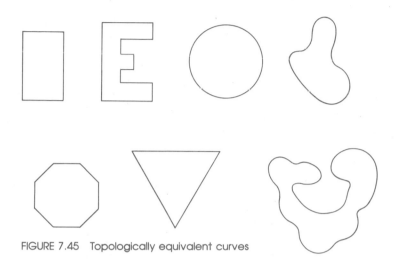

FIGURE 7.45 Topologically equivalent curves

if it has only one interior. Notice that, to pass from a point in the interior to a point in the exterior, it is necessary to cross over the given curve at least once. This property remains the same for any distortion and is therefore called an *invariant*.

In topology, objects and figures are classified according to their **genus,** that is, the number of cuts that can be made without cutting the object or figure into two pieces. Even though a study of genus is beyond the scope of this book, you can see from Figure 7.46 that the genus of an object is related to the number of holes in the object.

Problem Set 7.7

Which of the figures in Problems 1–4 are Saccheri quadrilaterals?

1.

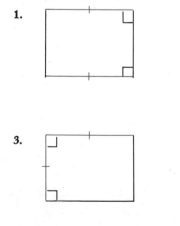

2.

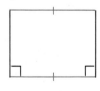

3.

4.

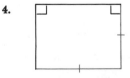

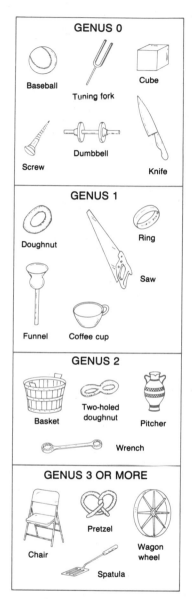

FIGURE 7.46 Genus of the surfaces of everyday objects

5. Let A be any point on a sphere, and let ℓ be a line passing through A. How many lines can you draw through A that are perpendicular to ℓ?

6. Let P be a point at the pole, and let m be a line passing through P. How many lines can you draw through P that are perpendicular to m?

7. Consider the spheres shown here.

    a. Why do the lines appear to be parallel?
    b. In Table 7.5 we said that there are no parallels on a circle. What's wrong with our reasoning in part a?

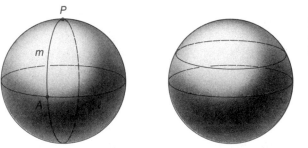

Remember that a line on a sphere is a great circle.

8. On a globe, locate San Francisco, Miami, and Detroit. Connect these cities with the shortest paths to form a triangle. Next, use a protractor to measure the angles. What is their sum?

9. Repeat Problem 8 for the cities of Tokyo, Seattle, and Honolulu.

10. Why do you think Euclidean geometry remains so prevalent today even though we know there are other valid geometries?

In Problems 11–15 determine whether each of the points A, B, and C is inside or outside the simple closed curve.

11.

12.

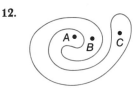

13.

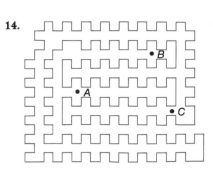

14.

15.

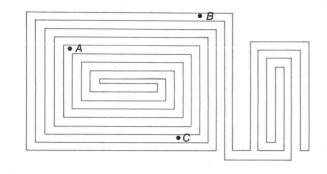

16. Let X be a point obviously outside the figure given in Problem 11. Draw AX. How many times does it cross the curve? Repeat for BX and CX.

17. Repeat Problem 16 for the figure given in Problem 12.

18. Repeat Problem 16 for the figure given in Problem 13.

19. Make a conjecture based on Problems 16–18. Use your conjecture for the figures given in Problems 14 and 15.

20. Walking north or south on the earth is defined as walking along meridians, and walking east or west is defined as walking along parallels. There is a point on the surface of the earth from which it is possible to walk a mile south, then a mile east, then a mile north and be right back where you started. Find one such point.

21. Find two additional, different points satisfying the conditions of Problem 20.

22. *He killed the noble Mudjokivis.*
Of the skin he made him mittens,
Made them with the fur side inside
Made them with the skin side outside.
He, to get the warm inside,
Put the skin side outside;
He, to get the cold side outside,
Put the warm side fur side inside.
That's why he put the fur side inside,
Why he put the skin side outside,
Why he turned them inside outside.—Author unknown

   a. If a right-handed mitten is turned inside out, as is suggested in the poem, will it still fit the right hand?

   b. Is the right-handed mitten topologically equivalent to the left-handed mitten?

This conjecture involves a theorem called the *Jordan Curve theorem*. It says that a simple closed curve divides the plane into two regions, an inside and an outside. Sometimes a simple closed curve is called a Jordan curve.

A Jordan curve has also been described as a Protestant curve: it doesn't cross itself.

Mind Bogglers

23. *Two-person problem.* Tie a piece of cord to each of your wrists. Then tie a second piece of cord to the wrists of a second person and through your cord, as shown at the top of the next page. Now see if you can separate yourselves without cutting the

(a)   (b)

cord, untying the knots, or taking the string off your wrists. Since the figure in (a) is topologically equivalent to the figure in (b), the problem can be solved.

24. *A five-color map?* One of the earliest and most famous problems in topology is the four-color problem. It was first stated in 1850 by the English mathematician Francis Guthrie. It states that any map on a plane or a sphere can be colored with at most four colors so that any two countries that share a common boundary are colored differently. All attempts to prove this conjecture had failed until Kenneth Appel and Wolfgang Haken of the University of Illinois announced their proof in 1976. The university honored their discovery by using the following postmark:

FOUR COLORS

SUFFICE

Some mathematicians were reluctant to accept this proof because of the necessity of computer verification. The proof was not "elegant" in the sense that it required the computer analysis of a large number of cases. Study the map in the following figure and determine for yourself if it is the *first five-color* map, providing a counterexample for the computerized "proof." Notice from the table that the region in question is bounded by four colors.

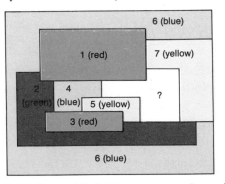

Is this the world's first five-color map?

Region	Blue	Yellow	Green	Red
1	✔	✔	✔	
2	✔	✔		✔
3	✔	✔	✔	
4		✔	✔	✔
5	✔		✔	✔
6		✔	✔	✔
7	✔	✔	✔	✔
?	✔	✔	✔	✔

Problems for Individual Study

25. *Perspective and projective geometry.* How are three-dimensional objects represented in two dimensions? What are different ways of projecting lines or surfaces on a plane?

Exhibit suggestions: Models of perspective; examples of how projective geometry and art are related.

References: Morris Kline, *Mathematics, a Cultural Approach* (Reading, Mass.: Addison-Wesley, 1962), chapters 10–11.

C. Stanley Ogilvy, *Excursions in Geometry* (New York: Oxford University Press, 1969), chapter 7.

26. The discovery and acceptance of non-Euclidean geometries had an impact on all of our thinking about the nature of scientific truth. Can we ever know truth in general? Write a paper on the nature of scientific laws, the nature of an axiomatic system, and the implications of non-Euclidean geometries. An excellent reference is *The Mathematical Experience* by Philip J. Davis and Reuben Hersh (Boston: Houghton Mifflin Company, 1981).

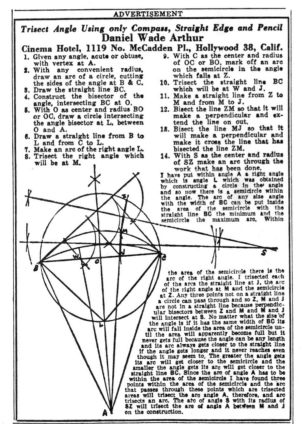

If someone claims to be able to trisect an angle with a straightedge and compass, a mathematician, *without ever looking at the construction,* will respond by saying that a mistake was made. This can be frustrating to someone who believes that he or she has accomplished the impossible. In fact, it was so frustrating to Daniel Wade Arthur that he was motivated to take out a paid advertisement in the *Los Angeles Times,* which is reproduced here. See Problem 27 on the next page.

27. Euclid clearly made a distinction between the definition of a figure and the proof that such a figure could be constructed. Two very famous problems in mathematics focus on this distinction:

1. *Trisecting an angle.* Using only a straightedge and compass, trisect a given angle.
2. *Squaring a circle.* Using only a straightedge and compass, construct a square with an area equal to the area of a given circle.

These problems have been *proved* to be impossible (as compared with unsolved problems that *might* be possible). Write a paper discussing the nature of an unsolved problem as compared with an impossible problem.

28. Make drawings of geometric figures on a piece of rubber inner tube. Demonstrate to the class various ways in which these figures can be distorted.

Chapter 7 Outline

   I. Geometry
      A. Geometry as a deductive system
         1. Undefined terms
         2. Axioms
         3. Theorems
      B. Transformations
         1. Definition
         2. Types of transformations
      C. Relationships
         1. Congruence
         2. Similarity
  II. Measuring Length and Area
      A. Length
         1. U.S. system (inch, foot, yard, and mile)
         2. Metric system (meter)
         3. Conversion factors
      B. Perimeter
         1. Polygons
            a. Types
            b. Regular
         2. Definition
         3. Formulas for perimeter of an equilateral triangle, a rectangle, and a square
      C. Circumference
         1. Circle
         2. Approximation for π
         3. Formula for circumference
      D. Relationship between kilometers and miles

    E. Area
        1. U.S. system
        2. Metric system
        3. Formulas for area of a square, a rectangle, a parallelogram, a triangle, and a circle

III. Angles
    A. Definition
    B. Measuring with a protractor
        1. Degree
        2. Congruent angles
    C. Classification of angles
        1. Right
        2. Acute
        3. Obtuse
        4. Straight
    D. Classification of pairs of angles
        1. Complementary angles
        2. Supplementary angles
        3. Vertical angles
        4. Alternate interior angles
        5. Alternate exterior angles
        6. Corresponding angles

IV. The Metric System—Volume, Mass, and Temperature
    A. Polyhedrons
    B. Volume
        1. U.S. system (cubic inch)
        2. Metric system (cubic centimeter)
    C. Capacity
        1. U.S. system (teaspoon, tablespoon, ounce, cup, pint, quart, and gallon)
        2. Metric system (liter)
        3. Conversion factors
    D. Mass
        1. U.S. system (ounce, pound, and ton)
        2. Metric system (gram)
    E. Metric conversions
        1. Metric prefixes (kilo, hecto, deka, deci, centi, milli)
        2. Changing metric units
    F. Temperature
        1. U.S. system (Fahrenheit)
        2. Metric system (Celsius)

V. Coordinate Geometry
    A. Map reading using ordered pairs
    B. Cartesian coordinate system
    C. Graphing a line

VI. Königsberg Bridge Problem
   A. Statement of problem
   B. Networks
      1. Vertices and arcs
      2. Traversible
      3. Odd vertex/even vertex
      4. Closed network
   C. Solution of problem
   D. Floor-plan problem
*VII. Euclid Does Not Stand Alone
   A. Projective geometry; perspective in art is achieved as a result of the study of projective geometry.
   B. The Saccheri quadrilateral
      1. A quadrilateral $ABCD$ with base angles A and B right angles and with sides AD and BC equal is a Saccheri quadrilateral.
      2. The angles C and D of a Saccheri quadrilateral might be acute, right, or obtuse depending on additional assumptions that are made.
      3. The assumptions we accept in determining the size of angles C and D of a Saccheri quadrilateral give rise to three geometries: Euclidean, hyperbolic, and elliptic.
   C. Euclidean geometry
      1. Probably the geometry you studied in high school; it conforms to the way we view the physical world.
      2. It is based on five postulates (see page 406).
      3. The plane serves as a model.
      4. Important results
         a. Given a point not on a line, there is one and only one line through the point and parallel to the given line (Euclid's fifth postulate).
         b. The sum of the angles of a triangle is 180°.
         c. Lines are infinitely long.
   D. Hyperbolic geometry
      1. Lobachevski made the assumption that the summit angles of a Saccheri quadrilateral are acute.
      2. The pseudosphere serves as a model.
      3. Important results
         a. Given a point not on a line, there are an infinite number of lines through the point that do not intersect the given line.
         b. The sum of the angles of a triangle is less than 180°.
         c. Lines are infinitely long.
   E. Elliptic geometry
      1. Riemann considered *re-entrant* lines and thus concluded that the base angles of a Saccheri quadrilateral are obtuse.

*Optional section.

> 2. The sphere serves as a model.
> 3. Important results
>   a. There are no parallels.
>   b. The sum of the angles of a triangle is more than 180°.
>   c. Lines are finite in length.
> F. Topology
>   1. Topology is the mathematics of distortion.
>   2. Topologically equivalent
>   3. Simple closed curves; interior/exterior

Chapter 7 Test

1. If a 3-cm cube is first painted green and then cut into nine 1-cm cubes, how many of those cubes will be painted on

   a. 4 sides?   b. 3 sides?   c. 2 sides?   d. 1 side?   e. 0 sides?

2. a. What are the three basic metric units?
   b. Give the meaning of each of the following prefixes:
      centi; deka; kilo; milli; deci

3. Find the measure of the requested angles.

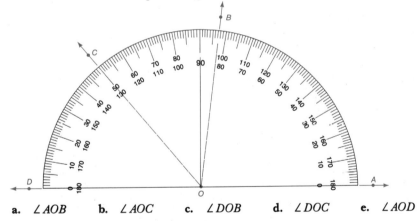

   a. $\angle AOB$ b. $\angle AOC$ c. $\angle DOB$ d. $\angle DOC$ e. $\angle AOD$

4. Classify the pairs of angles shown in the figure in the margin.

   a. $\angle 1$ and $\angle 5$ b. $\angle 5$ and $\angle 6$ c. $\angle 2$ and $\angle 4$
   d. $\angle 2$ and $\angle 8$ e. $\angle 1$ and $\angle 7$

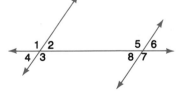

Choose the most appropriate answer for each of Problems 5–9.

5. The volume of a large upright freezer is about

   A. 8ℓ **B.** $.8 \text{ m}^3$ **C.** 2 m^3 **D.** 8 m^3 **E.** 154 kg

6. On a hot, summer day the temperature is about

   A. 0°C   **B.** 10°C   **C.** 40°C   **D.** 80°C   **E.** 100°C

Try to answer these questions without converting from one system to another—THINK METRIC!

7. The height of a tall basketball player might be

   A. 85 cm   **B.** 1.5 m   **C.** 72 ℓ   **D.** 4 m   **E.** 200 cm

8. If you had a fever your temperature might be

   A. 100°C   **B.** 50°C   **C.** 40°C   **D.** 30°C   **E.** 98.6°C

9. The length of a new pencil is about

   A. 2 km   **B.** 2 m   **C.** 2 cm   **D.** 2 dkm   **E.** 2 dm

10. Tell whether each of the following are measuring length, capacity, or mass.

    a. 15 kg   **b.** 38.2 m   **c.** 6 m³   **d.** 7 mℓ   **e.** 68 km

11. Name the metric unit you would most likely use to measure each of the following.

    a. Distance from New York to Chicago
    b. Distance around your waist
    c. Amount of gin to use in a martini
    d. Amount of space in the trunk of your car
    e. Weight of a pencil

12. Fill in the blanks.

    a. 4,300 m = _____ km   **b.** 1 ℓ = _____ m³
    c. 4.8 cm = _____ mm   **d.** 2.88 kg = _____ g
    e. 1 mile ≈ _____ km

13. Find the perimeter and area of the shaded regions.

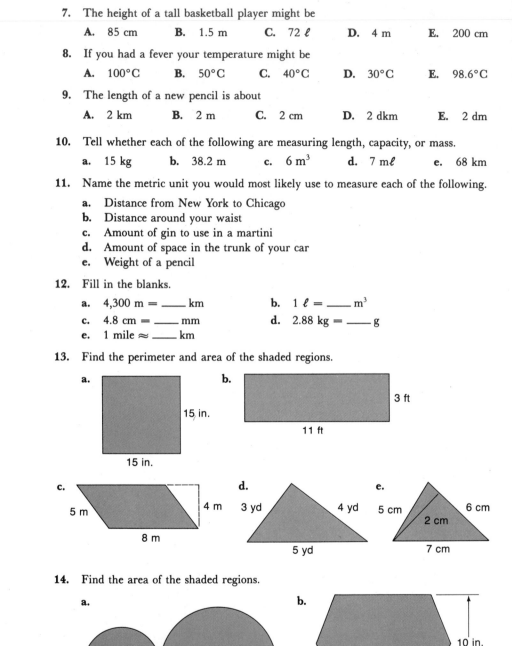

14. Find the area of the shaded regions.

   a.

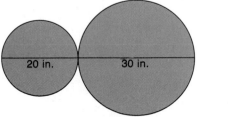

   b.

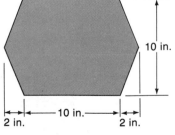

15. Find the volume of each box.

a.   b.

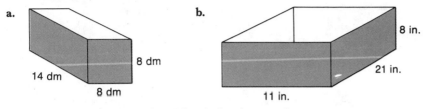

c. What is the capacity, in liters, for the box in part a?
d. What is the capacity, in gallons, for the box in part b?

16. Graph the following lines.

  a. $x + y = 5$ b. $2x - y = 5$ c. $x = y$

  d. $x = 3$ e. $y = {}^-30$

17. Indicate which of these networks are traversible. If the network is traversible, show how.

a.   b.

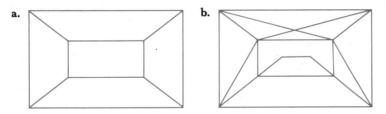

18. Indicate whether you can pass through all the rooms in the given floor plans by going through each door only once.

a.   b.

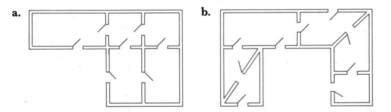

*19. What is a Saccheri quadrilateral? Discuss why this quadrilateral leads to different kinds of geometrics.

*20. Briefly discuss Euclidean, hyperbolic, and elliptic geometries.

Bonus Questions

21. a. Draw a map with seven regions that can be colored with three colors, no two bordering regions having the same color.
  b. Draw a map similar to the one in part a but that requires four colors.

22. Take a strip of paper 11″ × 1″, and give it four half-twists; join the edges together. How many edges and sides does this band have? Cut the band down the center. What is the result?

*Requires optional Section 7.7.

Date	Cultural History
1600	1600: Shakespeare's *Hamlet* 1609: Pocahontas saves John Smith 1611: King James bible published 1618: Beginning of the Thirty Years' War 1620: Pilgrims land at Plymouth 
1625	1636: Harvard College, the first American college 1637: Descartes, *Discourse on Method* 1643: Moliere founds *Théatre de la Comédie Française* 1646: Building of the Taj Mahal 1649: Cromwell abolishes English monarchy 
1650	1654: Coronation of Louis XIV of France 1656: Rembrandt declared bankrupt 1659: Birth of Alessandro Scarlatti 1665: Newton's experiments on gravitation 1668: La Fontaine's *Fables* 
1675	 1677: Spinoza's *Ethics* 1680: Stradavari makes the first cello 1685: J. S. Bach and Handel born 1689: Peter the Great tsar of Russia 1697: Yucatan remains of the Mayan culture obliterated by the Spanish 

Pocahontas

Taj Mahal

Rembrandt, *Lion Resting*

Spinoza

Bach, *Two-part Invention No. 5*

Mathematical History

1600: Galileo—physics, astronomy, projectiles

1610: Kepler—astronomy, continuity

1614: Napier—logarithms, Napier's rods

Kepler

Galileo

Mersenne

Oughtred

Napier

1630: Mersenne—number theory

1631: Oughtred—first table of natural logs

1635: Cavalieri—number theory
Fermat—number theory,
analytic geometry

1637: Descartes—analytic geometry

1640: Desargues—projective geometry

Fermat

Cavalieri

1650: Pascal—conics, probability,
computing machines,
Pascal's triangle
John Wallis—algebra,
imaginary numbers

Descartes

Pascal

1670: Sir Christopher Wren—architecture,
imaginary numbers

Wallis

Wren

1678: Ceva—nature of concurrency

1680: Sir Isaac Newton—calculus,
gravitation, series, hydrodynamics

1682: Gottfried Leibniz—calculus,
determinants, symbolic logic,
notation, computing machines

1690: Nicolaus Bernoulli—probability curves

Newton

Leibniz

8
The Nature
of
Counting

"Threes"

I think that I shall never c
A # lovelier than 3;
For 3 < 6 or 4,
And than 1 it's slightly more.

All things in nature
come in 3s,
Like ∴.s, trios, Q.E.D.'s;
While $s gain more dignity
If augmented 3 × 3—

A 3 whose slender curves
are pressed
By banks, for
compound interest;
Oh, would that, paying
loans or rent,
My rates were only 3%!

3^2 expands with
rapture free,
And reaches outward ∞;
3 complements each
x and y,
And intimately lives with π.

A ⊙'s # of °
Are best ÷ up by 3s,
But wrapped in
dim obscurity
Is the $\sqrt{-3}$.

Atoms are split by
men like me,
But only God is 1 in 3.

John Atherton

"Five trillion, four hundred eighty billion, five hundred twenty-three million, two hundred ninety-seven thousand, one hundred and sixty-two ..."

Reprinted with permission from The Saturday Evening Post Company, © 1976.

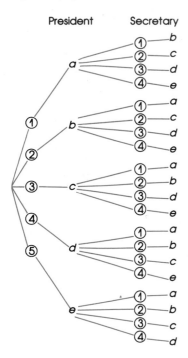

FIGURE 8.1 Tree diagram for election problem; a = Alfie, b = Bogie, c = Calvin, d = Doug, and e = Ernie

8.1 Counting Problems

In this chapter we'll investigate the nature of counting. It may seem like a simple topic for a math book, since everyone knows how to count. However, there are easier ways to count than that shown in the cartoon (see Problem 18). In this first section, we'll pose some counting problems and introduce the Fundamental Counting Principle

Consider a club with five members:

$$\{\text{Alfie, Bogie, Calvin, Doug, Ernie}\}$$

In how many ways could they elect a president and secretary? There are several ways to solve this problem. The first, and perhaps the easiest, is by making a tree diagram (see Figure 8.1). We see that there are 20 possibilities. This method is effective for "small" tree diagrams, but the technique quickly gets out of hand. For example, if we wished to see how many ways the club could elect a president, secretary, and treasurer, this technique would be very lengthy.

A second method of solution is by using boxes or "pigeon holes" representing each choice separately. Here we determine the number of ways of choosing a president and the number of ways of choosing a secretary.

Ways of choosing a president	Ways of choosing a secretary
5	4

↑
Since we have chosen a president, only 4 members remain.

If we multiply the numbers in the pigeon holes,

$$5 \cdot 4 = 20$$

and we see that the result is the same as that from the tree diagram. This method is much more satisfying and is an example of a general result called the **Fundamental Counting Principle.**

Fundamental Counting Principle

If task A can be performed in m ways, and, after task A is performed, a second task, B, can be performed in n ways, then task A followed by task B can be performed in $m \cdot n$ ways.

Example 1: How many skirt–blouse outfits can a woman wear if she has three skirts and five blouses?

Solution: There are two "tasks": (i) choosing a skirt, and (ii) choosing a blouse. We illustrate with two pigeon holes:

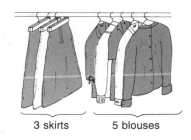

3 skirts   5 blouses

3	5
↑	↑
Skirts	Blouses

There are $3 \times 5 = 15$ skirt–blouse outfits.

Example 2: In how many ways could the given club choose a president, secretary, and treasurer?

Solution:

| 5 | × | 4 | × | 3 | = 60 |

↑	↑	↑
Ways of choosing a president	Ways of choosing a secretary	Ways of choosing a treasurer

Example 3: In how many ways could this club choose a president, vice-president, secretary, and treasurer?

Solution: $5 \cdot 4 \cdot 3 \cdot 2 = 120$

The State of California "ran out" of codes that could be issued on license plates. The scheme in California is practiced in many states; that is, license plates consist of three letters followed by three numerals. For example, CWB 072 is a license plate code. When the state ran out of available new numbers, a change had to take place. The decision was made to leave the old numbers in circulation and to issue new plates in the order of three numerals followed by three letters, such as

The problem is to determine how many plates were available before the switch.* The solution to this problem is found by using the Fundamental Counting Principle.

---

*Certain plates are never issued because the combinations of letters produce obscene or confusing words, such as CHP, which might be misinterpreted as a car of the California Highway Patrol. But in our discussion here, we will assume that all possible license plates were issued.

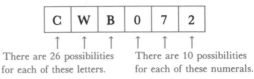

There are 26 possibilities   There are 10 possibilities
for each of these letters.   for each of these numerals.

Thus the total number of possible license plates that could be issued is:

$$26 \cdot 26 \cdot 26 \cdot 10 \cdot 10 \cdot 10 = 17{,}576{,}000$$

Example 4:   How many license plates can be formed if repetitions of letters or digits are not allowed?

Solution: The result is given by

$$26 \cdot 25 \cdot 24 \cdot 10 \cdot 9 \cdot 8 = 11{,}232{,}000$$

The Fundamental Counting Principle will not suffice in all counting situations. For example, consider that same club:

{Alfie, Bogie, Calvin, Doug, Ernie}

In how many ways could they elect a committee of two members?
One method of solution is easy (but tedious) and involves the enumeration of all possibilities.

{a,b}   {b,c}   {c,d}   {d,e}

{a,c}   {b,d}   {c,e}

{a,d}   {b,e}

{a,e}

We see that there are ten possible two-member committees. Do you see why we can't use the Fundamental Counting Principle for this committee problem in the same way we did for the election problem?

We have presented two different types of counting problems, the *election problem* and the *committee problem*. For the election problem, we found an easy numerical method of counting, using the Fundamental Counting Principle. For the committee problem, we did not; further investigation is necessary. Later sections of this chapter will illustrate a numerical method of counting for the committee problem and apply this method to still other counting problems. The task of this chapter, then, is to develop a method of "counting without counting."

Problem Set 8.1

When working these problems remember that the Fundamental Counting Principle will not work for some types of counting problems.

1.   State the Fundamental Counting Principle, and explain in your own words what it says.

2.   If a state issued license plates using the scheme of one letter followed by five digits, how many plates could it issue?

3. Answer Problem 2 if repetitions are not allowed.

4. Count the number of cubes in the structure shown in the margin. (Assume that the structure is solid.)

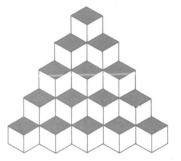

5. In how many ways can a group of 15 people elect a president, vice-president, and secretary?

6. How many two-member committees can be formed from a group of seven people?

7. How many squares (of any size) are in the design shown?

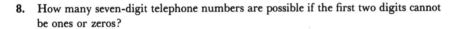

8. How many seven-digit telephone numbers are possible if the first two digits cannot be ones or zeros?

9. Foley's Village Inn offers the following menu in their restaurant:

Main Course	Dessert	Beverage
prime rib	ice cream	coffee
steak	sherbet	tea
chicken	cheesecake	milk
ham		Sanka
shrimp		

In how many different ways can someone order a meal consisting of one choice from each category?

10. Boats often relay messages by using flags on a flagpole. How many messages can be made using three different flags? (All three flags must be used.)

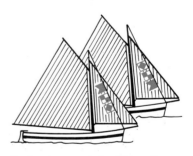

Each arrangement of these three flags is a "code" for some message. Two different messages are shown here. Can you find all other possible messages?

11. How many triangles are there in each figure?

a.

b.

c.
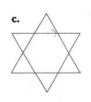

12. New York license plates consist of three letters followed by three numerals, and 245 letter arrangements are not allowed. How many plates can New York issue?

13. A typical Social Security number is 576-38-4459; how many Social Security numbers are possible if

  a. the first digit cannot be zero?
  b. neither of the first two digits can be a zero?

14. A club consists of four men and three women. In how many ways can this club elect a president, vice-president, and secretary, in that order, if

  a. the president must be a woman and the other two officers must be men?
  b. the president and the vice-president must be men and the secretary must be a woman?

15. How many triangles are there in the figure shown?

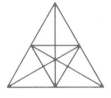

16. How many distinct arrangements are there of the letters in the word *math?* In the word *magnet?*

One face showing a 5

17. On page 183, we made the statement that 9^{9^9} bacteria would overflow the Milky Way, and the number of grains of sand on this and every other planet in our solar system would not be so large as this giant. How could we arrive at such a statement? We certainly could not obtain the result by direct counting. Give a convincing argument to support this statement.

18. The old lady jumping rope in the cartoon at the beginning of this section is counting one at a time. Assume that she jumps the rope 50 times per minute and that she jumps 8 hours a day, 5 days a week, 50 weeks a year. Estimate the length of time necessary for her to jump the rope the number of times indicated in the cartoon.

Two faces showing a total of 9

19. A die can be held so that one, two, or three of its faces can be seen at any one time. Is it possible to hold the die in different ways so that at different times the visible numbers add up to every number from 1 through 15?

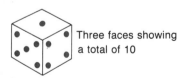
Three faces showing a total of 10

20. Outline a procedure that would allow the fellow in the cartoon to proclaim that he "did it without counting."

Mind Bogglers

21. Consider the contest from Problem 20. Estimate the number of beans without actually counting them. Assume that the container is a cylinder 15 in. tall and 10 in. across and that it is filled with kidney beans.

22. The marginal note on the next page presents an argument proving that at least two New Yorkers must have *exactly* the same number of hairs on their heads! For this problem, just *suppose* you *knew* the number of hairs on your head, as well as the *exact* number of hairs on anyone else's head. Now multiply the number of hairs on your head and the number of hairs on your neighbor's head. Take this result and

"I did it without counting. There are ---beans in the jar."

multiply by the number of hairs on the heads of each person in your town or country. Continue this process until you have done this for everyone in the entire world! Make a guess (you can use scientific notation if you like) about the size of this answer. [*Hint:* The author has worked this out and claims to know the exact answer!]

23. The advertisement shown here claims that the following 12 mix and match items give 122 different outfits:

   4 blouses
   2 slacks
   2 skirts
   1 sweater
   2 jackets
   1 scarf
   ___
   12 items total

**Nothing to Wear?
This Simple Wardrobe Will
Solve All Your Problems**

*12 Mix & Match Items
Give You 122 Outfits*

An example of reaching a conclusion about the number of objects in a set is given by M. Cohen and E. Nagel in *An Introduction to Logic* (London: Routledge & Kegan Paul Ltd., 1963). They conclude that there are at least two people in New York City who have the same number of hairs on their heads. This conclusion was reached not through counting the hairs on the heads of 8 million inhabitants of the city, but through studies revealing that: (1) the maximum number of hairs on the human scalp could never be as many as 5,000 per square centimeter, and (2) the maximum area of the human scalp could never reach 1,000 centimeters. We can now conclude that no human head could ever contain $5,000 \times 1,000 = 5,000,000$ hairs. Since this number is less than the population of New York City, it follows that at least two New Yorkers must have the same number of hairs on their heads!

Assume that the model must choose one top and one bottom item of clothing. She may or may not choose a sweater or a jacket or a scarf. How many different outfits are possible?

24. *Hexahexaflexagons.* Prepare a strip of paper as shown in figure a. Turn it over and mark the other side as shown in figure b.

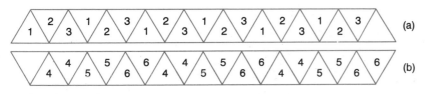

(a)

(b)

Strips for constructing a hexahexaflexagon. Make sure each of the numbered triangles is equilateral.

ℌistorical ℕote

In 1939 Arthur Stone, a 23-year-old graduate student from England, trimmed an inch from his American notebook sheets to make them fit into his English binder. For lack of something else to do, he began to fold the trimmed-off strips of paper in various ways. By rearranging them he came up with several interesting flexagons.

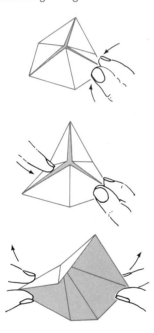

To "flex" your hexahexaflexagon, pinch together two of its triangles (top two figures). The inner edge may then be opened with the other hand (bottom). If the hexahexaflexagon cannot be opened, the adjacent pair of triangles is pinched. If it opens, turn it inside out, finding a side that was not visible before.

Starting from the left of figure b, fold the 4 onto the 4, the 5 onto the 5, 6 onto 6, 4 onto 4, and so on until your paper looks like the one shown in figures c and d.

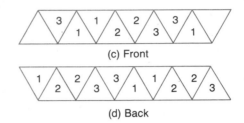

(c) Front

(d) Back

Continue by folding the 1 onto the 1 from the front, by folding the 1 onto the 1 from the back, and finally by bringing the 1 up from the bottom so that it rests on top of the 1 on the top. Your paper should now look like the one shown in figures e and f.

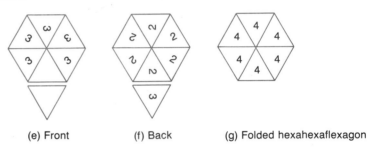

(e) Front   (f) Back   (g) Folded hexahexaflexagon

Paste the blank onto the blank, and the result is called a *hexahexaflexagon,* as shown in figure g.

With a little practice you'll be able to "flex" your hexahexaflexagon so that you can obtain a side with all 1s, all 2s, . . . , all 6s. After you have become fairly proficient at "flexing," count the number of flexes you require to obtain all six "sides." What do you think is the fewest flexes necessary to obtain all six sides? [*Note:* Hints on how to flex the hexahexaflexagon are given in the margin.]

Problem for Individual Study

25. *Paper folding.* How can all the constructions of Euclidean geometry be done by folding paper? What assumptions are made when paper folding is used to construct geometric figures? What polygons and polyhedrons can be formed by folding paper? What is a hexaflexagon? How can conic sections be formed by folding paper? What puzzles and tricks are based on paper folding?

Exhibit suggestions: Models, geometric constructions, polyhedrons, conic sections, puzzles formed by paper folding, the five regular solids.

References: Martin Gardner, *The Scientific American Book of Mathematical Puzzles and Diversions* (New York: Simon and Schuster, 1959), chapter 1.

Donovan Johnson, *Paper Folding for the Mathematics Class* (Washington, D.C.: National Council of Teachers of Mathematics, 1957).

T. Saundara Row, *Geometrical Exercises in Paper Folding,* rev. ed. (New York: Dover Publications, 1966).

8.2 Permutations

Consider the election problem of the previous section. We wish to elect a president, secretary, and treasurer from among

{Frank, George, Happy, Iggy, Jerry, Ken}

We list some possibilities:

President	Secretary	Treasurer
1. Frank	George	Happy
2. George	Frank	Happy
3. George	Frank	Iggy
.		
.		
.		

Certainly we see that direct listing would be too tedious. We seek an easier way. Is it possible to count the number of arrangements without actually counting them? We will look for some patterns in which the arrangement of the members of the set is important. That is, we will consider arrangements (1) and (2) as different; thus, the *order* in which we list the officers is important. Mathematicians call arrangements in which the order is important **permutations.**

We wish to choose a president, secretary, and treasurer. We say that this is a permutation of six objects (members of the club) taken three at a time (offices to be filled). We denote this by:

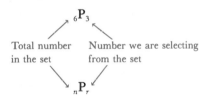

In general, $_nP_r$ is the number of permutations in which a total of n objects are under consideration and r objects are to be selected from the set. The problem, then, is to compute $_6P_3$ using the Fundamental Counting Principle.

President	Secretary	Treasurer
6	5	4

↑ Number of ways of selecting a president   ↑ Number of ways of selecting a secretary   ↑ Number of ways of selecting a treasurer

Thus, $_6P_3 = 6 \cdot 5 \cdot 4 = 120$.

Let's look for some patterns. Suppose the set had seven members. Then how many selections could we make? We see that we have

One more than $7 - 3$

$$_7P_3 = \underbrace{7 \cdot 6 \cdot 5}_{\text{Three factors}}$$

possibilities. What if the set contained ten members and we wished to elect four officers: a president, vice-president, secretary, and treasurer? Then we would have:

One more than $10 - 4$

$$_{10}P_4 = \underbrace{10 \cdot 9 \cdot 8 \cdot 7}_{\text{Four factors}}$$

Do you see a pattern? In general, we say that the number of permutations of n objects taken r at a time is given by:

One more than $n - r$

$$_nP_r = \underbrace{n \cdot (n - 1) \cdot (n - 2) \cdots (n - r + 1)}_{r \text{ factors}}$$

Example 1:   Evaluate the following.

a. $_5P_3 = 5 \cdot 4 \cdot 3 = 60$ **b.** $_6P_2 = 6 \cdot 5 = 30$
c. $_{10}P_3 = 10 \cdot 9 \cdot 8 = 720$ **d.** $_{10}P_2 = 10 \cdot 9 = 90$
e. $_{10}P_1 = 10$ **f.** $_{52}P_2 = 52 \cdot 51 = 2,652$

Example 2:   Can r be larger than n?
No, since n is the total number of objects.

Example 3:   Can $r = 0$?
You must be careful. This example doesn't exactly fit our definition, so we must interpret it separately. Now, $_6P_0$ means the number of ways you can permute six objects by taking none of them. There is only one way to take no objects, and that is by not taking any. Thus, we define

$$_nP_0 = 1$$

Example 4:   Can $r = n$?
Yes. Consider: $_5P_5 = 5 \cdot 4 \cdot 3 \cdot 2 \cdot 1 = 120$
                 $_6P_6 = 6 \cdot 5 \cdot 4 \cdot 3 \cdot 2 \cdot 1 = 720$

In our work with permutations, we will frequently encounter products such as

$$6 \cdot 5 \cdot 4 \cdot 3 \cdot 2 \cdot 1$$
$$10 \cdot 9 \cdot 8 \cdot 7 \cdot 6 \cdot 5 \cdot 4 \cdot 3 \cdot 2 \cdot 1$$
$$52 \cdot 51 \cdot 50 \cdot 49 \cdots 4 \cdot 3 \cdot 2 \cdot 1$$

Since these are rather lengthy things to write down, we use *factorial notation*.

Historical Note

Factorial notation was first used by Christian Kramp in 1808.

Definition of Factorial

For any counting number n,

$$n! = n(n - 1)(n - 2) \cdot \cdot \cdot 3 \cdot 2 \cdot 1$$

Also, $0! = 1$.

Notice that these numbers get big pretty fast. We need not wonder why the notation ! was chosen. For example,

$$52! \approx 8.065817517 \times 10^{67}$$

Example 5:   Find $5! - 4!$.

Solution: $5! = 5 \cdot 4 \cdot 3 \cdot 2 \cdot 1 = 120$
$\ \ \ 4! = 4 \cdot 3 \cdot 2 \cdot 1 = 24$
Thus, $5! - 4! = 120 - 24 = 96$.

Example 6:   Find $(5 - 4)!$.

Solution: Since $5 - 4 = 1$,   we have $1! = 1$.   Remember to work inside the parentheses first. Notice the difference between Examples 5 and 6.

Example 7:   Find $\dfrac{7!}{5!}$.

Solution: $7! = 7 \cdot 6 \cdot 5 \cdot 4 \cdot 3 \cdot 2 \cdot 1 = 5,040$
$\ \ \ 5! = 5 \cdot 4 \cdot 3 \cdot 2 \cdot 1 = 120$

Thus, $\dfrac{7!}{5!} = \dfrac{5,040}{120} = 42$.

A simpler solution arises by not multiplying the numbers out before dividing. That is, we could write:

$$\frac{7!}{5!} = \frac{7 \cdot 6 \cdot \cancel{5} \cdot \cancel{4} \cdot \cancel{3} \cdot \cancel{2} \cdot \cancel{1}}{\cancel{5} \cdot \cancel{4} \cdot \cancel{3} \cdot \cancel{2} \cdot \cancel{1}} = 7 \cdot 6 = 42$$

Example 8:   Find $\dfrac{10!}{7!}$.

Solution: $\dfrac{10!}{7!} = \dfrac{10 \cdot 9 \cdot 8 \cdot \cancel{7!}}{\cancel{7!}} = 10 \cdot 9 \cdot 8 = 720$

Notice that we could also write

$0! = 1$
$1! = 1$
$2! = 2 \cdot 1 = 2$
$3! = 3 \cdot 2 \cdot 1 = 6$
$4! = 4 \cdot 3 \cdot 2 \cdot 1 = 24$
$5! = 5 \cdot 4 \cdot 3 \cdot 2 \cdot 1 = 120$
$6! = 6 \cdot 5 \cdot 4 \cdot 3 \cdot 2 \cdot 1 = 720$
$7! = 7 \cdot 6! = 5,040$
$8! = 8 \cdot 7! = 40,320$
$9! = 9 \cdot 8! = 362,880$
$10! = 10 \cdot 9! = 3,628,800$

$n! = n(n - 1)!$

$$\frac{10!}{7!} = 10 \cdot 9 \cdot 8$$

directly by noticing the pattern that 10!/7! means count down (multiplying all the while) from 10 to 7 (but don't include 7).

Example 9:   Find $\dfrac{20!}{18!}$.

Solution: $\dfrac{20!}{18!} = 20 \cdot 19 = 380$

Using factorials, we can write some of our permutations a little more simply. For example, $_6P_6 = 6!$ and, in general, $_nP_n = n!$.

By looking for patterns, we can define $_nP_r$ in terms of factorial notation. Notice that

$$_7P_3 = 7 \cdot 6 \cdot 5 = \frac{7 \cdot 6 \cdot 5 \cdot 4!}{4!} \qquad \text{We multiplied by } 1 = \frac{4!}{4!}.$$

Thus,

$$_7P_3 = \frac{7!}{4!}$$

Also consider

$$_6P_3 = 6 \cdot 5 \cdot 4 = \frac{6!}{3!}$$

$$_{10}P_4 = 10 \cdot 9 \cdot 8 \cdot 7 = \frac{10!}{6!}$$

By studying this pattern, we see that it might be desirable to write these problems in terms of n and r.

$$_7P_3 = \frac{7!}{4!} = \frac{7!}{(7-3)!}$$

$$_6P_3 = \frac{6!}{(6-3)!}$$

$$_{10}P_4 = \frac{10!}{(10-4)!}$$

The general formula for this pattern is as follows.

Permutation Formula

$$_nP_r = \frac{n!}{(n-r)!}$$

Notice that this formula works even for

$$_nP_0 = \frac{n!}{(n-0)!} = \frac{n!}{n!} = 1$$

Example 10:   Find the number of license plates possible in a state, using only three letters if none of the letters can be repeated. This is a permutation of 26 objects taken 3 at a time. Thus the solution is given by

$$_{26}P_3 = \underbrace{26 \cdot 25 \cdot 24}_{3 \text{ factors}} = 15,600$$

Example 11:   Find the number of ways a baseball coach can arrange the batting order. This is a permutation of 9 objects taken 9 at a time. Thus,

$$_9P_9 = 9! = 362,880$$

Problem Set 8.2

Compute the results in Problems 1–20.

1. $7! - 5!$

2. $(7-5)!$

3. $\dfrac{10!}{8!}$

4. $\dfrac{12!}{9!}$

5. $6! - 5!$

6. $(10-4)!$

7. $\dfrac{10!}{6!}$

8. $\dfrac{10!}{4!6!}$

9. $\dfrac{52!}{3!(52-3)!}$

10. $\dfrac{12!}{3!(12-3)!}$

11. $6!$

12. $6 \cdot 5!$

13. $\dfrac{8!}{4!}$

14. $\left(\dfrac{8}{4}\right)!$

15. $8! - 4!$

16. $(8-4)!$

17. $\dfrac{8!}{5!3!}$

18. $\dfrac{10!}{5!}$

19. $\dfrac{9!}{6!3!}$

20. $\dfrac{52!}{5!47!}$

Evaluate each of the numbers in Problems 21–40.

21. $_9P_1$

22. $_9P_2$

23. $_9P_3$

24. $_9P_4$

25. $_9P_0$

26. $_5P_4$

27. $_{52}P_3$

28. $_7P_2$

29. $_4P_4$

30. $_{100}P_1$

31. $_{12}P_5$

32. $_5P_3$

33. $_8P_4$

34. $_8P_0$

35. $_gP_h$

36. $_{92}P_0$

37. $_{52}P_1$

38. $_7P_5$

39. $_{16}P_3$

40. $_nP_4$

41. A traveler must commute from Meridian to Centerville every day. There are four routes that he might take, as shown in the figure. This traveler would like to vary his trips as much as possible and always return on a different road. In how many different ways could he make the round trip?

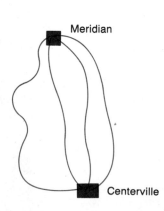
Meridian

Centerville

By "original position" we mean
that *all 52 cards are* in *exactly*
the same order as they were to
begin with.

42.   Write out a general formula for $_nP_r$ using only the numbers n, r, and the factorial. Show how you arrived at the answer.

43.   Give two original examples of arrangements of objects from everyday life that are permutations.

44.   A certain mathematics test consists of ten questions. Peppermint Patty wishes to answer the questions without reading them. In how many ways can she fill in the answer sheet if:

   a.   the possible answers are true and false?
   b.   the possible answers are true, false, and maybe?

Mind Bogglers

45.   *Radar puzzle.* Start on any one of the spaces marked R, and, by consecutive moves to adjacent squares, spell out the word *RADAR*. In how many ways can this be done?

$$
\begin{array}{ccccc}
R & A & D & A & R \\
A & D & A & R & A \\
D & A & R & A & D \\
A & R & A & D & A \\
R & A & D & A & R \\
\end{array}
$$

46.   *Bridge tournament.* Eight couples played bridge seven times. The people changed groupings every time they played, and, as it turned out, no man was ever paired with the same woman more than once as a partner or as an opponent. How were the couples distributed if no husband and wife were ever partners?

Problem for Individual Study

47.   *Card-shuffling problem.* Have you ever watched someone shuffle a deck of cards and wondered how well the deck was being mixed? Suppose we define a *perfect* riffle shuffle of a deck with an even number of cards.

   (1)   Divide the deck into two piles of equal size, putting the top half in your left hand and the bottom half in your right hand.
   (2)   Alternate the cards from the two stacks by selecting the bottom card in the right pile, then the bottom card in the left pile, and so on. When you finish this, the first and last cards will not have changed positions (see examples below).

If we repeatedly perform a perfect shuffle, how many shuffles are necessary to return a deck of 52 cards to its original position? [*Hint:* Start with simpler cases and look for a pattern.]

EXAMPLES:

Two cards (numbered 1 and 2)

Start: 1 2

Shuffle: 1 2

Original position after *one* shuffle

Four cards (numbered 1 to 4)

Start: 1 2 3 4

Shuffle: 1 3 2 4

Shuffle: 1 2 3 4

Original position after *two* shuffles

Ten cards (numbered 1 to 10)

Start: 1 2 3 4 5 6 7 8 9 10

Shuffle: 1 6 2 7 3 8 4 9 5 10

Shuffle: 1 8 6 4 2 9 7 5 3 10

Shuffle: 1 9 8 7 6 5 4 3 2 10

Shuffle: 1 5 9 4 8 3 7 2 6 10

Shuffle: 1 3 5 7 9 2 4 6 8 10

Shuffle: 1 2 3 4 5 6 7 8 9 10

Original position after *six* shuffles

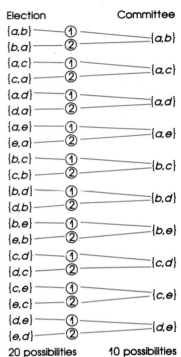

FIGURE 8.2 Comparison of the election and committee problems (two members)

8.3 Combinations

Let's compare the committee and election problems of Section 8.1. We saw in the preceding section, while considering the election problem, that the following arrangements were *different:*

President	Secretary	Treasurer
1. Alfie	Bogie	Calvin
2. Bogie	Alfie	Calvin

That is, the order was important. However, in the committee problem, the following three-member committees are the *same:*

1. Alfie	Bogie	Calvin
2. Bogie	Alfie	Calvin

In this case, the order is *not important.* When we list an arrangement of objects in which the order they are listed in is not important, we call the arrangement a **combination.**

Consider the example in Figure 8.2 comparing the election and committee problems. When we have a set of 5 elements taken 2 at a time, we see that we have twice as many permutations as combinations. That is,

$$_5C_2 = \frac{_5P_2}{2!}$$

Notice that we used the notation $_5C_2$ to mean the number of combinations of 5 objects taken 2 at a time.

Consider a set of 5 elements taken 3 at a time, as shown in Figure 8.3. We see that we have six times as many permutations as combinations. That is,

$$_5C_3 = \frac{_5P_3}{3!}$$

These examples suggest the following general rule.

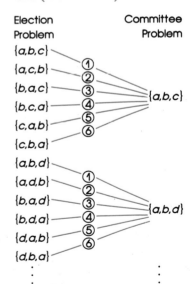

FIGURE 8.3 Comparison of the election and committee problems (three members)

Combination Formula

$$_nC_r = \frac{_nP_r}{r!}$$

That is, to find $_nC_r$, we proceed as follows:

1. Write $_nP_r$ in factored form.
2. Divide by $r!$.
3. Simplify the result.

Example 1: $_5C_3 = \dfrac{_5P_3}{3!} = \dfrac{5 \cdot 4 \cdot 3}{3 \cdot 2 \cdot 1} = 10$

Example 2: $_6C_2 = \dfrac{_6P_2}{2!} = \dfrac{6 \cdot 5}{2} = 15$

Example 3: $_{10}C_3 = \dfrac{_{10}P_3}{3!} = \dfrac{10 \cdot 9 \cdot 8}{3 \cdot 2 \cdot 1} = 120$

Example 4:   Find the number of ways different five-card hands can be drawn from an ordinary deck of cards.

$$_{52}C_5 = \frac{_{52}P_5}{5!} = \frac{52 \cdot 51 \cdot 50 \cdot 49 \cdot 48}{5 \cdot 4 \cdot 3 \cdot 2 \cdot 1} = 2{,}598{,}960$$

Example 5:   In how many ways can a diamond flush be drawn in poker? (A diamond flush is a hand of five diamonds.) This is a combination of 13 objects (diamonds) taken 5 at a time. Thus the solution is given by

$$_{13}C_5 = \frac{_{13}P_5}{5!} = \frac{13 \cdot 12 \cdot 11 \cdot 10 \cdot 9}{5 \cdot 4 \cdot 3 \cdot 2 \cdot 1} = 1{,}287$$

If n and r are large, then you can use the combination formula for $_nC_r$. However, for relatively small n and r (which will be most of the problems in this course) you should notice the following relationship:

$$_0C_0 = 1$$
$$_1C_0 = 1 \qquad _1C_1 = 1$$
$$_2C_0 = 1 \qquad _2C_1 = 2 \qquad _2C_2 = 1$$
$$_3C_0 = 1 \qquad _3C_1 = 3 \qquad _3C_2 = 3 \qquad _3C_3 = 1$$

Do you see a relationship between combinations and Pascal's triangle, which was first introduced in Section 1.1? Figure 8.4 shows Pascal's triangle down to row 15 so that you can use it for n and r up to 15.

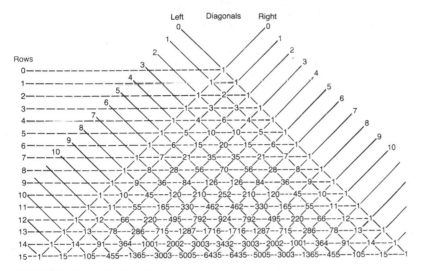

FIGURE 8.4 Pascal's triangle

In Examples 6–8, find the number of combinations using Pascal's triangle.

Example 6: $_8C_3$: row 8, diagonal 3: 56

Example 7: $_6C_5$: row 6, diagonal 5: 6

Example 8: $_{14}C_6$: row 14, diagonal 6: 3,003

In Problem 16 of Problem Set 8.1 we asked for the number of arrangements of the letters of the word *math*. The answer was found by using the Fundamental Counting Principle:

$$4 \cdot 3 \cdot 2 \cdot 1 = 24$$

We could say that this is a permutation of 4 objects taken 4 at a time:

$$_4P_4 = 4! = 24$$

However, suppose that the word you're considering has some repeated letters. For example, how many arrangements are there of the letters in the word *soon?* This is no longer a permutation problem because, although

$$_4P_4 = 4! = 24$$

you can find only 12 arrangements (as shown in the margin).

1. soon   7. oons
2. sono   8. onso
3. snoo   9. onos
4. oson   10. noos
5. osno   11. noso
6. oosn   12. nsoo

It seems that for a word with indistinguishable letters we need to divide by a factor, the number of arrangements of all letters, just as we did to find combinations. That is, if a word with n letters has one letter repeated a times, divide $_nP_n = n!$ by $a!$ For the word *soon*, for example, divide 4! by 2!:

$$\frac{4!}{2!} = \frac{24}{2} = 12$$

Example 9:   How many arrangements are there of the letters in the word *assist?*

Solution: There are six letters, and one letter (s) is repeated three times. Thus,

$$\frac{6!}{3!} = \frac{6 \cdot 5 \cdot 4 \cdot 3!}{3!} = 120$$

There are 120 arrangements of the letters in the word *assist*.

If there is more than one letter repeated, simply carry out the procedure indicated in Example 9 for *each* repeated letter. This procedure is shown in Example 10.

Example 10:   How many arrangements are there of the letters in the word *nonsense?*

Solution: The total number of letters is eight; the letter n is repeated three times; the letters s and e are each repeated two times. Thus, divide the total (8!) by 3! (for the letter n) *and* by 2! (for the letter s) *and also* by 2! (for the letter e):

$$\frac{8!}{3!2!2!} = \frac{8 \cdot 7 \cdot 6 \cdot 5 \cdot \cancel{4} \cdot \cancel{3} \cdot \cancel{2} \cdot \cancel{1}}{\cancel{3} \cdot \cancel{2} \cdot \cancel{1} \cdot \cancel{2} \cdot \cancel{1} \cdot \cancel{2} \cdot \cancel{1}} = 1,680$$

Examples 9 and 10 suggest a general result.

The number of permutations of n objects, of which a are alike, another b are alike, and another c are alike, is

$$\frac{n!}{a!b!c!}$$

Problem Set 8.3

Evaluate each of the expressions in Problems 1–20.

1. $_9C_1$	**2.** $_9C_2$	**3.** $_9C_3$	**4.** $_9C_4$	**5.** $_9C_0$
6. $_5C_4$	**7.** $_{52}C_3$	**8.** $_7C_2$	**9.** $_4C_4$	**10.** $_{100}C_1$
11. $_7C_3$	**12.** $_5C_5$	**13.** $_{50}C_{48}$	**14.** $_{25}C_1$	**15.** $_gC_h$

16. $_7C_5$ **17.** $_8C_0$ **18.** $_{10}C_2$ **19.** $_{12}C_5$ **20.** $_nC_4$

21. A bag contains 12 pieces of candy. In how many ways can 5 pieces be selected?

22. If the Senate is to form a new committee of 5 members, in how many different ways can the committee be chosen if all 100 Senators are available to serve on this committee?

23. How many permutations are there of the letters of the word *holiday?*

24. How many permutations are there of the letters of the word *annex?*

25. How many permutations are there of the letters of the word *eschew?*

26. How many permutations are there of the letters of the word *obfuscation?*

27. In how many ways can a dot and a dash be arranged into groups of three? For example, $\cdot\ \cdot\ \cdot$, $\cdot\ \cdot\ -$, $\cdot\ -\ \cdot$, and so on.

In Problems 28–33, suppose a jar contains 5 red and 6 green marbles.

28. In how many ways can 3 red marbles be drawn?

29. In how many ways can 4 green marbles be drawn?

30. In how many ways can 3 marbles be drawn?

31. In how many ways can 4 marbles be drawn?

32. In how many ways can 7 marbles be drawn?

33. In how many ways can 3 red marbles and 4 green marbles be drawn?

In Problems 34–38 you may need to refer to the following deck of cards:

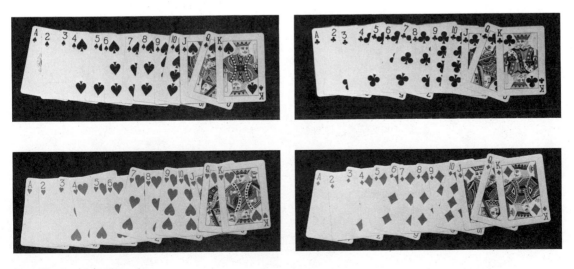

An ordinary deck of cards

34. In how many ways can three aces be drawn from a deck of cards?

35. In how many ways can two kings be drawn from a deck of cards?

36. In how many ways can a full house of three aces and two kings be obtained?

37. In how many ways can a heart flush be obtained? (A heart flush is a hand of five hearts.)

38. In how many ways can a flush be obtained?

39. Recall that $_nP_r$ can be written as $_nP_r = n!/(n-r)!$. Find a formula for $_nC_r$ in terms of n and r only (along with the factorial symbol).

40. Show that $_{n-1}C_{r-1} + {_{n-1}C_r} = {_nC_r}$.

Mind Bogglers

41. Prove that $_nC_r = {_nC_{n-r}}$.

42. *Calculator problem.* Suppose the population of the United States is 220 million. In how many ways could we elect a president and a vice-president? (Assume that everyone is eligible to be elected.) How many committees of two persons could we form?

Problems for Individual Study

43. Find a formula for the largest number in the nth row of Pascal's triangle.

*44. *Continuation of the card-shuffling problem.* If you tabulate the results of Problem 47 of Problem Set 8.2, you find:

Number of Cards	Number of Shuffles
4	2
6	4
8	3
10	6
12	10
14	12
16	4
.	
.	
.	

The pattern is not as obvious as we had hoped. So let's approach the problem another way. Label the first 10 cards:

$$1 \quad 2 \quad 3 \quad 4 \quad 5 \quad 6 \quad 7 \quad 8 \quad 9 \quad 10$$

---

*This problem requires Section 6.6.

Trace the position of the card we labeled "2." After successive shuffles the card labeled "2" is in the following positions:

2-3-5-9-8-6-2

This means that the second card moves to the third position after one shuffle, the fifth position after two shuffles, and so on. When it returns to the second position, all the cards will be in their original position, so we see that 10 cards require 6 shuffles. Find a pattern that predicts the position of the second card.

EXAMPLE: For 12 cards we would write: 2-3-5-9-6-11-10-8-4-7-2, which is 10 shuffles.

EXAMPLE: For 30 cards: 2-3-5-9-17-4-7-13-25-20-10-19-8-15-29-28-26-22-14-27-24-18-6-11-21-12-23-16-2, which is 28 shuffles.

8.4 Counting without Counting

This chapter is concerned with methods of counting that are more efficient than the old "one, two, three, . . ." technique. There are many ways of counting. We saw that some counting problems can be solved by using the Fundamental Counting Principle, some by using permutations, and some by using combinations. Other problems required that we combine some of these ideas. However, it is important to keep in mind that not all counting problems fall into one of these neat categories.

Keep in mind, therefore, as we go through this section, that, although some problems may be permutation problems and some may be combination problems, there are many counting problems that are neither.

In practice, we are usually required to decide whether a given counting problem is a permutation or a combination before we can find a solution. For the sake of review, recall the difference between the election and committee problems of the previous sections. The election problem is a permutation problem, and the committee problem is a combination problem. We summarize the definitions of permutation and combination.

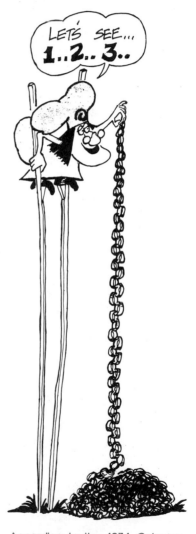

According to the 1974 *Guinness Book of World Records,* the longest recorded paper-link chain was 6,077 ft long and was made by the first- and second-grade children at Rose City Elementary School, Indiana.

Definition

A *permutation* of a set of objects is an arrangement of certain of these objects in a *specified order.*

Definition

A *combination* of a set of objects is an arrangement of certain of these objects *without regard to their order.*

Remember: Permutation: Order *is* important.
                        Combination: Order is *not* important.

Example 1:   Classify the following as permutations, combinations, or neither.

a. The number of three-letter "words" that can be formed using the letters {*m,a,t,h*}.
b. The number of ways you can change a $1 bill with 5 nickels, 12 dimes, and 6 quarters.
c. The number of ways a five-card hand can be drawn from a deck of cards.
d. The number of different five-numeral combinations on a combination lock.
e. The number of license plates possible in Florida.

Solutions:

a. Permutation, since *mat* is different from *tam*. (In the examples that follow, any combination of letters is considered a "word." Thus *atm* is also a "word.")
b. Combination, since "2 quarters and 5 dimes" is the same as "5 dimes and 2 quarters."
c. Combination, since the order in which you receive the cards is unimportant.
d. Permutation, since "4 to the L, 6 to the R, 3 to the L, . . ." is different from "6 to the R, 4 to the L, 3 to the L, . . ." We should not be misled by everyday usage of the word *combination*. We made a strict distinction between combination and permutation—one that is not made in everyday terminology. (The correct terminology would require that we call these "permutation locks.")
e. Neither; even though the *order* in which the elements are arranged is important, this does not actually fit the definition of a permutation because the objects are separated into two categories (pigeon holes). The arrangement of letters is a permutation and the arrangement of numerals is a permutation, but to count the actual number of arrangements for this problem would require permutations *and* the Fundamental Counting Principle.

Combination lock or permutation lock?

Sometimes it is easier to count what we are not counting than to enumerate all those items with which we are concerned. For example, suppose we wish to know the number of four-member committees that can be appointed from the club:

{Alfie,Bogie,Calvin,Doug,Ernie}

1. We could count directly:

    a. Alfie, Bogie, Calvin, Doug
    b. Alfie, Bogie, Calvin, Ernie
    c. Alfie, Bogie, Doug, Ernie
              ·
              ·
              ·

2. We could use the formula:

$$_5C_4 = \frac{_5P_4}{4!} = \cdots$$

3. We could count those we are not counting: For each four-member committee, there is one person left out. We can leave out Alfie, Bogie, Calvin, Doug, or Ernie. Therefore there are five four-member committees.

Example 2:   Suppose you flip a coin three times and keep a record of the results. In how many ways can you obtain at least one head?

Solution: Consider all possibilities:

$$
\begin{array}{ccc}
\text{H} & \text{H} & \text{H} \\
\text{H} & \text{H} & \text{T} \\
\text{H} & \text{T} & \text{H} \\
\text{H} & \text{T} & \text{T} \\
\text{T} & \text{H} & \text{H} \\
\text{T} & \text{H} & \text{T} \\
\text{T} & \text{T} & \text{H} \\
\text{T} & \text{T} & \text{T}
\end{array}
$$

You could count directly to obtain the answer, but you could also count what you are not counting by noticing that in only one out of eight possibilities do you obtain no heads. Thus there are $8 - 1 = 7$ possibilities in which you obtain at least one head.

Example 3:   How many Texas license plates of three letters and three numerals have a repeating letter or numeral?

Solution: The direct solution is difficult, but it's easy to count how many don't:

$$26 \times 25 \times 24 \times 10 \times 9 \times 8 = 11{,}232{,}000$$

From Example 4 of Section 8.1, there are 17,576,000 plates possible, so the number of plates that have a repeating letter is $17{,}576{,}000 - 11{,}232{,}000 = 6{,}344{,}000$. This figure is almost 36% of the total number of license plates in the state!

Example 4:   The principle of counting without counting is particularly useful when the results become more complicated. For example, suppose we wish to know the number of ways of obtaining at least one diamond on drawing five cards from an ordinary deck of cards. This is very difficult if we proceed directly, but we can compute the number of ways of not drawing a diamond:

$$_{39}C_5 = \frac{39 \cdot 38 \cdot 37 \cdot 36 \cdot 35}{5 \cdot 4 \cdot 3 \cdot 2 \cdot 1} = 575{,}757$$

From Example 4 of Section 8.3, there is a total of 2,598,960 possibilities so the number of ways of drawing at least one diamond is 2,598,960 − 575,757 = 2,023,203.

Another useful counting scheme occurs when you count the number of objects by first counting too much.* Suppose

$$U = \{1,2,3,4,5,6,7,8,9,10\}$$
$$A = \{2,4,6,8,10\}$$
$$B = \{1,3,5,7,9\}$$
$$C = \{1,2,3\}$$

If $|X|$ represents the number of elements in the set X, then $|U| = 10$, $|A| = 5$, $|B| = 5$, and $|C| = 3$. Furthermore, we see that $|A \cap B| = 0$, $|A \cap C| = 1$, $|B \cap C| = 2$. We wish to know about the number of elements in the union of these sets.

First, we work the problem by counting:

$$A \cup B = \{1,2,3,4,5,6,7,8,9,10\}$$
$$A \cup C = \{1,2,3,4,6,8,10\}$$
$$B \cup C = \{1,2,3,5,7,9\}$$

Thus,

$$|A \cup B| = 10$$
$$|A \cup C| = \ 7$$
$$|B \cup C| = \ 6$$

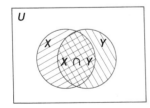

FIGURE 8.5 Venn diagram for the number of elements in the union of two sets

If these sets were very large, the counting method would not be satisfactory. Can we work the problem without counting? Consider the general Venn diagram shown in Figure 8.5. Notice that, if we count X and then count Y, the intersection has been counted twice. Therefore we write:

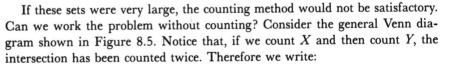

$$|X \cup Y| = \quad \underbrace{|X| + |Y|}_{\uparrow} \quad - \quad \underbrace{|X \cap Y|}_{\uparrow}$$

The elements in the intersection are counted twice here.

This corrects for the "error" introduced by counting those elements in the intersection twice.

Thus,

$$|A \cup B| = |A| + |B| - |A \cap B|$$
$$= 5 + 5 - 0$$
$$= 10$$

*Section 2.1 is required for the remainder of this section. This material can be omitted without any loss of continuity.

$$|A \cup C| = |A| + |C| - |A \cap C|$$
$$= 5 + 3 - 1$$
$$= 7$$
$$|B \cup C| = |B| + |C| - |B \cap C|$$
$$= 5 + 3 - 2$$
$$= 6$$

Example 5: What is the millionth positive integer that is not the square or cube of an integer?

Solution: We will solve this problem by counting what we are not counting; that is, we will count the squares and cubes and eliminate those from the rest of the numbers:

$$\cancel{1},2,3,\cancel{4},5,6,7,\cancel{8},\cancel{9},10,11,12,\ldots$$

Squares	Cubes
$1 = 1^2$	$1 = 1^3$
$2 = 2^2$	$8 = 2^3$
$9 = 3^2$	$27 = 3^3$
$16 = 4^2$	$64 = 4^3$
$25 = 5^2$	$125 = 5^3$
.	.
.	.
.	.
$1,000,000 = (10^3)^2$	$1,000,000 = (10^2)^3$

If $S = \{Squares\}$, then $|S| = 1,000$, and if $C = \{Cubes\}$, then $|C| = 100$. (Why?) We find $S \cap C$ as follows:

$$1 = 1^6$$
$$64 = 2^6$$
$$729 = 3^6$$
$$.$$
$$.$$
$$.$$
$$1,000,000 = 10^6$$

Thus, $|S \cap C| = 10$.

Now we can find

$$|S \cup C| = |S| + |C| - |S \cap C|$$
$$= 1,000 + 100 - 10$$
$$= 1,090$$

Since the next square and cube ($1,001^2 = 1,002,001$ and $101^3 = 1,030,301$, respectively) are both greater than 1,001,090, we see that there are no additional numbers to be excluded between 1,000,000 and 1,001,090. Thus the millionth number that is not a square or a cube is 1,001,090.

Problem Set 8.4

1. Explain the difference between a permutation and a combination.

2. Classify each of the following as permutations or combinations or neither.

   a. The number of arrangements of letters in the word *math*.
   b. At Mr. Furry's Dance Studio, every man must dance the last dance. If there are five men and eight women, in how many ways can dance couples be formed for the last dance?
   c. Martin's Ice Cream Store sells sundaes with chocolate, strawberry, butterscotch, or marshmallow toppings, nuts, and whipped cream. If you can choose exactly three of these, how many possible sundaes are there?
   d. Five people are to dine together at a rectangular table, but the hostess cannot decide on a seating arrangement. In how many ways can the guests be seated?
   e. In how many ways can three hearts be drawn from a deck of cards?

3. Classify each of the following as permutations or combinations or neither.

   a. A shipment of 100 TV sets is received. Six sets are to be chosen at random and tested for defects. In how many ways can the six sets be chosen?
   b. New students must register in person and be processed through three stages. At the first stage there are four tables (the student can go to any one of these), at the second stage there are six tables, and at the last stage there are five tables. In how many ways could the student be routed through the registration procedures?
   c. How many subsets can be formed from a set of 100 elements?
   d. A night watchman visits 15 offices every night. To prevent others from knowing when he will be at a particular office, he varies the order of his visits. In how many ways can this be done?
   e. A certain manufacturing process calls for the mixing of six chemicals. One liquid is to be poured into the vat, and then the others are to be added in turn. All possibilities must be tested to see which gives the best results. How many tests must be performed?

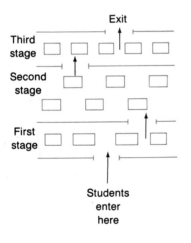

4. Classify each of the following as permutations or combinations or neither.

   a. There are three boys and three girls at a party. In how many ways can they be seated if they can sit down four at a time?
   b. If there are ten people in a club, in how many ways can they choose a dishwasher and a bouncer?
   c. What is the number of three-digit numbers that can be formed using the numerals {3,5,6,7} without repetition?
   d. In how many ways can you be dealt two cards from an ordinary deck of cards?
   e. In how many ways can five taxi drivers be assigned to six cars?

5. Classify each of the following as permutations or combinations or neither.

   a. How many arrangements are there of the letters in the word *gamble?*
   b. A student is asked to answer 10 out of 12 questions on an exam. In how many ways can she select the questions to be answered?
   c. In how many ways can a group of seven choose a committee of four?
   d. In how many ways can seven books be arranged on a bookshelf?

e. In how many ways can we choose two books to read from a bookshelf containing seven books?

6. Use the notation $_nP_r$ or $_nC_r$ to write each of the parts of Problem 2, showing the appropriate choice for n and r. For example, Problem 2a is a permutation of 4 objects (letters, in this case) taken 4 at a time. Thus the appropriate notation is $_4P_4$.

* 7. If $U = \{1,2,3,\ldots,100\}$, $A = \{\text{Multiples of 2}\}$, and $B = \{\text{Multiples of 3}\}$, find:

    a. $|A|$ **b.** $|B|$ **c.** $|A \cap B|$ **d.** $|A \cup B|$

8. A certain lock has five tumblers, and each tumbler can assume six positions. How many different possibilities are there?

9. Suppose you flip a coin and keep a record of the results. In how many ways could you obtain at least one head if you flip the coin four times?

10. You flip a coin five times and keep a record of the results. In how many ways could you obtain at least one head?

11. You flip a coin six times and keep a record of the results. In how many ways could you obtain at least one tail?

12. You flip a coin seven times and keep a record of the results. In how many ways could you obtain at least one tail?

13. You flip a coin n times and keep a record of the results. In how many ways could you obtain at least one head?

14. Answer the questions posed in Problem 2.

15. Answer the questions posed in Problem 4.

16. Answer the questions posed in Problem 5.

17. **a.** In how many ways could a club of 15 members choose a president, vice-president, secretary, and treasurer?
    b. In how many ways could a committee of 4 be appointed by this club?

18. Compute each of the following (you may leave your answer in factored form):

    a. $_{52}P_5$ **b.** $_{52}C_5$ **c.** $_nP_n$ **d.** $_nC_n$ **e.** $_nC_2$

19. Assume that this class has 30 members.

    a. In how many ways could we select a president, vice-president, and secretary?
    b. How may committees of three persons can we choose?

*20. In the text we found $|A \cup B| = |A| + |B| - |A \cap B|$. Show that

$$|A \cup B \cup C|$$
$$= |A| + |B| + |C| - |A \cap B| - |A \cap C| - |B \cap C| + |A \cap B \cap C|$$

Hint for Problem 20: Use the associative property to write $A \cup B \cup C$ as $(A \cup B) \cup C$. You will also need the distributive property:

$$(A \cup B) \cap C$$
$$= (A \cap C) \cup (B \cap C)$$

*21. What is the millionth positive integer that is not a square, cube, or fifth power? This problem requires the formula from Problem 20.

*These problems require Section 2.1.

22. On January 24, 1971, The Santa Rosa *Press Democrat* began the contest shown in the clipping. If we assume that the contest was to be solved by chance (no skill involved), how many possibilities for matching the baby to the adult are there?

Match the Mayors

HEY, here's your chance for some fun—and maybe a prize. Join The Press Democrat's MATCH THE MAYOR contest. During the next 10 days we'll print a picture of a well-known Mid-Empire mayor every day. And with it we'll include a picture of a baby—just like the one today.

Now the baby is one of the 10 mayors. And your job is to match the baby with the correct mayor. Do it correctly, through all 10, and you'll win a prize. And that's not so tough, but it's not so easy either. We'll tell you about the prizes tomorrow. So clip this first box today and you're on your way.

COL. W. R. LUCIUS
Healdsburg Mayor

Mind Bogglers

23. Answer the problem posed in Peppermint Patty's textbook.

24. The advertisement reproduced here appeared in several national periodicals. It

claims that the SEA graphic equalizer system can create 371,293 different sounds. Can this claim be substantiated from the advertisement and the knowledge of this chapter? If so, explain how; if not, tell why this number is impossible.

25. While the Dormouse lay sleeping in the middle of the table, the Mad Hatter and the March Hare sat down for tea. A number of places had been laid, and the Hatter and the Hare each moved alternately to a seat next to his old one. Each changed his direction of movement when, and only when, the other had just changed his or to avoid their both sitting on the same chair. They finished when both the Hatter and the Hare were again sitting in their initial positions. After how many moves was the tea party over?[†]

Problems for Individual Study

26. You have read about many mathematicians in the historical notes of this book. Write a news article about one of them as if you were a contemporary of the person you are writing about. You will need to do some additional research for this problem, but you can use one of the historical notes in this book as a starting point.

27. Prepare a list of black mathematicians from the history of mathematics.

28. Prepare a list of women mathematicians from the history of mathematics.

*8.5 Rubik's Cube and Instant Insanity

In the summer of 1974 an Hungarian architect invented a three-dimensional object that could rotate about *all three* axes (sounds impossible, doesn't it?). He wrote up the details of the cube and obtained a patent in 1975. The cube is now known worldwide as Rubik's Cube, and in the United States the Ideal Toy Corp. holds the manufacturing rights. In case you have not seen one of these cubes, it is shown in Figure 8.6.

When you purchase the cube it is arranged so that each face is showing a different color, but after a few turns it seems next to impossible to return to the start. In fact, Ideal claims there are 8.86×10^{22} possible arrangements of which there is only one correct solution. This claim is incorrect; there are 2,048 possi-

---

[†]This problem was proposed by R. N. Lloyd in *The American Mathematical Monthly*, January 1970, and was solved by R. W. Sielaff in the November 1970 *Monthly*.

*This is an optional section. The term "Rubik's Cube" is a trademark of Ideal Toy Corporation, Hollis, New York. The term "Rubik's Cube" as used in this book means the cube puzzle sold under any trademark.

The puzzle known as Rubik's
Cube was designed by Ernö
Rubik, an architect and teacher
in Budapest, Hungary. It was
also invented independently by
Terutoshi Ishige, an engineer in
Japan. Both applied for patents
in the mid-1970s. The cubes were
first manufactured in Hungary. In
1978, a Hungarian mathematics
professor brought several with
him to the International Congress
of Mathematicians in Helsinki,
Finland. This formally introduced
the cube to Europe and the rest
of the world. They have been
widely available in the United
States only since 1980.

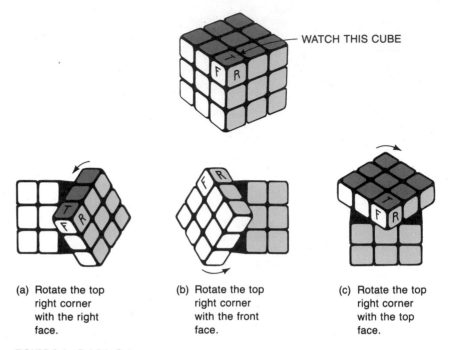

WATCH THIS CUBE

(a) Rotate the top
     right corner
     with the right
     face.

(b) Rotate the top
     right corner
     with the front
     face.

(c) Rotate the top
     right corner
     with the top
     face.

FIGURE 8.6   Rubik's Cube

ble solutions among the 8.86×10^{22} arrangements (actually, $8.85801027 \times 10^{22}$). This means there is one solution for each 4.3×10^{19} arrangements (actually, 43,252,003,274,489,856,000).

Example 1:   If it took you one-half second for each arrangement, how long would it take you to move the cube into all arrangements?

Solution: $4.3 \ \times 10^{19}$ Arrangements
       $= 2.15 \times 10^{19}$ Divide by 2 for the number of seconds.
       $\approx 3.58 \times 10^{17}$ Divide by 60 for the number of minutes.
       $\approx 5.97 \times 10^{15}$ Divide by 60 for the number of hours.
       $\approx 2.49 \times 10^{14}$ Divide by 24 for the number of days.
       $\approx 6.81 \times 10^{11}$ Divide by $365\frac{1}{4}$ for the number of years (counting leap years).
       $\approx 6.81 \times 10^{9}$ Divide by 100 for the number of centuries.
       $\approx 6.81 \times 10^{6}$ Divide by 1,000 for the number of millenniums.

Our problem ends here because you would require over

6,810,000 milleniums!

The age of the earth has not been estimated to be near this age. If a high-speed computer listed 1,000,000 arrangements per minute, it would take it over 800 milleniums to print out the possibilities. (It has only been 2 milleniums since Christ!)

Let's consider what we will call the standard position as shown in Figure 8.7. Label the faces Front (F), Right (R), Left (L), Back (B), Top (T), and Under (U), as shown.

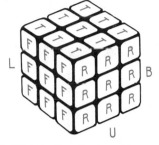

FIGURE 8.7   Standard position cube

Hold the cube in your left hand with T up and F toward you so that L is against your left palm. Now describe the results of the moves in Examples 2–6.

Example 2:   Rotate the right face 90° clockwise; denote this move by R.

Solution:

Return the cube to standard position. We denote this move by R^{-1}.

Example 3:   Rotate the right face 180° clockwise; denote this by R^2.

Solution:

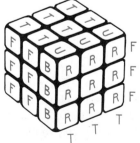

Return the cube to standard position by doing another R^2. Notice that

$$R^2 R^2 = R^4$$

which returns the cube to standard position.

Example 4:   Rotate the top 90° clockwise; call this T.

Solution:

Return the cube (T^{-1}).

Example 5: TR means rotate the top face 90° clockwise, *then* rotate the right face 90° clockwise.

Solution:

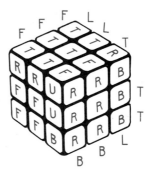

Example 6:   Describe the steps necessary to return the cube shown in Example 5 to standard position.

Solution: $R^{-1}T^{-1}$

Historical Note

Instant Insanity is not a new puzzle; it has appeared under several patents, each in a different form. The present one is a trademark of Parker Brothers, Inc. Other versions involve numbers, dots, groceries, or card suits on the faces. All these forms are mathematically equivalent.

Now, a rearrangement of the 54 colored faces of small cubes is a *permutation*. We discussed permutations of a square in Problem 25, Section 6.1, at which time we called them symmetries of a square. The same type of algebraic structure can be used to find a solution to Rubik's Cube. Although such a discussion is beyond the scope of this course, you can consult Problem 43 in Problem Set 8.5 if you are interested.

An older puzzle, simpler than Rubik's Cube, is called Instant Insanity. It provides four cubes colored red, white, blue, and green, as indicated in Figure 8.8. The puzzle is to assemble them into a $1 \times 1 \times 4$ block so that all four colors appear on each side of the block.

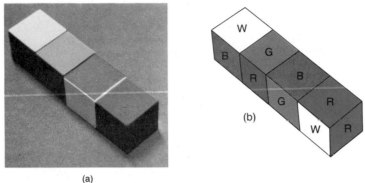

(a)

FIGURE 8.8

Example 7:   In how many ways can this Instant Insanity puzzle be arranged?

Solution: A common mistake is to assume that a cube can have 6 different arrangements (because of our experience with dice). In the case of Instant Insanity, we are interested not only in the top, but also with the other sides. There are 6 possible faces for the top and *then* 4 possible faces for the front. Thus, by the Fundamental Counting Principle, there are

$$6 \cdot 4 = 24 \text{ arrangements for one cube}$$

Now, for four cubes, again use the Fundamental Counting Principle to find

$$24 \cdot 24 \cdot 24 \cdot 24 = 331{,}776$$

This is *not* the number of *different* possibilities. You are asked to find this number in Problem 1 of Problem Set 8.5.

Problem Set 8.5

1. Example 7 shows that Instant Insanity has 331,776 possibilities (not all different). If you could make one different move per second, how long would it take to go through all of these arrangements?

2. Suppose a computer could print out 1,000,000 arrangements in Problem 1 every minute. How long would it take this compouter to print out all possibilities?

Show the result of the move indicated in Problems 3–26. Remember that R, F, L, B, T, and U mean rotate 90° clockwise the right, front, left, back, top, and under faces, respectively. Use the cube shown in Figure 8.7 as your starting point.

3. F 4. B 5. U 6. L 7. B^{-1} 8. F^{-1}

9. L^{-1} 10. T^{-1} 11. F^2 12. T^2 13. L^2 14. B^2

15. F^3 **16.** R^3 **17.** T^3 **18.** U^3 **19.** RL **20.** TU

21. FB^{-1} **22.** $F^{-1}B$ **23.** RT **24.** FT **25.** FT^{-1} **26.** $F^{-1}T^{-1}$

*In Problems 27–34, name the move or moves that will return the cube to standard
position, as shown in Figure 8.7.*

27. Problem 3   **28.** Problem 12   **29.** Problem 15

30. Problem 20   **31.** Problem 21   **32.** Problem 24

33. Problem 25   **34.** Problem 26

*35. Are two consecutive moves on Rubik's Cube commutative?

*36. Are two consecutive moves on Rubik's Cube associative?

37. In Example 6, $R^{-1}T^{-1}$ reversed the moves in Example 5. Would $T^{-1}R^{-1}$ also
reverse the moves in Example 5? Why or why not?

38. Find the number of *different* arrangements for the Instant Insanity blocks.

Mind Bogglers

39. The most difficult part of understanding a solution to Rubik's Cube is understanding the notation used by the author in stating the solution. In one solution, the
Ledbetter-Nering Algorithm (see the references in Problem 43), the authors describe a sequence of moves which they call THE MAD DOG. This series of moves
can be described using our notation by

$$(RTF)^5$$

Start with the cube shown in Figure 8.7 and show the result after carrying out
THE MAD DOG.

40. David Singmaster in his book *Notes on Rubik's "Magic Cube"* (see the references
in Problem 43) describes a sequence of moves to put the upper corners in place
(after certain other moves). Using our notation this set of moves is

$$FU^2F^{-1}U^2R^{-1}UR$$

Start with the cube shown in Figure 8.7 and show the result after carrying out this
sequence of moves.

41. In *The Simple Solution to Rubik's Cube* (see the references in Problem 43), James
Nourse describes a process to orient the bottom corner cubes. Using our notation,
this move is

$$R^{-1}U^{-1}RU^{-1}R^{-1}U^2RU^2$$

Start with the cube shown in Figure 8.7 and show the result after carrying out this
sequence of moves.

---

*These problems require Section 6.1.

Problems for Individual Study

42. *A numerical solution to Instant Insanity.* Let's associate numbers with the sides of the cubes of the Instant Insanity problem. Let

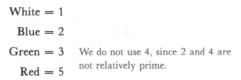

$$\text{White} = 1$$
$$\text{Blue} = 2$$
$$\text{Green} = 3 \quad \text{We do not use 4, since 2 and 4 are}$$
$$\text{Red} = 5 \quad \text{not relatively prime.}$$

Now, the product across the top must be 30, and the product across the bottom must also be 30 (why?). It follows that a solution must have a product of 900 for the faces on the top and bottom. Consider the four cubes:

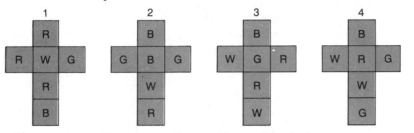

These cubes do not show the solution.

Here are the possible products of top and bottom for cube 1:

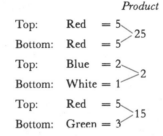

Product

| Top: | Red | = 5 | |
| Bottom: | Red | = 5 | 25 |

| Top: | Blue | = 2 | |
| Bottom: | White | = 1 | 2 |

| Top: | Red | = 5 | |
| Bottom: | Green | = 3 | 15 |

By a clever and systematic analysis of the products of top and bottom for cubes 2, 3, and 4, you will find that there are several ways of solving the top and bottom for a product of 900. Only one of these gives a product of 900 for front and back. Find the solution by using this method.

43. *A solution to Rubik's Cube.* Consult one of the references and learn to solve Rubik's Cube. Demonstrate your skill to the class. Nourse names the following categories:

20 minutes—WHIZ   5 minutes—EXPERT

10 minutes—SPEED DEMON   3 minutes—MASTER OF CUBE

References: Ledbetter and Nering, *The Solution to Rubik's Cube* (Rohnert Park, Ca.: Noah's Ark Enterprises, 1980).

James G. Nourse, *The Simple Solution to Rubik's Cube* (New York: Bantam Books, 1981).

David Singmaster, *Notes on Rubik's "Magic Cube,"* 5th ed. (Hillside, N.J.: Enslow Publishers, 1980).

Chapter 8 Outline

I. Counting Problems
  A. Fundamental Counting Principle
  B. Election problem
  C. License plate problem
  D. Committee problem
II. Permutations
  A. Order important (election problem)
  B. Notation ${}_nP_r$
  C. Permutation formula
    1. Factoral notation
    2. Simplifying expressions
III. Combinations
  A. Order not important (committee problem)
  B. Notation ${}_nC_r$
  C. Combination formula
  D. Pascal's triangle for combinations
  E. Counting permutations with indistinguishable elements
IV. Counting without Counting
  A. Distinguishing permutations, combinations, or neither
  B. Counting by counting what you are not counting
  C. Counting by counting too much
  D. Counting the union of two sets
***V.** Rubik's Cube and Instant Insanity
  A. Number of arrangements
  B. Permutations of the cube
  C. Instant Insanity

Chapter 8 Test

1. Simplify each of the given expressions.

a. $6! - 2!$ **b.** $6 - 2!$ **c.** $(6 - 2)!$ **d.** $\left(\frac{6}{2}\right)!$ **e.** $\frac{6!}{2!}$

2. Find the numerical value of each expression.

a. ${}_5C_3$ **b.** ${}_4P_2$ **c.** ${}_2P_0$ **d.** ${}_4C_2$ **e.** ${}_{12}C_8$

3. Discuss and define permutations and combinations.

4. **a.** How many 3-member committees can be chosen from a group of 12 people?
b. In how many ways can 5 people line up at a bank teller's window?

*Optional section.

5. How many permutations are there of the letters of the words *happy* and *college?*

6. The Wednesday Luncheon Club, consisting of ten members, decided to celebrate its first anniversary by having lunch at a fancy restaurant. When the members arrived and were ready to take their seats, they could not decide where to sit.

   Just when they were ready to leave because of their embarrassment at being unable to decide how to sit around the rectangular table, the manager came to the rescue and told them to sit down just where they were standing. Then the secretary of the club was to write down where they were sitting, and they were to return the following day and sit in a different order. If they continued this until they had tried all arrangements, the manager would give them anything on the menu, free of charge.

   It sounded like a very good deal, so they agreed. However, the day of the free luncheon never came. Can you explain why?

7. A jar contains four red and six white balls. Three balls are drawn at random. In how many ways can at least one red ball be drawn?

* 8. If $U = \{1,2,3, \ldots ,49,50\}$, $A = \{$Odd numbers$\}$, and $B = \{$Primes$\}$, find the value of:

   a. $|A|$ **b.** $|B|$ **c.** $|A \cap B|$ **d.** $|A \cup B|$

9. At Wendy's Old Fashioned Hamburgers, they advertise that you can have your hamburgers 256 ways. If they offer catsup, onion, mustard, pickles, lettuce, tomato, mayonnaise, and relish, is their claim correct? Explain why or why not.

†10. Use the cube shown in Figure 8.7 as your starting point.

   a. Show the result after *TRT.*
   b. What steps will return this cube back to the starting point?

Bonus Question

†11. One variation of the Instant Insanity cubes is a puzzle with five blocks instead of four.

   a. How many arrangements are possible?
   b. If you carry out one arrangement every second, how much time is required to show the number of arrangements named in part a?

---

*This problem requires Section 2.1.

†These problems require optional Section 8.5.

Date	Cultural History
1700	1705: Haley predicts return of 1682 comet 1706: Benjamin Franklin born 1710: Leibniz's *Théodicée* 1712: Last execution for witchcraft in England  Witch burning
1725	1726: Swift's *Gullivers Travels* 1738: Bach's *Mass in B Minor* 1740: Accession of Fredrick the Great Israel Baal Shem Toh founds Hasidhim 1742: Handel's *The Messiah*  Fredrick the Great
1750	1756–63: The Seven Years' War 1759: Voltaire's *Candide* 1762: Rousseau's *Social Contract* 1764: Paris Pantheon started Death of William Hogarth 1767: Watt's steam engine  Hogarth, *Physiognomy*
1775	1776: American Declaration of Independence 1779: Mozart's *Don Giovanni* 1789: French Revolution 1799: Napoleon rules France  French revolutionary broadside

Mathematical History

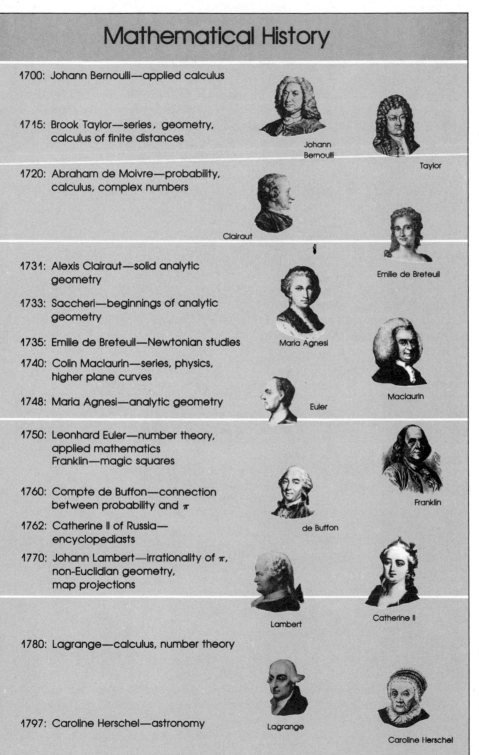

1700: Johann Bernoulli—applied calculus

1715: Brook Taylor—series, geometry, calculus of finite distances

1720: Abraham de Moivre—probability, calculus, complex numbers

Johann Bernoulli

Taylor

Clairaut

Emilie de Breteuil

1731: Alexis Clairaut—solid analytic geometry

1733: Saccheri—beginnings of analytic geometry

1735: Emilie de Breteuil—Newtonian studies

1740: Colin Maclaurin—series, physics, higher plane curves

1748: Maria Agnesi—analytic geometry

Maria Agnesi

Euler

Maclaurin

1750: Leonhard Euler—number theory, applied mathematics
Franklin—magic squares

1760: Compte de Buffon—connection between probability and π

1762: Catherine II of Russia— encyclopediasts

1770: Johann Lambert—irrationality of π, non-Euclidian geometry, map projections

Franklin

de Buffon

Lambert

Catherine II

1780: Lagrange—calculus, number theory

1797: Caroline Herschel—astronomy

Lagrange

Caroline Herschel

1799: Metric system adopted in France

9

The Nature of Probability

There once was a
breathy baboon
Who always breathed
down a bassoon,
For he said, "It appears
That in billions of years
I shall certainly hit
on a tune."

Sir Arthur Eddington

*9.1 Some Experiments in Probability

The nature of probability is best experienced by means of actual physical experiments. It is easy to say that the probability of obtaining a head on the toss of a coin is $\frac{1}{2}$, but what does that mean? We could define probability according to any scheme that is noncontradictory, but the definition would be "good" only insofar as it coincides with real-world events. In this section, we will perform experiments, make observations, and then form conjectures about the nature of probability. In the next section, we will formalize our conjectures and form mathematical models to fit the experiments.

Experiment 1: Tossing a Coin

In this experiment, we flip a coin 50 times. Make sure that, each time the coin is flipped, it rotates several times in the air and lands on a table or on the floor. Keep a record similar to that in Table 9.1. The figures in each table of this section are the result of an actual experiment.

TABLE 9.1   Outcomes of Tossing a Single Coin

Number of Throw	Outcome	Number of Throw	Outcome	Number of Throw	Outcome	Number of Throw	Outcome	Number of Throw	Outcome
1	H	11	H	21	H	31	H	41	T
2	T	12	T	22	T	32	T	42	H
3	H	13	T	23	T	33	H	43	H
4	H	14	T	24	H	34	H	44	T
5	T	15	H	25	H	35	H	45	H
6	T	16	T	26	H	36	T	46	H
7	T	17	T	27	T	37	T	47	H
8	T	18	H	28	T	38	T	48	T
9	H	19	H	29	H	39	T	49	H
10	H	20	T	30	T	40	T	50	T

TOTALS: Heads, 24; Tails, 26

In this section, no special knowledge is assumed. You are simply asked to perform some experiments and make some conjectures.

Repeat the experiment. The results are called **data.** We will study about the analysis of data in the next chapter. There are many ways to represent data, and you should choose the one most convenient for your purposes. For example, the findings in Table 9.1 could be recorded more conveniently by using tally marks, as in Table 9.2.

*Optional section.

TABLE 9.2 Outcomes of Tossing a Single Coin—Tally Method

Outcome	Number of Occurrences	Total	Percentage
Heads	卌 卌 卌 卌 IIII	24	48%
Tails	卌 卌 卌 卌 卌 I	26	52%

To find the percentage of occurrence, divide the total number of occurrences by the total number of trials: $24 \div 50 = .48$; then write the decimal as a percent.

Can you formulate any conjectures concerning these data? Find the percentage of heads and tails for your data. For our example, we have:

$$\frac{24}{50} \text{ heads} \qquad \frac{26}{50} \text{ tails}$$

$$.48 \text{ heads} \qquad .52 \text{ tails}$$

$$48\% \text{ heads} \qquad 52\% \text{ tails}$$

From our experience, we see that this is as we would expect; that is, heads have about a "50–50" chance of occurrence. If we write the percentage as a fraction, we say that the **probability** of heads occurring is about $\frac{1}{2}$. This means that, if we repeated the experiment a large number of times, we could expect heads to occur about $\frac{1}{2}$ the time.

Experiment 2: Tossing Three Coins

In this experiment, we flip three coins simultaneously 50 times. We are interested in the probability of obtaining 3 heads, 2 heads and 1 tail, 1 head and 2 tails, and 3 tails. The procedure is to flip the coins simultaneously and let them fall to the floor (it might be easier to flip them if you place them in a cup). Record the result in each case, as in Table 9.3.

TABLE 9.3 Outcomes of Flipping Three Coins Simultaneously

Outcome	First Trial	Second Trial	Third Trial	Total	Percentage of Occurrence
Three heads	卌 II	卌	卌 I	18	12%
Two heads and one tail	卌 卌 卌	卌 卌 卌 IIII	卌 卌 卌 卌 II	56	$37\frac{1}{3}\%$
Two tails and one head	卌 卌 卌 卌 IIII	卌 卌 卌 卌 II	卌 卌 III	59	$39\frac{1}{3}\%$
Three tails	IIII	IIII	卌 IIII	17	$11\frac{1}{3}\%$
TOTALS	50	50	50	150	100%

Repeat the experiment three times, and record the totals. Next, find the percentage of occurrence by dividing the number of times the event occurred by the total number of occurrences. Can you make any conjectures? It appears that these outcomes are *not* equally likely to occur. Can you make a conjecture about the probability of three heads occurring?

Notice some properties of probability. If we specify the probability of three heads as the percentage of occurrence (or as a fraction), we see that this percentage must be between 0% and 100% (why?). Therefore, if we express the probability as a fraction, the fraction will be between 0 and 1. The closer the fraction is to 0, the less likely the event is to occur; the closer it is to 1, the more likely the event is to occur. An event that can never occur has probability 0, and an event that is certain to occur has probability 1.

Experiment 3: Rolling One Die

In this experiment, we roll a single die* 50 times and record the results, as in Table 9.4. You can use the pattern in the margin for a die if you don't have dice from some game. Trace it out on a piece of cardboard and fold on the dashed lines.

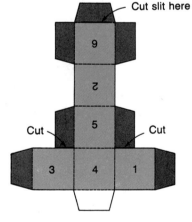

TABLE 9.4   Outcomes of Rolling One Die

Outcome	First Trial	Second Trial	Third Trial	Total	Percentage of Occurrence
1	卌 卌	卌卌	卌 卌	24	16%
2	卌 III	卌	卌 I	19	$12\frac{2}{3}$%
3	卌 卌	卌 卌 II	卌 II	28	$18\frac{2}{3}$%
4	卌 I	卌 卌 I	卌 卌 I	28	$18\frac{2}{3}$%
5	卌	卌 II	卌 卌 I	23	$15\frac{1}{3}$%
6	卌 卌 II	卌 卌 I	卌	28	$18\frac{2}{3}$%
TOTALS	50	50	50	150	100%

Repeat the experiment three times. Find the frequency of each outcome, and then compute the percentage of occurrence of each.

Can you make a conjecture about the probability of each? Notice that the average of each is close to $\frac{1}{6} = 16\frac{2}{3}$%. That is, for one die, it appears that all outcomes are equally likely.

*The word *die* is the singular form of *dice*. All dice used in this text are fair; that is, they are not loaded unless so designated.

Experiment 4: Rolling a Pair of Dice

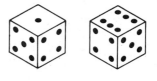

In this experiment, we will roll two dice. Here the possible sums are 2, 3, 4, 5, 6, 7, 8, 9, 10, 11, and 12. In each of these experiments we have made a list of all possible outcomes. This list of outcomes is called the **sample space** of the experiment. Roll the pair of dice 50 times.

TABLE 9.5   Outcomes of Rolling a Pair of Dice

Outcome	First Trial	Second Trial	Third Trial	Total	Percentage of Occurrence
2	│		│	2	$1\frac{1}{3}\%$
3	││││	││	││	7	$4\frac{2}{3}\%$
4	┼┼┼┼	┼┼┼┼	┼┼┼┼	14	$9\frac{1}{3}\%$
5	││	┼┼┼┼ ││	┼┼┼┼ │	15	10%
6	┼┼┼┼ ││││	┼┼┼┼ │││	┼┼┼┼ │	23	$15\frac{1}{3}\%$
7	┼┼┼┼ │││	┼┼┼┼ ┼┼┼┼ │	┼┼┼┼ ││	26	$17\frac{1}{3}\%$
8	┼┼┼┼ │	┼┼┼┼ │││	┼┼┼┼ ││││	23	$15\frac{1}{3}\%$
9	┼┼┼┼ │	┼┼┼┼	┼┼┼┼ ││	16	$10\frac{2}{3}\%$
10	││││	││││	││	10	$6\frac{2}{3}\%$
11	┼┼┼┼	││	┼┼┼┼	12	8%
12	│	│		2	$1\frac{1}{3}\%$
TOTALS	50	50	50	150	100%

Repeat the experiment three times. Now total to find the frequency of each outcome. Next, compute the percentage of occurrence for each.

Can you make a conjecture about the probability of each? In this problem, it is clear that there is no tendency for each outcome to be equally likely, as in the previous experiment. In fact, it seems as if there is a strong tendency for the middle outcomes to be more probable. Can you explain why?

Experiment 5: Tossing a Coin and Rolling a Die

In this experiment, we simultaneously toss a coin and roll a die. Repeat the experiment 50 times, and tabulate the results as shown in Table 9.6.

After performing this experiment, total and find the frequency and percentage of each outcome. Can you make a conjecture about the probabilities involved? We saw that the probability of heads should be $\frac{1}{2}$, and the probability of rolling a 3 on a die is $\frac{1}{6}$. How does the probability of obtaining 3H compare with the product of these probabilities?

Historical Note

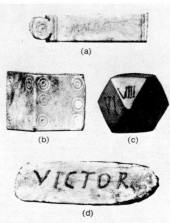

(a)

(b)   (c)

VICTOR

(d)

Some ancient Greek and Roman dice: (a) a strip of bone marked with *malest* (meaning "bad luck") on one side, (b) a bone die that is spotted very much like a modern die, (c) a stone with 14 faces marked with Roman numerals, and (d) a strip of ivory marked *victor* ("winner") on one side.

TABLE 9.6   Outcomes of Tossing a Coin and Rolling a Die

Outcome	First Trial	Second Trial	Third Trial	Total	Percentage of Occurrence
1H	卌	||||	|||	12	8%
1T	卌 |	|||	|||	12	8%
2H	|	卌 |||	||	11	$7\frac{1}{3}$%
2T	|	|||	卌 ||	11	$7\frac{1}{3}$%
3H	||||	卌	||||	13	$8\frac{2}{3}$%
3T	|||	||	卌 ||	12	8%
4H	卌 |||	卌	|	14	$9\frac{1}{3}$%
4T	卌 ||	|||	|	11	$7\frac{1}{3}$%
5H	|	|	卌 ||||	11	$7\frac{1}{3}$%
5T	卌 |||	卌	|||	16	$10\frac{2}{3}$%
6H	||||	卌	卌 |	15	10%
6T	||	卌 |	||||	12	8%
TOTALS	50	50	50	150	100%

(Probability of 3) · (Probability of H)

$$\frac{1}{6} \qquad \cdot \qquad \frac{1}{2} \qquad = \frac{1}{12}$$

This result follows when the events are **independent**—that is, when they are completely unrelated (the occurrence of one event does not affect the occurrence of the other event).

Suppose we reclassify this problem to determine the probability of obtaining an even number or a 3. This result can be tabulated as in Table 9.7.

TABLE 9.7   Outcomes of Obtaining an Even Number
or a 3 on Tossing a Coin and Rolling a Die

Outcome	First Trial	Second Trial	Third Trial	Total	Percentage of Occurrence
Success	30	37	32	99	66%
Failure	20	13	18	51	34%

If we obtain an even number or a 3 in Table 9.6, we'll classify the result as a success; otherwise, we'll call it a failure. Do the same for your data. Compute the total percentage of each. Notice that the event of rolling an even number and the event of obtaining a 3 are **mutually exclusive.** That is, if either one occurs, the other one cannot occur. In our example, if a 3 appears, then an even number cannot appear on that *same* roll. Also, if an even number comes up, then a 3 can't come up. Compare the result you obtain if you compute the following:

(Probability of an even number) + (Probability of 3)

$$\frac{1}{2} \qquad + \qquad \frac{1}{6} \qquad = \frac{2}{3}$$

Do you think this will always be true? To find the probability of mutually exclusive occurrences, we *add* the probabilities of each.

TABLE 9.8   Outcomes of Obtaining an Even Number
or a Head on Tossing a Coin and Rolling a Die

Outcome	First Trial	Second Trial	Third Trial	Total	Percentage of Occurrence
Success	33	40	37	110	$73\frac{1}{3}\%$
Failure	17	10	13	40	$26\frac{2}{3}\%$

On the other hand, suppose we reclassify the problem to determine the probability of obtaining an even number or a head. The result could be tabulated as in Table 9.8. Do the same for your data. If we obtain an even number or a head, we will call this result a success; otherwise, it will be a failure.

(Probability of even) + (Probability of head)

$$\frac{1}{2} \qquad + \qquad \frac{1}{2} \qquad = 1$$

Does this result fit the data obtained? Can you explain how the experiment of Table 9.7 differs from the experiment of Table 9.8? That is, for the probability of "an even or a 3," we can add the respective probabilities; but for the probability of "an even or a head," we apparently cannot. Can you make any conjectures concerning this experiment?

Experiment 6: Three-Card Problem

In this experiment, you need to prepare three cards that are identical except for color. One card is black on both sides, one is white on both sides, and one is black on one side and white on the other. One card is selected at random and then the side placed face up is also randomly chosen. You will see either a black or a white card; record the color. This is not the probability with which we are

Black on both sides

Black on one side, white on the other

White on both sides

concerned; rather, we are interested in predicting the probability of the *other* side being black or white. Record the color of the second side, as in Table 9.9. Repeat the experiment 50 times, and find the percentage of occurrence of black or white with respect to the known color.

TABLE 9.9   Outcomes of Three-Card Experiment

Color of Chosen Card	Number of Times	Outcome (Color of Second Side)	First Trial	Percentage of Occurrence*
White	‖‖‖ ‖‖‖ ‖‖‖ ‖‖‖ ‖‖‖ |	White	‖‖‖ ‖‖‖ ‖‖‖ ‖‖	65.38%
		Black	‖‖‖ ‖‖‖‖	34.62%
Black	‖‖‖ ‖‖‖ ‖‖‖ ‖‖‖ ‖‖‖‖	White	‖‖‖ ‖‖‖‖	37.5%
		Black	‖‖‖ ‖‖‖ ‖‖‖	62.5%

*This is with respect to the known color. For example, the first entry is found by dividing: $17 \div 26 \approx .6538$, or 65.38%.

Consider the following argument. The probability of a white on the underside of the shown card is $\frac{1}{2}$, since the chosen card must be one of two possibilities. Assume that the visible side is white. Then it cannot be the black-black card; it must be the white-white or the white-black. Thus the probability should be $\frac{1}{2}$. Does this argument match the data? If not, then either the data are biased, or the argument is not correct. You might wish to repeat the experiment.

The probabilities obtained in this section are **empirical probabilities.** In the next section, we will find **theoretical probabilities.** Our mathematical model will be good only insofar as these two probabilities are "about the same." That is, if we perform the experiment enough times, the difference between the empirical and the theoretical probability can be made as small as we please.

Problem Set 9.1

1–6.   Complete the six experiments as described in this section, and state any appropriate conclusions or conjectures.

*Problems 7–10 deal with **random numbers**. We are all familiar with the idea of "picking a number at random." What do we mean by this? Can you really pick random numbers? Problems 7–9 ask you to pick random numbers using different methods. You are then asked to check the "randomness" of the numbers you selected. **After** you have selected the numbers, look at page 471 and apply the tests of randomness. **Do not read these tests until after** you have selected your "random numbers."*

7.   Pick random numbers "from your head." Try writing down 100 random digits between 0 and 9 inclusive. Remember to write down the numbers as you think of them. You may not go back and change the numbers once they have been put down.

8. Pick random numbers using cards. From an ordinary deck of cards, remove all jacks, queens, and kings. Consider the ten as 0, the ace as 1, and the other cards as their face values. Shuffle the cards thoroughly. Select one card, note the result, and return it to the deck. Shuffle, and repeat 100 times.

9. Repeat Problem 8, except pick 20 cards, note the result, and return the cards to the deck. Shuffle, and repeat 5 times.

10. Pick random numbers using a table of random numbers. Consult a table of random numbers such as the table shown below. [There are whole books of random numbers—for example, Rand Corporation has a book entitled *A Million Random Digits with 100,000 Normal Deviants* (New York: The Free Press, 1955).] Go to any place on the table and, by reading across a row or down a column, pick 100 numbers.

11. Drop 16 balls into the device shown in the margin. As a ball hits a peg, let it fall randomly to the right or left as determined by the flip of a coin (heads—right; tails—left). Report on the final distribution of balls at the bottom.

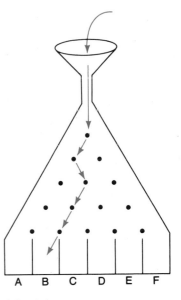

A hextat

12. Repeat the experiment in Problem 11 using the random number table. If a digit is even (0, 2, 4, 6, or 8) let it fall to the right; if a digit is odd (1, 3, 5, 7, or 9) let it fall to the left.

13. Repeat the experiment in Problem 12 for 64 balls.

Table of Random Numbers: This is a table of random digits, arranged into groups of 5 for ease of reading. There are books of random numbers, such as the Rand Corporation's *A Million Random Digits with 100,000 Normal Deviants* (New York: The Free Press, 1955).

76103	54833	70820	34913	98670	19279	09844	22971	92304	59293	35289	72050	01063	39217
48863	80541	05662	91234	68279	68524	38347	56276	46289	08490	30894	40350	38019	99324
45732	97777	95094	75664	28032	77949	29673	43353	12266	44999	17906	88336	39090	78957
06286	48453	55275	38581	05825	97943	33208	20674	15073	05128	14740	69565	73705	93054
08264	89736	27644	49229	53628	70723	50493	82657	86680	00443	05972	36225	93783	08026
55459	30688	81619	14591	32062	69629	68218	05069	63221	75975	21319	47239	98310	13552
82459	73179	00886	70194	31725	16658	98614	43381	30161	10846	22282	09708	24754	92896
67571	12653	88165	82561	07851	56660	71288	18642	72206	93517	42794	15872	19741	11757
15247	01518	55868	00166	76152	47436	69266	26621	05785	26612	41225	41310	87444	32098
58936	94717	62073	14288	42914	39416	23773	17905	09906	54660	12214	79899	50012	90279
24394	08154	42903	47442	78882	80560	28380	02828	56828	54497	61566	26977	24318	44348
18534	80284	07652	12150	77193	08640	91264	28554	09430	99046	20170	71345	50116	16327
11703	92464	94067	67479	93687	47464	77939	68513	74114	85589	89303	93201	20961	63648
41583	39061	12445	07282	69713	17034	02106	70092	63512	40342	90936	89590	17317	13310
77785	25025	94270	20441	94245	53364	40785	72264	45807	75318	35704	29122	86473	13771
32479	21037	67157	64614	00378	14772	30649	70212	02838	70342	27807	87898	45948	63632
69404	27495	56827	04026	43607	23164	99618	47578	33723	64703	32191	69755	73238	48101
60854	40778	20924	04486	12728	35928	17589	68749	97456	65785	25834	24803	99071	61100
52414	26291	55870	88153	99895	60996	97872	74404	10657	70924	11836	88313	51233	48771
29526	62514	00918	03119	31925	15446	93522	35609	26364	36632	90458	25664	42197	09952

Mind Bogglers

14. *The game of WIN.* Construct a set of nonstandard dice as shown.* (Blanks are counted as zero.) The game of WIN is for two players. Each player is to choose one of the dice, and then the players roll the dice simultaneously. The player with the larger number wins the game. By experimenting with the dice, which one would you choose? If you were to play the game of WIN, would you like to choose your die first or second?

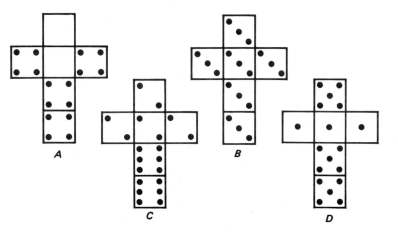

<table>
<tr><td>One way of using the table of random numbers to determine the winner is to select any digit in the table; if the digit is even then the game is won by the Marvels, and if the digit is odd then the game is won by the Stompers. For subsequent games, move in any direction one digit at a time from the first digit.</td></tr>
</table>

15. A tournament is organized between Shannon's Stompers and Melissa's Marvels. Suppose the teams each have an equal chance of winning, and the Marvels are leading the series three games to two when play is interrupted. If the first team to win five games wins the tournament, use the table of random numbers on page 469 to simulate the finish of the series and decide the winner. Record the name of the winner of the series. Repeat the experiment 25 times (or more if necessary), and record the frequency with which the Marvels win the series. Plot the relative frequencies after 5, 10, 15, 20, and 25 trials. Do these relative frequencies appear to be stabilizing around any particular number?

Problem for Individual Study

16. *Probability and chance.* What is probability? What is the risk involved in driving a car? What is Brownian motion? What are the odds of winning a lottery? What is the mathematical expectation in a carnival game? Does gambling pay? How is the law of disorder related to nuclear fission? How is probability used to derive new facts?

Exhibit suggestions: Illustrate with toys or models the probability of accident, crop damage from storms, winning bridge hands, slot-machine odds, coin tosses, dice totals, contest winnings, lock combinations, multiple births.

---

*These dice were designed by Bradley Efron of Stanford University.

Tests of Randomness

Do you think you can pick numbers at random? Well, check it out by looking at Problems 7–10. There are several tests for randomness. Here are two of them.

1. Each digit should occur approximately the same number of times. This means that, given enough such digits, the percent of occurrence for each should be close to 10%.

2. Are there any sequences of numbers? In naming 100 digits, we could expect random selection to have a sequence of several digits.

In many applications of science, it is important to obtain random numbers. Computers have been a great help in selecting random numbers for us.

9.2 Definition of Probability

Engraving by Darcis: *Le Trente-et-un*

Historical Note

The engraving depicts gambling in 18th century France. The mathematical theory of probability arose in France in the 17th century when a gambler, Chevalier de Méré, was interested in adjusting the stakes so that he could be certain of winning if he played long enough. In 1654 he sent some problems to Blaise Pascal, who in turn sent them to Pierre de Fermat. Together, Pascal and Fermat developed the first theory of probability. From these beginnings, the theory of probability developed into one of the most useful and important branches of mathematics, and today it also has applications in business, economics, genetics, insurance, medicine, physics, psychology, and the social sciences.

We will now formalize the results of the previous section and develop a mathematical model giving *theoretical probabilities*. That is, we'll formulate definitions, assume some properties, and thus create a model that will serve as a predictor of the actual percentage of occurrence for a particular experiment. Our model will be good only insofar as it conforms to experimental data.

First we should settle on some terminology. We will consider experiments and events. The **sample space** of an experiment is the set of all possible outcomes. In each of the experiments of the previous section, we tabulated the sample space under the column headed "Outcomes." Now, an **event** is simply a set of possible outcomes (a subset of the sample space). For example, in Experiment 5, we tossed a coin and rolled a die. The sample space is

$$\{1H,1T,2H,2T,3H,3T,4H,4T,5H,5T,6H,6T\}$$

and an event might be "obtaining an even number or a head." That is, the following subset of the sample space is considered to be this event:

$$\{1H,2H,2T,3H,4H,4T,5H,6H,6T\}$$

The event "obtaining a 5" is:

$$\{5H,5T\}$$

If the sample space can be divided into *mutually exclusive* and *equally likely* outcomes, we can define the probability of an event. Let's consider Experiment 1. We see here that a suitable sample space is:

$$S = \{Heads,Tails\}$$

Suppose we wish to consider the event of obtaining heads; we'll call it event A.

$$A = \{Heads\}$$

We wish to define the probability of event A, which we denote by $P(A)$. Notice that the outcomes in the sample space are **mutually exclusive.** That is, if one occurs, the others cannot occur. If we flip a coin, there are two possible outcomes, and *one and only one* outcome can occur on a toss. If each outcome in the sample space is **equally likely,** we define the probability of A as:

$$P(A) = \frac{\text{Number of successful results}}{\text{Number of possible results}} = \frac{|A|}{|S|}$$

A "successful" result is a result that corresponds to the probability we are seeking—in this case, {Heads}. Since we can obtain a head in only one way (success), and the total number of possible outcomes is two (cardinality of the sample space), then the probability of heads is given by this definition as:

$$P(\text{Heads}) = P(A) = \frac{1}{2}$$

• Life is a school of probability.
Walter Bagehot

This, of course, corresponds to our empirical results and to our experience of flipping coins. We now give a definition of probability.

Definition of Probability

If an experiment can occur in any of n mutually exclusive and equally likely ways, and if s of these ways are considered favorable, then the probability of the event E, denoted by $P(E)$, is:

$$P(E) = \frac{s}{n} = \frac{\text{Number of outcomes favorable to } E}{\text{Number of all possible outcomes}}$$

Example 1: Use the definition of probability to find the probability of white and the probability of black, using the spinner shown in the margin. Assume that the arrow will never lie on a border line.

Solution: $P(\text{White}) = \dfrac{2}{3}$ ←Two sections are white.
                             ←Three sections altogether

$P(\text{Black}) = \dfrac{1}{3}$

Example 2: Consider a jar containing marbles as shown. Suppose each marble has an equal chance of being picked from the jar. Find the indicated probabilities.

a. $P(\text{Black}) = \dfrac{4}{12}$ ←4 black marbles
                        ←12 marbles in jar

b. $P(\text{Red}) = \dfrac{7}{12}$

c. $P(\text{White}) = \dfrac{1}{12}$

Reduced fractions are used to state probabilities when the fractions are fairly simple. If, however, the fractions are not simple, the probabilities are usually stated as decimals, as shown by Example 3.

Example 3: Suppose that in a certain study, 46 out of 155 people showed a certain kind of behavior. Assign a probability to this behavior.

Solution: $P = \dfrac{46}{155}$ Press: 46 ÷ 155 =
                            Display: .29677419

$\approx .30$

Notice that Examples 1 and 2 are theoretical probabilities while Example 3 is an empirical probability. If, for example, we *actually* spun the dial in Exam-

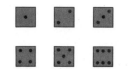

FIGURE 9.1 Sample space for tossing a single die

ple 1, or *actually* drew a marble from the jar in Example 2, and if we repeated the experiment a large number of times, we would expect the empirical probabilities and theoretical probabilities to be almost the same. The results obtained by applying this definition *must* be consistent with the results obtained by actually performing the experiment a large number of times.

Example 4: Compare the empirical and theoretical probabilities for Experiment 3 (rolling a single die).

Solution: The sample space is {1,2,3,4,5,6}, as shown in Figure 9.1.

Let's find the theoretical probability for one of these, say the probability of tossing a two:

$$P(Two) = \frac{Number\ of\ successful\ outcomes}{Number\ of\ all\ possible\ outcomes} = \frac{1}{6}$$

The probability of any other single occurrence is also $\frac{1}{6}$. Since $\frac{1}{6} \approx .1667$, we can compare the empirical and theoretical probabilities as shown in Table 9.10. These seem to be consistent.

TABLE 9.10 Comparison of Empirical and Theoretical Probabilities for Rolling a Single Die

Out-comes	Theoretical Probability	Empirical Probability*
1	.1667	.1600
2	.1667	.1267
3	.1667	.1867
4	.1667	.1867
5	.1667	.1533
6	.1667	.1867

*These figures vary according to the particular experiment and the number of trials. The figures used here are actual results of Experiment 3.

Example 5: Compare the empirical and theoretical probabilities for Experiment 4 (rolling a pair of dice).

Solution: Let's begin by reasoning in the same fashion as in Example 4. The sample space is {2,3,4,5,6,7,8,9,10,11,12}. Thus,

$$P(Two) = \frac{1}{11} \approx .0909$$

Suppose we say, as in Example 4, that the probability of any other single event will also be $\frac{1}{11}$. Table 9.11 compares this with the results we obtained for Experiment 4.

TABLE 9.11 Comparison of Empirical and Proposed Probabilities for Rolling a Pair of Dice

Out-comes	Theoretical Probability	Empirical Probability
2	.0909	.0133
3	.0909	.0467
4	.0909	.0933
5	.0909	.1000
6	.0909	.1533
7	.0909	.1733
8	..0909	.1533
9	.0909	.1067
10	.0909	.0667
11	.0909	.0800
12	.0909	.0133

The empirical probabilities do *not* substantiate our theoretical probability. The problem here is our assumption that the outcomes are equally likely. You *must* define a sample space so that the empirical and theoretical probabilities are consistent. With this in mind, consider another sample space for a pair of dice (see Figure 9.2).

Now,

$$P(Two) = \frac{1}{36}$$ ←—Number of equally likely possibilities in sample space

$$P(Three) = \frac{2}{36}$$ Do you see that there are two possibilities of obtaining a three in Figure 9.2?

$$P(Four) = \frac{3}{36}$$

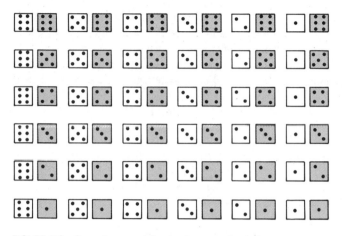

FIGURE 9.2 Sample space for tossing a pair of dice

After calculating the others (you will be asked to do some of these in Problem Set 9.2), we compare these theoretical probabilities with the empirical probabilities in Table 9.12. The results are consistent. Notice also that the sample space in Figure 9.2 is what you would expect from the Fundamental Counting Principle. Since each die can be arranged 6 ways, the total number of possibilities is $6 \cdot 6 = 36$.

Example 6 illustrates the use of the words *and* and *or* in probability problems.

Example 6: Suppose that a single card is selected from an ordinary deck of 52 cards.

a. What is the probability that it is a heart?
b. What is the probability that it is an ace?
c. What is the probability that it is a two *or* a king?
d. What is the probability that it is a two *or* a heart?
e. What is the probability that it is a two *and* a heart?
f. What is the probability that it is a two *and* a king?

Solution: In case you are not familiar with an ordinary deck of cards, the sample space is shown on page 441.

a. $P(\text{Heart}) = \dfrac{13}{52} = \dfrac{1}{4}$ **b.** $P(\text{Ace}) = \dfrac{4}{52} = \dfrac{1}{13}$

c. The word *or* is used in probability to mean that either the first event occurs *or* the second event occurs; that is, both events are counted as successes when applying the definition of probability. In this example, there are 4 twos and 4 kings in a deck, so

$$P(\text{Two } or \text{ King}) = \frac{8}{52} = \frac{2}{13}$$

TABLE 9.12 Revised Comparison of the Empirical and Theoretical Probabilities for Rolling a Pair of Dice

Outcomes	Theoretical Probability	Empirical Probability
2	.0278	.0133
3	.0556	.0467
4	.0833	.0933
5	.1111	.1000
6	.1389	.1533
7	.1667	.1733
8	.1389	.1533
9	.1111	.1067
10	.0833	.0667
11	.0556	.0800
12	.0278	.0133

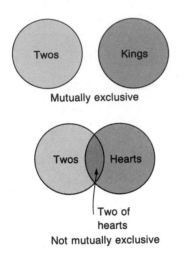

Mutually exclusive

Not mutually exclusive

Two of hearts

d. This seems to be very similar to part c, but there is one important difference. Look at the sample space and notice that although there are 4 twos and 13 hearts, the total number of successes is *not* $4 + 13 = 17$,   *but rather* 16. The reason for this is that you cannot count the same card twice; the two of hearts should *not* be counted *once* as a two *and again* as a heart. Thus,

$$P(\text{Two } or \text{ Heart}) = \frac{16}{52} = \frac{4}{13}$$

e. The word *and* is used in probability to mean that both events occur. This means, for this example, that when you draw the single card it must be *both* a two *and* a heart. By looking at the sample space, you can see that there is only one way of success. Thus,

$$P(\text{Two } and \text{ Heart}) = \frac{1}{52}$$

f. There is no way of drawing both a two *and* a king with a single draw, so

$$P(\text{Two } and \text{ King}) = \frac{0}{52} = 0$$

If an event *cannot* occur, then the probability is zero. The occurrence of albino twins in cows had never before been recorded, and, according to *The Dairyman*, it more than likely will never happen again. This is an example of a highly unlikely event, and its probability is very close to zero. But this is not the same as saying that the probability *is* zero.

Notice from Example 6f that the probability of an event that cannot occur is 0. In Problem Set 9.2 you are asked to show that the probability of an event that must occur is 1. These are the two extremes. All other probabilities will fall somewhere in between. The closer the probability is to 1, the more likely the event; the closer it is to 0, the less likely the event.

Procedure for Finding Probabilities

1. Describe and identify the sample space, and then count the number of elements (these should be equally likely); call this number n.
2. Count the number of occurrences that interest us; call this the number of successes and denote it by s.
3. Compute the probability of the event: $P(E) = \dfrac{s}{n}$

Problem Set 9.2

1. What is the difference between empirical and theoretical probabilities?

2. Define probability.

Use the spinner shown here for Problems 3–5 and find the requested probabilities. Assume that the pointer can never lie on a border line.

3. P(White) **4.** P(Black) **5.** P(Red)

Consider the jar containing marbles shown here. Suppose each marble has an equal chance of being picked from the jar. Find the probabilities in Problems 6–8.

6. P(White) **7.** P(Black) **8.** P(Red)

A single card is selected from an ordinary deck of cards. Find the probabilities in Problems 9–11.

9. P(Five of hearts) **10.** P(Five) **11.** P(Heart)

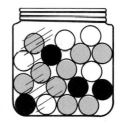

Give the probabilities in Problems 12–15 in decimal form (correct to two decimal places). A calculator may be helpful with these problems.

12. Last year, 1,485 calculators were returned to the manufacturer. If 85,000 were produced, assign a number to specify the probability that a particular calculator would be returned.

13. Last semester, a certain professor gave 13 As out of 285 grades. If grades were assigned randomly, what is the probability of an A?

14. Last year, in a certain city it rained on 85 days. What is the probability of rain on a day selected at random?

15. A certain campus club is having a raffle and they are selling 1,500 tickets. If the people on your floor of the dorm bought 285 of those tickets, what is the probability that someone on your floor will hold the winning ticket?

Suppose you toss a coin and roll a die. The sample space is shown here. Use this sample space to answer the questions in Problems 16–19.

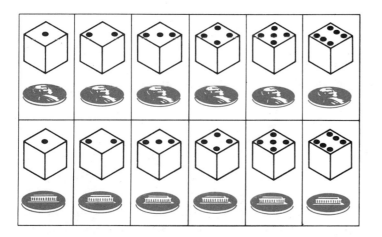

16. What is the probability of obtaining:

    a. A tail *and* a five?   **b.** A tail *or* a five?
    c. A head *and* a two?

17. What is the probability of obtaining:

  a. A tail?
  b. One, two, three, *or* four?
  c. A head *or* a two?

18. What is the probability of obtaining:

  a. A head *and* an odd number?
  b. A head *or* an odd number?

19. What is the probability of obtaining:

  a. A head *and* a five?
  b. A head *or* a five?

Use the sample space shown in Figure 9.1 to answer the questions in Problems 20–30. The requested probabilities in Problems 20–27 refer to the sum of the numbers shown on the tops of the dice.

20. P(Five)   **21.** P(Six)   **22.** P(Seven)

23. P(Eight)   **24.** P(Nine)   **25.** P(Four *or* Five)

26. P(Even number)   **27.** P(Eight *or* Ten)

28. Dice is a popular game in gambling casinos. Two dice are tossed, and various amounts are paid according to the outcome. If a seven or eleven occurs on the first roll, the player wins. What is the probability of winning on the first roll?

29. In dice, the player loses if the outcome of the first roll is a two, three, or twelve. What is the probability of losing on the first roll?

30. In dice, a pair of ones is called *snake eyes*. What is the probability of losing a dice game by rolling snake eyes?

31. Consider a die with only four sides, marked one, two, three, and four. Write out a sample space similar to the one shown in Figure 9.1 for rolling a pair of these dice.

32. Using the sample space you found in Problem 31, find the probability that the sum of the dice is the given number. Assume equally likely outcomes.

  a. P(Two)   **b.** P(Three)   **c.** P(Four)

33. Using the sample space you found in Problem 31, find the probability that the sum of the dice is the given number. Assume equally likely outcomes.

  a. P(Five)   **b.** P(Six)   **c.** P(Seven)

34. The game of Dungeons and Dragons uses nonstandard dice. Consider a die with eight sides marked one, two, three, four, five, six, seven, and eight. Write out a sample space similar to the one shown in Figure 9.1 for rolling a pair of these dice.

Mind Bogglers

35. Consider the game of WIN described in Problem 14 of Problem Set 9.1. Suppose

your opponent picks die *A*. Suppose you pick die *B*. Then we can enumerate the sample space as shown here.

A / B	0	0	4	4	4	4	
3	(3, 0)	(3, 0)	(3, 4)	(3, 4)	(3, 4)	(3, 4)	
3	(3, 0)	(3, 0)	(3, 4)	(3, 4)	(3, 4)	(3, 4)	
3	(3, 0)	(3, 0)	(3, 4)	(3, 4)	(3, 4)	(3, 4)	
3	(3, 0)	(3, 0)	(3, 4)	(3, 4)	(3, 4)	(3, 4)	← *A* wins
3	(3, 0)	(3, 0)	(3, 4)	(3, 4)	(3, 4)	(3, 4)	
3	(3, 0)	(3, 0)	(3, 4)	(3, 4)	(3, 4)	(3, 4)	

↑
B wins

We see that the probability of *A* winning is $\frac{24}{36}$, or $\frac{2}{3}$.

By enumeration of the sample space, find your probability of winning if you choose:

a. Die *C*   **b.** Die *D*

c. If your opponent picks die *A*, which die would you pick?

36. In WIN (see Problem 35), if your opponent picks die *B*, which would you pick?

37. In WIN (see Problem 35), if your opponent picks die *C*, which would you pick?

38. In WIN (see Problem 35), if your opponent picks die *D*, which would you pick?

39. By considering Problems 35–38, see if you can write a strategy for playing WIN. If you follow this strategy, what is your probability of winning?

40. Show that the probability of some event *E* falls between 0 and 1. Also, show that $P(E) = 0$ if the event cannot occur and $P(E) = 1$ if the event must occur.

Problems for Individual Study

41. *Friday-the-13th problem.* Find the probability that the 13th day of a randomly chosen month will be a Friday. We might begin by saying that the sample space is {Sunday, Monday, Tuesday, Wednesday, Thursday, Friday, Saturday}. That is, the 13th must be one of these. Thus, $P(\text{Fri}) = \frac{1}{7}$. This is *not correct*, since the outcomes listed are not equally likely. It turns out that the 13th of the month is more likely to be a Friday than any other day of the week. Show this.

    References: William Bailey, "Friday-the-Thirteenth," *Mathematics Teacher,* vol. LXII, no. 5 (1969), pp. 363–364.

                      C. V. Heuer, Solution to problem E1541 in *American Mathematical Monthly,* vol. 70, no. 7 (1963), p. 759.

                      G. L. Ritter, et al. "An Aid to the Superstitious," *Mathematics Teacher,* vol. 70, no. 5 (1977), pp. 456–457.

42. What is the Monte Carlo method?

9.3 Conditional Probability

At one time or another, we all become involved in a discussion of whether an expected child will be a boy or a girl, and many ideas of probability are used and misused in making these predictions. In this section, we will "keep it in the family" by considering applications of probability to family matters. We assume that each child is as likely to be a boy as a girl.

Let's begin by computing the probability of a family having two children of opposite sexes. The sample space might look like this:

2 boys

a boy and a girl

2 girls

But this sample space presents problems in computing probabilities, since the outcomes are not equally likely. Consider the following tree diagram:

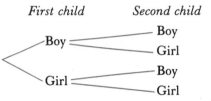

We see that there are four equally likely outcomes:

Boy —— Boy   Girl —— Boy

Boy —— Girl   Girl —— Girl

Thus the probability of having a boy and a girl with two children is:

$$\frac{\text{Number of favorable outcomes}}{\text{Total number of all possible outcomes}} = \frac{2}{4}$$

There is a probability of $\frac{1}{2}$ of having a boy and a girl family.

Example 1: Suppose a family wanted to have four children. What is the probability of having two boys and two girls?

Solution: In this problem, there are 16 possibilities (can you list them?). Of these possibilities, there are 6 ways of obtaining 2 boys and 2 girls: BBGG, BGBG, BGGB, GBGB, GBBG, GGBB. Thus the probability is

$$\frac{6}{16} = \frac{3}{8}$$

A couple with two boys wanted to have a girl. They knew a little about probability and reasoned that, since the chance of having three boys is only $\frac{1}{8}$, they really had a good chance of having a girl the next time. This is a common

A common fallacy is to say "$\frac{1}{2}$ boys and $\frac{1}{2}$ girls; thus the probability is $\frac{1}{2}$."

misconception. It is important to remember that the probability of a future event is unaffected by past occurrences. The probability of a girl on the next birth is the same as always—$\frac{1}{2}$—*regardless of the previous record.* The Emory Harrison family in Tennessee had 13 boys.

The Emory Harrison family

The probability of this occurring is

$$\frac{1}{8,192}$$

If the Harrisons were to have another baby, what would be the probability of this child being a boy? You *cannot* reason that the probability of 14 boys is $\frac{1}{16,384}$ and thus their chances of another boy are very slight. Actually, the probability of their having another boy is still $\frac{1}{2}$.

Frequently, we wish to compute the probability of an event but have additional information that will alter the sample space.

Example 2: Suppose a family has two children. What is the probability that the family has two boys? What is the probability that the family has two boys if we know that the older child is a boy?

Solution: The first answer is easy: P(2 boys) $= \frac{1}{4}$ (Can you explain why?) For the second answer, notice that we have altered the sample space.

Sample space

$$
\left.
\begin{array}{cc}
\text{B} & \text{B} \\
\text{B} & \text{G}
\end{array}
\right\}
\quad
\begin{array}{l}
\text{This is the altered} \\
\text{sample space.}
\end{array}
$$

$$\cancel{\text{G} \quad \text{B}}$$
$$\cancel{\text{G} \quad \text{G}}$$

We see that the probability is now $\frac{1}{2}$.

Example 3: Reconsider Example 2; what is the probability that a family with two children will have two boys if all we know is that at least one child is a boy?

Solution: Again, consider an altered sample space.

Sample space

B   B

B   G

G   B

~~G   G~~

Here we see that the probability is $\frac{1}{3}$!

These are problems involving **conditional probability.** We speak about the *probability of an event E, given that another event F has occurred.* We denote this by $P(E|F)$. This means that, rather than simply compute $P(E)$, we reevaluate $P(E)$ in light of the information that F has occurred. That is, we consider the altered sample space.

Example 4: Suppose we toss three coins. What is the probability that two or more heads are obtained, given that at least one head is obtained?

Solution: We consider the altered sample space.

Sample space

*HHH	*HHH	*Success
*HTH	HTT	
*THH	THT	
TTH	~~TTT~~	

The probability is $\frac{4}{7}$.

Example 5: Two cards are drawn from a standard deck of 52 cards. What is the probability that the first was an ace, given that the second was an ace?

Solution: Let $A1$ be the event that an ace is drawn on the first draw
   $A2$ be the event that an ace is drawn on the second draw
Find $P(A1|A2)$. The number of successful ways is 3 (there are 3 aces left) and the total number of possibilities is 51 (since there are 51 cards left, given the event $A2$).
   Thus, $P(A1|A2) = \frac{3}{51} = \frac{1}{17}$.

Example 6: In a life science experiment it is necessary to examine fruit flies and determine their sex and whether or not they have mutated after exposure to radiation. For 1,000 fruit flies examined, there were 643 females and 357 males. Also, 403 of the females were normal and 240 were mutated; of the males, 190 were normal and 167 were mutated. Use a calculator to find the following probabilities (correct to the nearest hundredth), assuming a single fruit fly is chosen at random from the sample of 1,000.

a. It is male.
b. It is a normal male.
c. It is normal, given that it is a male.
d. It is a male, given that it is normal.
e. It is mutated, given that it is a male.

Solution: It helps to construct a table when considering a problem such as this.

	Mutated	Normal	TOTALS
Male	167	190	357
Female	240	403	643
TOTALS	407	593	1,000

a. $P(\text{Male}) = \dfrac{357}{1,000} \approx .36$ Answers rounded to the nearest hundredth

b. $P(\text{Normal male}) = \dfrac{190}{1,000} \approx .19$

c. $P(\text{Normal}|\text{Male}) = \dfrac{190}{357} \approx .53$

d. $P(\text{Male}|\text{Normal}) = \dfrac{190}{593} \approx .32$

e. $P(\text{Mutated}|\text{Male}) = \dfrac{167}{357} \approx .41$

Problem Set 9.3

1. Suppose a family wants to have four children.

   a. What is the sample space?
   b. What is the probability of 4 girls? 4 boys?
   c. What is the probability of 1 girl and 3 boys? 1 boy and 3 girls?
   d. What is the probability of 2 boys and 2 girls?
   e. What is the sum of your answers in parts b through d?

2. In your own words, explain what we mean by "conditional probability."

3. What is the probability of a family of three children containing exactly two boys, given that at least one of them is a boy?

4. What is the probability of flipping a coin four times and obtaining exactly three heads, given that at least two are heads?

Suppose a single die is rolled. Find the probabilities requested in Problems 5–8.

5. 5, given that it was odd
6. 6, given that it was odd
7. Odd, given that it was 6
8. Odd, given that it was 5

Suppose a pair of dice are rolled. Consider the sum of the numbers on the top of the dice and find the probabilities requested in Problems 9–15.

9. 7, given that it was odd
10. 7, given that at least one die came up 2
11. 3, given that one die came up 2
12. 4, given that one die came up 2
13. 2, given that one die came up 2
14. 8, given that a double was rolled
15. A double, given that 8 was rolled

A single card is drawn from a standard deck of cards. Find the probabilities if the given information is known about the chosen card in Problems 16–21.

16. P(Face card | Jack)
17. P(Jack | Face card)
18. P(Two | Not a face card)
19. P(Heart | Not a spade)
20. P(Black | Jack)
21. P(Jack | Black)

Two cards are drawn from a standard deck of cards, and one of the two cards is noted. Find the probabilities of the second card, given the information about the noted card provided in Problems 22–27.

22. P(Ace | Two)
23. P(King | King)
24. P(Heart | Heart)
25. P(Heart | Spade)
26. P(Black | Red)
27. (Black | Black)

In Problems 28–35, use the information provided in Example 6. Give answers to the nearest hundredth.

28. P(Female)
29. P(Mutated male)
30. P(Mutated female)
31. P(Mutated | Female)
32. P(Normal female)
33. P(Normal | Female)
34. P(Female | Mutated)
35. P(Male | Mutated)

Mind Boggler

36. *Experiment.* Consider the birthdates of some of the mathematicians we've met in this book:

Abel	August 5, 1802
Cardano	September 24, 1501
Descartes	March 31, 1596
Euler	April 15, 1707

Fermat	August 20, 1601 (baptized)
Galois	October 25, 1811
Gauss	April 30, 1777
Newton	December 25, 1642
Pascal	June 19, 1623
Riemann	September 17, 1826

We do not know Fermat's birthdate, so we will use his baptismal date instead.

Birthday problem: If we select 23 persons out of a crowd, the chances are about even that 2 of them will have the same birthday.

Add to this list the birthdates of the members of your class. What is your *guess* as to the probability that at least two persons in this group will have exactly the same birthdate (not counting the year)? Be sure to make your guess *before* finding out the birthdates of your classmates. The answer, of course, depends on the number of persons on the list. We've listed 10 mathematicians, and you may have 20 persons in your class, giving 30 names on the list.

Problem for Individual Study

37. *Birthday problem.* Let's actually perform the experiment suggested by the *B.C.* cartoon on the facing page (except that we are concerned only with birthdates, not with mentalities). This problem is related to Problem 36. Instead of standing on one spot, let's consult an almanac and look up the birthdates of the presidents of the United States. What would you guess is the probability that two of them were born on the same day (not counting the year)? As it turns out, Polk and Harding were both born on November 2.

If 23 or more people are selected at random, the probability that 2 or more of them will have the same birthday is greater than 50%! This is a seemingly paradoxical situation that will fool most people. You might try it sometime when you are in a group of more than 23 persons. For example, if the class has 35 persons, the probability that 2 will have a birthday on the same day is almost 80%. For 2 persons, the probability is $\frac{1}{365}$; for more than 60 persons, it is almost 1 (after 60 the curve is indistinguishable from a straight line). The probability of 1 is not reached, of course, until there are 366 persons (see the graph in the margin).

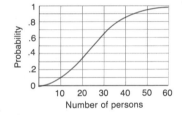

Probabilities for the birthday problem

a. Pick 23 names at random from a biological dictionary or a *Who's Who* and verify some of the probabilities of the table. The more you go past 23, the greater the probability of the same birthday.

b. Find the probabilities for part a. You can check almost any probability text for a discussion of this problem. One possibility is Emanuel Parzen's *Modern Probability Theory and Its Applications* (New York: Wiley, 1964), pp. 46–47.

9.4 Mathematical Expectation

Smiles toothpaste is giving away $10,000. All you must do to have a chance to win is send in a postcard with your name on it (the fine print says you do not need to buy a tube of toothpaste). Is it worthwhile to enter?

Suppose the contest receives one million postcards (a conservative estimate). We wish to compute your **expected value** (or your expectation) in entering this contest. We denote the expectation by E and compute:

$$E = \text{(Amount to win)} \cdot \text{(Probability of winning)}$$

$$= \$10,000 \cdot \frac{1}{1,000,000}$$

$$= \$.01$$

A game is **fair** if the expected value equals the cost of playing the game. Is this game fair? If the toothpaste company charges you 1¢ to play the game, it is fair. But how much does it really cost you to enter the contest? How much does the postcard cost? We see that it is not a fair game.

Example 1:   Suppose you are going to roll two dice. You will be paid $3.60 if you roll two 1s. You do not receive anything for any other outcome. How much should you be willing to pay for the privilege of rolling the dice?

Solution: You should be willing to pay the mathematical expectation of the game, which is computed as follows.

$$E = \text{(Amount of winnings)} \cdot \text{(Probability of winning)}$$

$$= \qquad \$3.60 \qquad \cdot \qquad \frac{1}{36}$$

$$= \$.10$$

You should be willing to pay 10¢ to play the game.

When we say that the expectation is 10¢, we certainly do *not* mean that you will win 10¢ every time you play. (Indeed, you will *never* win 10¢; it will be either $3.60 or nothing.) We mean that, if you were to play this game a very large number of times, you could expect to win an *average* of 10¢ per game.

Most games of chance will not be fair (the "house" must make a living). The question often is "Which games come closest to being fair?" We will consider this question in some of the problems.

Sometimes there is more than one payoff, as illustrated by Example 2.

Example 2:   A recent contest offered one grand prize worth $10,000, two second prizes worth $5,000 each, and ten third prizes worth $1,000 each.

Solution: The expectation can be computed by multiplying the amounts to be won by their respective probabilities and then adding the result. That is,

$$E = \text{(Amount of 1st prize)} \cdot P\text{(1st prize)}$$
$$+ \text{(Amount of 2nd prize)} \cdot P\text{(2nd prize)}$$
$$+ \text{(Amount of 3rd prize)} \cdot P\text{(3rd prize)}$$

Now, if we assume that there were one million entries and that winners' names were replaced in the pool after being drawn, we see that:

$$P\text{(1st prize)} = \frac{1}{1,000,000}$$

$$P(\text{2nd prize}) = \frac{2}{1,000,000}$$

$$P(\text{3rd prize}) = \frac{10}{1,000,000}$$

Hence,

$$E = \$10,000 \cdot \frac{1}{1,000,000} + \$5,000 \cdot \frac{2}{1,000,000} + \$1,000 \cdot \frac{10}{1,000,000}$$

$$= \$.01 + \$.01 + \$.01$$

$$= \$.03$$

Now we will give a formal definition of expectation.

𝕳istorical 𝕹ote

Definition of Mathematical Expectation

If an event has several possible outcomes with probabilities $p_1, p_2, p_3, \ldots$, and for each of these outcomes the amount that can be won is $a_1, a_2, a_3, \ldots$, the **mathematical expectation,** E, of the event is

$$E = a_1 \cdot p_1 + a_2 \cdot p_2 + a_3 \cdot p_3 + \cdots$$

Girolamo Cardano (1501–1576) was a doctor, mathematician, and astrologer. He investigated many interesting problems in probability and wrote a gambler's manual, probably because he was a compulsive gambler. He was the illegitimate son of a jurist and was at one time imprisoned for heresy for publishing a controversial horoscope of Christ's life. He had a firm belief in astrology and predicted the date of his death. However, when the day came, he was alive and well—so he drank poison toward the end of the day to make his prediction come true!

Example 3: A contest offered the following prizes (as indicated in the fine print):

Prize	Value	Probability of Winning
Grand prize trip	$1,500 = a_1$	$.000026 = p_1$
Weber Kettle	$110 = a_2$	$.000032 = p_2$
Magic Chef Range	$279 = a_3$	$.000016 = p_3$
Murray Bicycle	$191 = a_4$	$.000021 = p_4$
Lawn Boy Mower	$140 = a_5$	$.000026 = p_5$
Samsonite Luggage	$183 = a_6$	$.000016 = p_6$

What is the expected value for this contest?

Solution:

$$E = a_1 p_1 + a_2 p_2 + a_3 p_3 + a_4 p_4 + a_5 p_5 + a_6 p_6$$
$$= 1,500(.000026) + 110(.000032) + 279(.000016) + 191(.000021)$$
$$+ 140(.000026) + 183(.000016)$$
$$= .057563$$

The expected value is a little less than 6¢. Suppose we mail in our entry as specified in the rules. What is the cost of the stamp and the envelope?

Sometimes there are situations in which the amount you must pay to play a game has a role in the amount you will win, so that it is not feasible to compare the expectation with the amount you must pay to play. In such situations, you calculate the expenses for playing as part of the amount to win, and then you can have a positive, negative, or zero expectation. In this case, if the expectation is positive, then the game is in your favor; if it is negative, then the game is in your opponent's favor; and if the expectation is zero, then the game is said to be *fair*.

Example 4: Walt, who is a realtor, knows that if he takes a listing to sell a house, it will cost him $1,000. However, if he sells the house, he will receive 6% of the selling price. If another realtor sells the house, Walt will receive 3% of the selling price. If the house is unsold in 3 months, he will lose the listing and receive nothing. Suppose the probabilities for selling a particular $100,000 house are as follows: The probability that Walt will sell the house is .4; the probability that another agent will sell the house is .2; and the probability that the house will remain unsold is .4. What is Walt's expectation if he takes this listing?

Solution: First calculate

$$6\% \text{ of } \$100,000: \quad .06 \times \$100,000 = \$6,000$$
$$3\% \text{ of } \$100,000: \quad .03 \times \$100,000 = \$3,000$$

Probability that Walt sells the house

Probability that another agent sells the house

Probability that house is not sold

Notice that the expense is subtracted from the profit

$$E = (\$6,000 - \$1,000)(.4) + (\$3,000 - \$1,000)(.2) + (\$0 - \$1,000)(.4)$$
$$= \$5,000(.4) + \$2,000(.2) + (-\$1,000)(.4)$$
$$= \$2,000 + \$400 + (-\$400)$$
$$= \$2,000$$

Walt's expectation is $2,000.

A very popular casino game is called Keno. In Keno, a player tries to guess in advance which numbers will be selected from a pot containing 80 numbers. Twenty numbers are selected at random. The player may choose from 1 to 15 spots and is paid according to the amounts shown in Table 9.13.

***Example 5:** Suppose a person picks one number and is paid $2.10 if he wins. For the privilege of playing this game, he pays $.70. Is this a fair game?

*Examples 5 and 6 require Chapter 8.

YOU CAN WIN $50,000

KENO

1	2	3	4	5	6	7	8	9	10
11	12	13	14	15	16	17	18	19	20
21	22	23	24	25	26	27	28	29	30
31	32	33	34	35	36	37	38	39	40
41	42	43	44	45	46	47	48	49	50
51	52	53	54	55	56	57	58	59	60
61	62	63	64	65	66	67	68	69	70
71	72	73	74	75	76	77	78	79	80

TABLE 9.13 Keno Payoffs

MARK 1 SPOT 70¢ Minimum		
Catch	Play .70	Play 1.40
1	2.10	4.20

MARK 2 SPOTS 70¢ Minimum		
Catch	Play .70	Play 1.40
2	8.50	17.00

MARK 3 SPOTS 70¢ Minimum		
Catch	Play .70	Play 1.40
2	.70	1.40
3	30.00	60.00

MARK 4 SPOTS 70¢ Minimum		
Catch	Play .70	Play 1.40
2	.70	1.40
3	3.00	6.00
4	75.00	150.00

MARK 5 SPOTS 70¢ Minimum		
Catch	Play .70	Play 1.40
3	.70	1.40
4	6.50	13.00
5	580.00	1160.00

MARK 6 SPOTS 70¢ Minimum		
Catch	Play .70	Play 1.40
3	.40	.80
4	2.50	5.00
5	70.00	140.00
6	1300.00	2600.00

MARK 7 SPOTS 70¢ Minimum		
Catch	Play .70	Play 1.40
4	.70	1.40
5	16.00	32.00
6	260.00	520.00
7	6000.00	12000.00

MARK 8 SPOTS 70¢ Minimum		
Catch	Play .70	Play 1.40
5	6.30	12.60
6	63.00	126.00
7	1155.00	2310.00
8	12600.00	25000.00

MARK 9 SPOTS 70¢ Minimum		
Catch	Play .70	Play 1.40
5	2.10	4.20
6	31.50	63.00
7	234.50	469.00
8	3290.00	6580.00
9	12950.00	25000.00

MARK 10 SPOTS 70¢ Minimum		
Catch	Play .70	Play 1.40
5	1.40	2.80
6	14.00	28.00
7	99.40	198.80
8	700.00	1400.00
9	3150.00	6300.00
10	13300.00	25000.00

MARK 11 SPOTS 70¢ Minimum		
Catch	Play .70	Play 1.40
6	7.00	14.00
7	52.50	105.00
8	266.00	532.00
9	1400.00	2800.00
10	8750.00	17500.00
11	13650.00	25000.00

MARK 12 SPOTS 70¢ Minimum		
Catch	Play .70	Play 1.40
6	4.00	8.00
7	20.00	40.00
8	160.00	320.00
9	700.00	1400.00
10	1800.00	3600.00
11	12500.00	25000.00
12	25000.00	25000.00

MARK 13 SPOTS 70¢ Minimum		
Catch	Play .70	Play 1.40
6	1.50	3.00
7	11.00	22.00
8	55.00	110.00
9	500.00	1000.00
10	2500.00	5000.00
11	5000.00	10000.00
12	15000.00	25000.00
13	25000.00	25000.00

MARK 14 SPOTS 70¢ Minimum		
Catch	Play .70	Play 1.40
6	2.00	4.00
7	6.00	12.00
8	23.00	46.00
9	200.00	400.00
10	600.00	1200.00
11	1800.00	3600.00
12	8000.00	16000.00
13	25000.00	25000.00
14	25000.00	25000.00

MARK 15 SPOTS 70¢ Minimum		
Catch	Play .70	Play 1.40
6	1.00	2.00
7	6.00	12.00
8	16.00	32.00
9	60.00	120.00
10	160.00	320.00
11	2000.00	4000.00
12	8000.00	16000.00
13	20000.00	25000.00
14	25000.00	25000.00
15	25000.00	25000.00

Solution:

$$P(\text{One number}) = \frac{\text{Number of ways of choosing one number from the 20}}{\text{Number of ways of choosing one number from the 80}}$$

$$= \frac{{}_{20}C_1}{{}_{80}C_1} = \frac{20}{80} = \frac{1}{4}$$

Now we compute the expectation:

$$E = \$2.10 \cdot \frac{1}{4} = \$.53$$

No, it is not a fair game since the expected value is only 53¢ and the price to play is 70¢.

LET THE KING prohibit gambling and betting in his kingdom, for these are vices that destroy the kingdoms of princes.

The Code of Manu, *about 100 A.D.*

The game becomes more complicated for picking more numbers, as illustrated by Example 6.

Example 6:   What is the expectation for playing a three-spot Keno ticket?

Solution: There are four possibilities as follows:

Pick 0 numbers, win $0.   Pick 2 numbers, win $.70.

Pick 1 number, win $0.   Pick 3 numbers, win $30.00.

We compute the expected value as follows:

$$P(\text{Picking 3}) = \frac{_{20}C_3}{_{80}C_3} = \frac{20 \cdot 19 \cdot 18}{80 \cdot 79 \cdot 78} \approx .0139$$

$$P(\text{Picking 2}) = \frac{_{20}C_2 \cdot {_{60}C_1}}{_{80}C_3} = \frac{20 \cdot 19 \cdot 60 \cdot 3}{80 \cdot 79 \cdot 78} \approx .1388$$

We see that the expected value of one 70¢ play is as follows:

$$E \approx \$30.00 \cdot .0139 + \$.70 \cdot .1388$$

$$\approx \$.51$$

Thus we can expect to win only about 51¢ while paying 70¢ to play the game.

Probabilities are often stated in terms of odds. There are two ways of stating odds, **odds in favor** and **odds against.** If the odds against your winning are 2 to 5, then the odds in favor of your winning are 5 to 2. In general, we give the following definition of odds:

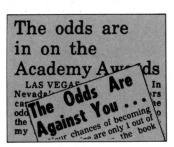

Definition of Odds

The *odds in favor of an event E,* where $P(E)$ is the probability that E will occur, is

$$P(E) \quad \text{to} \quad 1 - P(E) \qquad \text{or} \qquad \frac{P(E)}{1 - P(E)}$$

The *odds against an event E* is

$$1 - P(E) \quad \text{to} \quad P(E) \qquad \text{or} \qquad \frac{1 - P(E)}{P(E)}$$

Example 7:   If a contest has 1,000 entries and you purchase 10 tickets, what are the odds against your winning?

Solution:

$$P(\text{Winning}) = \frac{10}{1,000} = \frac{1}{100} \qquad 1 - P(\text{Winning}) = \frac{99}{100}$$

$$\textit{Odds against:} \quad \frac{\frac{99}{100}}{\frac{1}{100}} = \frac{99}{100} \times \frac{100}{1} = \frac{99}{1}$$

The odds against winning are 99 to 1.

In Example 7 we found the odds by first calculating the probability. Suppose, instead, you are given the odds and wish to calculate the probability.

Procedure for Finding the Probability, Given the Odds

If the odds in favor of an event E are s to b, then the probability of E is given by

$$P(E) = \frac{s}{b + s}.$$

Example 8: If the odds in favor of some event are 2 to 5, what is the probability?

Solution: In this example, $s = 2$ and $b = 5$, so $P(E) = \dfrac{2}{7}$.

Example 9: If the odds against you are 100 to 1, what is the probability?

Solution: First change (mentally) to odds in favor: 1 to 100.

Then, $P(E) = \dfrac{1}{101}$.

According to the *Encyclopaedia Britannica*, each game gives the casino a mathematical expectancy of winning. Over the course of a year the odds of a casino ending up without a profit have been calculated at about 1 chance in 50 billion!

The last example in this section relates to Experiment 6 of Section 9.1, but it is not necessary that you carried out that experiment to follow this example.

Example 10: Suppose your "friend" George shows you three cards. One is white on both sides, one is black on both sides, and the last is black on one side and white on the other. He mixes the cards and lets you select one at random and place it on the table. Suppose the upper side turns out to be black. It is not the white–white card; it must be the black–black or the black–white card. "Thus," says George, "I'll bet you $1 that the other side is black." Would you play?

Perhaps you hesitate. Now George says he feels generous. You need to pay him 75¢ if you lose, and he will pay you $1 if he loses. Would you play now?

Solution: Both these propositions can be answered by finding the expectation for the problem.

$$E = \text{(Amount if black)} \cdot P(\text{Black}) + \text{(Amount if white)} \cdot P(\text{White})$$

Notice: An *incorrect* assumption is that $P(\text{Black}) = P(\text{White}) = \frac{1}{2}$. Recall that the empirical results of Experiment 6 showed that $P(\text{Black}) \neq \frac{1}{2}$. What is it then? Consider the sample space. Let's distinguish between the front (side 1) and back (side 2) of each card. The sample space of *equally likely* events is:

	Card 1	Card 2	Card 3
Side showing:	B_1 B_2	B W	W_1 W_2
Side not showing:	B_2 B_1	W B	W_2 W_1

We see that

$$P(\text{Black face down}) = \frac{3}{6} = \frac{1}{2}$$

But we have additional information. We wish to compute the probability of black, *given* that a black card is face up. Alter the sample space to take into account the additional information.

	Card 1	Card 2	Card 3
Side showing:	B_1 B_2	B $\cancel{W}$	$\cancel{W_1}$ $\cancel{W_2}$
Side not showing:	B_2 B_1	W $\cancel{B}$	$\cancel{W_2}$ $\cancel{W_1}$

$$P(\text{A black is on the bottom given that a black is on top}) = \frac{2}{3}$$

Thus the desired expectation is:

$$E = \begin{pmatrix} \text{AMOUNT} \\ \text{TO LOSE} \end{pmatrix}\begin{pmatrix} \text{PROBABILITY} \\ \text{OF LOSING} \end{pmatrix} + \begin{pmatrix} \text{AMOUNT} \\ \text{TO WIN} \end{pmatrix}\begin{pmatrix} \text{PROBABILITY} \\ \text{OF WINNING} \end{pmatrix}$$

$$= (^-.75)\left(\frac{2}{3}\right) + (1)\left(\frac{1}{3}\right)$$

$$\approx {}^-.17$$

Since the expectation is negative, you should not play.

Problem Set 9.4

1. Suppose you roll two dice. You will be paid $5 if you roll a double. You will not receive anything for any other outcomes. How much should you be willing to pay for the privilege of rolling the dice?

2. A magazine subscription service is having a contest in which the prize is $80,000. If the company receives one million entries, what is the expectation of the contest?

3. Suppose you have 5 quarters, 5 dimes, 10 nickels, and 5 pennies in your pocket. You reach in and choose a coin at random so you can tip your barber. What is the barber's expectation? What tip is the barber most likely to receive?

4. A game involves tossing two coins and receiving $.50 if they are both heads. What is a fair price to pay for the privilege of playing?

5. Krinkles potato chips is having a "Lucky Seven Sweepstakes." The one grand prize is $70,000; 7 second prizes each pay $7,000; 77 third prizes each pay $700; and 777 fourth prizes each pay $70. How much is the expectation of this contest if there are 10 million entries?

6. A punch-out card contains 100 spaces. One space pays $100, 5 spaces pay $10, and the others pay nothing. How much should you pay to punch out one space?

7. In old gangster movies on TV, you often hear of "numbers runners" or the "numbers racket." The game, which is still played today, involves betting $1 and picking three digits, such as 245. Then, the next day some procedure for randomly selecting numbers is used. For example, the last three digits of the number of stocks sold on a particular day as reported in *The Wall Street Journal*. If the payoff is $500, what is the expectation for this numbers game?

8. What are the odds in favor of drawing an ace from an ordinary deck of cards?

State and national lotteries are often held to increase government income. The prizes are large, but, because of the large number of tickets sold, the expectation is low. The percentage returned to the gamblers is much smaller than with other forms of legal gambling.

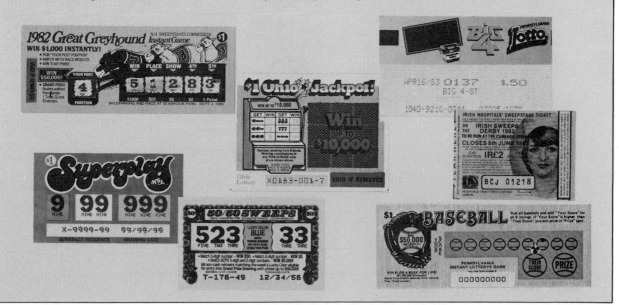

9. What are the odds in favor of drawing a heart from an ordinary deck of cards?

10. What are the odds against a family of four children containing four boys?

11. Suppose the odds that a man will be bald by the time he is 60 are 9 to 1. State this as a probability.

12. Suppose the odds are 33 to 1 that someone will lie to you at least once in the next seven days. State this as a probability.

A U.S. roulette wheel has 38 numbered slots (1–36, 0, 00), as shown in the figure. Some of the more common bets and payoffs are also shown. If the payoff is listed as 6 to 1, you would receive $7 ($6 plus $1 you bet) for each $1 you bet. Use the figure to find the expectation in Problems 13–22 for playing $1 one time. One play consists of the dealer spinning the wheel and a little white ball in opposite directions. As the ball slows to a stop it lands in one of the 38 numbered slots, which are colored black, red, or green.

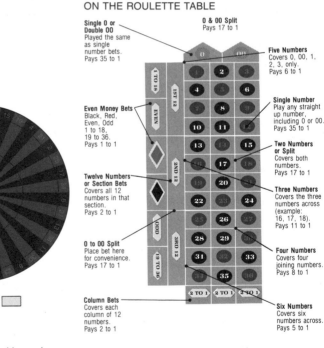

U.S. roulette wheel and board

13. Black

14. Odd

15. Single number bet

16. Double number bet

17. Three number bet

18. Four number bet

19. Five number bet

20. Six number bet

21. Twelve number bet

22. Column bet

23. A realtor who takes the listing on a house to be sold knows that she will spend $800 trying to sell the house. If she sells it herself, she will earn 6% of the selling price. If another realtor sells a house from her list, our realtor will earn only 3% of the price. If the house is unsold in 6 months, she will lose the listing. Suppose the probabilities are as follows:

	Probability
Sell by herself	.50
Sell by another realtor	.30
Not sell in 6 months	.20

What is the expected profit for listing a $85,000 house?

24. An oil drilling company knows that it will cost $25,000 to sink a test well. If oil is hit, the income for the drilling company will be $425,000. If natural gas only is hit, the income will be $125,000. If nothing is hit, they will have no income. If the probability of hitting oil is $\frac{1}{40}$ and if the probability of hitting gas is $\frac{1}{20}$, what is the expectation for the drilling company? Should they sink the test well?

25. In Problem 24, suppose the income for hitting oil is changed to $825,000 and the income for gas to $225,000. Now, what is the expectation for the drilling company? Should they sink the test well?

26. Suppose you roll one die. You are paid $5 if you roll a one, and you pay $1 otherwise. What is the expectation?

27. A game involves drawing a single card from an ordinary deck. If an ace is drawn, you receive $.50; if a heart is drawn, you receive $.25; if the queen of spades is drawn, you receive $1. If the cost of playing is $.10, should you play?

28. Consider the following game where a player rolls a single die. If a prime (2, 3, or 5) is rolled, the player wins $2. If a square (1 or 4) is rolled, the player wins $1. However, if the player rolls a perfect number (6), then it costs the player $11. Is this a good deal for the player or not?

29. Racetracks quote the approximate odds for a particular race on a large display board called a tote board. Some examples from a particular race are:

Horse Number	Odds
1	2 to 1
2	15 to 1
3	3 to 2
4	7 to 5
5	1 to 1

The odds stated are for the horse losing. Thus,

$$\text{P(Horse 1 losing)} = \frac{2}{2+1} = \frac{2}{3} \qquad \text{P(Horse 1 winning)} = 1 - \frac{2}{3} = \frac{1}{3}$$

What would be the probability of winning for each of these horses?

30. What are the odds in favor of rolling a 7 or 11 on a single roll of a pair of dice?

Prize	Number of Prizes Available	Approximate Probability of Winning
$1,000	13	.000005
100	52	.00002
10	520	.0002
1	28,900	.010989
Total	29,485	.011111

31. *Calculator problem.* A company with 210 franchises conducted the contest described in the table in the margin. What is the expectation for this game?

32. *Calculator problem.* A company held a bingo contest for which the following chances of winning were given:

Playing one card, your chances of winning are at least:

	One Card 1 Time	One Card 7 Times	One Card 13 Times
$25 prize	1 in 21,252	1 in 3,036	1 in 1,630
$3 prize	1 in 2,125	1 in 304	1 in 163
$1 prize	1 in 886	1 in 127	1 in 68
Any prize	1 in 609	1 in 87	1 in 47

What is the expectation for playing one card 13 times?

*33. Using Table 9.13, what is the expectation in playing Keno of picking one number and paying $1.40 to play?

*34. What is the expectation for playing Keno by picking two numbers and paying $.70 to play?

*35. What is the expectation for playing a three-spot ticket and paying $1.40 to play?

Mind Boggler

36. *St. Petersburg paradox.* Suppose you toss a coin and will win $1 if it comes up heads. If it comes up tails, you toss again. This time you will receive $2 if it comes up heads. If it comes up tails, toss again. This time you will receive $14 if it is heads. You continue in this fashion until you finally toss a head. Would you pay $100 for the privilege of playing this game? What is the mathematical expectation for this game?

Problem for Individual Study

37. Write a paper about the St. Petersburg paradox introduced in Problem 36.

Reference: Allen J. Ceasar, "The Saint Petersburg Paradox and Some Related Series," *Two-Year College Mathematics Journal,* November 1981.

*9.5 Probability Models

There are many ways to combine the simple probabilities you have found thus far in this chapter into new models in order to find probabilities of events that otherwise would be very difficult to find.

*Problems 33–35 and Section 9.5 require Chapter 8. Section 9.5 is optional.

The first model is to find the probability of the complementary event. Since any event either occurs or does not occur, and since we denote the occurrence of an event by E, the nonoccurrence of E is denoted by $\bar{E}$ and is called the **complement of E**. In symbols,

$$P(E) + P(\bar{E}) = 1$$

This is usually written in a more useful form:

$$P(\bar{E}) = 1 - P(E)$$

Example 1: If the probability of obtaining a pair in poker is .42, what is the probability of not obtaining a pair?

Solution: $P(\text{Pair}) = .42$, so

$$P(\text{Not a pair}) = 1 - P(\text{Pair})$$
$$= 1 - .42$$
$$= .58$$

Sometimes you need to use the ideas and combinations and permutations from the last chapter.

Example 2: What is the probability of drawing a hand of five cards and obtaining no aces?

Solution: The model for drawing a hand of cards is to use combinations, since the order in which the cards are drawn is not important.

$$P(\text{No aces}) = \frac{\text{Number of ways of drawing no aces}}{\text{Number of ways of drawing 5 cards}}$$

$$= \frac{{}_{48}C_5}{{}_{52}C_5} \begin{matrix} \leftarrow \text{There are 48 cards in the deck that are not aces.} \\ \leftarrow \text{There are 52 cards in the deck.} \end{matrix}$$

$$= \frac{\dfrac{48 \cdot 47 \cdot 46 \cdot 45 \cdot 44}{5 \cdot 4 \cdot 3 \cdot 2 \cdot 1}}{\dfrac{52 \cdot 51 \cdot 50 \cdot 49 \cdot 48}{5 \cdot 4 \cdot 3 \cdot 2 \cdot 1}}$$

$$\approx .6588$$

The distinction between using a combination model and a permutation model is illustrated by Examples 3 and 4. For these examples, suppose that two cards are drawn from a deck of cards. To draw **with replacement** means that the first card is drawn, the result noted, and then it is placed back into the deck before the second card is drawn. To draw **without replacement** means that one card is drawn, the result noted, and then a second card is drawn without replacing the first card.

Example 3: Find the probability of drawing a heart on the first draw and a spade on the second draw, given that the cards are drawn without replacement.

Solution: P(Spade on first draw and a heart on second draw)

$$= \frac{{}_{13}P_1 \cdot {}_{13}P_1}{{}_{52}P_2} \quad \text{Permutations since the order is important}$$

$$= \frac{13 \cdot 13}{52 \cdot 51}$$

$$= \frac{13}{204}$$

Example 4: Find the probability of two hearts drawn without replacement.

Solution: $$\text{P(Two hearts)} = \frac{{}_{13}C_2}{{}_{52}C_2} \quad \text{Combinations since the order is not important}$$

$$= \frac{\frac{13 \cdot 12}{2}}{\frac{52 \cdot 51}{2}}$$

$$= \frac{1}{17}$$

Notice that if the cards are drawn without replacement, the result of the first draw affects the second draw. If, on the other hand, the experiment is performed *with* replacement, the result of one draw has no effect on the result of the other draw. In such a case we say the two events are **independent.**

> **Multiplication Property of Probability**
>
> If events E and F are independent, then
>
> $$P(E \text{ and } F) = P(E) \cdot P(F)$$

Example 5: What is the probability of flipping a coin three times and receiving three tails?

Solution: $$\text{P(Three heads)} = \text{P(Heads)} \cdot \text{P(Heads)} \cdot \text{P(Heads)}$$

$$= \frac{1}{2} \cdot \frac{1}{2} \cdot \frac{1}{2}$$

$$= \frac{1}{8}$$

> **COINCIDENCE**
>
> IMPROBABLE-SOUNDING coincidences are often put forth as evidence of thought transference, or ESP. The Nobel laureate physicist Luis W. Alvarez once was startled by a coincidence that happened to him.
>
> Reading a newspaper one day, he came across a phrase that triggered associations and led to his thinking of a person from his college days, "very probably for the first time in thirty years." Five minutes later, in the same newspaper, he came across an obituary notice reporting the death of that person. Many people would have jumped to the conclusion that clairvoyance or precognition had been responsible.
>
> Not Alvarez. He realized that the connection was a coincidence, and, like the good scientist he is, he proceeded to calculate the chances of its happening. He found that the probability of thinking about a person five minutes before learning of that person's death is about 3 parts in 100,000 per year. Multiplying by the 100 million adults in the United States, an incredible 3,000 such experiences of the sort should occur every year, or about 10 per day. Seemingly improbable coincidences, in other words, are more likely to occur than we might assume.
>
> Alvarez concluded in a letter to *Science* entitled "A Pseudo Experience in Parapsychology," "With such a large sample to draw from, it is not surprising that some exceedingly astonishing coincidences are reported in the parapsychological literature as proof of extrasensory perception in one form or another."—Kendrick Frazier

Example 6: What is the probability of drawing two cards with replacement and receiving a heart and a spade?

Solution: P(Spade and a heart) = P(Spade) · P(Heart)

$$= \frac{1}{4} \cdot \frac{1}{4}$$

$$= \frac{1}{16}$$

Example 7: Consider a game consisting of three cuts with a deck of cards. You win if a face card turns up on at least one cut, but you lose if a face card does not (a face card is a jack, queen, or king). Should you play?

Solution: You should play if the probability of winning is bigger than .5 (why?). Use a tree diagram and the multiplication property of probability:

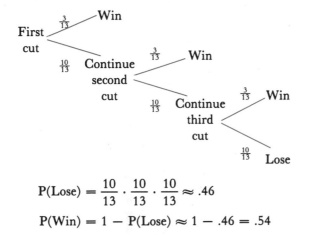

$$P(\text{Lose}) = \frac{10}{13} \cdot \frac{10}{13} \cdot \frac{10}{13} \approx .46$$

$$P(\text{Win}) = 1 - P(\text{Lose}) \approx 1 - .46 = .54$$

You should play; the expectation is

$$E = (1)(.54) + (^-1)(.46) = .08$$

The idea of independence is a hard idea for many gamblers to learn. If you have ever been to a casino and watched players bet on black or red on a roulette game, you have seen that if one color comes up several times in a row players start betting on the other color because it is "due." The spins of a roulette wheel are independent and it is a fallacy to think that one spin has any effect on any other spins. An example of this fallacy is given by Darrell Huff. On August 18, 1913, at a casino in Monte Carlo, black came up 26 times in a row on a roulette wheel. Had you bet $1 on black and continued for the entire run of blacks, you would have won $67,108,863. What actually happened confirms our statement that many people commit this fallacy. After about the 15th occurrence of black, there was a near-panic rush to bet on red. People kept doubling up on their bets

Gambling casinos like to take advantage of this fallacy in the thinking of their clients. There used to be a pair of dice displayed on a red velvet pillow behind glass in the lobby of the Desert Inn in Las Vegas. This pair of dice made 28 straight passes (winning tosses) in 1950. The man rolling them won only $750; if he had bet $1 and let his winnings double for each pass, he would have won $268,435,455.

in the belief that, after black came up the 20th time, there was not a chance in a million of another repeat. The casino came out ahead by several million francs. Remember: for each spin of the wheel, the probability of black remains the same, *regardless* of past performance.

Problem Set 9.5

Use the table of probabilities of poker hands and the formula $P(\bar{E}) = 1 - P(E)$ *to find the probabilities in Problems 1–6 correct to three decimal places. In a poker hand, five cards are drawn from a deck of 52 cards.*

1. P(Not obtaining a flush)

2. P(Not obtaining a straight)

3. P(Not obtaining two pairs)

4. Let $T = \{$Three of a kind$\}$; $P(\bar{T})$

5. Let $F = \{$Four of a kind$\}$; $P(\bar{F})$

6. Let $H = \{$Full house$\}$; $P(\bar{H})$

7. **a.** Find P(Club flush) and leave your answer in fractional form.
    b. Find P(Flush) and leave your answer in fractional form.
    c. Does your answer correspond to the entry for a flush in the table?

8. **a.** Find P(Ace, king, queen, jack, and 10 of spades) and leave your answer in fractional form.
    b. Find P(Royal flush) and leave your answer in fractional form.
    c. Does your answer correspond to the entry for a royal flush in the table?

9. **a.** Find the probability of obtaining a full house of three aces and a pair of twos and leave your answer in fractional form.
    b. Does your answer correspond to the entry for a full house in the table?

10. A news story reported that a man in Rochester, New York, took two chances at Russian roulette and killed himself on the second try. What is the probability that he would kill himself on the first try, assuming that there was only one bullet in his six-shooter? What is the probability that he would kill himself on the second try (assuming we know that he didn't kill himself on the first try)?

11. What is the probability of flipping a coin five times and obtaining five tails?

12. What is the probability of flipping a coin five times and obtaining at least one tail?

13. A game consists of three cuts with a deck of 52 cards. You win $1 if a heart turns up on at least one cut, but you lose $1 otherwise. Should you play?

14. A game consists of four cuts with a deck of 52 cards. You win $1 if a face card turns up on at least one cut, but lose $1 if a face card does not turn up. Should you play?

15. One roulette system is to bet $1 on black. If black comes up on the first spin of the wheel, you win $1 and the game is over. If black does not come up, double your bet ($2). If you win, your net winnings for two spins is still $1. Continue this doubling procedure until you eventually win; in every case your net winnings amount to $1. What is the fallacy with this betting "system"?

A club flush hand is a hand of five clubs; a flush is five of one suit—clubs, spades, hearts, or diamonds.

A royal flush is an ace, king, queen, jack, and 10 of one suite—clubs, spades, hearts, or diamonds.

Table of Probabilities of Poker Hands

Hand		Number of Favorable Events	Probability
Royal flush		4	.00000153908
Other straight flush		36	.00001385169
Four of a kind		624	.00240096038
Full house		3,744	.00144057623
Flush		5,108	.00196540155
Straight		10,200	.00392464782
Three of a kind		54,912	.02112845138
Two pair		123,552	.04753901561
One pair		1,098,240	.42256902761
Other hands		1,302,540	.50117739403
Total		2,598,960	1.00000000000

16. *Slot machine problem.* Suppose a slot machine has three identical independent wheels, each with 13 symbols as follows:

1 bar	2 lemons
2 bells	3 plums
2 cherries	3 oranges

Find the following probabilities and leave your answers in fractional form.

a. What is the probability of a jackpot (three bars)?
b. What is the probability of a cherry on the first wheel?
c. What is the probability of cherries on the first two wheels?
d. What is the probability of cherries on the first two wheels and a bar on the third wheel?
e. What is the probability of three cherries?
f. What is the probability of three oranges?
g. What is the probability of three plums?
h. What is the probability of three bells?

17. *Calculator problem.* Suppose a slot machine has three independent wheels, as shown in the margin. Answer the questions posed in Problem 16. Leave your answers in decimal form.

18. *Calculator problem: slot machine expectations.* Using Problem 17, compute the expectation for playing a slot machine that gives the following payoffs:

First one cherry	2 coins
First two cherries	5 coins
First two cherries and a bar	10 coins
Three cherries	10 coins
Three oranges	14 coins
Three plums	14 coins
Three bells	20 coins
Three bars (jackpot)	50 coins

In Problems 19–22, assume a jar has five red marbles and three black marbles and two marbles are drawn with replacement. Find the requested probabilities.

19. P(Two red marbles)

20. P(Two black marbles)

21. P(One red and one black marble)

22. P(Red marble on the first draw and black on the second draw)

In Problems 23–26, repeat the experiment of Problems 19–22, except draw without replacement.

23.	Problem 19	24.	Problem 20
25.	Problem 21	26.	Problem 22

In Problems 27–32 suppose that in an assortment of 20 electronic calculators there are 5 with defective switches.

27. If a machine is selected at random, what is the probability that it has a defective switch?

28. If two are selected at random without replacement, what is the probability that they both have defective switches?

29. Repeat Problem 28 with replacement.

30. If three are selected at random with replacement, what is the probability that they are all defective switches?

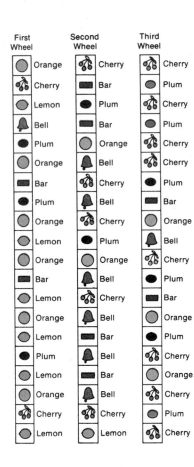

First Wheel / Second Wheel / Third Wheel

31. Repeat Problem 30 without replacement.

32. If three are selected at random, what is the probability that at least one has a defective switch?

Mind Bogglers

33. A man begins a stroll at point X. He walks north one block, at which time he flips a coin. If it's a head, he makes a right turn; if it's a tail, he makes a left turn. He walks four blocks, flipping the coin at each corner. What is the probability that he ends up at the starting point?

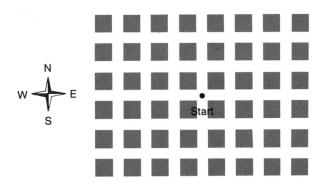

34. *Calculator problem.* Chevalier de Méré used to bet that he could get at least one 6 in four rolls of a die. He also bet that, in 24 tosses of a pair of dice, he would get at least one 12. He found that he won more often than he lost with the first bet, but not with the second. He did not know why, so he wrote to Pascal seeking the probabilities of these events. What are the probabilities for winning in these two games?

See the Historical Note on page 471.

Problem for Individual Study

35. *Computer problem: gambler's financial ruin.* In all casinos, the payoffs are adjusted so that the "house" has the advantage. This means that your expectation is always less than the amount you must pay to play the game. Thus, if you continue to play long enough, regardless of your original bankroll, you will meet your financial ruin. The BASIC program for finding the time to a gambler's ruin is shown.

All that is necessary is a beginning bankroll and the computed expectation for the game (both the bankroll and expectation must be greater than 0). Notice that the program makes the assumption that a $1 bet is made every minute continuously—with no time off. If you have ever been in a gambling casino, you will know that some people try to do precisely that—bet continuously, with no time off for anything!

```
 5  PRINT
 6  PRINT
 7  PRINT
10  PRINT "THIS PROGRAM WILL FIND THE LENGTH OF TIME TO A GAMBLER'S";
13  PRINT " RUIN."
15  PRINT "THE PROGRAM ASSUMES A $1 BET PER MINUTE."
```

"How did you do with your new blackjack system?"

```
 17  PRINT
 20  PRINT "WHAT IS THE GAMBLER'S ORIGINAL BANKROLL";
 22  INPUT B
 24  PRINT "WHAT IS THE EXPECTATION FOR THE GAME";
 26  INPUT E
 30  PRINT
 35  IF E > 1 GOTO 200
 40  IF E = 1 GOTO 250
 45  LET M = B/(1 - E)
 50  LET H = M/60
 55  LET D = H/24
100  LET A1 = D - INT(D)
105  LET A2 = 24*A1
110  LET A3 = A2 - INT(A2)
115  LET A4 = 60*A3
120  PRINT "FINANCIAL RUIN IN";
125  PRINT INT(D);
130  PRINT "DAYS,";
135  PRINT INT(A2);
140  PRINT "HOURS, AND ";
145  PRINT INT(A4);
150  PRINT "MINUTES."
155  PRINT "GOOD LUCK. BUT AS YOU SEE, FINANCIAL RUIN IS INEVITABLE."
157  PRINT
158  PRINT
159  PRINT
160  GOTO 300
200  PRINT "YOU HAVE FOUND THE PERFECT GAME! IF YOU PLAY LONG ENOUGH";
203  PRINT " YOU WILL"
205  PRINT "BE A WEALTHY PERSON."
206  PRINT
207  PRINT
208  PRINT
210  GOTO 300
250  PRINT "THIS IS A FAIR GAME. YOU CAN BREAK EVEN IF YOU PLAY";
255  PRINT "LONG ENOUGH."
260  PRINT
270  PRINT
280  PRINT
290  PRINT
300  END
```

THIS PROGRAM WILL FIND THE
LENGTH OF TIME TO A GAMBLER'S
RUIN.
THE PROGRAM ASSUMES A $1 BET
PER MINUTE.

WHAT IS THE GAMBLER'S ORIGINAL
BANKROLL? 100
WHAT IS THE EXPECTATION FOR
THE GAME? .9474

FINANCIAL RUIN IN 1 DAYS,
7 HOURS, AND 41 MINUTES.
GOOD LUCK. BUT AS YOU SEE,
FINANCIAL RUIN IS INEVITABLE.

THIS PROGRAM WILL FIND THE
LENGTH OF TIME TO A GAMBLER'S
RUIN.
THE PROGRAM ASSUMES A $1 BET
PER MINUTE.

WHAT IS THE GAMBLER'S ORIGINAL
BANKROLL? 100
WHAT IS THE EXPECTATION FOR
THE GAME? 1

THIS IS A FAIR GAME. YOU CAN
BREAK EVEN IF YOU PLAY LONG
ENOUGH.

THIS PROGRAM WILL FIND THE
LENGTH OF TIME TO A GAMBLER'S
RUIN.
THE PROGRAM ASSUMES A $1 BET
PER MINUTE.

WHAT IS THE GAMBLER'S ORIGINAL
BANKROLL? 100
WHAT IS THE EXPECTATION FOR
THE GAME? 1.25

YOU HAVE FOUND THE PERFECT
GAME! IF YOU PLAY LONG ENOUGH
YOU WILL BE A WEALTHY PERSON.

Some sample runs of this program are shown in the margin. Using a computer, verify the entries in the table.

Time to Gambler's Ruin

Game	Expectation per Game ($1 bet)	Time Required to Lose $100 by Betting $1 per Minute		
		Days	Hours	Minutes
Roulette (with 0 and 00)				
1. Black or Red bet	$.9474	1	7	41
2. Odd or Even bet	.9474	1	7	41
3. Single Number bet	.9211	0	21	7
4. Two Number bet	.8947	0	15	49
5. Three Number bet	.8684	0	12	39
6. Four Number bet	.8421	0	10	33
7. Five Number bet	.7895	0	7	55
8. Six Number bet	.7895	0	7	55
9. Twelve Number bet	.6316	0	4	31
Dice				
1. Eleven (or Three)	.8333	0	9	59
2. Twelve (or Two)	.8333	0	9	59
3. Craps	.7778	0	7	30
4. Seven	.6667	0	5	0
5. Any Doubles	.6667	0	5	0
6. Field	.4444	0	3	0
7. Under Seven (or over)	.4167	0	2	51
Slot Machine (Problem 17, Problem Set 9.5)	.6410	0	4	38
Keno				
1. One Spot	.7500	0	6	40
2. Two Spot	.7515	0	6	42
3. Three Spot	.7412	0	6	26
4. Four Spot	.7502	0	6	40
5. Five Spot	.7458	0	6	33
6. Six Spot	.7492	0	6	38
7. Seven Spot	.7477	0	6	36
8. Eight Spot	.7345	0	6	16
9. Nine Spot	.7453	0	6	32
10. Ten Spot	.7498	0	6	39
11. Eleven Spot	.7445	0	6	31
12. Twelve Spot	.7448	0	6	31
13. Thirteen Spot	.7490	0	6	38
14. Fourteen Spot	.7468	0	6	34
15. Fifteen Spot	.7444	0	6	31

Chapter 9 Outline

*I. Some Experiments in Probability
   A. Empirical probability
   B. Theoretical probability
II. Definition of Probability
   A. Terminology
      1. Sample space
      2. Event
      3. Mutually exclusive
      4. Equally likely
   B. If an experiment can occur in any of n mutually exclusive and equally likely ways, and if s of those ways are considered favorable, then the *probability* of the event E is given by

$$P(E) = \frac{s}{n} = \frac{\text{Number of outcomes favorable to } E}{\text{Number of all possible outcomes}}$$

III. Conditional Probability
   A. Altered sample space
   B. $P(E|F)$ is the probability of the event E, given the information that an event F has occurred.
IV. Mathematical Expectation
   A. Definition
   B. Fair game
   C. Odds
      1. In favor
      2. Against
*V. Probability Models
   A. Probability of the complement of an event
   B. Models using permutations and combinations
   C. Probabilities of independent events

Chapter 9 Test

1. Define the (theoretical) probability of an event.

2. A card is selected from an ordinary deck of cards. What is the probability that it is:

   a. A diamond?   **b.** A diamond and a 2?   **c.** A diamond or a 2?

*Optional sections.

3. What is the probability that Clumsy Carp will take a fall, according to the *B.C.* cartoon?

4. A die is rolled. What is the probability that it comes up:

   a. 3? **b.** 3 or 4? **c.** 3 and 4?

5. Two dice are tossed. What is the probability that the sum is 6 or at least one of the two faces is odd?

6. Two dice are thrown. What is the probability of at least one being a 6? What is the probability of both being a 6?

7. A lottery offers a prize of a color TV (value $500), and 800 tickets are sold. What is the expectation if you buy three tickets? If the tickets cost $1 each, should you buy them? How much should you be willing to pay for the three tickets?

8. If Ferdinand the Frog has twice as much chance of winning the jumping contest as the other two champion frogs:

   a. What is the probability that Ferdinand will lose?
   b. What are the odds in favor of Ferdinand's winning?

9. What is the probability (to the nearest hundredth) of drawing 5 cards (without replacement) from a deck of 52 cards and not receiving even a pair or better in poker? (See the table on page 501.)

*10. A box contains three orange and two purple balls, and two are chosen at random. What is the probability of obtaining two orange balls if we know that the first drawn was orange and we draw the second ball:

   a. After replacing the first ball?
   b. Without replacing the first ball?

Bonus Questions

11. Consider the following set of octahedral dice.

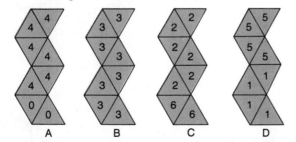

   Suppose your opponent chooses die D. Which die would you choose, and why, if the game consists of each player rolling their chosen die once and the higher number on top winning?

12. Describe a strategy for playing the game described inn Problem 11 if your opponent always has the first choice of dice. What is your probability of winning if you use your chosen strategy?

*Requires optional Section 9.5.

Date	Cultural History

1800

1803: Fulton's first steamboat
1804: Beethoven's *Eroica* symphony
1808: Goethe's *Faust*, Part I
1810: Goya's *The Disasters of War*

1815: Battle of Waterloo

1821: Rosetta Stone deciphered

Goya, *The Carnivorous Vulture*

1825

1824–27: Simon Bolivar dictator of Peru

1828: Dumas' *The Three Musketeers*

1830: Simon Bolivar liberates South America

1836: The first telegraph

1848: Marx's *Communist Manifesto*

Marx

1850

1851: Melville's *Moby Dick*
1855: Whitman's *Leaves of Grass*
1859: Darwin's *Origin of the Species*
1861: American Civil War
1862: Pasteur's germ theory of infection
1866: Alfred Nobel invents dynamite
Dostoevsky's *Crime and Punishment*

Darwin's Orangoutan

1875

1876: Bell invents telephone
1879: Edison's first light bulb
1880: Rodin's *The Thinker*

1886: Rimbaud's *Illuminations*

1888: Eastman develops box camera

1898: Spanish-American War

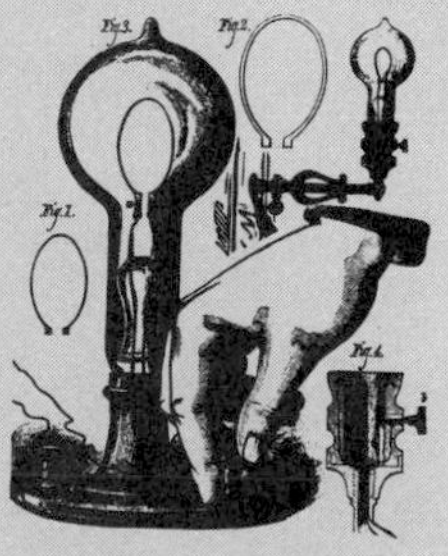

Edison's first light

Mathematical History

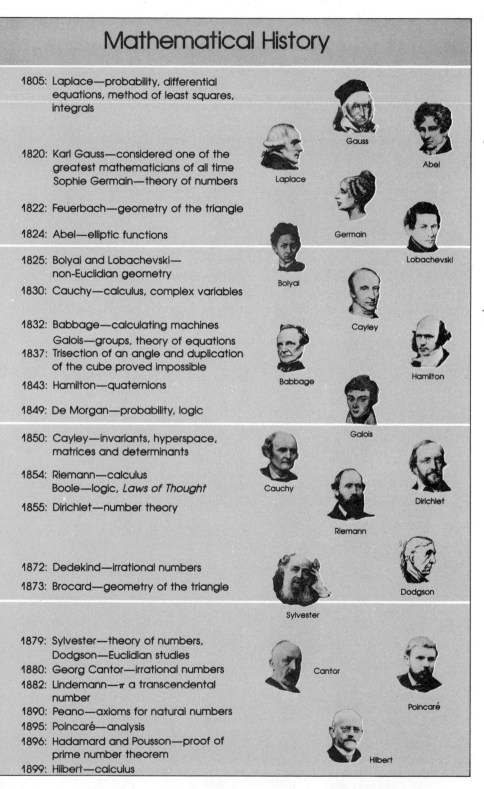

1805: Laplace—probability, differential equations, method of least squares, integrals

1820: Karl Gauss—considered one of the greatest mathematicians of all time
Sophie Germain—theory of numbers

1822: Feuerbach—geometry of the triangle

1824: Abel—elliptic functions

1825: Bolyai and Lobachevski— non-Euclidian geometry

1830: Cauchy—calculus, complex variables

1832: Babbage—calculating machines
Galois—groups, theory of equations

1837: Trisection of an angle and duplication of the cube proved impossible

1843: Hamilton—quaternions

1849: De Morgan—probability, logic

1850: Cayley—invariants, hyperspace, matrices and determinants

1854: Riemann—calculus
Boole—logic, *Laws of Thought*

1855: Dirichlet—number theory

1872: Dedekind—irrational numbers

1873: Brocard—geometry of the triangle

1879: Sylvester—theory of numbers,
Dodgson—Euclidian studies

1880: Georg Cantor—irrational numbers

1882: Lindemann—π a transcendental number

1890: Peano—axioms for natural numbers

1895: Poincaré—analysis

1896: Hadamard and Pousson—proof of prime number theorem

1899: Hilbert—calculus

Laplace · Gauss · Abel · Germain · Lobachevski · Bolyai · Cayley · Babbage · Hamilton · Galois · Cauchy · Dirichlet · Riemann · Sylvester · Dodgson · Cantor · Poincaré · Hilbert

There are three kinds of lies:
lies, damned lies,
and statistics.

Benjamin Disraeli

> • Statistical thinking will one day be as necessary for efficient citizenship as the ability to read and write.
>
> —*H. G. Wells*

10.1 Frequency Distributions

Undoubtedly, you have some idea about what is meant by the term *statistics.* For example, we hear about:

1. The latest statistics on the cost of living
2. Statistics on population growth
3. The Gallup poll's use of statistics to predict election outcomes
4. The Nielsen ratings, which show that one show has 30% more viewers than another
5. Baseball statistics

We could go on and on, but we see that there are two main uses for the word **statistics.** First, we use the term to mean a mass of data, including charts and tables. Second, the word refers to *statistical methods,* which are techniques used in the collection of numerical facts, called *data,* the analysis and interpretation of these data, and, finally, the presentation of these data. In this chapter we'll introduce some of these statistical methods.

In the last chapter, we performed experiments and gathered data. For example, in Experiment 4, we rolled a pair of dice 50 times. The outcomes were:

3	2	6	5	3	8	8	7	10	9
7	5	12	9	6	11	8	11	11	8
7	7	7	10	11	6	4	8	8	7
6	4	10	7	9	7	9	6	6	9
4	4	6	3	4	10	6	9	6	11

The next step was to organize the data in a convenient way. We used a **frequency distribution,** as shown in Table 10.1.

Example 1: The waiting time in 1982 for a marriage license by state was as follows:

TABLE 10.1 Frequency Distribution for an Experiment of Rolling a Pair of Dice 50 Times

Outcome	Tally	Frequency
2	|	1
3	|||	3
4	⊬	5
5	||	2
6	⊬ ||||	9
7	⊬ |||	8
8	⊬ |	6
9	⊬ |	6
10	||||	4
11	⊬	5
12	|	1

Alabama—None
Arizona—None
California—None
Connecticut—4 days
Florida—3 days
Hawaii—None
Illinois—1 day
Iowa—3 days
Kentucky—3 days
Maine—5 days
Massachusetts—3 days
Minnesota—5 days
Missouri—3 days
Nebraska—2 days
New Hampshire—5 days
New Mexico—None
North Carolina—None

Ohio—5 days
Oregon—3 days
Rhode Island—None
South Dakota—None
Texas—None
Vermont—3 days
Washington—3 days
Wisconsin—5 days
Alaska—3 days
Arkansas—3 days
Colorado—None
Delaware—24 hours
Georgia—3 days
Idaho—3 days
Indiana—3 days
Kansas—3 days
Louisiana—72 hours

Maryland—48 hours
Michigan—3 days
Mississippi—3 days
Montana—8 days
Nevada—None
New Jersey—72 hours
New York—None
North Dakota—None
Oklahoma—None
Pennsylvania—3 days
South Carolina—24 hours
Tennessee—3 days
Utah—None
Virginia—None
West Virginia—3 days
Wyoming—None

Prepare a frequency distribution for these data.

Solution: We make three columns. The first column will contain the number of days' wait (notice that some times are given in hours; these must be converted to days), the second will have the tally, and the third, the frequency.

Wait for Marriage License (Days)	Tally	Frequency (Number of States)
0	ЖЖ ЖЖ ЖЖ II	17
1	III	3
2	II	2
3	ЖЖ ЖЖ ЖЖ ЖЖ I	21
4	I	1
5	ЖЖ	5
6		0
7		0
8	I	1
TOTAL		50

A disadvantage of a frequency distribution is that it lacks visual appeal. Data classified into groups are often represented by means of **bar graphs.** To construct a bar graph, draw a horizontal axis and a vertical axis, as shown in Figure 10.1. We have labeled the vertical axis "Frequency" and the horizontal axis "Outcomes of Experiment of Rolling a Pair of Dice." The data are from Table 10.1.

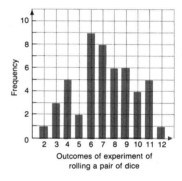

Outcomes of experiment of rolling a pair of dice

FIGURE 10.1

Example 2:   The numbers of years several leading batters played in the major leagues is as follows:

Ty Cobb	24	George Sisler	16	Willie Keeler	19
Rogers Hornsby	23	Nap Lajoie	21	Ted Williams	19
Ed Delehanty	16	Cap Anson	22	Tris Speaker	22
Dan Brouthers	19	Eddie Collins	25	Billy Hamilton	14
Sam Thompson	15	Paul Waner	20	Harry Heilmann	17
Al Simmons	20	Stan Musial	22	Babe Ruth	22
Joe DiMaggio	13	Henie Manush	17	Jesse Burkett	16
Jimmy Foxx	20	Honus Wagner	21	Bill Terry	14
Lou Gehrig	17				

Give the frequency distribution and draw the bar graph for these data.

Solution:

Number of Years	Tally	Frequency	Number of Years	Tally	Frequency
13	|	1	20	|||	3
14	||	2	21	||	2
15	|	1	22	||||	4
16	|||	3	23	|	1
17	|||	3	24	|	1
18		0	25	|	1
19	|||	3			

The bar graph for this distribution is given in Figure 10.2a.

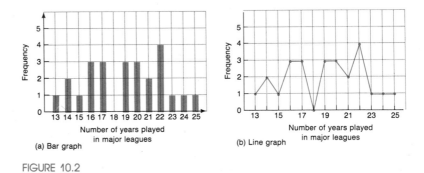

FIGURE 10.2

Another type of graph is the **line graph,** which uses dots that are marked and then connected in order. The line graph for Figure 10.2a is shown in Figure 10.2b.

Example 3:   The monthly normal rainfall for Honolulu and New Orleans is given in Figure 10.3.

a. During which month is there the most rain in New Orleans? In Honolulu?
b. During which month is there the least rain in New Orleans? In Honolulu?

Solution:

a. It rains most in New Orleans in July and in Honolulu in January.
b. The least rain falls in New Orleans in October and in Honolulu in June.

There are other kinds of graphs, some of which are discussed in Problem Set 10.1.

• Factual science may collect statistics and make charts. But its predictions are, as has been well said, but past history reversed.

—*John Dewey*

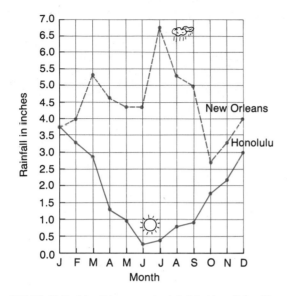

FIGURE 10.3 Monthly normal rainfall for two U.S. cities

Problem Set 10.1

1. The heights of 30 students are as follows (figures rounded to the nearest inch):

   66 68 64 70 67 67 68 64 65 66
   64 70 72 71 69 64 63 70 71 63
   68 67 67 65 69 65 67 66 69 69

   Give the frequency distribution, and draw a bar graph to represent these data.

2. The wages of the employees of a small accounting firm are as follows:

$10,000	$15,000	$15,000	$15,000
$20,000	$20,000	$25,000	$40,000
$60,000	$ 8,000	$ 8,000	$ 8,000
$ 8,000	$ 8,000	$ 6,000	$ 4,000

   Give the frequency distribution, and draw a bar graph to represent these data.

3. The line graph in the margin shows the expenses for two salespeople of the Lead-well Pencil Company for each month of last year.

   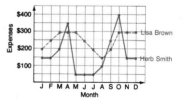

   a. During which month did Herb incur the most expense?
   b. During which month did Lisa incur the least expense?

4. The purchasing power of the dollar (1967 = $1.00) is shown (to the nearest cent):

1960	$1.13	1962	$1.10	1964	$1.07
1966	$1.03	1968	$ .96	1970	$ .86
1972	$ .80	1974	$ .68	1976	$ .59
1978	$ .49	1980	$ .38	1982	$ .32

   The purchasing power of the dollar in 1940 was $2.38. In 1950 it was $1.39.

   Show these figures by using a line graph.

5. The amount of time it takes for three leading pain relievers to reach your blood-stream is as follows:

Brand A   480 seconds   Brand B   490 seconds   Brand C   500 seconds

a. Make a bar graph using the scale shown in figure a.
b. Make a bar graph using the scale shown in figure b.

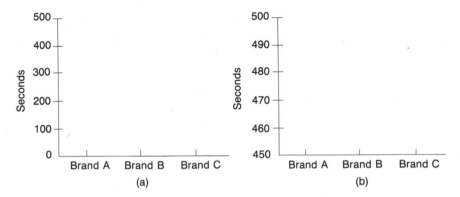

(a)   (b)

c. If you were an advertiser working on a promotion campaign for Brand A, which graph would seem to give your product a more distinct advantage?
d. Consider the graph in an Anacin advertisement, as shown in the margin. Does it tell you anything at all about the effectiveness of the three pain relievers? Discuss.

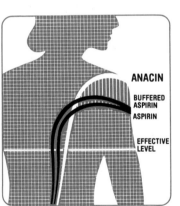

6. The percentage of claims paid by five leading insurance companies is as follows:

Company A   96.2%   Company D   96.1%
Company B   94.1%   Company E   95.9%
Company C   97.6%

a. Make a bar graph using the scale shown in figure a.
b. Make a bar graph using the scale shown in figure b.

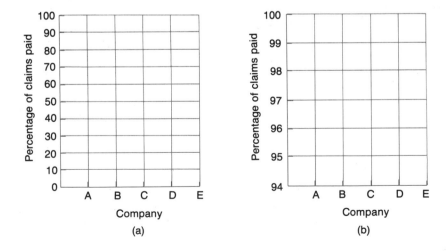

(a)   (b)

c. If you were an advertiser working on a promotion campaign for Company C, which graph would seem to give your client a more distinct advantage?

7. Comment on the graph and advertisement statement for Brand C oil shown in the margin.

8. Refer to the table of random numbers, on page 469.

   a. Select 100 random digits. Make a frequency distribution for the digits $0, 1, 2, \ldots, 8, 9$. Draw a line graph.
   b. Repeat part a for a different 100 random digits. Draw a line graph.
   c. Combine the frequencies for the digits from parts a and b, and draw a line graph.

9. Following is a list of holidays in 1984. Find the frequency distribution for the number of holidays celebrated on each day of the week, and draw a bar graph to represent the data.

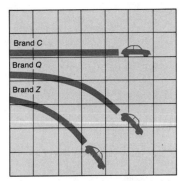

To prove that Brand C is better suited for smaller, hotter, higher-revving engines we tested Brand C against Brand Q and Brand Z. As the graph above plainly shows, only Brand C didn't break down.

January		
1 (Sun)	New Year's Day	
8 (Sun)	Battle of New Orleans (Louisiana)	
16 (Mon)	Arbor Day (Florida)	
19 (Thur)	Robert E. Lee's Birthday	

January
1 (Sun) New Year's Day
8 (Sun) Battle of New Orleans (Louisiana)
16 (Mon) Arbor Day (Florida)
19 (Thur) Robert E. Lee's Birthday

February
12 (Sun) Lincoln's Birthday
14 (Tue) Admission Day (Arizona)
20 (Mon) Washington's Birthday

March
2 (Fri) Texas Independence Day (Texas)
　　　　 Town Meeting Day (Vermont)
25 (Sun) Maryland Day (Maryland)
26 (Mon) Kuhio Day (Hawaii)
　　　　 Seward's Day (Alaska)

April
2 (Mon) Pascua Florida Day (Florida)
13 (Fri) Jefferson's Birthday (Missouri)
19 (Thur) Patriot's Day (Maine & Massachusetts)
20 (Fri) Good Friday
21 (Sat) San Jacinto Day (Texas)

22 (Sun) Arbor Day (Nebraska)
26 (Thur) Fast Day (New Hampshire)
　　　　 Confederate Memorial Day (Alabama & Mississippi)
30 (Mon) Arbor Day (Utah)

May
4 (Fri) Rhode Island Independence (Rhode Island)
10 (Thur) Confederate Memorial Day (North Carolina & South Carolina)
20 (Sun) Mecklenburg Day (North Carolina)
28 (Mon) Memorial Day

June
3 (Sun) Confederate Memorial Day (Kentucky & Louisiana)
4 (Mon) Jefferson Davis's Birthday (Alabama & Mississippi)
14 (Thur) Flag Day (Pennsylvania)
20 (Wed) West Virginia Day (West Virginia)

July
4 (Wed) Independence Day
24 (Tue) Pioneer Day (Utah)

August
6 (Mon) Colorado Day (Colorado)
13 (Mon) Victory Day (Rhode Island)
16 (Thur) Bennington Battle Day (Vermont)
17 (Fri) Admission Day (Hawaii)
27 (Mon) Lyndon Johnson's Birthday (Texas)
30 (Thur) Huey B. Long's Birthday (Louisiana)

September
3 (Mon) Labor Day
9 (Sun) Admission Day (California)
12 (Wed) Defenders Day (Maryland)

October
8 (Mon) Columbus Day
31 (Wed) Nevada Day (Nevada)

November
1 (Thur) All Saints Day (Louisiana)
6 (Tue) Election Day
11 (Sun) Veterans Day
22 (Thur) Thanksgiving

December
10 (Mon) Wyoming Day (Wyoming)
19 (Wed) Hanukkah
25 (Tue) Christmas

10. Listed below and on the facing page are the home-run champions for the American and National Leagues for every year from 1901 to 1982. Divide the data into five even categories, and construct a line graph for the number of players in each category. Do the National League and the American League separately, but put the graphs on the same coordinate system.

National League

Year	Player	No. of Home Runs	Year	Player	No. of Home Runs	Year	Player	No. of Home Runs	Year	Player	No. of Home Runs
1901	Crawford	16	1922	Hornsby	42	1941	Camilli	34	1961	Cepeda	46
1902	Leach	6	1923	Williams	41	1942	Ott	30	1962	Mays	49
1903	Sheckard	9	1924	Fournier	27	1943	Nicholson	29	1963	Aaron ⎱	44
1904	Lumley	9	1925	Hornsby	39	1944	Nicholson	33		McCovey ⎰	
1905	Odwell	9	1926	Wilson	21	1945	Holmes	28	1964	Mays	47
1906	Jordan	12	1927	Wilson ⎱	30	1946	Kiner	23	1965	Mays	52
1907	Brain	10		Williams ⎰		1947	Kiner ⎱	51	1966	Aaron	44
1908	Jordan	12	1928	Wilson ⎱	31		Mize ⎰		1967	Aaron	39
1909	Murray	7		Bottomley ⎰		1948	Kiner ⎱	40	1968	McCovey	36
1910	Beck ⎱	10	1929	Klein	43		Mize ⎰		1969	McCovey	45
	Schulte ⎰		1930	Wilson	56	1949	Kiner	54	1970	Bench	45
1911	Schulte	21	1931	Klein	31	1950	Kiner	47	1971	Stargell	48
1912	Zimmerman	14	1932	Klein ⎱	38	1951	Kiner	42	1972	Bench	40
1913	Cravath	19		Ott ⎰		1952	Kiner ⎱	37	1973	Stargell	44
1914	Cravath	19	1933	Klein	28		Sauer ⎰		1974	Schmidt	36
1915	Cravath	24	1934	Ott ⎱	35	1953	Mathews	47	1975	Schmidt	38
1916	Robertson ⎱	12		Collins ⎰		1954	Kluszewski	49	1976	Schmidt	38
	Williams ⎰		1935	Berger	34	1955	Mays	51	1977	Foster	52
1917	Robertson ⎱	12	1936	Ott	33	1956	Snider	43	1978	Foster	40
	Cravath ⎰		1937	Ott ⎱	31	1957	Aaron	44	1979	Kingman	48
1918	Cravath	8		Medwick ⎰		1958	Banks	47	1980	Schmidt	48
1919	Cravath	12	1938	Ott	36	1959	Mathews	46	1981	Schmidt	31
1920	Williams	15	1939	Mize	28	1960	Banks	41	1982	Kingman	37
1921	Kelly	23	1940	Mize	43						

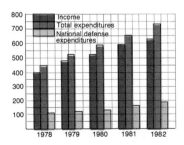

Government income and expenditures for selected years

11. Answer the following questions by referring to the bar graph.

a. In which year was the ratio of national defense expenditures to total expenditures the smallest?

b. In which year was the ratio of national defense expenditures to total income the greatest?

c. In which year was the ratio of total expenditures over income the greatest?

American League

Year	Player	No. of Home Runs	Year	Player	No. of Home Runs	Year	Player	No. of Home Runs	Year	Player	No. of Home Runs
1901	Lajoie	13	1923	Ruth	41	1945	Stephens	24	1967	Yastrzemski Killebrew	44
1902	Seybold	16	1924	Ruth	46	1946	Greenberg	44	1968	Howard	44
1903	Freeman	13	1925	Meusel	33	1947	Williams	32	1969	Killebrew	49
1904	Davis	10	1926	Ruth	47	1948	DiMaggio	39	1970	Howard	44
1905	Davis	8	1927	Ruth	60	1949	Williams	43	1971	Melton	33
1906	Davis	12	1928	Ruth	54	1950	Rosen	37	1972	Allen	37
1907	Davis	8	1929	Ruth	46	1951	Zernial	33	1973	Jackson	32
1908	Crawford	7	1930	Ruth	49	1952	Doby	32	1974	Allen	32
1909	Cobb	9	1931	Gehrig Ruth	46	1953	Rosen	43	1975	Jackson Scott	36
1910	Stahl	10				1954	Doby	32			
1911	Baker	9	1932	Foxx	58	1955	Mantle	37	1976	Nettles	32
1912	Baker	10	1933	Foxx	48	1956	Mantle	52	1977	Rice	39
1913	Baker	12	1934	Gehrig	49	1957	Sievers	42	1978	Rice	46
1914	Baker Crawford	8	1935	Foxx Greenberg	36	1958	Mantle	42	1979	Thomas	45
						1959	Colavito Killebrew	42	1980	Jackson Oglivie	41
1915	Roth	7	1936	Gehrig	49						
1916	Pipp	12	1937	DiMaggio	46	1960	Mantle	40	1981	Amas Evans Grich Murray	31
1917	Pipp	9	1938	Greenberg	58	1961	Maris	61			
1918	Ruth Walker	11	1939	Foxx	35	1962	Killebrew	48			
			1940	Greenberg	41	1963	Killebrew	45			
1919	Ruth	29	1941	Williams	37	1964	Killebrew	49	1982	Thomas Jackson	39
1920	Ruth	54	1942	Williams	36	1965	Conigliaro	32			
1921	Ruth	59	1943	York	36	1966	Robinson	49			
1922	Williams	39	1944	Etten	22						

12. Another type of graph is the **circle graph,** which is useful in displaying percentages or parts of a whole. For example, the amount of electricity used in a typical "all-electric" home is shown by the circle graph in the margin. If in a certain month a home used 1,100 kwh (kilowatt-hours), the number of kwh used by the water heater is found by looking at the graph and finding that the water heater consumes about 11% of the electricity:

$$1{,}100 \text{ kwh} \times .11 = 121 \text{ kwh}$$

a. Find the amount of electricity used by the clothes dryer in this home.
b. Find the amount of electricity used by the refrigerator in this home.

13. *Calculator problem.* According to the UCLA General Catalog, a student's principal expenses for a year are:

Registration fee	$300	Books & supplies	$ 200
Educational fee	300	Room & board	1,325
Miscellaneous fees	25	Miscellaneous	500
		TOTAL	$2,650

Make a circle graph showing these data.

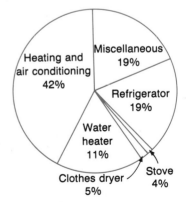

Circle graph: Distribution of the electricity used by a typical home during one month

14. Another graphical technique, which is sometimes misused, is the **pictogram.** Consider the marital status of persons age 65 and older in 1978. The raw data are shown in the table and are illustrated by the bar graph.

1978 Marital Status of Persons Age 65 and Older (in Millions)

	Married	Widowed	Divorced	*Never Married*
Women	5.4	7.2	.4	.8
Men	7.5	1.4	.3	.6

Figures rounded to the nearest 100,000.

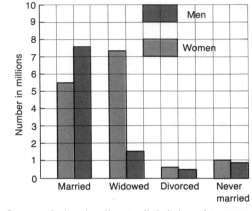

Bar graph showing the marital status of persons age 65 and older in 1978

However, this information might also be shown as a pictogram, as in figure a.

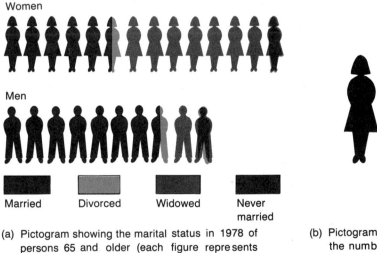

(a) Pictogram showing the marital status in 1978 of persons 65 and older (each figure represents approximately 1 million people)

(b) Pictogram showing the number of widowed persons age 65 and older in 1978

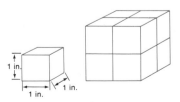

Some writers wish to exaggerate the differences, so they use the size of the objects as the measuring point in drawing the pictogram, as shown in figure b. This is very misleading, since the figures in figures a and b are viewed as three dimensional, and the differences seem much larger than they actually are. Let's look at a simpler case. Consider the "pictogram" shown in the margin.

a. How do the heights of the objects compare?
b. How do the areas of the faces compare?
c. How do the volumes compare?

Mind Boggler

15. Read the SAAB advertisement carefully. There is a fallacy in the statistics given about the car. See if you can find it.

There are two cars built in Sweden. Before you buy theirs, drive ours.

When people who know cars think about Swedish cars, they think of them as being strong and durable. And conquering some of the toughest driving conditions in the world.

But, unfortunately, when most people think about buying a Swedish car, the one they think about usually isn't ours. (Even though ours doesn't cost any more.)

Ours is the SAAB 99E. It's strong and durable. But it's also a lot different from their car.

Our car has Front-Wheel Drive for better traction, stability and handling.

It has a 1.85 liter, fuel-injected, 4-cylinder, overhead cam engine as standard in every car. 4-speed transmission is standard too. Or you can get a 3-speed automatic (optional).

Our car has four-wheel disc brakes and a dual-diagonal braking system so you stop straight and fast every time.

It has a wide stance. (About 55 inches.) So it rides and handles like a sports car.

Outside, our car is smaller than a lot of "small" cars. 172″ overall length, 57″ overall width.

Inside, our car has bucket seats up front and a full five feet across in the back so you can easily accommodate five adults.

It also has more headroom than a Rolls Royce and more room from the brake pedal to the back seat than a Mercedes 280. And it has factory air conditioning as an option.

There are a lot of other things that make our car different from their car. Like roll cage construction and a special "hot seat" for cold winter days.

So before you buy their car, stop by your nearest SAAB dealer and drive our car. The SAAB 99E. We think you'll buy it instead of theirs. **SAAB 99E**

Phone 800-243-6000 toll-free for the name and location of the SAAB dealer nearest you. In Connecticut, call 1-800-942-0655.

Problems for Individual Study

16. Collect examples of good statistical graphs and examples of misleading graphs. Use some of the leading newspapers and national magazines.

17. Read *How to Lie with Statistics* by Darrell Huff (New York: Norton, 1954), and make a presentation to the class. Illustrate your presentation with graphs and examples from the book.

10.2 Descriptive Statistics

In the last section, we organized data into a frequency distribution and then discussed their representation as a graph. However, there are some properties of data that can help us to interpret masses of information.

Violet is using one of these properties in the cartoon. Do you suppose her dad necessarily bowls better on Monday nights than on Thursday nights? Don't be too hasty to say yes unless you first look at the scores making up these averages.

	Monday night	*Thursday night*
Game 1:	175	180
Game 2:	150	130
Game 3:	160	161
Game 4:	180	185
Game 5:	160	163
Game 6:	183	185
Game 7:	287	186
TOTALS	1,295	1,190

To find the averages used by Violet, we divide the total scores by the number of games:

$$\text{\textit{Monday night}} \qquad \text{\textit{Thursday night}}$$
$$\frac{1,295}{7} = 185 \qquad \frac{1,190}{7} = 170$$

However, if we consider the games separately, we see that Violet's dad did better on Thursday in five out of seven games. Are there any other properties of the bowling scores that would tell us this fact?

The average used by Violet is only one kind of statistical measure that can be used. It is the measure that most of us think of when we hear someone use the word *average*. It is called the *mean*. In statistics, however, there are other averages besides the mean. All these terms are now defined.

Definition

Given a set of data, three common **measures of central tendency** (or **averages**) are:

1. The **mean:** the number found by adding the data and then dividing by the number of data. The mean is usually denoted by $\bar{x}$.
2. The **median:** the middle number when the numbers are arranged in order of size. If there are two middle numbers (in the case of an even number of data), the median is the mean of these two middle numbers.
3. The **mode:** the value that occurs most frequently. If there is no number that occurs more than once, there is no mode. It is possible to have more than one mode.

If we consider these other measures for the bowling scores, we find:

The median

Monday night		Thursday night
150		130
160		161
160	The middle	163
175	← number is the →	180
180	median.	185
183		185
287		186

For the median, arrange the data in order from lowest to highest.

The mode

Monday night		Thursday night
150		130
160⎤		161
160⎦	The mode is the	163
175	number that occurs	180
180	most frequently.	⎧185
183		⎩185
287		186

When we compare the three measures of the bowling scores, we find the following:

	Monday night	Thursday night
Mean	185	170
Median	175	180
Mode	160	185

We are no longer convinced that Violet's dad "did better" on Monday nights than on Thursday nights.

Example 1: Find the mean, median, and mode for the following sets of numbers.

a. 3, 5, 5, 8, 9   **b.** 4, 10, 9, 8, 9, 4, 5   **c.** 6, 5, 4, 7, 1, 9

Solution.

a. Mean: $\dfrac{\text{Sum of the terms}}{\text{Number of terms}} = \dfrac{3 + 5 + 5 + 8 + 9}{5}$

$$= \frac{30}{5}$$

$$= 6$$

There is a rather nice physical example that illustrates the idea of the mean. Consider a seesaw that consists of a plank and a movable support (called a fulcrum).

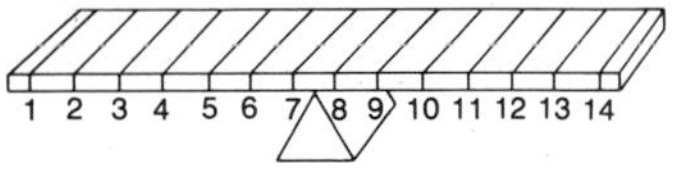

We assume that the plank has no weight and is marked off into units as shown. Now let's place some 1-pound weights in the positions of the numbers in our distribution. The balance point is the mean. In Example 1a, we place weights at 3, 5 (two weights), 8, and 9.

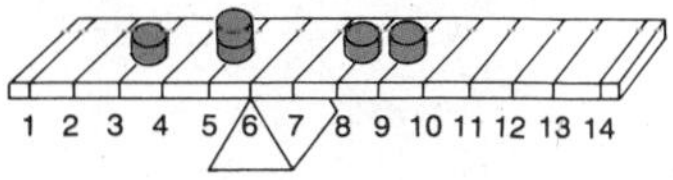

If we place the fulcrum at the mean, 6, the seesaw will balance.

Median: Arrange in order: 3, 5, 5, 8, 9
          The middle term, 5, is the median.
Mode: The term that occurs most frequently is the mode, which is 5.

b. Mean: $\dfrac{4 + 10 + 9 + 8 + 9 + 4 + 5}{7} = \dfrac{49}{7}$

$= 7$

Median: 4, 4, 5, 8, 9, 9, 10
          The median is 8.
Mode: This set of data is *bimodal,* with modes 4 and 9. *Bimodal* means
          that the set of data has two modes.

c. Mean: $\dfrac{6 + 5 + 4 + 7 + 1 + 9}{6} = \dfrac{32}{6}$

≈ 5.33

Median: 1, 4, 5, 6, 7, 9

$\dfrac{5 + 6}{2} = \dfrac{11}{2} = 5.5$

Mode: There is no mode.

TABLE 10.2

Days' Wait for a Marriage License	Frequency (Number of States)
0	17
1	3
2	2
3	21
4	1
5	5
6	0
7	0
8	1
Total	50

Suppose our data are presented in a frequency distribution. For example, consider the days one must wait for a marriage license, as shown in Table 10.2. Suppose we're interested in the average wait for a marriage license in the United States.

1. *Mean:*   To find the mean, we could, of course, add all 50 individual numbers. But, instead, notice that

0 occurs 17 times, so we write $0 \cdot 17$;

1 occurs 3 times, so we write $1 \cdot 3$;

2 occurs 2 times, so we write $2 \cdot 2$;

3 occurs 21 times, so we write $3 \cdot 21$;

$\cdot$
$\cdot$
$\cdot$

Thus the mean is

$$\bar{x} = \frac{0 \cdot 17 + 1 \cdot 3 + 2 \cdot 2 + 3 \cdot 21 + 4 \cdot 1 + 5 \cdot 5 + 6 \cdot 0 + 7 \cdot 0 + 8 \cdot 1}{50}$$

$$= \frac{0 + 3 + 4 + 63 + 4 + 25 + 0 + 0 + 8}{50}$$

$$= \frac{107}{50} = 2.14$$

2. *Median:* Since there are 50 values, the mean of the 25th and 26th largest values is the median. From Table 10.2, we see that the 25th term is 3 and the 26th term is 3, so the median is

$$\frac{3+3}{2} = \frac{6}{2} = 3$$

3. *Mode:* The mode is the value that occurs most frequently, which is 3.

The measures we've been discussing can help us interpret information, but they do not give the whole story. For example, consider the two sets of data given in Figure 10.4.

	First example	*Second example*
	8	2
	9	9
	9	9
	9	12
	10	13
Mean:	9	9
Median:	9	9
Mode:	9	9

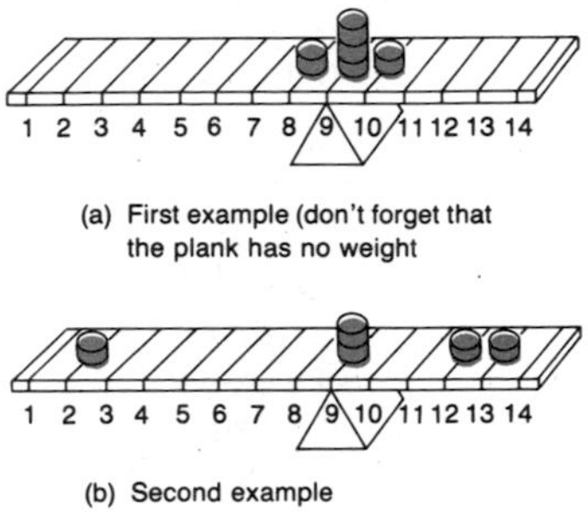

(a) First example (don't forget that the plank has no weight

(b) Second example

FIGURE 10.4 Comparison of three statistical measures for two sets of data

Notice that the two examples have the same mean, median, and mode, but the second set of data is more spread out than the first set. The amount that the data is spread out is called the **dispersion.** There are three measures of dispersion that we'll consider: the *range,* the *variance,* and the *standard deviation.*

The simplest measure of dispersion is the range.

Definition

The **range** in a set of data is the difference between the largest and the smallest numbers in the set.

The ranges for the examples given in Figure 10.4 are:

$$\text{First example:} \quad 10 - 8 = 2$$
$$\text{Second example:} \quad 13 - 2 = 11$$

Historical Note

The Wright brothers wrote to the
United States Weather Bureau
seeking a location that offered
privacy and a consistent wind
so that they could conduct
experiments. Kitty Hawk, North
Carolina, was suggested. Yet the
reported winds represented an
annual average of calms and
gales. Many days were spent in
waiting and frustration because
the winds were unsuitable. If the
Weather Bureau's report had
included information about
dispersion, the Wright brothers
might have recognized that the
acceptable average was produced
by unacceptable extremes.

Notice that the range is determined only by the largest and the smallest numbers in the set; it does not give us any information about the other numbers. It
thus seems reasonable to invent another measure of dispersion that takes into
account all the numbers in the data.

Definition

The **variance,** denoted by **var,** is a measure of dispersion which is found
as follows:

1. Determine the mean of the numbers.
2. Subtract each number from the mean.
3. Square each of these differences.
4. Find the sum of the squares of these differences.
5. Divide this sum by 1 less than the number of pieces of data.

Example 2:   Find the variance for each set of data:

a. 8, 9, 9, 9, 10   **b.** 2, 9, 9, 12, 13

Solution: Remember that the mean, median, and mode for both these examples is the same (it is 9). The range for Example a is 2 and for Example b is 11.
Now we want to find the second measure of dispersion, called *variance.* The
procedure is lengthy, so make sure you follow each step carefully.

Step 1. *Find the mean.*

a. $\bar{x} = 9$ **b.** $\bar{x} = 9$

Step 2. *Subtract each number from the mean.*

a.

Data	Difference from the Mean
8	1
9	0
9	0
9	0
10	⁻1

b.

Data	Difference from the Mean
2	7
9	0
9	0
12	⁻3
13	⁻4

Step 3. *Square each of these differences.* Notice that some of the differences in Step 2 are positive and others are negative. Remember, we wish to find a measure of total dispersion. But if we add all these differences, we won't obtain the total variability. Indeed, if we simply add the differences for either example, the sum is zero. But we don't wish to say there is no dispersion. To resolve this difficulty with positive and negative differences, we square each difference, so the result will always be nonnegative.

a.

Data	*Difference from the Mean*	*Square of the Difference*
8	1	1
9	0	0
9	0	0
9	0	0
10	$^-1$	1

b.

Data	*Difference from the Mean*	*Square of the Difference*
2	7	49
9	0	0
9	0	0
12	$^-3$	9
13	$^-4$	16

Step 4. *Find the sum of these squares.*

a. $1 + 0 + 0 + 0 + 1 = 2$ **b.** $49 + 0 + 0 + 9 + 16 = 74$

Step 5. *Divide this sum by 1 less than the number of terms.* In each of these examples there are five pieces of data, so to find the variance, divide by 4:

a. $\text{var} = \frac{2}{4} = .5$ **b.** $\text{var} = \frac{74}{4} = 18.5$

The larger the variance, the more dispersion there is in the original data.

A third measure of dispersion, and by far the most commonly used, is called the *standard deviation*. This is found by taking the square root of the variance.

Definition

The **standard deviation,** denoted by σ, is the square root of the variance. That is,

$$\sigma = \sqrt{\text{var}}$$

Historical Note

Even though the origins of statistics are ancient, the development of modern statistical techniques began in the 16th century. Adolph Quetelet (1796–1874) was the first person to apply statistical methods to accumulated data. He correctly predicted (to his own surprise) the crime and mortality rates from year to year. Florence Nightingale (1820–1910), Gregor Mendel (1822–1884), and Sir Francis Galton (1822–1911) were all proponents of the use of statistics in their work. Since the advent of the computer, statistical methods have come within the reach of almost everyone. Today, statistical methods are strongly influencing agriculture, biology, business, chemistry, economics, education, electronics, medicine, physics, psychology, and sociology.

TABLE 10.3   Squares and Square Roots

n	n^2	$\sqrt{n}$	$\sqrt{10n}$	n	n^2	$\sqrt{n}$	$\sqrt{10n}$
1	1	1.000	3.162	51	2601	7.141	22.583
2	4	1.414	4.472	52	2704	7.211	22.804
3	9	1.732	5.477	53	2809	7.280	23.022
4	16	2.000	6.325	54	2916	7.348	23.238
5	25	2.236	7.071	55	3025	7.416	23.452
6	36	2.449	7.746	56	3136	7.483	23.664
7	49	2.646	8.367	57	3249	7.550	23.875
8	64	2.828	8.944	58	3364	7.616	24.083
9	81	3.000	9.487	59	3481	7.681	24.290
10	100	3.162	10.000	60	3600	7.746	24.495
11	121	3.317	10.488	61	3721	7.810	24.698
12	144	3.464	10.954	62	3844	7.874	24.900
13	169	3.606	11.402	63	3969	7.937	25.100
14	196	3.742	11.832	64	4096	8.000	25.298
15	225	3.873	12.247	65	4225	8.062	25.495
16	256	4.000	12.649	66	4356	8.124	25.690
17	289	4.123	13.038	67	4489	8.185	25.884
18	324	4.243	13.416	68	4624	8.246	26.077
19	361	4.359	13.784	69	4761	8.307	26.268
20	400	4.472	14.142	70	4900	8.367	26.458
21	441	4.583	14.491	71	5041	8.426	26.646
22	484	4.690	14.832	72	5184	8.485	26.833
23	529	4.796	15.166	73	5329	8.544	27.019
24	576	4.899	15.492	74	5476	8.602	27.203
25	625	5.000	15.811	75	5625	8.660	27.386
26	676	5.099	16.125	76	5776	8.718	27.568
27	729	5.196	16.432	77	5929	8.775	27.749
28	784	5.292	16.733	78	6084	8.832	27.928
29	841	5.385	17.029	79	6241	8.888	28.107
30	900	5.477	17.321	80	6400	8.944	28.284
31	961	5.568	17.607	81	6561	9.000	28.460
32	1024	5.657	17.889	82	6724	9.055	28.636
33	1089	5.745	18.166	83	6889	9.110	28.810
34	1156	5.831	18.439	84	7056	9.165	28.983
35	1225	5.916	18.708	85	7225	9.220	29.155
36	1296	6.000	18.974	86	7396	9.274	29.326
37	1369	6.083	19.235	87	7569	9.327	29.496
38	1444	6.164	19.494	88	7744	9.381	29.665
39	1521	6.245	19.748	89	7921	9.434	29.833
40	1600	6.325	20.000	90	8100	9.487	30.000
41	1681	6.403	20.248	91	8281	9.539	30.166
42	1764	6.481	20.494	92	8464	9.592	30.332
43	1849	6.557	20.736	93	8649	9.644	30.496
44	1936	6.633	20.976	94	8836	9.695	30.659
45	2025	6.708	21.213	95	9025	9.747	30.822
46	2116	6.782	21.448	96	9216	9.798	30.984
47	2209	6.856	21.679	97	9409	9.849	31.145
48	2304	6.928	21.909	98	9604	9.899	31.305
49	2401	7.000	22.136	99	9801	9.950	31.464
50	2500	7.071	22.361	100	10000	10.000	31.623

The most efficient way of finding square roots is to use a key marked $\boxed{\sqrt{}}$ on your calculator. However, if you don't have a calculator, or if your calculator doesn't have this key, you can use Table 10.3.

Example 3:   Find the standard deviation for the data:

a. 8, 9, 9, 9, 10   **b.** 2, 9, 9, 12, 13

Solution: We did most of the work in finding the variance in Example 2. Now, all we need to do is find the square root of the answers to Example 2.

a. $\sqrt{.5} \approx .71$ **b.** $\sqrt{18.5} \approx 4.30$

Problem Set 10.2

Find the mean, median, mode, range, variance, and standard deviation for each set of values in Problems 1–12.

1. 1, 2, 3, 4, 5

2. 17, 18, 19, 20, 21

3. 103, 104, 105, 106, 107

4. 765, 766, 767, 768, 769

5. 4, 7, 10, 7, 5, 2, 7

6. 15, 13, 10, 7, 6, 9, 10

7. 3, 5, 8, 13, 21

8. 1, 4, 9, 16, 25

9. 79, 90, 95, 95, 96

10. 70, 81, 95, 79, 85

11. 1, 2, 3, 3, 3, 4, 5

12. 0, 1, 1, 2, 3, 4, 16

13. Compare Problems 1–4. What do you notice about the mean and standard deviation?

14. By looking at Problems 1–4 and discovering a pattern, find the mean and standard deviation of the set of numbers being juggled in the illustration in the margin.

15. Find the mean, median, and mode of the following salaries of the Moe D. Lawn Landscaping Company:

Salary	Frequency
$ 5,000	4
8,000	3
10,000	2
15,000	1

16. G. Thumb, the leading salesperson for the Moe D. Lawn Landscaping Company, turned in the following summary of sales for the week of October 23–28. Find the mean, median, and mode.

Date	Number of clients contacted by G. Thumb
Oct. 23	12
Oct. 24	9
Oct. 25	10
Oct. 26	16
Oct. 27	10
Oct. 28	21

17. Find the mean, median, and mode of the test scores shown in the margin.

217,852
217,851
217,850
217,849
217,853

Test score	Frequency
90	1
80	3
70	10
60	5
50	2

18. A class obtained the following scores on a test:

Score	Frequency
90	1
80	6
70	10
60	4
50	3
40	1

Find the mean, median, mode, and range for the class.

19. A class obtained the following scores on a test:

Score	Frequency
90	2
80	4
70	9
60	5
50	3
40	1
30	2
0	4

Find the mean, median, mode, and range for the class.

20. Suppose a variance is zero. What can you say about the data?

21. A professor gives five exams. Two students' scores have the same mean, although one student seemed to do better on all tests except one. Give an example of such scores.

22. A professor gives six exams. Two students' scores have the same mean, although one student's scores have a small standard deviation and the other student's scores have a large standard deviation. Give an example of such scores.

23. The salaries for the executives of a certain company are shown in the margin. Find the mean, median, and mode. Which measure seems to best describe the average executive salary for the company?

Position	Salary
President	$80,000
1st VP	30,000
2nd VP	30,000
Supervising Manager	24,000
Accounting Manager	20,000
Personnel Manager	20,000

24. The number of miles driven on each of five tires was 17,000, 19,000, 19,000, 20,000, and 21,000 miles. Find the mean, range, and standard deviation for these mileages.

25. What is the average length of the words in the first paragraph of this section? Find the mean, median, and mode. Which average seems more descriptive of the data? Find the range and standard deviation of the word size in this problem.

26. Repeat Problem 25 for the Historical Note on page 524.

27. Roll a single die until all six numbers occur at least once. Repeat the experiment 20 times. Find the mean, median, mode, and range of the number of tosses.

28. Roll a pair of dice until all 11 numbers occur at least once. Repeat the experiment 20 times. Find the mean, median, mode, and range of the number of tosses.

CALCULATOR PROBLEMS

Find the variance and standard deviation for the data indicated in Problems 29–34.

29. Problem 17   **30.** Problem 18   **31.** Problem 19

32. Problem 15   **33.** Problem 16   **34.** Problem 23

Many calculators have keys for finding the mean, variance, and standard deviation.

35. *Computer problem.* Below is a BASIC program for finding the mean, variance, and standard deviation. A sample output (using the data from Problem 5) is shown in the margin.

```
102  REM: COMPUTES THE MEAN, VARIANCE, AND STANDARD DEVIATION
104  LET T = 0
105  LET A = 0
106  LET B = 0
107  PRINT "INPUT DATA (TYPE 9999 TO INDICATE END OF DATA)"
110  PRINT "WHAT IS YOUR DATA";
112  INPUT N
114  IF N = 9999 THEN 120
115  LET T = T + 1
116  LET A = A + N
118  LET B = B + N*N
119  GOTO 112
120  LET M = A/T
122  LET X = B*T - A*A
124  LET D = SQR(X)/SQR(T*T - T)
126  PRINT "THE MEAN IS "M
130  PRINT
140  PRINT "THE VARIANCE IS "X/(T*T - T)
141  PRINT
145  PRINT "THE STANDARD DEVIATION IS "D
998  DATA 9999
999  END
```

```
THIS PROGRAM COMPUTES THE
MEAN, VARIANCE, AND STANDARD
DEVIATION

INPUT DATA (TYPE 9999 TO
INDICATE END OF DATA)

WHAT IS YOUR DATA? 4
? 7
? 10
? 7
? 5
? 2
? 7
? 9999

THE MEAN IS 6

THE VARIANCE IS 6.666667

THE STANDARD DEVIATION IS
2.581989
```

The computer, of course, is most useful for problems that have a large number of data. Use this program to find the mean, variance, and standard deviation of the heights given in Problem 1 of Problem Set 10.1 (page 513).

36. *Computer problem.* Find the mean, variance, and standard deviation of the number of home runs hit by the yearly champions in the National League, as given in Problem 10 of Problem Set 10.1 (pages 516–517).

Mind Bogglers

37. If you roll a die 36 times, the expected number of times for rolling each of the numbers is summarized in the table on the next page. A graph of the data in this table is shown in the figure. Find the mean, variance, and standard deviation for this model.

Experiment: Rolling a Pair of Dice
36 Times

Outcome	Expected Frequency
2	1
3	2
4	3
5	4
6	5
7	6
8	5
9	4
10	3
11	2
12	1

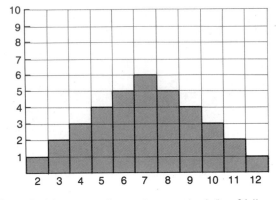

Expected frequency from rolling a pair of dice 36 times

A calculator may be helpful for
Problems 37–39.

38. *Continuation of Problem 37.* Roll a pair of dice 36 times. Construct a table and a graph similar to the ones shown for Problem 37. Find the mean, variance, and standard deviation.

39. *Continuation of Problems 37 and 38.* Compare the results of Problems 37 and 38. If this is a class problem, you might wish to pool the entire class's results for Problem 38 before making the comparison.

Problems for Individual Study

40. Prepare a report or exhibit showing how statistics are used in baseball.

41. Prepare a report or exhibit showing how statistics are used in educational testing.

42. Prepare a report or exhibit showing how statistics are used in psychology.

43. Prepare a report or exhibit showing how statistics are used in business. Use a daily report of the New York Stock Exchange's transactions. What inferences can you make from the information reported?

44. Investigate the work of Quetelet, Galton, Pearson, Fisher, and Nightingale. Prepare a report or an exhibit of their work in statistics.

10.3 The Normal Curve

The cartoon suggests that there are more children with above-normal intelligence than with normal intelligence. But what do we mean by *normal* or *normal intelligence?*

   Suppose we survey the results of 20 children's scores on an IQ test. The scores (rounded to the nearest 5 points) are: 115, 90, 100, 95, 105, 105, 95, 105, 95, 125, 120, 110, 100, 100, 90, 110, 100, 115, and 80. The mean is 103, the standard deviation is 10.93, and the frequency is shown in Figure 10.5.

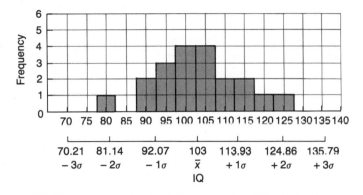

FIGURE 10.5 Frequencies of IQs for a sample of 20 children

If we consider 100,000 IQ scores instead of only 20, we might obtain the frequency distribution shown in Figure 10.6. As we can see, the frequency distribution approximates a curve. If we connect the end points of the bars in Figure 10.5, we obtain a curve that is very close to something called a *normal frequency curve,* or simply a **normal curve,** as shown in Figure 10.7. It is a theoretical distribution that extends indefinitely in both directions, but for all practical purposes, the entire distribution falls within three standard deviations on either side of the mean.

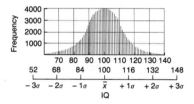

FIGURE 10.6 Frequencies of IQs for a sample of 100,000 persons

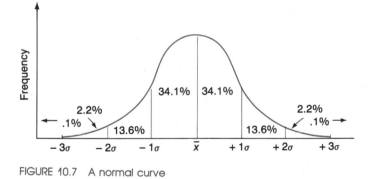

FIGURE 10.7 A normal curve

The distribution represented by the normal curve is very common and occurs in a wide variety of circumstances. For example, the heights of horses, the nodes on a jellyfish, baseball averages, the size of peas, the density of stars, the life span of light bulbs, chest size, and intelligence quotients all are normally distributed. That is, if we obtain the frequency distribution of a large number of measurements (as with the IQ example), the corresponding graph tends to look like a normal curve.

The interesting and useful property of the normal curve is that roughly 68% of all values lie within one standard deviation above and below the mean. About

96% lie within two standard deviations, and virtually all (99.8%) values lie within three standard deviations of the mean. These percentages are the same regardless of the particular mean or standard deviation. Thus, for 1,000 people taking an IQ test, about 34%, or 340 persons, would have scores between 100 and 116. These results are summarized in Table 10.4.

TABLE 10.4   Sample of 1,000 IQ Scores

Range	Percentage of Scores	Expected Number of Scores
Below 52	.1	1
52–68	2.2	22
69–84	13.6	136
85–100	34.1	341
101–116	34.1	341
117–132	13.6	136
133–148	2.2	22
Above 148	.1	1
TOTAL		1,000

Example 1:   A teacher claims to grade "on a curve." This means the teacher believes that the scores on a given test are normally distributed. If 200 students take the exam, with mean 73 and standard deviation 9, what are the grades the teacher would give?

Solution: First draw a normal curve with mean 73 and standard deviation 9, as shown in Figure 10.8. The range of 73 to 82 will contain about 34% of the class, and the range of 82 to 91 will contain about 14% of the class. Finally, about 2% of the class will score higher than 91. The teacher would therefore give grades according to the following table:

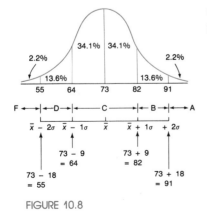

FIGURE 10.8

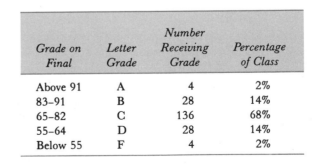

Grade on Final	Letter Grade	Number Receiving Grade	Percentage of Class
Above 91	A	4	2%
83–91	B	28	14%
65–82	C	136	68%
55–64	D	28	14%
Below 55	F	4	2%

Example 2: The Eureka Light Bulb Company tested a new line of light bulbs and found them to be normally distributed, with a mean life of 98 hours and a standard deviation of 13.

a. What percentage of bulbs will last fewer than 72 hours?
b. What is the probability that a bulb selected at random will last more than 111 hours?

Solution: Draw a normal curve with mean 98 and standard deviation 13, as shown in Figure 10.9.

a. About 2% (2.2%) will last under 72 hours.
b. We see that about 16% (15.8%) of the bulbs last longer than 111 hours, so

P(Bulb will last longer than 111 hours) ≈ .16

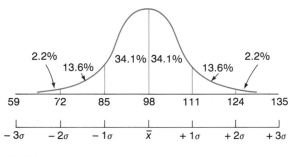

FIGURE 10.9

Problem Set 10.3

THINK
METRIC

In Problems 1–5, suppose that people's heights (in centimeters) are normally distributed, with a mean of 170 and a standard deviation of 5. We take a sample of 50 persons.

1. How many would you expect to be between 165 and 175 cm tall?

2. How many would you expect to be taller than 160 cm?

3. How many would you expect to be taller than 175 cm?

4. If a person is selected at random, what is the probability that he or she is taller than 165 cm?

5. What is the variance for this sample?

In Problems 6–10, suppose that, for a certain exam, a teacher grades on a curve. It is known that the mean is 50 and the standard deviation is 5. There are 45 students in the class.

6. How many students should receive a C?

7. How many students should receive an A?

8. What score would be necessary to obtain an A?

9. If an exam paper is selected at random, what is the probability that it will be a failing paper?

10. What is the variance for this exam?

11. Suppose the breaking strength of a rope (in pounds) is normally distributed, with a mean of 100 pounds and a standard deviation of 16. What is the probability that a certain rope will break with a force of 132 pounds?

12. The diameter of an electric cable is normally distributed, with a mean of .9 inch and a standard deviation of .01. What is the probability that the diameter will exceed .91 inch?

13. Suppose the annual rainfall in Ferndale, California, is known to be normally distributed, with a mean of 35.5 inches and a standard deviation of 2.5. About 2.2% of the time, the rainfall will exceed how many inches?

14. In Problem 13, what is the probability that the rainfall will exceed 30.5 inches in Ferndale?

15. The diameter of a pipe is normally distributed, with a mean of .4 inch and a variation of .0004. What is the probability that the diameter will exceed .44 inch?

16. The breaking strength (in pounds) of a certain new synthetic is normally distributed, with a mean of 165 and a variance of 9. The material is considered defective if the breaking strength is under 159 pounds. What is the probability that a sample chosen at random will be defective?

17. Suppose the neck size of men is normally distributed, with a mean of 15.5 inches and a standard deviation of .5. A shirt manufacturer is going to introduce a new line of shirts. How many of each of the following sizes should be included in a batch of 1,000 shirts?

    a. 14   **b.** 14.5   **c.** 15   **d.** 15.5
    e. 16   **f.** 16.5   **g.** 17

18. A package of Toys Galore Cereal is marked "Net Wt. 12 oz." The actual weight is normally distributed, with a mean of 12 oz and a variance of .04.

    a. What percentage of the packages will weigh under 12 oz?
    b. What weight will be exceeded by 2.3% of the packages?

19. Instant Dinner comes in packages that are normally distributed, with a standard deviation of .3 oz. If 2.3% of the dinners weigh more than 13.5 oz, what is the mean weight?

Mind Boggler

d = Distance between boards

Toothpick of length ℓ

20. *Buffon's needle problem.* Toss a toothpick onto a hardwood floor 1,000 times (see description in the margin), or toss 1,000 toothpicks, one at a time, onto the floor. Let ℓ be the length of the toothpick and d be the distance between the parallel lines determined by the floorboards (see the figure).

a. Make a guess as to the probability, p, that a toothpick will cross a line. Do this before you begin the experiment.

b. Perform the experiment and find p empirically.

c. By direct measurement, find ℓ and d.

d. Calculate the numbers 2ℓ and pd.

e. Divide 2ℓ by pd. What is the result?

Problems for Individual Study

21. Select something that you think might be normally distributed (for example, the ring size of students at your college). Next, select a sample of 100 people, and make the appropriate measurements (in this example, ring size). Calculate the mean and standard deviation. Are your data normally distributed? Make a presentation of your findings to the class.

22. *Statistics.* How can statistics be summarized? What are the measures of central tendency? What are measures of dispersion? How is the relationship between sets of measures determined? What is a normal curve? What is quality control? How is the accuracy of a sample measured? How are statistics used to draw conclusions?

Exhibit suggestions: Sample distributions of original data with analysis and graphs; examples of statistics found in advertisements and newspapers; examples of misuses of statistics; models of random-sampling devices and Gauss' probability board.

Equipment needed: A box of toothpicks of uniform length and a large sheet of paper with equidistant parallel lines. A hardwood floor works as well as a sheet of paper. The length of the toothpicks should be less than the perpendicular distance between the parallel lines. To find p, you will divide the number crossing a line by the number of toothpicks tossed (1,000 in this case).

*10.4 Sampling

The first sections of this chapter dealt with the accumulation of data, measures of central tendency, and dispersion. However, a more important part of statistics is its ability to help us make predictions about a population based on a sample from that population. A **sample** is a small group of items chosen to represent a larger group. The larger group is called a **population.** Sampling necessarily involves some error, and therefore statistics is also concerned with estimating the error involved in predictions based on samples.

In national polls, everyone can't be interviewed; instead, predictions are based on the responses of a small sample. The inferences drawn from a poll can, of course, be wrong. In 1936 the *Literary Digest* predicted that Alfred Landon would defeat Franklin D. Roosevelt—who was subsequently reelected by a landslide. (The magazine ceased publication the following year.) In 1948 the *Chicago Daily Tribune* drew an incorrect conclusion from its polls and declared with a headline that Dewey was elected president. As recently as 1976, the *Milwaukee Sentinel* printed the erroneous headline shown here, which was based on the result of its polls.

In an attempt to minimize error in their predictions, statisticians follow very careful procedures:

*Optional section.

JIMMY CARTER holds up an early edition of the *Milwaukee Sentinel* which declared in banner headlines that Carter had been defeated by Morris Udall in the Wisconsin primary. Late election returns, however, proved the newspaper wrong. Carter caught up and eventually defeated Udall.

Step 1. Propose some hypothesis about a population.

Step 2. Gather a sample from the population.

Step 3. Accept or reject the hypothesis. (It is also important to be able to estimate the error involved with making this decision.)

Suppose a friend hands you a coin to flip and wants to bet on the outcome. Now, this friend is known to be of dubious integrity, and you suspect that the coin is "rigged." You decide to test this hypothesis by taking a sample. You flip the coin twice, and it is heads both times. You say "Aha, I knew it was rigged." Your friend says "Don't be silly. Any coin can come up heads twice in a row."

You decide to go along with your friend for the time being, and accept the hypothesis that the coin is fair. But you decide to perform an experiment of flipping the coin 100 times. The result is:

Heads: 55   Tails: 45

Do you accept or reject the hypothesis "The coin is fair"? The expected number of heads is 50, but certainly a fair coin might well produce the results above.

As you can readily see, there are two types of error possible:

Type I: Rejection of the hypothesis when it is true

Type II: Acceptance of the hypothesis when it is false

How, then, can we proceed to minimize the possibility of making either error?

Let's carry our example further and repeat the experiment of flipping the coin 100 times:

Trial number	Number of heads
1	55
2	52
3	54
4	57
.	
.	
.	

If the coin is fair and we repeat the experiment a large number of times, the distribution would be normal, with a mean of 50 and a standard deviation of 5, as shown in Figure 10.10. Suppose you are willing to accept the coin as fair only if the number of heads falls between 45 and 55 ($\pm 1\sigma$). You know you will be correct 68% of the time. But your friend says "Wait, you're rejecting a lot of fair coins! If you accept coins between 40 and 60 ($\pm 2\sigma$), you'll be correct 96% of the time."

You say "But suppose the coin really favors heads, so that the mean is 60 with a standard deviation of 5. I'd be accepting all those coins in the shaded region of Figure 10.11."

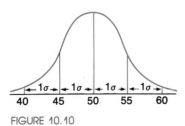

FIGURE 10.10

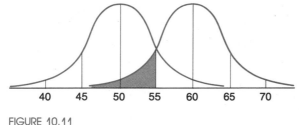

FIGURE 10.11

As you can see, a decrease in Type I error increases the Type II error, and vice versa. Deciding which type of error to minimize depends on the stakes involved as well as on some statistical calculations that are beyond the scope of this course. Consider a company that produces two types of valves. The first type is used in jet aircraft, and the failure of this valve might cause many deaths. A sample of the valves is taken, and the company must accept or reject the entire shipment. Under these circumstances, they would rather reject many good valves than accept a bad one.

On the other hand, the second valve is used in toy airplanes; the failure of this valve would cause the crash of the model. In this case, the company would not care to reject too many good valves, so they would minimize the probability of rejecting the good valves.

Problem Set 10.4

Problems for Individual Study

1. You are interested in knowing the number and ages of children (0–18 years) in a part (or all) of your community. You will need to sample 50 families and find out the number of children in each family and the age of each child. It is important that you select the 50 families at random.

   Step 1. Determine the geographical boundaries of the area with which you are concerned.
   Step 2. Consider various methods for selecting the families at random. For example, could you: (a) select the first 50 homes at which someone is at home when you call? (b) select 50 numbers from the phone book?
   Step 3. Consider different ways of asking the question. Can the way the family is approached affect the response?
   Step 4. Gather your data.
   Step 5. Organize your data. Construct a frequency distribution for the children, with integral values from 0 to 18.
   Step 6. Find out the number of families actually living in the area you've selected. If you can't do this, assume that the area has 1,000 families.

   a. What is the average number of children per family?
   b. What percentage of the children are in the first grade (age 6)?
   c. If all the children of ages 12–15 are in junior high, how many are in junior high for the geographical area you are considering?

d. See if you can actually find out the answers to parts b and c, and compare these answers with your projections.

e. What other inferences can you make from your data?

2. Five identical boxes must be prepared for this problem, with contents as follows:

Examples of how to prepare for Problem 2:

1. Use shoe boxes and marbles, beads, or poker chips.

2. Or use paper cups and beads. Cut a window out of each cup, and use a clear plastic wrap to make a cover for the cups.

3. Or use cardboard salt shakers and some small beads (a little larger than the openings at the top).

Box	Contents
1	15 red, 15 white
2	30 red, 0 white
3	25 red, 5 white
4	20 red, 10 white
5	10 red, 20 white

Select one of the boxes at random so that you do not know its contents.

Step 1. Shake the box.

Step 2. Select one marker, note the result, and return it to the box.

Step 3. Repeat the first two steps 20 times.

a. What do you think is inside the box you've sampled? That is, from the sample, do you think you can predict whether you have box 1, or box 2, or . . . ?

b. Could you have guessed the contents by repeating the experiment five times? Ten times? Do you think you should have more than 20 observations? Discuss.

Chapter 10 Outline

I. Frequency Distributions
  A. Constructing tables
  B. Bar graphs
  C. Line graphs
  D. Circle graphs
  E. Pictograms

II. Descriptive Statistics
  A. Measures of central tendency

The *mean* is the most sensitive average. It reflects the entire distribution and is probably the most important of the averages.

The *median* gives the middle value. It is especially useful when there are a few extraordinary values that distort the mean.

The *mode* is the average that measures "popularity." It is possible to have no mode or more than one mode.

  1. The **mean** is the number found by adding the values and then dividing by the number of values.
  2. The **median** is the middle number when the numbers are arranged in order of size.
  3. The **mode** is the value that occurs most frequently.
  B. Measures of dispersion
  1. The **range** in a set of data is the difference between the largest and the smallest numbers in the set.
  2. The **variance** is the sum of the squares of the differences of each number from the mean, divided by one less than the number of values.

3. The **standard deviation** is the square root of the variance.

III. The Normal Curve

    A. The normal curve is shown below.

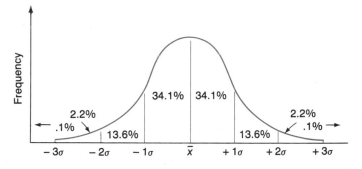

    B. The normal distribution

        1. About 68% of all values lie within one standard deviation of the mean.

        2. About 96% of all values lie within two standard deviations of the mean.

        3. About 99.8% of all values lie within three standard deviations of the mean.

*IV. Sampling

    A. Statistical predictions

        1. Propose some hypothesis about a population.

        2. Gather a sample from the population.

        3. Accept or reject the hypothesis.

    B. The accuracy of the predictions depends on the way the sample is chosen as well as on the sample size.

    C. Two types of sampling error

        1. Reject the hypothesis when it is true.

        2. Accept the hypothesis when it is false.

Chapter 10 Test

1. A writer, wishing to emphasize the tremendous shrinking of the dollar over the years, selects 1930 as his base year and constructs a graph for 1930 to 1980 [see, for example, *Building Your Fortune with Silver* by Robert Preston (Salt Lake City: Hawkes Publications, 1973)]. Another writer, wishing to minimize the effects, selects 1973 as her base year and depicts the years 1970 to 1973. A third writer selects for his base a "neutral year"—say 1967—as is done in most current almanacs (see, for example, Problem 4 in Problem Set 10.1). Discuss how the selection of different base years might support different viewpoints. Illustrate your discussion by drawing different graphs using the same data shown in the margin to support the following *opposing* viewpoints:

Year	Defense Expenditure (in Millions of Dollars)
1970	$ 78,360
1975	$ 87,471
1980	$136,138

*Optional section.

Viewpoint 1: There has been slow, steady growth in defense expenditures (about 7.4% per year for 1970–1980).

Viewpoint 2: Defense expenditures are out of hand (over $57,778,000,000 *more* was spent in 1980 than in 1970).

2. Given the data: 23, 24, 25, 26, 27, find the following.

   a. Mean   **b.** Median   **c.** Mode   **d.** Range   **e.** Variance
   f. Standard deviation

3. Given the data: 7, 8, 8, 7, 8, 9, 6, 9, 8, 6, 8, 9, 10, 7, 7, find the following.

   a. Mean   **b.** Median   **c.** Mode   **d.** Range
   e. Draw a bar graph to represent the data.

4. Find the mean, median, and mode for each of the following applications, and tell which measure of central tendency seems to be the most representative for each.

   a. A small grocery store stocked several sizes of Copycat Cola last year. The sales figures were as shown in the margin. If the store manager decides to cut back the variety and stock only one size, which measure of central tendency will be most useful in making this decision?

   b. A student's scores in a certain math class are 72, 73, 74, 85, and 91. Which measure of central tendency is most representative of the student's scores?

   c. The earnings of the employees of a small real estate company for the past year were (in thousands of dollars) 12, 18, 15, 9, 11, 22, 18, 17, and 10. Which measure of central tendency is more representative of the employees' earnings?

Size	Number of Cases Sold
6 oz	5
10 oz	10
12 oz	35
16 oz	50

5. **a.** Comment on a sales manager who tells the staff that no one should have sales below the average for the week. Is it possible for the staff to comply?

   b. An advertisement claims "Four out of five dentists surveyed recommend Brightdent Sugarless Gum for their patients who chew gum." State why you would or would not accept this as a valid claim to product superiority.

6. Find the range, variance, and standard deviation for the student's scores in Problem 4b.

7. The heights of 10,000 women students at an Eastern college are known to be normally distributed, with a mean of 67 inches and a standard deviation of 3.

   a. How many women would you expect to be between 64 and 70 inches tall?
   b. How many women would you expect to be shorter than 5′1″?

8. **a.** Two instructors gave the same Math 1A test in their classes. Both classes had the same mean, but one class had a standard deviation twice as large as that of the other class. Which class do you think would be easier to teach, and why?

   b. A student received a score of 91 on a history test for which the mean was 86 and the standard deviation 5. She also received a score of 79 on a mathematics test for which the mean was 75 and the standard deviation 2. On which test did she have the better score in comparison to her classmates?

9. **a.** Determine the upper and lower scores for the middle 68% of a normally distributed test in which $\bar{x} = 73$ and $\sigma = 8$.

   b. In a random sample of 1,000 children, IQ scores were normally distributed, with a mean of 100 and a standard deviation of 16. How many children could you expect to have an IQ greater than 132?

*10. How would you obtain a sample of one-syllable words used in this text? How would you apply those results to the whole book?

Bonus Question

11. Suppose two veteran baseball players are negotiating their contracts for the coming season. They are equal in every way and agree that the higher salary will go to the player showing the better batting average during his career. Which player would get the higher salary, given the following records:[†]

	Player A				Player B		
	At Bat	Hits	Pct		At Bat	Hits	Pct
1971	378	114	.301	1971	170	53	.312
1972	160	43	.269	1972	288	83	.288
1973	425	145	.341	1973	105	38	.362
1974	143	35	.245	1974	413	103	.249
1975	425	128	.301	1975	110	38	.345
1976	195	53	.272	1976	525	148	.282
1977	563	178	.316	1977	213	70	.329
1978	258	74	.287	1978	350	103	.294
1979	303	115	.380	1979	150	58	.387
1980	260	68	.262	1980	325	90	.277
TOTALS	3,110	953	.306		2,649	784	.296

[*]Requires optional Section 10.4.

[†]My thanks to Bill Leonard for this problem.

Date	Cultural History

Date

Cultural History

1900

1903: First powered aircraft

1908: Henry Ford's first model T

1912: Picasso's cubism

1914: World War I begins

1915: Panama-Pacific Exposition

1917: Russian Revolution

1920: U.S. women gain right to vote

1924: Lenin dies, succeeded by Stalin

Early monoplane

1925

1927: Lindbergh's solo transatlantic flight

1929: Stock market crash

1930: Gandhi leads march to the salt sea

1933: Hitler becomes chancellor of Germany

1939: World War II begins
William Carlos Williams
Collected Poems

1941: Japan bombs Pearl Harbor

1942: First controlled nuclear chain reaction

1945: United Nations formed

1946: First electronic computer

ТЫ

ЗАПИСАЛСЯ
ДОБРОВОЛЬЦЕМ?

Russian revolutionary poster

1950

1950: Korean War begins

1953: Watson and Crick discover double
helix structure of DNA

1955: Salk polio vaccine
Death of Charles Parker

1957: Sputnik I

1963: Vietnam War begins

1967: First human heart transplant

1969: First man on the moon

Ration stamps, WW II

1975

1976: Physicists discover the
"Charmed Quark"

1977: Viking mission lands on Mars

1980: Smallpox declared extinct

1981: Voyager II sends back pictures
from Saturn

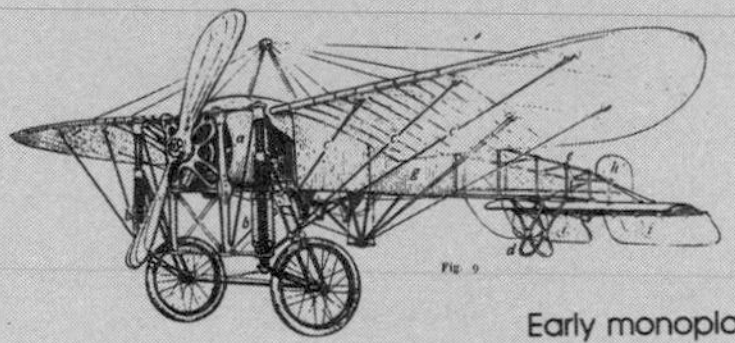

Gandhi

Mathematical History

1900: Russell and Whitehead—*Principia Mathematica,* logic
Cezanne orients paintings around the cone, sphere and rectilinear solid

1906: Fréchet—abstract spaces

1916: Einstein—general theory of relativity

1917: Hardy and Ramanujan—analytic number theory

Cezanne, *Large Bathers*

1930: Emmy Noether—algebra

1931: Gödel's theorem
Bourbaki—*Elements*

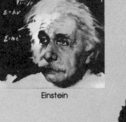

Einstein

Emmy Noether

1950: Norbert Wiener—cybernetics

1952: John von Neumann—game theory

1955: Homological algebra

1957: Datatron introduces first medium-priced computer ($325,000–$600,000)

1963: Cohen—continuum hypothesis

von Neumann

1976: Appel and Haken solve four-color problem

1977: Apple II personal computer introduced (price, $1799)

1980: Rubik's cube sweeps the world

Rubik's cube

The Nature of Mathematical Modeling

Now, when all these studies reach the point of intercommunion and connection with one another, and come to be considered in their mutual affinities, then, I think, but not till then, will the pursuit of them have a value for our objects; otherwise there is no profit in them.

Plato

*11.1 Introduction to Matrices

Mathematics may or may not describe the physical world around us. That is, mathematics can be thought of as a logical system and its structure investigated without any reference to familiar phenomena. This logical system or structure is called a **mathematical model** for the application at hand. We had a good example of mathematical modeling in Chapter 9 when we invented a theoretical definition of probability and then set about checking its consistency with the empirical results of performing experiments. The experiment, flipping a coin and obtaining 51% heads describes a real-world occurrence. The mathematical procedure of calculating the probability of heads as $\frac{1}{2}$ is a mathematical, or theoretical, model. Now, the model will be good only insofar as it is consistent with other real-world occurrences.

One of the biggest problems in applying mathematics to the real world is developing a means of systematizing and handling the great number of variables and large amounts of data that are inherent in a real-life situation. One step in handling large amounts of data is the creation of a mathematical model involving what are called **matrices.**

Definition

A **matrix** is a rectangular array of numbers. (The plural is *matrices.*)

You are already familiar with matrices from your everyday experiences. Table 11.1 shows a car rental chart given in the form of a matrix.

TABLE 11.1   Example of a Matrix:Costs of Car Rentals in Europe

Country	*Fiat Panda*	*Ford Fiesta*	*Opel Kadett*	*Renault R14*	*Volkswagen Microbus*
Austria	US $149	US $179	US $219	US $269	US $289
Belgium	130	143	222	273	N/A
Denmark	164	212	269	408	N/A
France	189	214	257	326	386
Great Britain	160	174	206	215	243
Ireland	176	185	198	222	233
Holland	156	183	213	259	305
Italy	179	213	259	353	408
Luxembourg	130	143	222	273	N/A
Spain	156	188	206	247	306
Sweden	178	205	246	281	315

*This chapter requires Section 4.5.

In mathematics, we enclose the array of numbers in brackets and denote the entire array by using a capital letter. The size of the matrix is called the **order** and is specified by naming the number of rows (rows are horizontal) first, and then the number of columns (columns are vertical). A matrix with order $m \times n$ (read m by n) means that the matrix has m rows and n columns. The order of the matrix in Table 11.1 is 11×5.

Example 1: Name the order of each given matrix.

$$A = \begin{bmatrix} 1 & 6 & 3 \\ 4 & 8 & ^-2 \end{bmatrix} \qquad B = \begin{bmatrix} 6 & 1 \\ 4 & ^-3 \\ 2 & 0 \\ ^-1 & 4 \end{bmatrix} \qquad C = \begin{bmatrix} 5 \\ 8 \\ 1 \end{bmatrix}$$

$$D = \begin{bmatrix} 4 & 8 & 6 & 2 \end{bmatrix} \qquad E = \begin{bmatrix} 6 & 8 & ^-2 \\ 5 & 4 & ^-1 \\ 0 & ^-5 & 7 \end{bmatrix}$$

Solution: The order of A is 2×3; B is 4×2; C is 3×1; D is 1×4; and E is 3×3.

If the matrix has only one column (for example, matrix C), then it is called a **column matrix**; if it has only one row (matrix D, for example), then it is called a **row matrix**; and if the number of rows and columns are the same, it is called a **square matrix**. The numbers in a matrix are called the **entries** or **components** of a matrix. If all the entries of a matrix are zero, it is called a **zero matrix**.

Two matrices are said to be **equal** if they have the same order and if the corresponding entries are equal. In Examples 2–4, find x and y, if possible, so that the pairs of matrices are equal.

Example 2: $\begin{bmatrix} 2 & x \\ y & ^-3 \end{bmatrix} = \begin{bmatrix} 2 & 5 \\ ^-1 & ^-3 \end{bmatrix}$ $\quad x = 5$ and $y = ^-1$ by inspection

Example 3: $\begin{bmatrix} 6 & 4 \\ x & y \end{bmatrix} = \begin{bmatrix} 6 & 0 \\ 4 & ^-7 \end{bmatrix}$ $\quad$ No values of x or y will make these matrices equal since $4 \neq 0$ in the upper right entries.

Example 4: $\begin{bmatrix} x & 8 \\ y & ^-4 \end{bmatrix} = \begin{bmatrix} 2 & 8 & ^-1 \\ 3 & ^-4 & 0 \end{bmatrix}$ $\quad$ No values of x and y are possible since the matrices must have the same order to be equal.

Example 5: Suppose a company produces two models of wireless telephones and manufactures them at four factories. The number of units produced at each factory can be summarized by the following matrix:

Factories

	A	B	C	D
Standard model	16	25	15	8
Deluxe model	10	0	14	3

Assume the entries represent thousands of units, and answer the following questions by reading the matrix.

a. How many standard models are produced at factory D?
b. How many standard models are produced at factory A?
c. How many deluxe models are produced at factory C?
d. At which factory are no deluxe models produced?

Solution: **a.** 8,000   **b.** 16,000   **c.** 14,000   **d.** Factory B

Sometimes entries 1 and 0 are used to summarize information in matrix form, as illustrated by Example 6.

Example 6:   Use matrix notation to summarize the following information: The United States has diplomatic relations with the U.S.S.R. and Mexico, but not with Cuba. Mexico has diplomatic relations with the United States and the U.S.S.R., but not Cuba. The U.S.S.R. has diplomatic relations with the United States, Mexico, and Cuba. And finally, Cuba has diplomatic relations with the U.S.S.R., but not with the United States and Mexico.

Solution: Let 1 be used if the row entry and the corresponding column entry have diplomatic relations and 0 if they do not. Then:

	U.S.	U.S.S.R.	Cuba	Mexico
U.S.	0	1	0	1
U.S.S.R.	1	0	1	1
Cuba	0	1	0	0
Mexico	1	1	0	0

The solution to Example 6 is called a **communication matrix.** We will use communication matrices in the next section.

If two matrices have the same order, we define the **sum** of the matrices as the matrix each of whose components is the sum of all the corresponding components of the given matrices. If the matrices are not of the same order, then we say the matrices are **not conformable** for addition.

Example 7:   Let

$$0 = [0 \quad 0 \quad 0] \qquad A = [4 \quad 2 \quad 6] \qquad B = \begin{bmatrix} 6 \\ 1 \end{bmatrix} \qquad C = [1 \quad 7 \quad 2]$$

$$D = \begin{bmatrix} 1 \\ 2 \end{bmatrix} \qquad E = \begin{bmatrix} 2 & 1 & 0 \\ 4 & 7 & 3 \\ -2 & 0 & 1 \end{bmatrix} \qquad F = \begin{bmatrix} 6 & 1 & 2 \\ 3 & -10 & 4 \\ 1 & 3 & -2 \end{bmatrix}$$

Find (whenever possible):

a. $A + C$ **b.** $B + D$ **c.** $A + 0$ **d.** $A + B$ **e.** $E + F$

Solution:

a. $A + C = [4 + 1 \quad 2 + 7 \quad 6 + 2] = [5 \quad 9 \quad 8]$

b. $B + D = \begin{bmatrix} 6 + 1 \\ 1 + 2 \end{bmatrix} = \begin{bmatrix} 7 \\ 3 \end{bmatrix}$

c. $A + 0 = [4 + 0 \quad 2 + 0 \quad 6 + 0] = [4 \quad 2 \quad 6]$

d. Not conformable

e. $E + F = \begin{bmatrix} 2 + 6 & 1 + 1 & 0 + 2 \\ 4 + 3 & 7 + (^-10) & 3 + 4 \\ ^-2 + 1 & 0 + 3 & 1 + (^-2) \end{bmatrix} = \begin{bmatrix} 8 & 2 & 2 \\ 7 & ^-3 & 7 \\ ^-1 & 3 & ^-1 \end{bmatrix}$

Example 8: Let S_{1982} and S_{1983} represent the sales (in thousands of dollars) for a company in 1982 and 1983, respectively:

$$S_{1982} = \begin{bmatrix} 150 & 200 & 350 \\ 100 & 150 & 50 \end{bmatrix} \begin{array}{l} \text{Wholesale} \\ \text{Retail} \end{array}$$

with columns Chicago, Los Angeles, New York

$$S_{1983} = \begin{bmatrix} 175 & 300 & 400 \\ 110 & 100 & 100 \end{bmatrix} \begin{array}{l} \text{Wholesale} \\ \text{Retail} \end{array}$$

with columns Chicago, Los Angeles, New York

What is the combined sales (in thousands of dollars) for the company for the given years?

Solution: The answer is found by

$$S_{1982} + S_{1983} = \begin{bmatrix} 150 + 175 & 200 + 300 & 350 + 400 \\ 100 + 110 & 150 + 100 & 50 + 100 \end{bmatrix}$$

$$= \begin{bmatrix} 325 & 500 & 750 \\ 210 & 250 & 150 \end{bmatrix}$$

For Example 8, we might be interested in finding the increase (or decrease) in sales between 1982 and 1983. We see that the difference in wholesale sales in Chicago is found by subtracting: $175 - 150 = 25$. The difference for the other components is found similarly. The **difference** of matrices A and B of the same order is found by subtracting the corresponding entries.

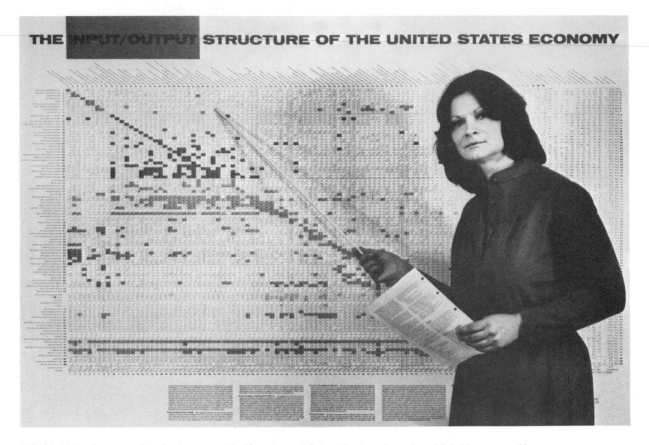

THE INPUT/OUTPUT STRUCTURE OF THE UNITED STATES ECONOMY

FIGURE 11.1   An example of a large matrix (Courtesy of Scientific American, Inc. All rights reserved.)

Can you see how the sales figures in Example 8 could be enlarged to summarize all the data for a large company like Sears? Figure 11.1 shows a chart in matrix form summarizing the input/output structure for the entire U.S. economy, based on a model developed by Wassily Leontief. When he began in the 1930s, such matrices had to be analyzed by hand; now they are being handled by computer.

Example 9:   Find the change in sales between 1982 and 1983 for the company with sales given by S_{1982} and S_{1983} in Example 8.

Solution: The answer is found by

$$S_{1983} - S_{1982} = \begin{bmatrix} 175 - 150 & 300 - 200 & 400 - 350 \\ 110 - 100 & 100 - 150 & 100 - 50 \end{bmatrix}$$

$$= \begin{bmatrix} 25 & 100 & 50 \\ 10 & ^{-}50 & 50 \end{bmatrix}$$

The negative entry indicates a decrease in sales.

Continuing with the same example, suppose we are interested in finding the profit generated by the sales in each category. It has been found that 3% of the gross sales for a given year is actual profit. In 1983, wholesale sales in Chicago generated a profit that is found by multiplying .03 (3%) by 175 (the wholesale sales in thousands of dollars in Chicago in 1983):

$$.03 \cdot 175 = 5.25$$

To find the profit in each category, we multiply the matrix

$$S_{1983} = \begin{bmatrix} 175 & 300 & 400 \\ 110 & 100 & 100 \end{bmatrix}$$

by .03. This operation of multiplying each entry of a matrix A by a number c is denoted by cA, and is called **scalar multiplication.**

Example 10: Find the profit (in thousands of dollars) for 1983 in Example 8.

Solution: Calculate $(.03)S_{1983}$.

$$(.03)S_{1983} = \begin{bmatrix} (.03)(175) & (.03)(300) & (.03)(400) \\ (.03)(110) & (.03)(100) & (.03)(100) \end{bmatrix}$$

$$= \begin{bmatrix} 5.25 & 9 & 12 \\ 3.3 & 3 & 3 \end{bmatrix}$$

You now know how to represent information in matrix form, how to add and subtract matrices, and how to multiply a matrix and a number (scalar multiplication). In the next section we'll consider how to multiply two matrices. We conclude this section with a numerical example combining scalar multiplication and subtraction.

Example 11: Find $3A - 2B$ for

$$A = \begin{bmatrix} 2 & 1 & 0 \\ 4 & 7 & 3 \\ -2 & 0 & 1 \end{bmatrix} \qquad B = \begin{bmatrix} 6 & 1 & 2 \\ 3 & -10 & 4 \\ 1 & 3 & -2 \end{bmatrix}$$

Solution:

$$3A - 2B = \begin{bmatrix} 3 \cdot 2 & 3 \cdot 1 & 3 \cdot 0 \\ 3 \cdot 4 & 3 \cdot 7 & 3 \cdot 3 \\ 3 \cdot (-2) & 3 \cdot 0 & 3 \cdot 1 \end{bmatrix} - \begin{bmatrix} 2 \cdot 6 & 2 \cdot 1 & 2 \cdot 2 \\ 2 \cdot 3 & 2 \cdot (-10) & 2 \cdot 4 \\ 2 \cdot 1 & 2 \cdot 3 & 2 \cdot (-2) \end{bmatrix}$$

$$= \begin{bmatrix} 6 - 12 & 3 - 2 & 0 - 4 \\ 12 - 6 & 21 + 20 & 9 - 8 \\ -6 - 2 & 0 - 6 & 3 + 4 \end{bmatrix}$$

$$= \begin{bmatrix} -6 & 1 & -4 \\ 6 & 41 & 1 \\ -8 & -6 & 7 \end{bmatrix}$$

Problem Set 11.1

1. State the order for each of the given matrices.

$$A = \begin{bmatrix} 6 & 1 & 4 \\ 7 & 9 & 2 \\ 1 & 5 & 3 \end{bmatrix} \quad B = \begin{bmatrix} 2 \\ 1 \\ 5 \end{bmatrix} \quad C = \begin{bmatrix} 4 & 9 \\ 1 & 6 \\ 7 & 5 \end{bmatrix} \quad D = \begin{bmatrix} 4 \\ 0 \\ 1 \\ 3 \end{bmatrix}$$

$$E = \begin{bmatrix} 4 & 1 & 7 \\ 9 & 6 & 5 \end{bmatrix} \quad F = \begin{bmatrix} 5 & 0 & 1 & 2 \end{bmatrix} \quad G = \begin{bmatrix} 6 & 9 \end{bmatrix} \quad H = \begin{bmatrix} 6 & 5 \\ 9 & 2 \end{bmatrix}$$

2. If $\begin{bmatrix} 2 & 4 & 6 \\ 5 & 9 & 3 \end{bmatrix} = \begin{bmatrix} 2 & x & y \\ z & 9 & 3 \end{bmatrix}$, what is x, y, and z?

3. **a.** On a calendar, does the month of January ever form a matrix? Why or why not?
    b. If the dates on the calendar for February form a matrix, on what day of the week does the 13th fall?

4. Suppose a company supplies four parts for General Motors. These parts are manufactured at four factories. Summarize the following production information using a matrix. Factory A produces 25 units of part 1, 42 units of part 2, and 193 units of part 3; factory B produces 16 units of part 1, 39 units of part 2, and 150 units of part 3; factory C produces 50 units of each part; and factory D produces 320 units of part 4 only.

5. A study of the eating habits of chimpanzees and three Old World monkeys (langur, macaque, baboon) showed the following: All eat insects; the macaque, baboon, and chimpanzee eat nuts; the macaque and baboon eat roots; only the chimpanzee eats honey; only the baboon eats scorpions; only the baboon and chimpanzee eat mammals. Summarize the above information in matrix form.

6. A company with five offices in the Bay Area operates its own delivery service for office mail. In the figure in the margin, arrows represent the direction of the communication. Note that A can send mail directly to offices B, D, and E but not to C. Express this delivery network in matrix form.

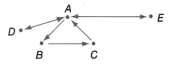

In Problems 7–22, find the indicated matrices given.

$$A = \begin{bmatrix} 2 & 0 & 7 \\ 4 & ^-1 & 3 \\ 1 & 3 & 0 \end{bmatrix} \quad B = \begin{bmatrix} 1 & 3 & 0 \\ 1 & 3 & ^-2 \\ ^-5 & 2 & 1 \end{bmatrix} \quad C = \begin{bmatrix} 4 \\ 1 \\ 6 \end{bmatrix} \quad D = \begin{bmatrix} ^-2 \\ 1 \\ 3 \end{bmatrix}$$

$$E = \begin{bmatrix} 2 & 4 & 7 \end{bmatrix} \quad F = \begin{bmatrix} 6 & ^-1 & 0 \end{bmatrix}$$

7. $A + B$	8. $B + A$	9. $A - B$	10. $B - A$
11. $C + D$	12. $C - D$	13. $E + F$	14. $E - F$
15. $2A + B$	16. $3A - 4B$	17. $2C + 3D$	18. $C - 2D$
19. $3E - 2D$	20. $3E - 2F$	21. $E + 3F$	22. $3B + 2C$

23. Suppose that one company supplies all the gas, electricity, and water to customers in a certain geographical area. Also suppose there are three types of consumers in this region: the general public, industry, and government. The *demand* matrix for each type of consumer is:

	Gas	Electricity	Water	
General public	[2	3	1]	$= P$
Industry	[4	3	2]	$= I$
Government	[2	2	1]	$= G$

What is the total demand? (It is $P + I + G$.)

Mind Bogglers

Let A, B, and C be square matrices of order 3, and let a and b be real numbers. For Problems 24–29, give an example to show that each of the following is false, or else give an example for which the property holds.

24. $A + B = B + A$

25. $A - B = B - A$

26. $(A - B) - C = A - (B - C)$

27. $(A + B) + C = A + (B + C)$

28. $a(B + C) = aB + aC$

29. $(a + b)C = aC + bC$

*30. What is the identity element for $\{A, B, C\}$ and the operation of addition?

11.2 Matrix Multiplication

Matrix multiplication is defined in a rather unusual way, so we'll begin by considering an example.

The Seedy Vin Company produces three kinds of wines: riesling (white, semidry), charbono (red, dry), and rosé (rose, dry). The company receives orders from distributors for these wines, which we can represent as a row vector:

$D = $ [Demand for riesling   Demand for charbono   Demand for rosé]

Suppose the demand on a particular day is

$$D = [15 \quad 10 \quad 20]$$

In order to produce each of these wines, the most important raw materials are grapes and labor. The numbers in the matrix below give the amounts of raw materials, expressed in convenient units, for each type of wine.

	Grapes	Labor	
Riesling	8	7	
Charbono	6	6	$= R$
Rosé	5	4	

*This problem requires Section 6.1.

Suppose the Little Old Winemaker wishes to know how much raw material he needs to satisfy the demand. First, to find the quantity of grapes needed, we would calculate the sum of the following: (1) the demand for riesling (15) times the amount of grapes needed for riesling (8); (2) the demand for charbono (10) times the amount of grapes needed for charbono (6); (3) the demand for rosé (20) times the amount of grapes needed for rosé (5). This can be shown as

$$15 \cdot 8 + 10 \cdot 6 + 20 \cdot 5 = 280$$

which can be obtained if we place matrix D to the left of matrix R and write

$$DR = [15 \quad 10 \quad 20] \begin{bmatrix} 8 & 7 \\ 6 & 6 \\ 5 & 4 \end{bmatrix}$$

The result found is defined as the entry of the first row, first column of the product.

$$DR = [\ 15 \quad 10 \quad 20\ ] \begin{bmatrix} 8 & 7 \\ 6 & 6 \\ 5 & 4 \end{bmatrix} = [\ 280 \qquad]$$

Next, to find the quantity of labor needed to meet the demand, we would take the corresponding sum of products using the amount of labor instead of the amount of grapes needed:

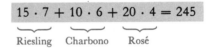

$$15 \cdot 7 + 10 \cdot 6 + 20 \cdot 4 = 245$$

Riesling   Charbono   Rosé

This is defined to be the entry of the first row, second column of the product (repeat the same procedure as with the first column):

$$DR = [\ 15 \quad 10 \quad 20\ ] \begin{bmatrix} 8 & 7 \\ 6 & 6 \\ 5 & 4 \end{bmatrix} = [280 \quad 245\ ]$$

This says that 280 units of grapes and 245 units of labor are needed.

Suppose that the grapes cost $20 per unit and labor costs $10 per unit. We can write the cost as a column vector:

$$C = \begin{bmatrix} 20 \\ 10 \end{bmatrix}$$

Next, we find the cost for each type of wine.

1. *Riesling:* Calculate the number of units of grapes (8) times the cost per unit of grapes (20), plus the number of units of labor (7) times the cost per unit of

labor (10):

$$8 \cdot 20 + 7 \cdot 10 = 230$$

This can be shown by placing the matrix R to the left of matrix C and defining this result as the entry in the first row, first column.

$$RC = \begin{bmatrix} 8 & 7 \\ 6 & 6 \\ 5 & 4 \end{bmatrix} \begin{bmatrix} 20 \\ 10 \end{bmatrix} = \begin{bmatrix} 230 \\ \\ \end{bmatrix}$$

2. *Charbono:* We find the sum

$$6 \cdot 20 + 6 \cdot 10 = 180$$

and define this result as the entry in the second row, first column of the product RC.

$$RC = \begin{bmatrix} 8 & 7 \\ 6 & 6 \\ 5 & 4 \end{bmatrix} \begin{bmatrix} 20 \\ 10 \end{bmatrix} = \begin{bmatrix} 230 \\ 180 \\ \end{bmatrix}$$

3. *Rosé:* We find the sum

$$5 \cdot 20 + 4 \cdot 10 = 140$$

and define the result as the entry in the third row, first column of the product RC.

$$RC = \begin{bmatrix} 8 & 7 \\ 6 & 6 \\ 5 & 4 \end{bmatrix} \begin{bmatrix} 20 \\ 10 \end{bmatrix} = \begin{bmatrix} 230 \\ 180 \\ 140 \end{bmatrix}$$

This criss-cross definition of multiplication can be illustrated as one mathematical step. The preceding illustrations are repeated as Examples 1 and 2, showing the mathematical process in one step.

Example 1: $DR = \begin{bmatrix} 15 & 10 & 20 \end{bmatrix} \begin{bmatrix} 8 & 7 \\ 6 & 6 \\ 5 & 4 \end{bmatrix}$

$$= \begin{bmatrix} 15 \cdot 8 + 10 \cdot 6 + 20 \cdot 5 & 15 \cdot 7 + 10 \cdot 6 + 20 \cdot 4 \end{bmatrix}$$
$$= \begin{bmatrix} 280 & 245 \end{bmatrix}$$

Example 2: $RC = \begin{bmatrix} 8 & 7 \\ 6 & 6 \\ 5 & 4 \end{bmatrix} \begin{bmatrix} 20 \\ 10 \end{bmatrix} = \begin{bmatrix} 8 \cdot 20 + 7 \cdot 10 \\ 6 \cdot 20 + 6 \cdot 10 \\ 5 \cdot 20 + 4 \cdot 10 \end{bmatrix}$

$$= \begin{bmatrix} 230 \\ 180 \\ 140 \end{bmatrix}$$

How can you decide on the order of the answer?

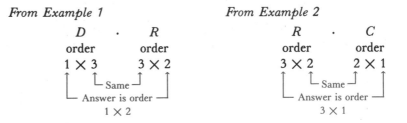

From Example 1

$$\underset{\text{order}}{D} \quad \cdot \quad \underset{\text{order}}{R}$$

$$1 \times 3 \qquad 3 \times 2$$

Same

Answer is order

$$1 \times 2$$

From Example 2

$$\underset{\text{order}}{R} \quad \cdot \quad \underset{\text{order}}{C}$$

$$3 \times 2 \qquad 2 \times 1$$

Same

Answer is order

$$3 \times 1$$

Multiplication can be extended to larger matrices as shown by Examples 3 and 4.

Example 3: Notice the numerals that are in color in this example; make sure you can follow each step.

$$AB = \begin{bmatrix} 1 & 2 \\ 3 & 4 \end{bmatrix} \begin{bmatrix} 4 & ^-1 & 2 & 6 \\ 1 & 3 & ^-1 & 2 \end{bmatrix}$$

$$= \begin{bmatrix} 1 \cdot 4 + 2 \cdot 1 & 1 \cdot (^-1) + 2 \cdot 3 & 1 \cdot 2 + 2 \cdot (^-1) & 1 \cdot 6 + 2 \cdot 2 \\ 3 \cdot 4 + 4 \cdot 1 & 3 \cdot (^-1) + 4 \cdot 3 & 3 \cdot 2 + 4 \cdot (^-1) & 3 \cdot 6 + 4 \cdot 2 \end{bmatrix}$$

$$= \begin{bmatrix} 6 & 5 & 0 & 10 \\ 16 & 9 & 2 & 26 \end{bmatrix}$$

If A has order $m \times p$ and B has order $n \times q$, then A and B are said to be *conformable* for multiplication if $n = p$. If not, then the matrices are said to be *nonconformable* for multiplication.

Example 4: Let

$$A = \begin{bmatrix} 1 & 2 \\ 3 & 4 \end{bmatrix} \quad B = \begin{bmatrix} ^-1 & 2 \\ 1 & 3 \end{bmatrix} \quad C = \begin{bmatrix} 1 & 2 & 3 \\ 4 & 5 & 6 \end{bmatrix} \quad D = \begin{bmatrix} 1 & 0 & 2 & 1 \\ 2 & 1 & 0 & 0 \\ 3 & 2 & 1 & 0 \end{bmatrix}$$

Find: **a.** AB **b.** BA **c.** CD **d.** DC

Solution:

a. $AB = \begin{bmatrix} 1 & 2 \\ 3 & 4 \end{bmatrix} \begin{bmatrix} ^-1 & 2 \\ 1 & 3 \end{bmatrix} = \begin{bmatrix} 1 \cdot (^-1) + 2 \cdot 1 & 1 \cdot 2 + 2 \cdot 3 \\ 3 \cdot (^-1) + 4 \cdot 1 & 3 \cdot 2 + 4 \cdot 3 \end{bmatrix}$

$\quad = \begin{bmatrix} 1 & 8 \\ 1 & 18 \end{bmatrix}$

b. $BA = \begin{bmatrix} ^-1 & 2 \\ 1 & 3 \end{bmatrix} \begin{bmatrix} 1 & 2 \\ 3 & 4 \end{bmatrix} = \begin{bmatrix} (^-1) \cdot 1 + 2 \cdot 3 & (^-1) \cdot 2 + 2 \cdot 4 \\ 1 \cdot 1 + 3 \cdot 3 & 1 \cdot 2 + 3 \cdot 4 \end{bmatrix}$

$\quad = \begin{bmatrix} 5 & 6 \\ 10 & 14 \end{bmatrix}$

c. $CD = \begin{bmatrix} 1 & 2 & 3 \\ 4 & 5 & 6 \end{bmatrix} \begin{bmatrix} 1 & 0 & 2 & 1 \\ 2 & 1 & 0 & 0 \\ 3 & 2 & 1 & 0 \end{bmatrix}$

$\qquad = \begin{bmatrix} 1+4+9 & 0+2+6 & 2+0+3 & 1+0+0 \\ 4+10+18 & 0+5+12 & 8+0+6 & 4+0+0 \end{bmatrix}$

$\qquad = \begin{bmatrix} 14 & 8 & 5 & 1 \\ 32 & 17 & 14 & 4 \end{bmatrix}$

d. $DC = \begin{bmatrix} 1 & 0 & 2 & 1 \\ 2 & 1 & 0 & 0 \\ 3 & 2 & 1 & 0 \end{bmatrix} \begin{bmatrix} 1 & 2 & 3 \\ 4 & 5 & 6 \end{bmatrix}$

These cannot be multiplied since the matrices are not conformable for multiplication.

Remember, matrix multiplication requires that the *first* of the two matrices being multiplied has the same number of columns as the *second* matrix has rows. If not, then they are not conformable for multiplication.

We see from Examples 4a and 4b that $AB \neq BA$. That is, matrix multiplication is not commutative. Thus, *you must be careful not to switch the order of the matrix factors.*

We conclude this section with an example of communication matrices. Example 6 of Section 11.1 writes a matrix representing the diplomatic relations of the United States, U.S.S.R., Cuba, and Mexico. Let A represent the communication matrix from that example.

$$A = \begin{bmatrix} 0 & 1 & 0 & 1 \\ 1 & 0 & 1 & 1 \\ 0 & 1 & 0 & 0 \\ 1 & 1 & 0 & 0 \end{bmatrix}$$

If we consider AA to equal A^2, this will give us the channels of communication that are open to the various countries if they are willing to speak through an intermediary.

$$A^2 = \begin{bmatrix} 0 & 1 & 0 & 1 \\ 1 & 0 & 1 & 1 \\ 0 & 1 & 0 & 0 \\ 1 & 1 & 0 & 0 \end{bmatrix} \begin{bmatrix} 0 & 1 & 0 & 1 \\ 1 & 0 & 1 & 1 \\ 0 & 1 & 0 & 0 \\ 1 & 1 & 0 & 0 \end{bmatrix}$$

$$= \begin{bmatrix} 0+1+0+1 & 0+0+0+1 & 0+1+0+0 & 0+1+0+0 \\ 0+0+0+1 & 1+0+1+1 & 0+0+0+0 & 1+0+0+0 \\ 0+1+0+0 & 0+0+0+0 & 0+1+0+0 & 0+1+0+0 \\ 0+1+0+0 & 1+0+0+0 & 0+1+0+0 & 1+1+0+0 \end{bmatrix} = \begin{bmatrix} 2 & 1 & 1 & 1 \\ 1 & 3 & 0 & 1 \\ 1 & 0 & 1 & 1 \\ 1 & 1 & 1 & 2 \end{bmatrix}$$

We see that the United States can communicate with Cuba (entry in the first row, third column) through an intermediary. The zeros indicate that the U.S.S.R. and Cuba cannot communicate through intermediaries (they can communicate only directly). Matrix A^3 would tell how many ways the countries could communicate if they used two intermediaries.

Problem Set 11.2

In Problems 1–16, find the indicated matrices, if possible, for:

$$A = \begin{bmatrix} 1 & 0 & 2 \\ 3 & ^-1 & 2 \\ 4 & 1 & 0 \end{bmatrix} \quad B = \begin{bmatrix} 1 & 4 & 0 \\ 3 & ^-1 & 2 \\ ^-2 & 1 & 5 \end{bmatrix} \quad C = \begin{bmatrix} 8 & 1 & 6 \\ 3 & 5 & 7 \\ 4 & 9 & 2 \end{bmatrix}$$

1. $A + B$
2. $2B - C$
3. AB
4. AC
5. $A(2B)$
6. BA
7. CA
8. $(B + C)A$
9. $A(B + C)$
10. BC
11. CB
12. $AB + AC$
13. $A(BC)$
14. $(AB)C$
15. $B(AC)$
16. $(AB)C$

In Problems 17–28, find the indicated matrices, if possible, for:

$$A = \begin{bmatrix} 1 & 2 \\ 4 & 0 \\ ^-1 & 3 \\ 2 & 1 \end{bmatrix} \quad B = \begin{bmatrix} 4 & 2 \\ ^-1 & 3 \end{bmatrix} \quad C = \begin{bmatrix} 1 & 0 & 0 & 0 \\ 0 & 1 & 0 & 0 \\ 0 & 0 & 1 & 0 \\ 0 & 0 & 0 & 1 \end{bmatrix} \quad D = \begin{bmatrix} 4 & 1 & 3 & 6 \\ ^-1 & 0 & ^-2 & 3 \end{bmatrix}$$

17. AB
18. BA
19. B^2
20. CA
21. BD
22. DB
23. $(B + C)A$
24. $BA + CA$
25. CD
26. A^2
27. C^3
28. B^3

29. When will matrix addition be conformable?

30. When will matrix multiplication be conformable?

31. In Problem 6 of Problem Set 11.1, you wrote the following matrix to represent the possible direct deliveries for the network shown in the margin.

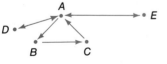

		To				
		A	B	C	D	E
	A	0	1	0	1	1
	B	0	0	1	0	0
From	C	1	0	0	0	0
	D	1	0	0	0	0
	E	1	0	0	0	0

Use matrix multiplication to find the deliveries possible if one intermediary is used.

32. Use matrix multiplication to find the deliveries of Problem 31 if two intermediaries are used.

33. In the wine example in this section, we found DR and RC.

  a. Find $D(RC)$.
  b. Attach meaning to the result of part a, given the meaning of the matrices D, R, and C as explained in the text.

34. The Seedy Vin Company produces riesling, charbono, and rosé wines. There are three procedures for producing each wine (the procedures for production affect the cost of the final product). One procedure allows an outside company to bottle the wine; a second allows the wine to be produced and bottled at the winery; a third allows the wine to be estate bottled. The amount of wine produced by the Seedy Vin Company by each method is shown by the matrix

$$W = \begin{matrix} & \text{Riesling} & \text{Charbono} & \text{Rosé} \\ & \begin{bmatrix} 2 & 1 & 3 \\ 4 & 3 & 6 \\ 1 & 2 & 4 \end{bmatrix} & & \begin{array}{l} \text{Outside bottling} \\ \text{Produced and bottled at winery} \\ \text{Estate bottled} \end{array} \end{matrix}$$

Suppose that the cost for each method of production is given by the row vector

$$C = [\text{Outside} \quad \text{Winery} \quad \text{Estate}] = [1 \quad 4 \quad 6]$$

Each unit of production cost varies by a dollar amount. Suppose that it is given by the column vector

$$D = \begin{bmatrix} \text{Riesling} \\ \text{Charbono} \\ \text{Rosé} \end{bmatrix} = \begin{bmatrix} 40 \\ 60 \\ 30 \end{bmatrix}$$

Find the cost of producing each of the three types of wines. [*Hint:* Find CW.]

35. Find the dollar amount for producing each unit of the different types of wines in Problem 34 for the three methods of production. [*Hint:* Find WD.]

36. Find the following products for the matrices of Problem 34 by calculating

  a. $(CW)D$ **b.** $C(WD)$
  c. Compare your answers for a and b. What do these products represent in the context given in Problem 34?

Mind Bogglers

37. Sociologists often study the dominance of one group over another. Suppose in a certain society there are four classes which we will call: abigweel, aweel, upancomer, pon.

  Abigweel dominate aweel, upancomer, and pon.
  Aweel dominate upancomer and pon.
  Upancomer dominate pon.

  a. Write a dominance matrix, D, representing these relationships. Use 1 to show dominance and use 0 otherwise.

b. We say that an individual has two-stage dominance over another if 1 appears in the matrix D^2 for those individuals. Write the two-stage dominance matrix.

c. We define the *power* of an individual as the sum of the entries in the appropriate row of the matrix $P = D + D^2$. Determine the power of abigweel, aweel, upancomer, and pon.

d. Rank the abigweels, aweels, upancomers, and pons.

38. In the newly formed *World Organization Opposed to Polluting the Sea* there are eight member countries. In appointing representatives to WOOPS the member countries' governing officers overlooked their appointee's language qualifications. Thus each appointee cannot speak to all of the other representatives directly. It is also possible that some representatives are acting under a "closed-mouth" policy. This policy allows a representative to listen to one speaking to him but he, in turn, will *not* speak or respond to this particular representative. Suppose:

The Spanish representative speaks to the German representative.

The German representative speaks to the Spanish and the American representatives.

The Italian representative speaks to the American and the French representatives.

The Chinese representative speaks to the Russian representative.

The American representative speaks to the German and the Italian representatives.

The Nigerian representative speaks to the Russian representative.

The Russian representative speaks to the Chinese, the Nigerian, and the French representatives.

The French representative speaks to the Russian representative.

(In this problem the French representative is the only one operating under the "closed-mouth" policy.) Finally, let's assume that a message becomes hopelessly garbled if it is translated *more* than five times.

The German representative wishes to argue the validity of throwing empty beer bottles overboard from his country's freighters. Can his statement be routed so that all representatives of WOOPS receive his message? Which countries *can* and which *cannot* get messages through to all representatives?

11.3 Systems of Equations

In Section 6.2, we solved linear equations such as

$$2x + 3 = 5$$

A **solution** to this equation is a replacement for the variable that makes this equation true; namely, $x = 1$:

$$2(1) + 3 = 5 \quad \text{is true}$$

A **system of equations** is two or more equations that are presented together for solution by using a brace:

$$\begin{cases} 2x + 3y = 8 \\ 3x + 2y = 7 \end{cases}$$

A **solution** to a system is a set of replacements for all the variables so that all the equations are true simultaneously, or at the same time; namely, $x = 1$ and $y = 2$:

$$2(1) + 3(2) = 2 + 6 = 8 \quad \text{is true}$$
$$\text{and}$$
$$3(1) + 2(2) = 3 + 4 = 7 \quad \text{is true}$$

We will now use matrices to find a solution to a system of equations.

Two matrices A and B of the same order are said to be **equivalent,** written $A \sim B$ or $B \sim A$, if A can be changed into B by one or more of the following **elementary row operations:**

1. Interchange two rows.
2. Multiply any row by a nonzero real number.
3. Add a multiple of any row to another row.

Example 1: **a.** $\begin{bmatrix} 2 & 4 \\ 6 & 8 \end{bmatrix} \sim \begin{bmatrix} 6 & 8 \\ 2 & 4 \end{bmatrix}$ Interchange two rows.

b. $\begin{bmatrix} 6 & ^-1 \\ 3 & ^-2 \end{bmatrix} \sim \begin{bmatrix} 12 & ^-2 \\ 3 & ^-2 \end{bmatrix}$ Multiply row 1 by 2.

c. $\times 2 \begin{bmatrix} 8 & 6 \\ ^-4 & ^-3 \end{bmatrix} \sim \begin{bmatrix} 0 & 0 \\ ^-4 & ^-3 \end{bmatrix}$ Multiply row 2 by 2 and add it to row 1; the way we will show this is with the little arrow and numeral, as shown at the left.

We'll now use elementary row operations to solve a system.

Example 2: $\begin{cases} 4x + 3y = 6 \\ x - 5y = 13 \end{cases}$

Solution: Write the matrix:

$$\begin{bmatrix} 4 & 3 & | & 6 \\ 1 & ^-5 & | & 13 \end{bmatrix}$$ We call this matrix the **augmented matrix.**

Coefficients of   Constant
the variables   terms

Use the elementary row operations to transform the left side of the augmented matrix to the form

$$\begin{bmatrix} 1 & 0 \\ 0 & 1 \end{bmatrix}$$

if possible. Notice that as we transform the left side, we carry out the same operations on the right side.

┌─ **Step 1.** *Obtain a 1 here.*

$$\begin{bmatrix} 4 & 3 & \vdots & 6 \\ 1 & -5 & \vdots & 13 \end{bmatrix} \sim \begin{bmatrix} ① & -5 & \vdots & 13 \\ 4 & 3 & \vdots & 6 \end{bmatrix}$$ Interchange rows.

$$\sim \begin{bmatrix} 1 & -5 & \vdots & 13 \\ ⓪ & 23 & \vdots & -46 \end{bmatrix}$$ Multiply row 1 by $^-4$ and add it to row 2, leaving the result in row 2.

└─ **Step 2.** *Obtain a 0 here under the 1.*

$$\begin{bmatrix} 1 & -5 & \vdots & 13 \\ 0 & ① & \vdots & -2 \end{bmatrix}$$ Multiply row 2 by $\frac{1}{23}$ (or divide by 23).

└─ **Step 3.** *Obtain a 1 here.*

┌─ **Step 4.** *Obtain a 0 here above the 1.*

$$\begin{bmatrix} 1 & ⓪ & \vdots & 3 \\ 0 & 1 & \vdots & -2 \end{bmatrix}$$ Multiply row 2 by 5 and add it to row 1.

The answer is found in this column: $x = 3$ and $y = {}^-2$.

Example 3:   Solve the system

$$\begin{cases} x + 2y - z = 0 \\ 2x + 3y - 2z = 3 \\ {}^-x - 4y + 3z = {}^-2 \end{cases}$$

Solution:

$${}^{-2}\begin{bmatrix} 1 & 2 & -1 & \vdots & 0 \\ 2 & 3 & -2 & \vdots & 3 \\ -1 & -4 & 3 & \vdots & -2 \end{bmatrix} \sim \begin{bmatrix} 1 & 2 & -1 & \vdots & 0 \\ 0 & -1 & 0 & \vdots & 3 \\ 0 & -2 & 2 & \vdots & -2 \end{bmatrix}$$ Add $^-2$ times the first row to the second row and add the first row to the third row.

$$\sim \begin{bmatrix} 1 & 2 & -1 & \vdots & 0 \\ 0 & 1 & 0 & \vdots & -3 \\ 0 & -2 & 2 & \vdots & -2 \end{bmatrix}$$ Multiply the second row by $^-1$.

$$\sim \begin{bmatrix} 1 & 2 & -1 & \vdots & 0 \\ 0 & 1 & 0 & \vdots & -3 \\ 0 & 0 & 2 & \vdots & -8 \end{bmatrix}$$ Add 2 times the second row to the third row.

X Y Z

WHO NEEDS 'EM !

$$\begin{bmatrix} 1 & 2 & -1 & \vdots & 0 \\ 2 & 3 & -2 & \vdots & 3 \\ -1 & -4 & 3 & \vdots & -2 \end{bmatrix}$$

$$\sim \begin{bmatrix} 1 & 2 & -1 & | & 0 \\ 0 & 1 & 0 & | & -3 \\ 0 & 0 & 1 & | & -4 \end{bmatrix} \quad \text{Multiply the third row by } \tfrac{1}{2}.$$

$$\sim \begin{bmatrix} 1 & 2 & 0 & | & -4 \\ 0 & 1 & 0 & | & -3 \\ 0 & 0 & 1 & | & -4 \end{bmatrix} \quad \text{Add the third row to the first row.}$$

$$\sim \begin{bmatrix} 1 & 0 & 0 & | & 2 \\ 0 & 1 & 0 & | & -3 \\ 0 & 0 & 1 & | & -4 \end{bmatrix} \quad \begin{array}{l}\text{Multiply the second} \\ \text{row by } ^-2 \text{ and add it} \\ \text{to the first row.}\end{array}$$

The solution is $x = 2$, $y = {}^-3$, and $z = {}^-4$.

Example 4: A rancher has to mix three types of feed for her cattle. The following analysis shows the amounts per bag (100 lb) of grain.

Grain	Protein	Carbohydrates	Sodium
A	7 lb	88 lb	1 lb
B	6 lb	90 lb	1 lb
C	10 lb	70 lb	2 lb

How many bags of each type of grain should she mix to provide 71 lb of protein, 854 lb of carbohydrates, and 12 lb of sodium?

Solution: Let a, b, and c be the number of bags of grains A, B, and C, respectively, that are needed for the mixture. Then,

	Protein	Carbohydrates	Sodium
A:	$7a$	$88a$	a
B:	$6b$	$90b$	b
C:	$10c$	$70c$	$2c$
TOTALS:	71	854	12

Thus,

$$\begin{cases} 7a + 6b + 10c = 71 \\ 88a + 90b + 70c = 854 \\ a + b + 2c = 12 \end{cases}$$

$$\begin{bmatrix} 7 & 6 & 10 & | & 71 \\ 88 & 90 & 70 & | & 854 \\ 1 & 1 & 2 & | & 12 \end{bmatrix} \sim \begin{bmatrix} 1 & 1 & 2 & | & 12 \\ 88 & 90 & 70 & | & 854 \\ 7 & 6 & 10 & | & 71 \end{bmatrix} \sim \begin{bmatrix} 1 & 1 & 2 & | & 12 \\ 0 & 2 & -106 & | & -202 \\ 0 & -1 & -4 & | & -13 \end{bmatrix} \sim \begin{bmatrix} 1 & 1 & 2 & | & 12 \\ 0 & 1 & -53 & | & -101 \\ 0 & -1 & -4 & | & -13 \end{bmatrix}$$

$$\sim \begin{bmatrix} 1 & 1 & 2 & | & 12 \\ 0 & 1 & -53 & | & -101 \\ 0 & 0 & -57 & | & -114 \end{bmatrix} \sim \begin{bmatrix} 1 & 1 & 2 & | & 12 \\ 0 & 1 & -53 & | & -101 \\ 0 & 0 & 1 & | & 2 \end{bmatrix} \sim \begin{bmatrix} 1 & 1 & 0 & | & 8 \\ 0 & 1 & 0 & | & 5 \\ 0 & 0 & 1 & | & 2 \end{bmatrix} \sim \begin{bmatrix} 1 & 0 & 0 & | & 3 \\ 0 & 1 & 0 & | & 5 \\ 0 & 0 & 1 & | & 2 \end{bmatrix}$$

Mix three bags of grain A, five bags of grain B, and two bags of grain C.

Problem Set 11.3

For Problems 1–5, let

$$A = \begin{bmatrix} -2 & 2 & 4 & | & 8 \\ -1 & 3 & -2 & | & 3 \\ 1 & 5 & 3 & | & -2 \end{bmatrix}$$

1. Interchange two rows to obtain a 1 in the first position of the first row.

2. Multiply each member of row 2 by 2.

3. Multiply each member of row 1 by $\frac{1}{2}$.

4. Multiply row 3 by 2 and add the result to row 1.

5. Multiply row 3 by $^-1$ and add the result to row 2.

For Problems 6–10, let

$$B = \begin{bmatrix} 6 & -2 & 4 & | & 6 \\ 1 & -2 & 9 & | & -7 \\ -3 & 4 & 0 & | & 3 \end{bmatrix}$$

6. Interchange two rows to obtain a 1 in the first position of the first row.

7. Multiply each member of row 3 by $^-3$.

8. Multiply each member of row 1 by $\frac{1}{2}$.

9. Multiply row 2 by 3 and add the result to row 3.

10. Multiply row 2 by $^-6$ and add the result to row 1.

Solve the systems in Problems 11–28 using the matrix method.

11. $\begin{cases} x + y = 3 \\ 2x + 3y = 8 \end{cases}$
12. $\begin{cases} x - y = 2 \\ 3x + 2y = 11 \end{cases}$
13. $\begin{cases} x - 2y = {}^-1 \\ 4x - y = 17 \end{cases}$

14. $\begin{cases} 2x - 3y = 7 \\ x + 2y = 0 \end{cases}$
15. $\begin{cases} 4x + y = 10 \\ x - 3y = 9 \end{cases}$
16. $\begin{cases} 3x - y = {}^-8 \\ x + 2y = 2 \end{cases}$

17. $\begin{cases} 2x + 3y = 7 \\ 5x - 4y = {}^-17 \end{cases}$
18. $\begin{cases} 3x - 2y = 7 \\ 4x - 5y = 0 \end{cases}$
19. $\begin{cases} 2x - 4y = 0 \\ 3x - 7y = {}^-3 \end{cases}$

20. $\begin{cases} x + y + z = 6 \\ 2x - y + z = 3 \\ x - 2y - 3z = {}^-12 \end{cases}$
21. $\begin{cases} 2x - y + z = 3 \\ x - 3y + 2z = 7 \\ x - y - z = {}^-1 \end{cases}$

22. $\begin{cases} x + y + z = 4 \\ x + 3y + 2z = 4 \\ x - 2y + z = 7 \end{cases}$
23. $\begin{cases} x + 2z = 13 \\ 2x + y = 8 \\ {}^-2y + 9z = 41 \end{cases}$

24. $\begin{cases} x + 2z = 7 \\ x + y = 11 \\ -2y + 9z = -3 \end{cases}$

25. $\begin{cases} 4x + y + 2z = 7 \\ x + 2y = 0 \\ 3x - y - z = 7 \end{cases}$

26. $\begin{cases} 6x + y + 20z = 27 \\ x - y = 0 \\ y + z = 2 \end{cases}$

27. $\begin{cases} 2x - y + 4z = 13 \\ 3x + 6y = 0 \\ 2y - 3z = 3 + 3x \end{cases}$

28. $\begin{cases} 3x - 2y + z = 5 \\ 5x - 3y = 24 \\ 2y + z = -5 \end{cases}$

29. In order to control a certain type of crop disease, it is necessary to use 23 gallons of chemical A and 34 gallons of chemical B. The dealer can order commercial spray I, each container of which holds 5 gallons of chemical A and 2 gallons of chemical B, and commercial spray II, each container of which holds 2 gallons of chemical A and 7 gallons of chemical B. How many containers of each type of commercial spray should be used to attain exactly the right proportion of chemicals needed?

30. In order to manufacture a certain alloy, it is necessary to use 33 kg of metal A and 56 kg of metal B. It is cheaper for the manufacturer if she buys and mixes an alloy, each bar of which contains 3 kg metal A and 5 kg of metal B along with another alloy, each bar of which contains 4 kg of metal A and 7 kg of metal B. How much of the two alloys should she use in order to produce the alloy desired?

Mind Bogglers

31. A candy maker mixes chocolate, milk, and almonds to produce three kinds of candy—I, II, and III—with the following proportions:

   I: 7 lb chocolate, 5 gal milk, and 1 oz almonds

   II: 3 lb chocolate, 2 gal milk, and 2 oz almonds

   III: 4 lb chocolate, 3 gal milk, and 3 oz almonds

   If 67 pounds of chocolate, 48 gallons of milk, and 32 ounces of almonds are available, how much of each kind of candy can be produced?

32. The total number of registered Democrats, Republicans, and Independents in a certain community is 100,000. Voter turnout in a recent election was tabulated as follows:

   50 percent of the Democrats voted.

   60 percent of the Republicans voted.

   70 percent of the Independents voted.

   55,200 votes were cast (assume there were no write-in votes; everyone voted Democratic, Republican, or Independent).

   If the registered Democrats outnumber the registered Independents by 9 to 1, how many registered Democrats, Republicans, and Independents are there in the community?

*11.4 Identity and Inverse Properties for Matrices

In Section 6.1, we introduced an algebraic structure that focused on various properties of real numbers. We can also investigate these properties for matrices. Recall that an identity for an operation in a given set is that element of the set that does not change another element when this other element is operated on by the identity element. The identity for addition of matrices is the zero matrix. For multiplication, consider the diagonal consisting of elements in the first row, first column; second row, second column; third row, third column; and so on. This diagonal is called the **main diagonal.**

Main diagonal (shown with arrows)

$$\begin{bmatrix} 1 & 0 \\ 0 & 1 \end{bmatrix} \quad \begin{bmatrix} 1 & 0 & 0 \\ 0 & 1 & 0 \\ 0 & 0 & 1 \end{bmatrix} \quad \begin{bmatrix} 1 & 0 & 0 & 0 \\ 0 & 1 & 0 & 0 \\ 0 & 0 & 1 & 0 \\ 0 & 0 & 0 & 1 \end{bmatrix}$$

The square matrix, I, of order $n \times n$, consisting of 1s on the main diagonal and zeros elsewhere, is called the **identity matrix** of order n since

$$IA = AI = A$$

for every conformable matrix A. For example,

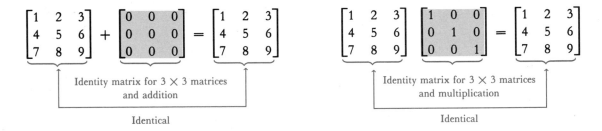

$$\begin{bmatrix} 1 & 2 & 3 \\ 4 & 5 & 6 \\ 7 & 8 & 9 \end{bmatrix} + \begin{bmatrix} 0 & 0 & 0 \\ 0 & 0 & 0 \\ 0 & 0 & 0 \end{bmatrix} = \begin{bmatrix} 1 & 2 & 3 \\ 4 & 5 & 6 \\ 7 & 8 & 9 \end{bmatrix} \qquad \begin{bmatrix} 1 & 2 & 3 \\ 4 & 5 & 6 \\ 7 & 8 & 9 \end{bmatrix} \begin{bmatrix} 1 & 0 & 0 \\ 0 & 1 & 0 \\ 0 & 0 & 1 \end{bmatrix} = \begin{bmatrix} 1 & 2 & 3 \\ 4 & 5 & 6 \\ 7 & 8 & 9 \end{bmatrix}$$

Identity matrix for 3 × 3 matrices
and addition

Identity matrix for 3 × 3 matrices
and multiplication

Identical   Identical

Recall the idea of inverse from Section 6.1. The inverse of an element for an operation in a given set is that element of the set that, when operated on by the given element, produces the identity for that operation. The inverse matrix for addition is simply the matrix whose entries are opposites of the original matrix. However, it is the inverse for multiplication that is of particular interest to us. Given a square matrix A, we denote its inverse by A^{-1} as indicated in the following definition.

*This section requires Section 6.1.

<table>
<tr><td>

Inverse of a Matrix

If A is a square matrix, and if there exists a matrix A^{-1} such that

$$A^{-1}A = AA^{-1} = I$$

then A^{-1} is called the **inverse of A for multiplication**.

</td></tr>
</table>

INVERSE PROPERTIES IN
THE SET OF REAL NUMBERS

Addition

$$a + (^-a) = (^-a) + a = 0$$

The sum of a real number and its
opposite (additive inverse) is 0
(identity for addition).

Multiplication

$$a \cdot a^{-1} = a^{-1} \cdot a = 1$$

The product of a nonzero real
number and its reciprocal
(multiplicative inverse) is 1
(identity for multiplication).

Usually in the context of matrices, when we talk simply of the inverse of A
we mean the inverse of A for multiplication.

Example 1: Verify that the inverse of $A = \begin{bmatrix} 2 & 1 \\ 3 & 2 \end{bmatrix}$ is $B = \begin{bmatrix} 2 & ^-1 \\ ^-3 & 2 \end{bmatrix}$.

Solution:
$$AB = \begin{bmatrix} 2 & 1 \\ 3 & 2 \end{bmatrix}\begin{bmatrix} 2 & ^-1 \\ ^-3 & 2 \end{bmatrix} \qquad BA = \begin{bmatrix} 2 & ^-1 \\ ^-3 & 2 \end{bmatrix}\begin{bmatrix} 2 & 1 \\ 3 & 2 \end{bmatrix}$$

$$= \begin{bmatrix} 4-3 & ^-2+2 \\ 6-6 & ^-3+4 \end{bmatrix} \qquad = \begin{bmatrix} 4-3 & 2-2 \\ ^-6+6 & ^-3+4 \end{bmatrix}$$

$$= \begin{bmatrix} 1 & 0 \\ 0 & 1 \end{bmatrix} \qquad = \begin{bmatrix} 1 & 0 \\ 0 & 1 \end{bmatrix}$$

$$= I \qquad = I$$

Thus, $B = A^{-1}$.

Example 2: Show that A and B are inverses, where

$$A = \begin{bmatrix} 0 & 1 & 2 \\ ^-1 & 1 & 2 \\ 1 & ^-2 & ^-5 \end{bmatrix} \quad \text{and} \quad B = \begin{bmatrix} 1 & ^-1 & 0 \\ 3 & 2 & 2 \\ ^-1 & ^-1 & ^-1 \end{bmatrix}$$

Solution:
$$AB = \begin{bmatrix} 0 & 1 & 2 \\ ^-1 & 1 & 2 \\ 1 & ^-2 & ^-5 \end{bmatrix}\begin{bmatrix} 1 & ^-1 & 0 \\ 3 & 2 & 2 \\ ^-1 & ^-1 & ^-1 \end{bmatrix}$$

$$= \begin{bmatrix} 0+3-2 & 0+2-2 & 0+2-2 \\ ^-1+3-2 & 1+2-2 & 0+2-2 \\ 1-6+5 & ^-1-4+5 & 0-4+5 \end{bmatrix}$$

$$= \begin{bmatrix} 1 & 0 & 0 \\ 0 & 1 & 0 \\ 0 & 0 & 1 \end{bmatrix}$$

$$= I$$

$$BA = \begin{bmatrix} 1 & ^-1 & 0 \\ 3 & 2 & 2 \\ ^-1 & ^-1 & ^-1 \end{bmatrix} \begin{bmatrix} 0 & 1 & 2 \\ ^-1 & 1 & 2 \\ 1 & ^-2 & ^-5 \end{bmatrix}$$

$$= \begin{bmatrix} 0+1+0 & 1-1+0 & 2-2+0 \\ 0-2+2 & 3+2-4 & 6+4-10 \\ 0+1-1 & ^-1-1+2 & ^-2-2+5 \end{bmatrix}$$

$$= \begin{bmatrix} 1 & 0 & 0 \\ 0 & 1 & 0 \\ 0 & 0 & 1 \end{bmatrix}$$

$$= I$$

Since $AB = I = BA$, then $B = A^{-1}$.

If a given matrix has an inverse, we say that it is **nonsingular.** We can use the elementary row operations presented in the last section to find the inverse of a given square matrix. In this case, augment the matrix A with the matrix I instead of the constant terms, as done in the last section. That is, write

$$[A \mid I]$$

where I is the identity matrix of the same order as A. Our goal is to perform elementary row operations to change the matrix A into the identity matrix (if possible). If we can do this, the result in the second part (where we began with I) will be the inverse of A.

Example 3: Find the inverse of the matrix A where

$$A = \begin{bmatrix} 1 & 2 \\ 1 & 4 \end{bmatrix}$$

Solution: We write the augmented matrix $[A \mid I]$.

$$\begin{bmatrix} 1 & 2 & \vdots & 1 & 0 \\ 1 & 4 & \vdots & 0 & 1 \end{bmatrix}$$

Now we want to perform elementary row operations to make the left-hand side

$$\begin{matrix} 1 & 0 \\ 0 & 1 \end{matrix}$$

(if possible).

$$\begin{bmatrix} 1 & 2 & \vdots & 1 & 0 \\ 1 & 4 & \vdots & 0 & 1 \end{bmatrix} \sim \begin{bmatrix} 1 & 2 & \vdots & 1 & 0 \\ 0 & 2 & \vdots & ^-1 & 1 \end{bmatrix}$$ Multiply row 1 by $^-1$ and add to row 2.

$$\sim \begin{bmatrix} 1 & 2 & \vdots & 1 & 0 \\ 0 & 1 & \vdots & -\frac{1}{2} & \frac{1}{2} \end{bmatrix}$$ Multiply row 2 by $\frac{1}{2}$.

$$\sim \begin{bmatrix} 1 & 0 & | & 2 & ^-1 \\ 0 & 1 & | & -\frac{1}{2} & \frac{1}{2} \end{bmatrix}$$ Multiply row 2 by $^-2$ and add to row 1.

The right-hand side of this augmented matrix is the inverse of A. Thus,

$$A^{-1} = \begin{bmatrix} 2 & ^-1 \\ -\frac{1}{2} & \frac{1}{2} \end{bmatrix}$$

Example 4: Find the inverse of the matrix $\begin{bmatrix} 1 & 2 \\ 0 & 0 \end{bmatrix}$.

Solution: We write the augmented matrix $[A \mid I]$.

$$\begin{bmatrix} 1 & 2 & | & 1 & 0 \\ 0 & 0 & | & 0 & 1 \end{bmatrix}$$

We want to make the left-hand side   1   0
                                          0   1

This is impossible since there are no elementary row operations that will put it into the required form. Thus, there is no inverse.

Example 5: Find the inverse of the matrix

$$\begin{bmatrix} 0 & 1 & 2 \\ 2 & ^-1 & 1 \\ ^-1 & 1 & 0 \end{bmatrix}$$

Solution: We write the augmented matrix $[A \mid I]$:

$$\begin{bmatrix} 0 & 1 & 2 & | & 1 & 0 & 0 \\ 2 & ^-1 & 1 & | & 0 & 1 & 0 \\ ^-1 & 1 & 0 & | & 0 & 0 & 1 \end{bmatrix}$$

$$\begin{bmatrix} 0 & 1 & 2 & | & 1 & 0 & 0 \\ 2 & ^-1 & 1 & | & 0 & 1 & 0 \\ ^-1 & 1 & 0 & | & 0 & 0 & 1 \end{bmatrix} \sim \begin{bmatrix} ^-1 & 1 & 0 & | & 0 & 0 & 1 \\ 2 & ^-1 & 1 & | & 0 & 1 & 0 \\ 0 & 1 & 2 & | & 1 & 0 & 0 \end{bmatrix}$$ Interchange row 1 and row 3.

$$\sim \begin{bmatrix} 1 & ^-1 & 0 & | & 0 & 0 & ^-1 \\ 2 & ^-1 & 1 & | & 0 & 1 & 0 \\ 0 & 1 & 2 & | & 1 & 0 & 0 \end{bmatrix}$$ Multiply row 1 by $^-1$.

$$\sim \begin{bmatrix} 1 & ^-1 & 0 & | & 0 & 0 & ^-1 \\ 0 & 1 & 1 & | & 0 & 1 & 2 \\ 0 & 1 & 2 & | & 1 & 0 & 0 \end{bmatrix}$$ Multiply row 1 by $^-2$ and add to row 2.

$$\sim \begin{bmatrix} 1 & ^-1 & 0 & | & 0 & 0 & ^-1 \\ 0 & 1 & 1 & | & 0 & 1 & 2 \\ 0 & 0 & 1 & | & 1 & ^-1 & ^-2 \end{bmatrix}$$ Multiply row 2 by $^-1$ and add to row 3.

(continued)

$$\sim \begin{bmatrix} 1 & ^-1 & 0 & | & 0 & 0 & ^-1 \\ 0 & 1 & 0 & | & ^-1 & 2 & 4 \\ 0 & 0 & 1 & | & 1 & ^-1 & ^-2 \end{bmatrix}$$
Multiply row 3 by $^-1$ and add to row 2.

$$\sim \begin{bmatrix} 1 & 0 & 0 & | & ^-1 & 2 & 3 \\ 0 & 1 & 0 & | & ^-1 & 2 & 4 \\ 0 & 0 & 1 & | & 1 & ^-1 & ^-2 \end{bmatrix}$$
Add row 2 to row 1.

$$A^{-1} = \begin{bmatrix} ^-1 & 2 & 3 \\ ^-1 & 2 & 4 \\ 1 & ^-1 & ^-2 \end{bmatrix}$$

By studying these examples we see that there is a standard procedure by which we find the inverse if it exists. We summarize the procedure. Given $[A \mid I]$:

Step 1. Obtain a 1 in the first position on the main diagonal. This can be done by interchanging rows or by multiplying a row by a constant. If this is not possible (the first column has all zeros), then the matrix does not have an inverse.

Step 2. Use the 1 in the first position on the main diagonal to obtain zeros for all the entries in the first column under the 1.

Step 3. Obtain a 1 in the second position on the main diagonal (if possible).

Step 4. Use the 1 in the second position on the main diagonal to obtain zeros for all the entries in the second column under the 1.

Step 5. Obtain a 1 in the third position on the main diagonal (if possible). Use this to obtain zeros for all entries below this 1. Continue this procedure until the main diagonal contains all 1s and all entries below those 1s are zeros.

Step 6. Use the 1 in the last position on the main diagonal to obtain zeros in all entries in the last column above the 1. Continue until all the entries above the 1s on the diagonal are zero.

Problem Set 11.4

In Problems 1–10, use multiplication to determine if the given matrices are inverses.

1. $\begin{bmatrix} 1 & 2 \\ 2 & 3 \end{bmatrix}$, $\begin{bmatrix} ^-3 & 2 \\ 2 & ^-1 \end{bmatrix}$

2. $\begin{bmatrix} 2 & ^-5 \\ ^-1 & 2 \end{bmatrix}$, $\begin{bmatrix} ^-2 & ^-5 \\ ^-1 & ^-2 \end{bmatrix}$

3. $\begin{bmatrix} 3 & 5 \\ 4 & 7 \end{bmatrix}$, $\begin{bmatrix} 7 & ^-5 \\ ^-4 & 3 \end{bmatrix}$

4. $\begin{bmatrix} 4 & 7 \\ 5 & 9 \end{bmatrix}$, $\begin{bmatrix} 9 & ^-7 \\ ^-5 & 4 \end{bmatrix}$

5. $\begin{bmatrix} 4 & 3 \\ 2 & 2 \end{bmatrix}$, $\begin{bmatrix} 1 & -\frac{3}{2} \\ -1 & 2 \end{bmatrix}$ 6. $\begin{bmatrix} 2 & 3 \\ 2 & 1 \end{bmatrix}$, $\begin{bmatrix} -\frac{1}{4} & \frac{3}{4} \\ \frac{1}{2} & -\frac{1}{2} \end{bmatrix}$

7. $\begin{bmatrix} 0 & 1 & 0 \\ 1 & -1 & 0 \\ -1 & 2 & 1 \end{bmatrix}$, $\begin{bmatrix} -1 & -1 & 0 \\ -1 & 0 & 0 \\ 1 & -1 & -1 \end{bmatrix}$ 8. $\begin{bmatrix} 1 & 0 & 0 \\ 0 & 1 & 1 \\ 2 & 0 & 1 \end{bmatrix}$, $\begin{bmatrix} 1 & 0 & 0 \\ 2 & 1 & -1 \\ -2 & 0 & 1 \end{bmatrix}$

9. $\begin{bmatrix} 3 & -2 & 4 \\ 2 & 1 & 2 \\ 5 & 3 & 5 \end{bmatrix}$, $\begin{bmatrix} -1 & 22 & -8 \\ 0 & -5 & 2 \\ 1 & -19 & 7 \end{bmatrix}$

10. $\begin{bmatrix} 6 & -1 & -5 \\ -7 & 1 & 5 \\ -10 & 2 & 11 \end{bmatrix}$, $\begin{bmatrix} -1 & -1 & 0 \\ -27 & -16 & -5 \\ 4 & 2 & 1 \end{bmatrix}$

Find the inverse of the given matrix in Problems 11–25, if possible.

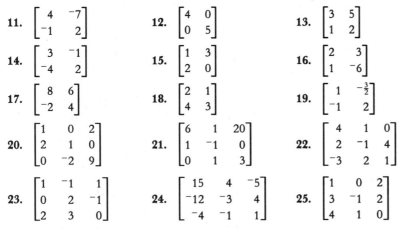

11. $\begin{bmatrix} 4 & -7 \\ -1 & 2 \end{bmatrix}$ 12. $\begin{bmatrix} 4 & 0 \\ 0 & 5 \end{bmatrix}$ 13. $\begin{bmatrix} 3 & 5 \\ 1 & 2 \end{bmatrix}$

14. $\begin{bmatrix} 3 & -1 \\ -4 & 2 \end{bmatrix}$ 15. $\begin{bmatrix} 1 & 3 \\ 2 & 0 \end{bmatrix}$ 16. $\begin{bmatrix} 2 & 3 \\ 1 & -6 \end{bmatrix}$

17. $\begin{bmatrix} 8 & 6 \\ -2 & 4 \end{bmatrix}$ 18. $\begin{bmatrix} 2 & 1 \\ 4 & 3 \end{bmatrix}$ 19. $\begin{bmatrix} 1 & -\frac{3}{2} \\ -1 & 2 \end{bmatrix}$

20. $\begin{bmatrix} 1 & 0 & 2 \\ 2 & 1 & 0 \\ 0 & -2 & 9 \end{bmatrix}$ 21. $\begin{bmatrix} 6 & 1 & 20 \\ 1 & -1 & 0 \\ 0 & 1 & 3 \end{bmatrix}$ 22. $\begin{bmatrix} 4 & 1 & 0 \\ 2 & -1 & 4 \\ -3 & 2 & 1 \end{bmatrix}$

23. $\begin{bmatrix} 1 & -1 & 1 \\ 0 & 2 & -1 \\ 2 & 3 & 0 \end{bmatrix}$ 24. $\begin{bmatrix} 15 & 4 & -5 \\ -12 & -3 & 4 \\ -4 & -1 & 1 \end{bmatrix}$ 25. $\begin{bmatrix} 1 & 0 & 2 \\ 3 & -1 & 2 \\ 4 & 1 & 0 \end{bmatrix}$

Mind Bogglers

Find the inverse of the given matrix in Problems 26 and 27.

26. $\begin{bmatrix} 1 & 0 & 0 & 1 \\ 0 & 2 & 0 & 0 \\ 0 & 0 & 0 & 1 \\ 2 & 0 & 1 & 0 \end{bmatrix}$ 27. $\begin{bmatrix} 0 & 1 & 2 & 0 \\ 0 & 0 & 0 & 1 \\ 1 & 1 & 3 & 0 \\ 2 & 4 & 0 & 0 \end{bmatrix}$

28. In algebra we know that $(b + c)a = ba + ca$. Does this property hold for matrices A, B, and C? That is, does $(B + C)A = BA + CA$ hold for any conformable matrices A, B, and C?

29. In algebra we know that $(a + b)^2 = a^2 + 2ab + b^2$. Does this property hold for matrices A, B, and C? That is, does $(A + B)^2 = A^2 + 2AB + B^2$ for any conformable matrices A and B?

30. Why is it impossible for a matrix that is not square to have an inverse?

*11.5 Game Theory

• It has long been an axiom of mine that the little things are infinitely the most important.

Sir Arthur Conan Doyle in The Adventures of Sherlock Holmes. A Case of Identity.

An interesting application of matrices is in the analysis of conflict situations, called **game theory.** For example, a game of cards or chess is played between friends; union and management sit down at the bargaining table; a presidential election is held; opposing armies meet on a battlefield; an oilman digs for oil (his opponent is nature).

Game theory is a way of analyzing certain types of conflict. Probability was invented to help professional gamblers understand games of chance and take advantage of the odds. More difficult games require more than a knowledge of odds; they require certain strategies. The outcome in the games of chess and poker, for example, depends upon more than mere chance. A good player will play according to some set of strategies in anticipation of what his opponent will do.

Let's begin our study of game theory by considering a rather simple game. Suppose a friend of ours, Linda, likes to play games. She is an incredible person but lacks the mathematical skill to analyze the various games she plays. For this reason, we will act as her consultant in matters relating to game theory. Her opponent, therefore, becomes our opponent.

The first game Linda decides to play is a variation of the game of two-finger Morra. The game consists of two persons simultaneously holding out either one or two fingers from a closed fist. Linda's opponent makes her the following offer, "Let's play two-finger Morra. If the sum of the fingers is two, I'll pay you 5¢. If it is four, you pay me 5¢. If the sum is odd we break even."

This is an example of a two person **zero-sum game.** If the payoffs for a game are such that a win for one player results in a corresponding loss for the other player, then we describe this situation as a zero-sum game. The game of two-finger Morra can be summarized by the 2×2 *game matrix* shown below:

Number of fingers shown: Opponent

$$\text{Linda} \quad \begin{array}{c} 1 \\ 2 \end{array} \begin{bmatrix} \overset{\displaystyle 1}{5} & \overset{\displaystyle 2}{0} \\ 0 & {}^{-}5 \end{bmatrix}$$

A game matrix is constructed so that each element represents a payoff to the first player. The preceding matrix shows the payoffs to Linda. We see that if Linda and her opponent both hold out one finger she will win 5¢, whereas if they both hold out two fingers she will win ⁻5¢ (which is interpreted as a 5¢ loss for her). Linda's opponent, however, views negative entries as winning payoffs and positive entries as losing payoffs.

We would like to develop a **strategy.** Each row of the game matrix represents a strategy for Linda, while each column represents a strategy for her oppo-

*This is an optional section; it requires Sections 9.3 and 9.4.

nent. If we examine the game matrix, we see that Linda's best strategy is to show 1 finger since the *worst* she can do is break even. On the other hand the best strategy for her opponent is to show 2 fingers, since the worst he can do is break even. The payoff when they both play their best strategies is called the **value of the game.** We see that the value of this game is 0, and whenever the value of a game is 0 we call the game **fair.**

Example 1: Determine the best strategies and value of the game of three-finger Morra, where player II agrees to pay player I the difference between the fingers shown.

Solution: Write the game matrix:

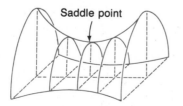

$$
\begin{array}{c}
 & \text{Player II} \\
 & \begin{array}{ccc} 1 & 2 & 3 \end{array} \\
\text{Player I} \quad \begin{array}{c} 1 \\ 2 \\ 3 \end{array} &
\left[\begin{array}{ccc}
0 & ^-1 & ^-2 \\
1 & 0 & ^-1 \\
2 & 1 & 0
\end{array} \right]
\end{array}
$$

We see that player I can reason as follows: "If I play 1 finger, then I stand to lose 2; if I play 2 fingers, I stand to lose 1; if I play 3 fingers, the *worst* I can do is break even." We see that player I looks for the *minimum entry* in each row. On the other hand, player II looks for the *maximum entry* of each column (since the negative values indicate a win for player II). He reasons, "If I play column 1, I stand to lose 2; if I play column 2, I can lose 1; but if I play column 3, the *worst* I can do is break even." In this game we see that player I would play 3 fingers and player II would also play 3 fingers. The value of the game is 0.

Games such as this are said to be **strictly determined** since a knowledge of our opponent's strategy would not alter our own strategy. In a strictly determined game the smallest entry in some row is the largest entry in some column. This point is called a **saddle point.** The reason it is called a saddle point is because it can be described by thinking of the surface of a saddle. The same point is a maximum for one person and at the same time a minimum for another. There is such a point on a saddle as we can see in Figure 11.2. The **value** of a strictly determined game is the value of this saddle point.

Saddle point

FIGURE 11.2 Saddle point

Is the game determined by each matrix given in Examples 2–4 strictly determined?

Example 2:

$$\begin{bmatrix} 1 & ^-2 & ^-3 \\ 2 & 4 & 1 \\ ^-3 & ^-4 & ^-2 \end{bmatrix}$$

Example 3:

$$\begin{bmatrix} 2 & 1 & ^-3 \\ ^-2 & 0 & 2 \\ 3 & ^-1 & 1 \end{bmatrix}$$

Example 4:

$$\begin{bmatrix} 0 & 5 & ^-2 & 3 \\ 3 & 4 & 3 & 5 \\ 2 & 0 & 1 & ^-1 \\ 3 & 5 & 3 & 6 \end{bmatrix}$$

Procedure

Step 1. Place an * by the minimum of each row.

Step 2. Check to see if each element marked by an * is the maximum in its column. If so, circle it. This is a saddle point and the game is strictly determined.

Solution for Example 2:

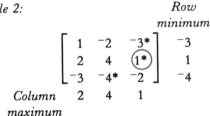

The saddle point is circled and the value of the game is 1. The game is strictly determined.

Solution for Example 3:

                                    Row
                                  minimum
$$\begin{bmatrix} 2 & 1 & ^-3* \\ ^-2* & 0 & 2 \\ 3 & ^-1* & 1 \end{bmatrix} \quad \begin{matrix} ^-3 \\ ^-2 \\ ^-1 \end{matrix}$$
Column   3   1   2
maximum

There is no saddle point. If this is the case, we say the game is **nonstrictly determined.**

Solution for Example 4:

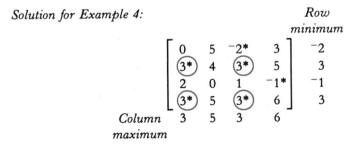

The saddle points are circled and the value of the game is 3. We see that a game can have more than one saddle point, but all saddle points must be numerically the same. The game is strictly determined.

Let's consider a game that is nonstrictly determined. Linda's friend says he is tired of playing two-finger Morra and has thought of another game. He says each person needs a dime and a nickel and will then select one of their two coins and put it in their hand. On a given signal they will open their hands together. If they match, Linda will win both coins. If they don't match, the opponent will win both coins. We advise Linda to play his silly game, provided we can first write out the game matrix.

$$
\begin{array}{cc}
 & \text{Opponent} \\
 & \text{Nickel} \quad \text{Dime} \\
\text{Linda} \quad
\begin{matrix} \text{Nickel} \\ \text{Dime} \end{matrix}
&
\begin{bmatrix} 5 & {}^-5 \\ {}^-10 & 10 \end{bmatrix}
\end{array}
$$

We quickly check for a saddle point and see that there is none, so this is not a strictly determined game. Game theory will not help Linda play this game if she plays only once, but if she plays over and over we can try to develop a **mixed** strategy that tells her how often to play a nickel and how often to play a dime.

Let p be the probability that Linda will choose row 1, and let q be the probability that her opponent will choose column 1 (that is, both choose nickels). The p and q are numbers between 0 and 1 (inclusive). Also, since Linda must choose either row 1 or row 2, the probability she will choose row 2 is $1 - p$. Likewise, column 2 for her opponent will be chosen with probability $1 - q$. Now, if we represent Linda's strategy by the row matrix

$$
[p \quad 1 - p]
$$

and her opponent's strategy by the column matrix

$$
\begin{bmatrix} q \\ 1 - q \end{bmatrix}
$$

we mean that Linda will choose row 1 of the game matrix with probability p and her opponent will choose column 1 with probability q.

Since Linda and her opponent are making their choices independently, the probability that the payoff to Linda is 5¢ is pq:

Linda picks row 1 ⌐ ⌐ Opponent picks column 1
$$
pq
$$

If Linda's opponent plays column 1, then Linda's expectation is

$$
5p - 10(1 - p)
$$

Historical Note

John von Neumann (1903–1957)

Game theory is primarily
concerned with the logic
of strategy and was first
envisioned by the great German
mathematician Leibniz. However,
the theory of games as we know
it today was developed in the
1920s by John von Neumann
and Emile Borel. It gained wide
acceptance in 1944 in a book
*Theory of Games and Economic
Behavior* by von Neumann
and Oskar Morgenstern. Many
stories about von Neumann
are passed around among
mathematicians. One such story
is told by Leo Moser about a
type of geometry invented by von
Neumann in which points play
no role. His original name for
this geometry was "pointless
geometry."

If her opponent plays column 2, then her expectation is

$$^-5p + 10(1 - p)$$

These expectations can be summarized by using matrix multiplication:

$$[p \quad 1 - p]\begin{bmatrix} 5 & ^-5 \\ ^-10 & 10 \end{bmatrix} = [5p - 10(1 - p) \quad ^-5p + 10(1 - p)]$$
$$= [5p - 10 + 10p \quad ^-5p + 10 - 10p]$$
$$= [15p - 10 \quad ^-15p + 10]$$

In 1928 John von Neumann proved that if the game is not strictly determined, then the best strategy for the first player (Linda) is to choose p so that the entries of this product matrix are equal. That is,

$$15p - 10 = ^-15p + 10$$
$$30p = 20$$
$$p = \frac{2}{3}$$

Thus, Linda's best strategy for this game is to pick row 1 two-thirds of the time and row 2 one-third of the time. For her opponent, find

$$\begin{bmatrix} 5 & ^-5 \\ ^-10 & 10 \end{bmatrix}\begin{bmatrix} q \\ 1 - q \end{bmatrix} = [5q - 5(1 - q) \quad ^-10q + 10(1 - q)]$$
$$= [5q - 5 + 5q \quad ^-10q + 10 - 10q]$$
$$= [10q - 5 \quad ^-20q + 10]$$

Solve:

$$10q - 5 = ^-20q + 10$$
$$30q = 15$$
$$q = \frac{1}{2}$$

The best strategy for Linda's opponent is to choose each row one-half of the time.

Von Neumann also proved that if both players select the optimum strategy, then the expectation of winning for the row player is the same as the expectation of losing for the column player. This is called *Von Neumann's Minimax Theorem.* This common expectation is called the **value** of this nonstrictly determined game. Thus, to calculate the value of the game, you can use either player's optimum strategy:

$$[\tfrac{2}{3} \quad \tfrac{1}{3}]\begin{bmatrix} 5 & ^-5 \\ ^-10 & 10 \end{bmatrix}\begin{bmatrix} \tfrac{1}{2} \\ \tfrac{1}{2} \end{bmatrix}$$

Use the product in any row or column to find the value (we choose the column 1 product):

$$\frac{2}{3}(5) + \frac{1}{3}(^-10) = \frac{10}{3} - \frac{10}{3} = 0$$

The value of this game is 0, which means that it is a fair game.

Optimal Strategy Theorem

The zero-sum two-person game

$$\begin{bmatrix} a & b \\ c & d \end{bmatrix}$$

with strategies

$$[p \quad 1 - p] \quad \text{and} \quad \begin{bmatrix} q \\ 1 - q \end{bmatrix}$$

has optimal strategy for the row player such that the entries of the product

$$[p \quad 1 - p] \begin{bmatrix} a & b \\ c & d \end{bmatrix}$$

are equal, provided the game matrix has no saddle point. The optimal strategy for the column player is such that the entries of the product

$$\begin{bmatrix} a & b \\ c & d \end{bmatrix} \begin{bmatrix} q \\ 1 - q \end{bmatrix}$$

are equal, provided the game matrix has no saddle point.

Example 5: Find the optimum strategies and value for the game

$$\begin{bmatrix} ^-10 & 5 \\ 8 & ^-10 \end{bmatrix}$$

Solution: There is no saddle point. Next, find the products.

Row player:

$$[p \quad 1 - p] \begin{bmatrix} ^-10 & 5 \\ 8 & ^-10 \end{bmatrix} = [^-10p + 8(1 - p) \quad 5p - 10(1 - p)]$$

Solve:
$$^-10p + 8(1 - p) = 5p - 10(1 - p)$$
$$^-10p + 8 - 8p = 5p - 10 + 10p$$
$$^-18p + 8 = 15p - 10$$
$$^-33p = ^-18$$
$$p = \frac{18}{33}$$

Column player:

$$\begin{bmatrix} -10 & 5 \\ 8 & -10 \end{bmatrix} \begin{bmatrix} q \\ 1 - q \end{bmatrix} = [\,-10q + 5(1 - q) \qquad 8q - 10(1 - q)]$$

Solve:
$$-10q - 5(1 - q) = 8q - 10(1 - q)$$
$$-10q + 5 - 5q = 8q - 10 + 10q$$
$$-15q + 5 = 18q - 10$$
$$-33q = -15$$
$$q = \frac{15}{33}$$

Thus the optimum strategies are $[\frac{18}{33} \quad \frac{15}{33}]$ and $\begin{bmatrix} \frac{15}{33} \\ \frac{18}{33} \end{bmatrix}$. The value of the game is

$$-10\left(\frac{18}{33}\right) + 8\left(1 - \frac{18}{33}\right) = -\frac{60}{33}$$

The game is biased in favor of the second player since the value of the game is negative.

Sometimes the original game is not 2×2, but can be reduced to a 2×2 game matrix by eliminating certain rows or columns. Consider the following game matrix:

$$\text{Opponent}$$
$$\text{Linda} \begin{bmatrix} 2 & 4 & -3 \\ -1 & 3 & 2 \\ 0 & -1 & -4 \end{bmatrix}$$

In finding the optimum strategies, we first look for a saddle point and see that there is none. Suppose Linda is playing this game with an opponent. Certainly she would never choose row 3, since every entry in row 3 is smaller than the corresponding entries in row 1. She can therefore eliminate row 3 from any further consideration. We say that row 1 *dominates* row 3. *We can delete dominated rows.* The result is a 2×3 *subgame* of the original game.

$$\text{Opponent}$$
$$\text{Linda} \begin{bmatrix} 2 & 4 & -3 \\ -1 & 3 & 2 \end{bmatrix}$$

On the other hand, Linda's opponent wants to make his choices as small as possible (negative values are good for him). He would not pick column 2 as a strategy since each entry in column 1 is smaller than the corresponding entries in column 2. We say that column 2 dominates column 1 and the opponent would *delete the dominating* column from further consideration. The result is a 2×2 subgame of the original game.

SOCIAL APPLICATION
OF MATHEMATICS

There are three very readable
articles on mathematical models
in the social sciences in the
classic book, *The World of
Mathematics* by James R.
Newman (New York: Simon and
Schuster, 1956). These articles
give game theory examples
in economics, gambling, and
sociology. (See pp. 1264–1313,
Vol. 2.)

Opponent

$$\text{Linda} \begin{bmatrix} 2 & -3 \\ -1 & 2 \end{bmatrix}$$

We see that the following principle can be applied to any matrix game.

Dominating Principle

In a given matrix, dominated rows can be eliminated (eliminate the smaller row). Dominating columns can be eliminated (eliminate the larger column).

We have already discussed how to find the optimum strategies for the resulting 2×2 game matrix. We see (the calculations are left for the reader) that the optimum strategies for the subgame are

$$[\tfrac{3}{8} \quad \tfrac{5}{8}] \quad \text{and} \quad \begin{bmatrix} \tfrac{5}{8} \\ \tfrac{3}{8} \end{bmatrix}$$

and for the original game the strategies are

$$[\tfrac{3}{8} \quad \tfrac{5}{8} \quad 0] \quad \text{and} \quad \begin{bmatrix} \tfrac{5}{8} \\ 0 \\ \tfrac{3}{8} \end{bmatrix}$$

Problem Set 11.5

Determine the optimal strategies and the value of the game for Problems 1–18.

1. $\begin{bmatrix} 4 & 0 \\ 3 & -2 \end{bmatrix}$ 2. $\begin{bmatrix} 0 & -3 \\ -2 & 0 \end{bmatrix}$ 3. $\begin{bmatrix} 1 & -1 \\ -1 & 1 \end{bmatrix}$

4. $\begin{bmatrix} 1 & -2 \\ 0 & 2 \end{bmatrix}$ 5. $\begin{bmatrix} 3 & -2 \\ 1 & -3 \end{bmatrix}$ 6. $\begin{bmatrix} 10 & 0 \\ -10 & 5 \end{bmatrix}$

7. $\begin{bmatrix} 5 & -1 \\ 3 & -2 \end{bmatrix}$ 8. $\begin{bmatrix} 1.5 & -.5 \\ .5 & 2.5 \end{bmatrix}$ 9. $\begin{bmatrix} -1 & 0 \\ \tfrac{1}{4} & -\tfrac{1}{4} \end{bmatrix}$

10. $\begin{bmatrix} 2 & 1 & 3 \\ -2 & 0 & 3 \\ 4 & -2 & -3 \end{bmatrix}$ 11. $\begin{bmatrix} 2 & 3 & 3 \\ 1 & 0 & -1 \\ 0 & 0 & 4 \end{bmatrix}$ 12. $\begin{bmatrix} 2 & 1 & 3 \\ 0 & 1 & -2 \\ -1 & 2 & -3 \end{bmatrix}$

13. $\begin{bmatrix} 4 & -2 & 3 \\ 3 & -3 & 1 \\ 2 & 0 & -1 \end{bmatrix}$ 14. $\begin{bmatrix} -1 & -8 & 5 \\ 1 & 5 & -4 \\ -5 & 7 & 2 \end{bmatrix}$ 15. $\begin{bmatrix} 5 & -3 & -4 \\ 4 & -1 & 0 \\ 3 & -3 & 6 \end{bmatrix}$

16. $\begin{bmatrix} 0 & 1 & 2 & 0 \\ 1 & 2 & 2 & 1 \\ 3 & 0 & {}^-1 & 0 \\ 2 & {}^-1 & 0 & {}^-2 \end{bmatrix}$ **17.** $\begin{bmatrix} 2 & {}^-1 & 4 & 3 \\ {}^-4 & {}^-5 & {}^-2 & 1 \\ {}^-1 & 0 & 3 & 0 \end{bmatrix}$ **18.** $\begin{bmatrix} 9 & 10 & 11 & 9 \\ 10 & 11 & 11 & 10 \\ 12 & 9 & 8 & 9 \\ 11 & 8 & 9 & 8 \end{bmatrix}$

19. A friend suggests the following game. He will hide a quarter, dime, or a half-dollar in one hand behind his back. You are to guess which coin he has. If you guess correctly you get the coin. If you guess incorrectly he gets the difference between your guess and the coin held. Write the game matrix and tell if the game is strictly determined.

20. A friend suggests the following four-finger Morra. If the sum is even, you win an amount equal to the sum of the fingers shown; if the sum is odd your opponent wins the amount equal to the sum of the fingers shown. Write the game matrix and tell if the game is strictly determined.

21. A general with two regiments is trying to capture a city. He can attack from the north with both regiments, from the south with both regiments or from the north with one regiment and the south with one regiment. The defending forces also have two regiments, which they can deploy for protection in the north, in the south, or one in the north and one in the south. The general of the attacking forces will gain 1 point if the attack succeeds, $^-1$ if the attack fails, and 0 if the forces are held at a standoff. Whichever force has more regiments wins the battle and if one regiment is deployed against one regiment the armies are held at a standoff. Write the game matrix and state whether this is a strictly determined game.

Historical Note

Plan of the Battle of Waterloo, showing the relative position of the armies at noon, June 15, 1815. Napoleon maintained that he could mathematically eliminate chance from a battle by carefully calculating every move and countermove beforehand. But events proved that chance cannot be eliminated.

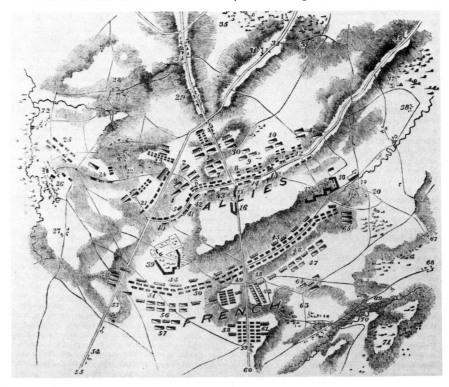

22. A friend proposes the following game. We each flip an imaginary coin and call out heads or tails without knowing our opponent's choice. If they match, I receive $1 and if they don't match, I lose $1. What are the strategies for me and my opponent and what is the value of the game?

23. In a presidential campaign the Democratic and Republican candidates can either campaign in the urban or rural areas. The units assigned to each choice are in gains or losses of thousands of votes.

Republican

		Urban	Rural
Democratic	Urban	-5	3
	Rural	4	2

What are the strategies and what is the value of the game?

24. In submarine warfare, the submarine can attack enemy ships from shallow depths or can launch rockets toward enemy ships from deep water. The shallow depth attack gives more accurate results, but is also more dangerous for the submarine. The surface ships can also drop depth charges set for shallow detonation or drop charges set for deep detonation. If the submarine avoids the depth charges we credit them with 50 points if they are deep and 80 points if they are shallow (since they are more effective if they are shallow). If they are hit with a depth charge they are credited with -100 points. What are the strategies and what is the value of the game?

Mind Bogglers

25. In Conan Doyle's story, "The Final Problem," Sherlock Holmes was pursuing his archenemy Professor Moriarty. The Professor is out to kill Holmes, whose only chance for escape is to flee to the Continent by means of taking a train from London to Dover. Just as the train is leaving London, the two men see each other and the Professor is left at the station. Holmes knows that if he meets Moriarty again it means certain death. Let us assign this occurrence the value of -100 (for Holmes). Holmes can stay on the train until he reaches Dover or he can get off at Canterbury. What should he do? If he eludes Professor Moriarty and makes it to Dover, we will assign this occurrence the value of 50 (for Holmes). If he eludes Moriarty but only reaches Canterbury, we will call this a draw and assign this occurrence the value of 0. On the other hand, Professor Moriarty can catch Holmes by chartering a special train, but must decide whether to go to Canterbury or to Dover. Find the strategies for Sherlock Holmes and Professor Moriarty. Assume that the results are for repeated plays of the "game."

26. Shannon is planning to build a carry-out restaurant in a certain city. He can build uptown, downtown, or in the suburbs. A competitor also plans to build a carry-out restaurant in the same city. The payoff matrix in thousands of dollars is shown below:

Opponent

		Uptown	Downtown	Suburbs
	Uptown	4	1	-3
Shannon	Downtown	-2	0	-5
	Suburbs	4	2	3

What are the strategies and what is the value of the game?

Chapter 11 Outline

I. Introduction to Matrices
  A. Mathematical models
  B. Matrix
    1. Definition
    2. Order
    3. Associated terminology
  C. Sum of matrices
  D. Scalar multiplication
II. Matrix Multiplication
  A. Definition
  B. Conformable
  C. Multiplication is not commutative
  D. Communication matrices
III. Systems of Equations
  A. Definition
  B. Equivalent matrices
  C. Elementary row operations
  D. Solving systems
    1. Augmented matrices
    2. Procedure
IV. Identity and Inverse Properties for Matrices
  A. Identity matrix
  B. Inverse matrix
  C. Procedure for finding inverse matrix
*V. Game Theory
  A. Introduction
    1. Game theory
    2. Zero-sum game
    3. Strategies
    4. Fair game
  B. Strictly determined games
    1. Saddle point
    2. Column maximum/row minimum
  C. Nonstrictly determined games
    1. Mixed strategy
    2. Value of a game
    3. Optimum strategy theorem
  D. Dominating principle
    1. Eliminate dominated rows
    2. Eliminate dominating columns

*Optional section.

Chapter 11 Test

In Problems 1–5, let

$$A = \begin{bmatrix} 4 & ^-2 \\ 3 & 0 \end{bmatrix} \quad B = \begin{bmatrix} ^-1 & 3 \\ ^-2 & 1 \end{bmatrix} \quad C = \begin{bmatrix} 1 & ^-2 & 2 \\ 0 & 4 & ^-15 \\ 3 & ^-5 & 5 \end{bmatrix}$$

$$D = \begin{bmatrix} ^-2 & 1 & 3 \\ 4 & 5 & 1 \\ 1 & 0 & 2 \end{bmatrix} \quad E = \begin{bmatrix} 1 & 0 & 3 & 4 \end{bmatrix}$$

1. Name the order of each matrix.

2. Find, if possible:

    a. $C + D$ b. $3A - 2B$

    c. For what value of x does $\begin{bmatrix} x & y \\ 3 & z \end{bmatrix} = A?$

3. Find, if possible:

    a. AB b. BA c. $(AB)C$

4. Find, if possible:

    a. A^2 b. B^2 c. CD

5. Find, if possible:

    a. A^{-1} b. C^{-1}

6. Solve: $\begin{cases} 3x - 2y = 7 \\ 2x + 3y = 22 \end{cases}$ 7. Solve: $\begin{cases} x - y + z = ^-4 \\ 2x + y - 3z = ^-8 \\ 3x - 2y + 5z = ^-1 \end{cases}$

8. A manufacturer must mix two alloys to produce a steel brace with the proper strength. The minimum acceptable standards call for 30 units of alloy I and 17 units of alloy II. The acceptable mixed batches consist of:

	Alloy I	Alloy II
Batch A:	7 units	3 units
Batch B:	2 units	5 units

    How many units of batches A and B should be mixed in order to produce the minimum acceptable alloy?

* 9. Is the following game strictly determined?

$$
\begin{bmatrix}
1 & 0 & ^-1 & 2 \\
0 & ^-2 & 3 & ^-1 \\
1 & 2 & ^-1 & 0 \\
^-1 & 0 & 1 & ^-2
\end{bmatrix}
$$

*10. Suppose we are planning to build a grocery store in a certain city. We can build uptown, downtown, or in the suburbs. Unfortunately, a competitor also plans to build a grocery store in the same city. If we both decide to build uptown, the difference in sales is $2,000 to our advantage. If we both build downtown, we will trail by $3,000, and if we both build in the suburbs, we will trail by $4,000. If we build uptown, our advantage is $4,000 if our competitor builds downtown and $^-$2,000 if he builds in the suburbs. If we build downtown, our advantage will be $1,000 if our competitor builds uptown and $^-$3,000 if he builds in the suburbs. If we build in the suburbs, our advantage will be $3,000 if our competitor builds uptown and $1,000 if he builds downtown. Write the game matrix for this problem and state whether it is a strictly determined game.

Bonus Question

*11. Consider the following game. I hold a two of spades and an eight of hearts. My opponent holds a two of diamonds and an eight of clubs. We each select a card and place it face up on the table. If they are both the same color, I win the amount equal to the sum of the values of the cards on the table. If they are different colors, my opponent wins the amount equal to the sum of the values of the cards. What are the strategies and what is the value of the game?

*Requires optional Section 11.5.

1. These steps are given on pages 4–5.   **3.** Answers vary.   **5.** They are in diagonal 3.
Note to student: Problems 7–12 are tricky; they are given here to get you to read the question carefully and think about what is being asked.
7. 12 (a dozen stamps, cards, pennies, and so on all have twelve items)
9. Cemeteries only bury the dead.
11. If it is a full 9 innnings, there are 54 outs (6 outs per inning).
13. 2 down, 3 over; 10 paths (from Pascal's triangle)
15. 7 down, 4 over; 330 paths (from Pascal's triangle)
17. There are intermediate paths, so number the vertices; 37 paths
19. 11 (by numbering paths—if you took a path down to Leavenworth, I think you would be backtracking when you reach McAllister, so basically your choices are to come down Taylor or Jones Streets)
21. 20 miles (the fly flies for 1 hour at 20 mph)

Problem Set 1.2, pages 16–20

1.

4	9	2
3	5	7
8	1	6

3.

4	9	5	16
15	6	10	3
14	7	11	2
1	12	8	13

5.

6	7	2
1	5	9
8	3	4

7.

13	2	3	16
8	11	10	5
12	7	6	9
1	14	15	4

9.

1	14	15	4
12	7	6	9
8	11	10	5
13	2	3	16

11.

8	1	6
3	5	7
4	9	2

13.

21	7	8	18
10	16	15	13
14	12	11	17
9	19	20	6

15.

8	1	6
3	5	7
4	9	2

Don't forget to drop down one space after each group of three (after 3 and again after 6)

17.

20	27	4	11	18
26	8	10	17	19
7	9	16	23	25
13	15	22	24	6
14	21	28	5	12

19.

27	34	11	18	25
33	15	17	24	26
14	16	23	30	32
20	22	29	31	13
21	28	35	12	19

21.

32	41	50	3	12	21	30
40	49	9	11	20	29	31
48	8	10	19	28	37	39
7	16	18	27	36	38	47
15	17	26	35	44	46	6
23	25	34	43	45	5	14
24	33	42	51	4	13	22

Answers to the Odd-Numbered Problems

Problem Set 1.3, pages 23–26

1. 19   **3.** 22   **5.** 36   **7.** 29   **9.** 38   **11. a.** $9 \times 54321 - 1$ **b.** 488888   **c.** 8888888888
13. a. 142,857   **b.** 285,714   **c.** 428,571   **d.** 571,428   **e.** 714,285   **f.** 857,142
g. The answers all exhibit the sequence   **15. a.** 5555555505   **b.** 6666666606   **c.** 7777777707   **d.** 8888888808

only with a different starting place.

17.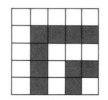

19. 5, 1, 6, 2, 7, 3, 8, 4, 9, 5, 1, 6, 2, 7, 3, 8, 4, 9, 5, 1, 6, . . .   **21.** 3, 6, 9, 3, 6, 9, . . .   **23.** 1111111110
25. 8888888888   **27.** $50 \times 50 = 2{,}500$
29. a. It should be approximately 90°.   **b.** They should be approximately the same (due to slight inaccuracies in measurement).   **c.** They are always 90°.

Problem Set 1.4, pages 32–36

1. Arithmetic; $d = 4$,   next term is 57   **3.** Arithmetic; $d = 6$,   next term is 43   **5.** Not arithmetic
7. Arithmetic; $d = 3$,   next term is 131   **9.** Geometric; $r = 3$,   next term is 162
11. Geometric; $r = 4$,   next term is 256   **13.** Not geometric   **15.** Not geometric
17. Arithmetic; $d = 3$,   next term is 17   **19.** Neither; the difference between terms increases by 1; next term is 22
21. Geometric; $r = 2$; next term is 96   **23.** Neither; subtract 1, subtract 2, subtract 3, . . . ;   next term is 85
25. Neither; one 5, two 5s, . . . , each separated by a 2;   next term is 2
27. Neither; ⊓   **29.** Answers vary; they are numbers in the sequence 1, 1, 2, 3, 5, 8, 13, 21, . . .

Problem Set 1.5, pages 40–42

1. a. 13   **b.** 987   **c.** 9,227,465   **d.** 89   **3. a.** 34   **b.** 4,181   **c.** 1,346,269   **d.** 196,418
5. a. 233   **b.** 2,584   **c.** 3,524,578   **d.** 75,025   **7. a.** 1   **b.** 3   **c.** 8   **d.** 21   **9. a.** 2   **b.** 4   **c.** 7   **d.** 12
11. $F_5 \times F_6$ **13.**   **15.** Answers vary; about 1.6.   **17.** Answers vary; about 1.6.
19. Answers vary; about 1.6.   **21.** Answers vary.
23. Answers vary; should look similar to Figure 1.13.

Chapter 1 Test, pages 43–45

1. Answers vary; see the description on pages 4–5
2. Turn the chessboard so that it forms a triangle with the rook at the top as suggested by the hint. To get the appropriate square, look at 7 blocks down and 7 blocks over, as illustrated by the figure at the right. The number of paths is 3,432.

3. Answers may vary.   **a.**

5	20	24	15	1
3	8	17	14	23
10	19	13	7	16
22	12	9	18	4
25	6	2	11	21

b.

21	4	16	23	1
11	18	7	14	15
2	9	13	17	24
6	12	19	8	20
25	22	10	3	5

c.

25	6	2	11	21
22	12	9	18	4
10	19	13	7	16
3	8	17	14	23
5	20	24	15	1

4.

18	25	2	9	16
24	6	8	15	17
5	7	14	21	23
11	13	20	22	4
12	19	26	3	10

It is a different magic square.   **5.** Answers vary; see pages 20–23

6. 12,345,678,987,654,321   **7. a.** 12   **b.** 543   **8. a.** 10   **b.** 12   **c.** 65   **9. a.** Part a   **b.** $r = \frac{1}{10}$
10 a. 55   **b.** Answers vary; see pages 40–41.   **11.** 1

Chapter 2: Problem Set 2.1, page 53

1. Answers vary; see pages 48–49.   **3.** Answers vary; see pages 48–49.   **5.** { }, {m}, {y}, {m,y}
7. { }, {y}, {o}, {u}, {y,o}, {y,u}, {o,u}, {y,o,u}
9. { }, {m}, {a}, {t}, {h}, {m,a}, {m,t}, {m,h}, {a,t}, {a,h}, {t,h}, {m,a,t}, {m,a,h}, {m,t,h}, {a,t,h}, {m,a,t,h}
11. {2,6,8,10}   **13.** {3,4,5}   **15.** {2,3,5,6,8,9}   **17.** {1,3,4,5,6,7,10}   **19.** {1,2,3,4,5,6,7,9,10}
21. {1,2,3,4,5,6}   **23.** {1,2,3,5,6,7}   **25.** {3}   **27.** {5,6,7}   **29.** {1,2,4,6}
31. 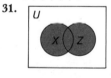   **33.** 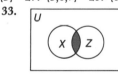   **35.**

Problem Set 2.2, pages 59–61

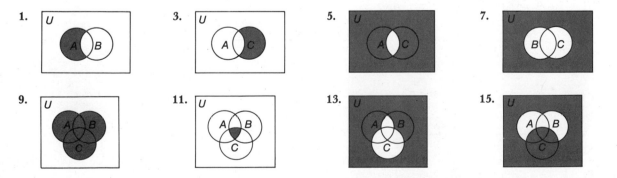

17. ∅   **19.** A or {2,4,6,8}   **21.** {4,6}   **23.** $\overline{A}$ or {1,3,5,7,9,10}   **25.** F   **27.** T   **29.** F   **31.**

33. a. 12   **b.** 7   **c.**

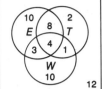

E = Proposition 8
T = Proposition 13
W = Proposition 2

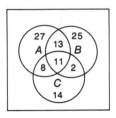

Problem Set 2.3, pages 68–70

1. a. Statement   **b.** Statement   **c.** Not a statement   **d.** Statement
3. a. Statement   **b.** Statement   **c.** Not a statement   **d.** Statement   **5.** Answers vary; see Table 2.1, page 64.
7. Answers vary; see Table 2.3, page 65.
9. a. Some counting numbers are not divisible by 1.   **b.** No apples are rotten.   **c.** All integers are odd.
d. Some triangles are squares.
11. a. $p \lor \sim q$ **b.** $p \land \sim q$ **c.** $p \land \sim q$ **d.** $\sim p \land q$
13. Answers may vary.   **a.** Paul is peculiar and likes to read math texts.   **b.** Paul is not peculiar and likes to read math texts.   **c.** Paul is peculiar or does not like to read math texts.   **d.** Paul is not peculiar and does not like to read math texts.
15. a. T   **b.** F   **c.** F   **d.** F
17. a. $(p \lor q) \lor (r \land \sim q)$ **b.** $\sim(\sim p) \lor (p \land q)$ **c.** $(p \land q) \lor (p \land \sim r)$ **d.** $(\sim p \lor q) \land \sim p$

$\quad\;$ (F ∨ T) ∨ (T ∧ ∼T)$\qquad$ ∼(∼F) ∨ (F ∧ T)$\qquad$ (F ∧ T) ∨ (F ∧ ∼T)$\qquad$ (∼F ∨ T) ∧ ∼F

$\qquad\quad$ T $\quad$ ∨ $\quad$ F$\qquad\qquad$ F $\quad$ ∨ $\quad$ F$\qquad\qquad$ F $\quad$ ∨ $\quad$ F$\qquad\qquad$ (T ∨ T) ∧ $\;$ T

$\qquad\qquad\quad$ T$\qquad\qquad\qquad\qquad$ F$\qquad\qquad\qquad\qquad$ F$\qquad\qquad\qquad\quad$ T $\quad$ ∧ $\;$ T

$\qquad\qquad\qquad\qquad\qquad\qquad\qquad\qquad\qquad\qquad\qquad\qquad\qquad\qquad\qquad\qquad\qquad\qquad$ T

19. Inductive reasoning
21. $e \land d \land t$ Where e: W. C. Fields is eating; d: W. C. Fields is drinking; t: W. C. Fields is having a good time.
23. $\sim t \land \sim m$ Where t: Jack will go tonight; m: Rosamond will go tomorrow.
25. $(j \lor i) \land \sim p$ Where j: The decision will depend on judgment; i: the decision will depend on intuition;
p: the decision will depend on who paid the most.   **27.** T   **29.** F   **31.** T

Problem Set 2.4, pages 78–79

1.

p	q	$\sim p$	$\sim p \lor q$
T	T	F	T
T	F	F	F
F	T	T	T
F	F	T	T

3.

p	q	$p \land q$	$\sim(p \land q)$
T	T	T	F
T	F	F	T
F	T	F	T
F	F	F	T

5.

r	$\sim r$	$\sim(\sim r)$
T	F	T
F	T	F

7.

p	q	$\sim q$	$p \land \sim q$
T	T	F	F
T	F	T	T
F	T	F	F
F	F	T	F

9.

p	q	$\sim p$	$\sim p \land q$	$\sim q$	$(\sim p \land q) \lor \sim q$
T	T	F	F	F	F
T	F	F	F	T	T
F	T	T	T	F	T
F	F	T	F	T	T

11.

p	q	$\sim p$	$\sim p \lor q$	$q \land p$	$(\sim p \lor q) \land (q \land p)$
T	T	F	T	T	T
T	F	F	F	F	F
F	T	T	T	F	F
F	F	T	T	F	F

13.

p	q	$p \to q$	$p \vee (p \to q)$
T	T	T	T
T	F	F	T
F	T	T	T
F	F	T	T

15.

p	q	$p \wedge q$	$(p \wedge q) \to p$
T	T	T	T
T	F	F	T
F	T	F	T
F	F	F	T

17.

p	q	$p \wedge q$	$\sim q$	$p \to \sim q$	$(p \wedge q) \wedge (p \to \sim q)$
T	T	T	F	F	F
T	F	F	T	T	F
F	T	F	F	T	F
F	F	F	T	T	F

19.

p	q	$\sim q$	$p \to \sim q$	$\sim p$	$q \to \sim p$	$(p \to \sim q) \to (q \to \sim p)$
T	T	F	F	F	F	T
T	F	T	T	F	T	T
F	T	F	T	T	T	T
F	F	T	T	T	T	T

21.

p	q	$p \vee q$	$p \wedge (p \vee q)$	$[p \wedge (p \vee q)] \to p$
T	T	T	T	T
T	F	T	T	T
F	T	T	F	T
F	F	F	F	T

23.

p	q	r	$p \vee q$	$\sim r$	$(p \vee q) \wedge (\sim r)$	$[(p \vee q) \wedge \sim r] \wedge r$
T	T	T	T	F	F	F
T	T	F	T	T	T	F
T	F	T	T	F	F	F
T	F	F	T	T	T	F
F	T	T	T	F	F	F
F	T	F	T	T	T	F
F	F	T	F	F	F	F
F	F	F	F	T	F	F

25. p is T and q is T. Using truth tables, we see that the statement in Problem 1 is true, that in Problem 2 is false, and that in Problem 3 is false.

27. a. If there is time enough, then everything happens to everybody sooner or later.

  b. If we make a proper use of those means which the God of Nature has placed in our power, then we are not weak.

  c. If it is a useless life, then there is an early death.

  d. If it is work, then it is noble.

  e. If only you can find it, then everything has got a moral. (Don't confuse the "if only" of this problem with the "only if" connective discussed later in this chapter.)

29. a. T **b.** T **c.** T **d.** T **e.** F

31. If I get paid, then I will go Saturday. *Converse:* If I go Saturday, then I will get paid. *Inverse:* If I do not get paid, then I will not go Saturday. *Contrapositive:* If I do not go Saturday, then I do not get paid.

33. *Statement:* $\sim p \to \sim q$ *Converse:* $\sim q \to \sim p$ *Inverse:* $p \to q$ *Contrapositive:* $q \to p$

35. *Statement:* $\sim t \to \sim s$ *Converse:* $\sim s \to \sim t$ *Inverse:* $t \to s$ *Contrapositive:* $s \to t$

Problem Set 2.5, pages 84–86

1. Tautology **3.** Not a tautology **5.** Not a tautology

7.

p	q	$\sim q$	$\sim q \to p$
T	T	F	T
T	F	T	T
F	T	F	T
F	F	T	F

9.

p	q	$p \vee q$	$p \wedge q$	$\sim(p \wedge q)$	$(p \vee q) \wedge \sim(p \wedge q)$
T	T	T	T	F	F
T	F	T	F	T	T
F	T	T	F	T	T
F	F	F	F	T	F

11. $\sim(p \vee q)$ is equivalent to $\sim p \wedge \sim q$ by De Morgan's Law.

13. Let h: I will buy a new house
  p: All provisions of the sale are clearly understood
  h unless p: $\sim p \to \sim h$.

15. Let m: It is a man
  i: It is an island
  No m is i: $m \to \sim i$.

17. Let n: Be nice to people on your way up
  m: You will meet people on your way down
  n because m: $(n \wedge m) \wedge (m \to n)$.

19. Let f: The majority, by mere force of numbers, deprives a minority of a clearly written constitutional right
  r: Revolution is justified
  If f then r: $f \to r$.

21. Paul did not go out for baseball and he did not go out for soccer. **23.** Sally is on time or she did not miss the boat.

25. You are out of Schlitz and you have beer. **27.** $x - 5 = 4$ or $x \neq 1$.

29.

p	q	$\sim p$	$\sim q$	$p \to q$	$\sim p \to \sim q$	$\sim q \to \sim p$	$p \to q \Leftrightarrow \sim q \to \sim p$
T	T	F	F	T	T	T	T
T	F	F	T	F	T	F	T
F	T	T	F	T	F	T	T
F	F	T	T	T	T	T	T

31.

p	q	$p \to q$	$\sim p$	$\sim p \vee q$	$(p \to q) \Leftrightarrow (\sim p \vee q)$
T	T	T	F	T	T
T	F	F	F	F	T
F	T	T	T	T	T
F	F	T	T	T	T

Problem Set 2.6, pages 90–94

1. Direct reasoning **3.** Transitive **5.** Direct reasoning **7.** Transitive **9.** Indirect reasoning

11.

p	q	$p \to q$	$(p \to q) \wedge \sim q$	$[(p \to q) \wedge \sim q] \Rightarrow \sim p$
T	T	T	F	T
T	F	F	F	T
F	T	T	F	T
F	F	T	T	T

13. I am lazy.

15. We do not interfere with the publication of false information.

17. If you climb the highest mountain, then you will feel great. **19.** $a = 0$ or $b = 0$

21. Let p: We win first prize
  e: We will go to Europe
  i: We are ingenious

Given argument: **1.** $p \to e$
  2. $i \to p$
  3. i

Rearrange the premises:
  2. $i \to p$ Given
  1. $p \to e$ Given
  $\therefore i \to e$ Transitive law
  3. i Given
  $\therefore e$ Direct reasoning

Conclusion: We will go to Europe.

23. If I am tired, then I did not understand the material.

25. Let p: Puppies
   n: Nice creatures
   a: This animal

Given argument:
1.	$p \rightarrow n$	
2.	$a \rightarrow p$	
3.	$n \rightarrow \sim d$	

Then we have:
2.	$a \rightarrow p$	Given
1.	$p \rightarrow n$	Given
	$\therefore a \rightarrow n$	Transitive law
3.	$n \rightarrow \sim d$	Given
	$\therefore a \rightarrow \sim d$	Transitive

Conclusion: This animal is not dangerous.

27. Let s: Sane people (assume $\sim s \leftrightarrow$ Lunatic)
   l: People who can do logic
   j: People who are fit to serve on a jury
   y: Your sons

Given argument:
1.	$s \rightarrow l$	
2.	$j \rightarrow \sim(\sim s)$	
3.	$y \rightarrow \sim l$	

Then we have:
2'.	$j \rightarrow s$	Double negative of (2)
1.	$s \rightarrow l$	Given
	$\therefore j \rightarrow l$	Transitive law
3'.	$l \rightarrow \sim y$	Contrapositive of (3)
	$\therefore j \rightarrow \sim y$	Transitive law

Conclusion: None of your sons are fit to serve on a jury.

29. Let p: The butler was present
   s: The butler was seen
   q: The butler was questioned
   r: The butler replied
   h: The butler was heard
   d: The butler was on duty

Given argument:
1.	$p \rightarrow s$	
2.	$s \rightarrow q$	
3.	$q \rightarrow r$	
4.	$r \rightarrow h$	
5.	$\sim h$	
6.	$(\sim s \wedge \sim q) \rightarrow d$	
7.	$d \rightarrow p$	
	$\therefore q$	

We do not need all of the premises:
4.	$r \rightarrow h$	Given
5.	$\sim h$	Given
	$\therefore \sim r$	Indirect reasoning
3.	$q \rightarrow r$	Given
	$\therefore \sim q$	Indirect reasoning

Thus, it is not valid. A correct conclusion is:
The butler was not questioned.

Problem Set 2.7, pages 98–99

1. Answers vary.

3.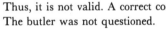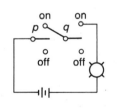

5.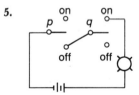

7.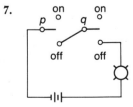

9.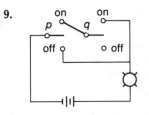

11.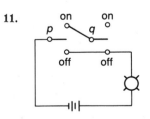

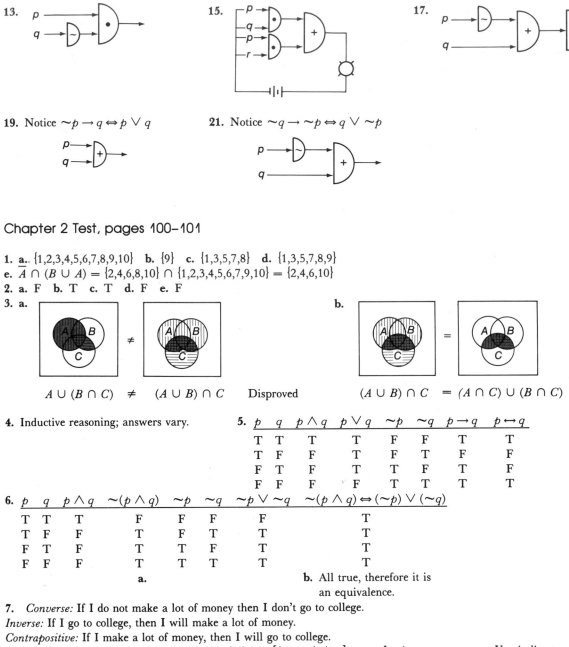

13.

15.

17.

19. Notice $\sim p \rightarrow q \Leftrightarrow p \vee q$

21. Notice $\sim q \rightarrow \sim p \Leftrightarrow q \vee \sim p$

Chapter 2 Test, pages 100–101

1. a. {1,2,3,4,5,6,7,8,9,10} **b.** {9} **c.** {1,3,5,7,8} **d.** {1,3,5,7,8,9}
e. $\overline{A} \cap (B \cup A) = \{2,4,6,8,10\} \cap \{1,2,3,4,5,6,7,9,10\} = \{2,4,6,10\}$
2. a. F **b.** T **c.** T **d.** F **e.** F
3. a.

$A \cup (B \cap C) \quad \neq \quad (A \cup B) \cap C$ Disproved

b.

$(A \cup B) \cap C \quad = \quad (A \cap C) \cup (B \cap C)$ Proved

4. Inductive reasoning; answers vary.

5.

p	q	$p \wedge q$	$p \vee q$	$\sim p$	$\sim q$	$p \rightarrow q$	$p \leftrightarrow q$
T	T	T	T	F	F	T	T
T	F	F	T	F	T	F	F
F	T	F	T	T	F	T	F
F	F	F	F	T	T	T	T

6.

p	q	$p \wedge q$	$\sim(p \wedge q)$	$\sim p$	$\sim q$	$\sim p \vee \sim q$	$\sim(p \wedge q) \Leftrightarrow (\sim p) \vee (\sim q)$
T	T	T	F	F	F	F	T
T	F	F	T	F	T	T	T
F	T	F	T	T	F	T	T
F	F	F	T	T	T	T	T

a.

b. All true, therefore it is an equivalence.

7. *Converse:* If I do not make a lot of money then I don't go to college.
Inverse: If I go to college, then I will make a lot of money.
Contrapositive: If I make a lot of money, then I will go to college.

8. a.

$p \rightarrow q$
p
$\therefore q$

p	q	$p \rightarrow q$	$(p \rightarrow q) \wedge p$	$[(p \rightarrow q) \wedge p] \rightarrow q$
T	T	T	T	T
T	F	F	F	T
F	T	T	F	T
F	F	T	F	T

b. Answers vary. **c.** Yes; indirect reasoning

Always true and is therefore an implication.

9. a. Mr. Gent obtained a tan.   **b.** You do not want peace.   **c.** If we go on a picnic, then we must go to the store.
d. I do not attend to my duties.   **e.** No prune is an artificial sweetener.

10. a.   **b.**

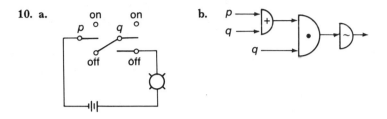

11. Let a = Ace did it, b = Lucky did it, c = Alfie did it, d = Ears did it, f = Maurice did it.

   1. $\sim(a \vee b) \rightarrow d$ Given; 3 tell truth
   2. $d \rightarrow (a \vee b)$ Given; compare statements
   3. $\sim(a \vee b) \rightarrow (a \vee b)$ Transitive, statements 1 and 2
   4. $a \vee b$ Logically equivalent to statement 3
   5. $(a \vee b) \rightarrow \sim c$ Given; compare statements
   6. $\sim c$ Direct reasoning from statements 4 and 5
   7. $\sim b \rightarrow f$ Given; 3 tell truth
   8. $f \rightarrow \sim d$ Given; compare statements
   9. $\sim b \rightarrow \sim d$ Transitive, statements 7 and 8
 10. $\sim b \rightarrow d$ Given; 3 tell truth
 11. $\sim d \rightarrow b$ Contrapositive of statement 10
 12. $\sim b \rightarrow b$ Transitive, statements 9 and 11
 13. b Logically equivalent to statement 12
 14. $\sim a \rightarrow d$ Given; 3 tell truth
 15. $d \rightarrow \sim f$ Contrapositive of statement 8
 16. $\sim f \rightarrow a$ Given; 3 tell truth
 17. $\sim a \rightarrow a$ Transitive, statements 14, 15, and 16
 18. a Logically equivalent to statement 17
 19. $a \wedge b$ Combine statements 13 and 18
   Therefore, from statement 19, Alfie did it.

Chapter 3:   Problem Set 3.1, pages 111–113

1. a. 27   **b.** 63   **c.** 243   **d.** 432   **e.** 504
3. It introduced the idea of the punched card and a program to complete a task.   **5.** A billionth of a second
7. and **9.** Answers vary, but you should support your opinions whenever possible.

Problem Set 3.2, pages 118–120

1. Arithmetic, algebraic, and RPN; answers vary.   **3.** 57334 (HEELS)   **5.** 57334.4614 (HIGH˙HEELS)
7. 53045 (SHOES)   **9.** 4914 (HIGH)   **11.** 40 (OH)   **13.** 510714 (HILOIS)
15. $2.814749767 \cdot 10^{14}$ or 281,474,976,710,656   **17.** 30,814 minutes or 21 days, 9 hours, and 34 minutes   **19.** .618
21. The result is the original number.   **23.** .1666 . . .   **25.** .142857142857 . . .   **27.** .0036900369 . . .
29. a. .142857142857 . . .   **b.** .285714285714 . . .   **c.** .428571428571 . . .   **d.** .571428571428 . . .
e. .714285714285 . . .   **f.** .857142857142 . . .

Problem Set 3.3, pages 127–130

1. Answers vary; see pages 121–122.   **3.** Answers vary; see page 127.
5. CRT; magnetic tape; punched cards; mark-sense cards; floppy disk (any three)   **7.** Answers vary; consider a truth table.
9. 25   **11.** 13   **13.** 11   **15.** 23   **17.** 99   **19.** 184   **21.** 1111_{two} **23.** 101110_{two} **25.** 111111_{two} **27.** 100000000_{two}
29. 1001100111_{two} **31.** 1100100011_{two}

Problem Set 3.4, pages 138–142

1. Step 1: Analyze the problem.   **Step 2:** Prepare plan or flowchart.
Step 3: Put the flowchart into a language that the machine can "understand."   **Step 4:** Test program.
Step 5: Debug program.

3. a.   **b.**   **c.**   **d.**   **e.**

5. Answers vary. To debug a program means to run the program under a variety of circumstances in order to test that it is running properly. When an error is detected, it is corrected and then checked again.
7. GOOD MORNING,
    CAN YOU DO THIS PROBLEM CORRECTLY?
    5
9. a. $35x^2 - 13x + 2$ **b.** $6x - 7$
11. a. $6.29^2 - 7$ **b.** $13x^2 + \dfrac{15}{2}$ **13.** $\left(\dfrac{16.34}{12.5}\right)(42.1^2 - 64)$ **15.** $\dfrac{5}{2}x + \dfrac{135}{2}$
17. a. 5*Y↑2 – 6*X↑2 + 11   **b.** 14*Y↑2 + 12*X↑2 + 3
19. a. (5 – X)*(X + 3)↑2   **b.** 6*(X + 3)*(2*X – 7)↑2   **21.** (2*X – 3)*(3*X↑2 + 1)
23. (2/3)*X↑2 + (1/3)*X – 17
25. 10   PRINT "I WILL DEMONSTRATE MY COMPUTATIONAL SKILL."   **27.** 10   PRINT(1 + .08)↑2
    20   END   20   END
29. 10   PRINT(23.5↑2 – 5*61.1)/2   **31.**   1   **33.** Answers vary.
    20   END   3
                                                   5
                                                   7
                                                   9
                                                  1 1

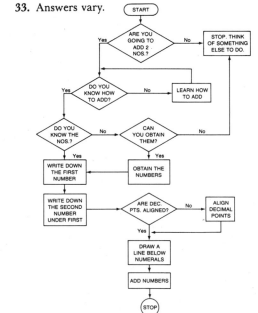

Problem Set 3.5, pages 148–151

1. (1) To do calculations, as in 10 PRINT 4 + 5;
(2) To type out text, as in 10 PRINT "HELLO";
(3) To generate a line feed, as in 10 PRINT.

3. LET or INPUT **5.** $6 + 5 = 11$
$6 - 5 = 1$
$6*5 = 30$ $6/3 = 2$

7. 9
```
HELLO, I LIKE YOU.   -   5   84
4 + 5 = 9
```
9. 19

11. No operation symbol between the factors in line 20.

13.
```
I WILL CALCULATE IQ.
WHAT IS MENTAL AGE? [input 12 here]
WHAT IS CHRONOLOGICAL AGE? [input 10 here]

IQ IS 120
```

15.
```
10   PRINT "WHAT IS X";
20   INPUT X
30   PRINT 5*X↑2 + 17*X - 128
40   END
```

17. Answers vary; change line 30 to:
```
30   PRINT "A =";P*R↑2
```
and delete line 40.

19.
```
10   PRINT "WHAT ARE YOUR SALES FOR THIS WEEK";
20   INPUT S
30   PRINT .05*X + 100
40   END
```

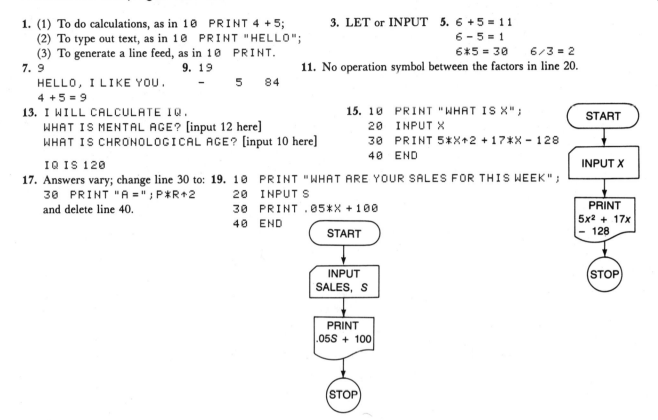

Problem Set 3.6, pages 156–160

1. A loop is a repetition of one or more lines in a program.   **3.** HI FRIEND.

5.
```
10   PRINT "HI FRIEND."
20   END
```
7. A IS BETWEEN 1 AND 100 INCLUSIVE.   **9.** A IS LESS THAN 1.

11.
NUMBER	NUMBER SQUARED
1	1
2	4
4	16
3	9

13.
1	5
5	13
2	7
4	11

15.
1
4
9

17.
```
10   LET N = 100
20   PRINT N;
30   PRINT "BOTTLES OF BEER ON THE WALL."
40   IF N = 2 THEN GOTO 70
50   LET N = N - 1
60   GOTO 20
70   PRINT "1 BOTTLE OF BEER ON THE WALL."
80   PRINT "THAT'S ALL FOLKS!"
90   END
```

19.
```
10   PRINT "FIRST 100 FIBONACCI NUMBERS"
20   LET A = 1
30   LET B = 1
40   LET N = 1
50   PRINT A
60   PRINT B
70   LET C = A + B
80   PRINT C
90   LET A = B
100  LET B = C
110  LET N = N + 1
120  IF N < 98 GOTO 70
130  END
```

21.
```
10   LET N = 1
20   PRINT N↑2
30   IF N > 50 GOTO 60
40   LET N = N + 1
50   GOTO 20
60   END
```

Chapter 3 Test, pages 162–163

1. Answers vary (see Section 3.1).　**2. a.** 44.1　**b.** 3.009375　**3. a.** 7735.05 (SO'SELL)　**b.** 3,857,620

4. *Input:* means of putting information into the computer. *Storage:* the part of the computer that saves or stores information. *CPU:* that part of the computer in which the arithmetic operations are carried on. *Control:* the part of the computer that directs the flow of data to and from the various components of the computer. It is also the part in which the steps of the program are followed. *Output:* means of getting results from the computer.

5. Computer programming refers to the process by which we give a computer a series of step-by-step instructions to complete a particular task or problem. The level of a program and type of instructions the computer can "understand" may vary greatly.

6. a. 29　**b.** 123　**c.** 1100_{two}　**d.** 110100_{two}　**7. a.** $5(x + 2)^2$　**b.** 3*(X + 4)*(2*X − 7)↑2

8.

```
10   PRINT "WHAT IS X";
20   INPUT X
30   PRINT "18*X↑2 + 10 IS";18*X↑2 + 10;
40   PRINT "WHEN X = ";X
50   END
```

9. THE MISSISSIPPI IS WET.

10.
```
10   PRINT "WHAT IS THE RADIUS";
20   INPUT R
30   PRINT "THE AREA IS";
40   PRINT 3.1416*R↑2
50   END
```

11. Answers vary.
```
10   LET N = 1
20   PRINT "1/";N;"=";1/N
30   LET N = N + 1
40   IF N < 101 GOTO 20
50   END
```

Chapter 4: Problem Set 4.1, pages 171–174

1. Answers vary; a number is an idea or a concept, while a numeral is a symbol for a number.

3. a. Yes, since it is not a positional system　**b.** No; answers vary.

5. a. ∩∩ |||||| **b.** ∩∩∩ ||||| **c.** 99 9∩∩| **d.** 9999 ∩∩∩∩ |||||| **e.** 99∩∩∩∩ ∩∩∩∩∩||

7. Answers vary.　**9. a.** 1,256　**b.** $\dfrac{1}{200}$　**11. a.** 24　**b.** 261　**13. a.** $123\frac{1}{2}$　**b.** $\dfrac{1}{100}$　**15. a.** 152　**b.** 4,881

17. ℒ 999∩||||||||| **19.** 99||| **21.** ⊲⊲⊲⊲⊲ **23.** ▾▾⊲▾▾ **25.** Answers vary. **27.** One
99||||

Problem Set 4.2, pages 178–179

1. a. $b^n = \underbrace{b \cdot b \cdot b \cdots b}_{n \text{ factors}}$ **b.** $b^0 = 1$ **c.** $b^{-n} = 1/b^n$ **3. a.** .01 **b.** 300 **5. a.** 1,000,000 **b.** .007

7. a. 521,658 **b.** 60,004,001 **9.** 3028.5462

11. a. $(7 \times 10^2) + (4 \times 10^1) + (1 \times 10^0)$ **b.** $(7 \times 10^5) + (2 \times 10^4) + (8 \times 10^3) + (4 \times 10^2) + (7 \times 10^0)$

13. a. $(5 \times 10^2) + (2 \times 10^1) + (1 \times 10^0)$ **b.** $(6 \times 10^3) + (2 \times 10^2) + (4 \times 10^1) + (5 \times 10^0)$

15. a. $(4 \times 10^1) + (7 \times 10^0) + (3 \times 10^{-2}) + (2 \times 10^{-3}) + (1 \times 10^{-4}) + (5 \times 10^{-5})$ **b.** $(1 \times 10^5) + (1 \times 10^{-3})$

17. a. $(6 \times 10^5) + (7 \times 10^4) + (8 \times 10^3) + (1 \times 10^{-2})$

b. $(5 \times 10^4) + (7 \times 10^3) + (2 \times 10^2) + (8 \times 10^1) + (5 \times 10^0) + (9 \times 10^{-1}) + (3 \times 10^{-2}) + (6 \times 10^{-3}) + (1 \times 10^{-4})$

19.

Problem Set 4.3, pages 185–187

1. a. One million **b.** 10 **c.** 6 **d.** 1,000,000 or $10 \times 10 \times 10 \times 10 \times 10 \times 10$

3. a. 3.2×10^3 **b.** 3.2 03 **c.** 3.2E+3 **5. a.** 5.629×10^3 **b.** 5.629 03 **c.** 5.629E+3

7. a. 3.5×10^{10} **b.** 3.5 10 **c.** 3.5E+10 **9. a.** 3.5×10^{-14} **b.** 3.5 −14 **c.** 3.5E−14

11. a. 6×10^{-9} **b.** 6 −09 **c.** 6E−9 **13. a.** 6.8×10^{-11} **b.** 6.8 −11 **c.** 6.8E−11 **15.** 2.2×10^8; 2.5×10^{-6}

17. 3×10^{10} **19.** 5.9×10^7 **21. a.** 72,000,000,000 **b.** .0021 **23. a.** 6.81 **b.** .00006 9 **25. a.** 40,700 **b.** 707.8

27. a. .00000 3217 **b.** .00000 00000 889 **29.** 907 **31.** 333,000 **33.** .00000003 **35.** 1,280,000

37. $1973 = 1.973 \times 10^3$; 1.6 million $= 1.6 \times 10^6$; half-million $= 5 \times 10^5$; half-billion $= 5 \times 10^8$; $63,000 = 6.3 \times 10^4$

39. 5.157×10^{15}

41. There are approximately 17 pennies per inch; $\frac{1,000,000}{17} \approx 58,823$ in. or 4,902 ft, or .92 mi. Thus, a million pennies would be about a mile high.

43. 186,282 mi/sec $\approx 1.9 \times 10^5$ mi/sec

$$= 1.9 \times 10^5 \times 60 \times 60 \times 24 \times 365 \times 25 \text{ mi/year}$$
$$= (1.9 \times 10^5)(6 \times 10)(6 \times 10)(2.4 \times 10)(3.6 \times 10^2)$$
$$= (1.9)(6)(6)(2.4)(3.6) \times 10^5 \times 10 \times 10 \times 10 \times 10^2$$
$$\approx 590 \times 10^{10}$$
$$\approx 5.9 \times 10^{12}$$

45. a. 3 ft = 36 in., and on the scale shown, 1 in. is about 26 light-years, so 3 ft is about 936 light-years away.

b. $(9.4 \times 10^2)(5.9 \times 10^{12}) = (9.4)(5.9) \times 10^2 \times 10^{12} \approx 55.46 \times 10^{14} \approx 5.5 \times 10^{15}$

Problem Set 4.4, pages 193–198

1. Commutative **3.** Commutative **5.** Both **7.** Commutative **9.** Commutative
11. Answers vary; commutative property because he switched the order of his words. **13.** No **15.** Answers vary.
17. a. ^-i **b.** $^-1$ **c.** i **d.** $^-1$ **19.** Reasons vary. **a.** Yes **b.** Yes **c.** Yes
21. Reasons vary. **a.** Yes **b.** Yes **c.** No **23.** Yes
25. a.

b.

Since the shaded parts are the same,
$(X \cup Y) \cup Z = X \cup (Y \cup Z)$.

$(X \cup Y) \cup Z$ $X \cup (Y \cup Z)$

27. Yes; proof is similar to that shown in the answer to Problem 25.
29. No; yes, the set of odd numbers is closed for multiplication. **31. a.** A **b.** F **c.** A **d.** F **e.** R
33. It is closed for both ↑ and ↓.
35. It is the same as the program in the text but with the following changes:

```
     30   IF A - B = B - A GOTO 80
     100  PRINT "THE SET IS COMMUTATIVE FOR SUBTRACTION."
37.  10   LET A = 0
     20   LET B = 0
     30   LET C = 0
     40   IF (A*B)*C = A*(B*C) GOTO 100
     50   PRINT "NOT ASSOCIATIVE, SINCE I HAVE FOUND A";
     55   PRINT "COUNTER-EXAMPLE WHEN"
     60   PRINT "A = ";A;
     70   PRINT ", B = ";B;
     80   PRINT ", AND C = C";C
     90   STOP
     100  PRINT "CHECKING A = ";A;", B = ";B;", AND C = ";C
     120  PRINT "(A*B)*C = ";(A*B)*C;", AND A*(B*C) = ";A*(B*C)
     130  LET C = C + 1
     140  IF C < 2 GOTO 40
     150  LET B = B + 1
     160  IF B < 2 GOTO 30
     170  LET A = A + 1
     180  IF A < 2 GOTO 20
     190  PRINT "THE SET IS ASSOCIATIVE FOR MULTIPLICATION."
     200  END
```

Problem Set 4.5, pages 209–212

1. a. You moved to the *right* 7 units. **b.** $^+6$ **c.** $^-8$ **3. a.** $^+4$ **b.** $^-4$ **c.** $^-14$ **d.** $^+135$ **e.** $^-19$ **f.** $^-1$
5. a. $^+3$ **b.** $^-3$ **c.** $^+10$ **d.** $^+5$ **e.** $^+10$ **f.** $^+10$ **7. a.** $^-47$ **b.** $^-35$ **c.** $^-4$ **d.** $^+7$ **e.** $^+1$ **f.** 0
9. a. $^-18$ **b.** $^-20$ **c.** $^+2$ **d.** $^+2$ **e.** $^-26$ **f.** $^+2$ **11. a.** $^-3$ **b.** $^+42$ **c.** $^+132$ **d.** $^+170$ **e.** $^-1$ **f.** $^+40$
13. a. 0 **b.** $^-13$ **c.** $^+6$ **d.** $^+7$ **e.** $^+44$ **f.** $^+15$ **15. a.** $^-56$ **b.** $^-75$ **c.** $^+30$ **d.** $^-14$ **e.** $^+2$ **f.** $^+18$
17.

19. a. An operation * is commutative for a set S if $a*b = b*a$ for all elements a and b in S.
b. Yes for addition; no for subtraction   **c.** Yes for multiplication; no for division
21. a. An operation * is associative for a set S if $(a*b)*c = a*(b*c)$ for all elements a, b, and c in S.
b. Yes for addition; no for subtraction   **c.** Yes for multiplication; no for division
23. a. $^+16$ **b.** $^-1$ **c.** $^+1$ **d.** $^+1$ **25.** Yes. Consider the set $\{^-1, 0, ^+1\}$.

Problem Set 4.6, pages 218–220

1. a. $\frac{1}{3}$ **b.** $\frac{2}{3}$ **c.** $\frac{1}{5}$ **3. a.** 2 **b.** 2 **c.** 46 **5. a.** $\frac{3}{5}$ **b.** $\frac{2}{3}$ **c.** $\frac{1}{8}$ **7.** $\frac{13}{9}$ **9.** $\frac{-92}{105}$ **11.** $\frac{63}{25}$ **13.** $\frac{1}{9}$ **15.** $\frac{1}{15}$
17. $\frac{239}{1,000}$ **19.** $\frac{10}{21}$ **21.** $\frac{30}{7}$ **23.** $\frac{2}{(-7)}$ or $\frac{-2}{7}$ **25.** $\frac{2}{3}$ **27.** 21 **29.** $\frac{-625}{686}$

31.

Sets	5	$^-7$	0	$\frac{4}{5}$	$\frac{-1}{2}$
Natural numbers	✔	X	X	X	X
Whole numbers	✔	X	✔	X	X
Integers	✔	✔	✔	X	X
Rational numbers	✔	✔	✔	✔	✔

33. Given any two rationals a/b and c/d where $c \neq 0$. To show $a/b \div c/d$ is rational.
Now, $a/b \div c/d = ad/bc$ by the definition of division. ad is an integer because a and d are integers; also, bc is a nonzero integer because b and c are nonzero integers.
Therefore, ad/bc is a rational number, by the definition of a rational number.
35. Closure, associative, and commutative

Problem Set 4.7, pages 227–229

1.

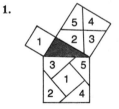

3. Answers vary; about 1.41.   **5. a.** 8   **b.** 10   **c.** 25   **7. a.** 807   **b.** 169   **c.** 400
9. Rational; 5   **11.** Irrational; 5.477   **13.** Irrational; 7.071   **15.** Rational; 20
17. Irrational; 31.623   **19.** Rational; 44   **21.** 24 ft   **23.** 10 ft   **25.** $\sqrt{18}$
27. $\sqrt{325}$; 18 ft; 37 ft   **29.** 15 ft
31. Answers are not unique.   **a.** $11 = 3^2 + 1^2 + 1^2 + 0^2$
b. $39 = 6^2 + 1^2 + 1^2 + 1^2$ **c.** $143 = 11^2 + 3^2 + 3^2 + 2^2$
d. $1,000 = 30^2 + 10^2 + 0^2 + 0^2$ **e.** $1,980 = 44^2 + 6^2 + 2^2 + 2^2$

Problem Set 4.8, pages 234–237

1.

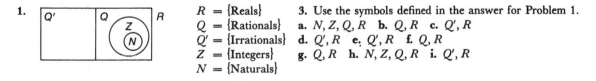

$R = \{\text{Reals}\}$ **3.** Use the symbols defined in the answer for Problem 1.
$Q = \{\text{Rationals}\}$ **a.** N, Z, Q, R **b.** Q, R **c.** Q', R
$Q' = \{\text{Irrationals}\}$ **d.** Q', R **e.** Q', R **f.** Q, R
$Z = \{\text{Integers}\}$ **g.** Q, R **h.** N, Z, Q, R **i.** Q', R
$N = \{\text{Naturals}\}$

5. a. 1.5 **b.** .7 **7. a.** .4 **b.** .47 **9. a.** $.8\overline{3}$ **b.** $.\overline{285714}$ **11. a.** $.1\overline{2}$ **b.** $2.1\overline{6}$ **13. a.** $.\overline{6}$ **b.** $2.\overline{153846}$
15. a. $^-.8$ **b.** $^-.6$ **17. a.** $\frac{1}{2}$ **b.** $\frac{4}{5}$ **19. a.** $\frac{2}{25}$ **b.** $\frac{3}{25}$ **21. a.** $\frac{9}{20}$ **b.** $\frac{117}{500}$ **23. a.** $\frac{987}{10}$ **b.** $\frac{63}{100}$
25. a. $\frac{153}{10}$ **b.** $\frac{139}{20}$ **27. a.** 1 **b.** Use division   **c.** Yes   **d.** 1   **e.** This division is correct.   **29.** Answers vary.

Chapter 4 Test, pages 239–241

1. a. The position in which the individual digits are listed is relevant. Examples will vary.
b. Addition is easier in a simple grouping system. Examples will vary.
c. It uses ten symbols; it is positional; it has a place-holder symbol (0); and it uses 10 as its basic unit for grouping.
2. a. $(4 \times 10^2) + (3 \times 10^1) + (6 \times 10^0) + (2 \times 10^{-1}) + (1 \times 10^{-5})$ **b.** 4,020,005.62   **3.** Answers vary.
4. a. .000579   **b.** 401,000   **c.** .1   **d.** 43,210,000
5. a. 3.4×10^{-3} **b.** 4.0003×10^6 **c.** 1.74×10^4 **d.** 5   or 5×10^0
6. a. $b^n = \underbrace{b \cdot b \cdots b}_{n \text{ factors}}$ **b.** $b^{-n} = 1/b^n$ **c.** $b^0 = 1$ **d.** Answers vary.

7. a. $^-7$ **b.** 23   **c.** $^-17$ **d.** $^-3$ **e.** $^-8$ **f.** $^-20$ **g.** 9   **h.** 15   **i.** $^-14$ **j.** $^-4$
8. a. $^-11$ **b.** $\frac{36}{5}$ **c.** $\frac{71}{63}$ **d.** $\frac{-1}{36}$ **e.** $\frac{35}{3}$ **9. a.** $\frac{12}{5}$ **b.** $\frac{1}{15}$ **c.** $\frac{31}{30}$ **d.** $\frac{15}{8}$ **e.** $\frac{7}{4}$
10. a. .625   **b.** $\frac{333}{500}$ **c.** 2 (Use the commutative and associative properties to simplify first.)
d. 7 (Use the commutative and associative properties to simplify first.)   **e.** $\dfrac{3}{5} \times \left(\dfrac{25}{97} + \dfrac{72}{97} \right) = \dfrac{3}{5} \times 1 = \dfrac{3}{5}$

11. a. $\sqrt{16,200}$; about 127 ft   **b.** Same as a.   **12.** 7 and 10 are the only happy integers between 1 and 10 (inclusive).

Chapter 5:   Problem Set 5.1, pages 251–253

1. $400; $469.33; $69.33 more   **3.** $720; $809.86; $89.86 more
5. $12,000; $43,231.47; $31,231.47 more   **7.** $1,400; $1,469.33; $69.33 more
9. $2,720; $2,809.86; $89.86 more   **11.** $17,000; $48,231.47; $31,231.47 more
13. 9%; 5; 1.538624; $1,538.62; $538.62   **15.** 8%; 3; 1.259712; $629.86; $129.86
17. 2%; 12; 1.268242; $634.12; $134.12   **19.** 4.5%; 40; 5.816365; $29,081.83; $24,081.83
21. 5%; 40; 7.039988; $35,199.95; $30,199.95   **23.** 2%; 60; 3.281031; $13,124.12; $9,124.12
25. 4%; 5; 1.216653; $1,520.82; $270.82   **27.** $2.02   **29.** $767.44   **31.** $18,849.41   **33.** $269,205.03
35. $29,960   **37.** 6.17%
39. Effective yield of $5\frac{1}{4}\%$ compounded daily is 5.39%. Therefore, the $5\frac{1}{2}\%$ compounded annually is the better interest rate.
41. The difference is almost $23 million.

Problem Set 5.2, pages 258–260

1. $432 interest; $45.34 payments   **3.** $330 interest; $76.25 payments   **5.** $720 interest; $130.00 payments   **7.** 15%
9. 18%   **11.** 8%   **13.** About 22%   **15.** About 20%   **17.** About 26%   **19.** $7.50   **21.** $9.86   **23.** $4.50
25. $.75   **27.** $50.05
29. Five payments of $5.00 plus a last payment of $7.50; total payments of $32.50, with interest of $2.50.
APR is almost 30%.

Problem Set 5.3, pages 264–266

1. $4,998   **3.** $6,591   **5.** $7,051   **7.** $8,492   **9.** $10,457   **11.** $738   **13.** $1,016   **15.** $810   **17.** $191   **19.** $96
21. $6,023 to $6,310   **23.** $7,987 to $8,368   **25.** $8,254 to $8,647   **27.** $9,117 to $9,551   **29.** $11,093 to $11,622
31. $16\frac{1}{2}\%$ **33.** $14\frac{1}{2}\%$

Problem Set 5.4, pages 272–274

1. $302   **3.** $346   **5.** $814   **7.** $2,425   **9.** $5,320   **11.** $17,000   **13.** $970   **15.** $1,596   **17.** $5,950   **19.** 11.86%
21. 14.64%   **23.** 15.96%   **25.** $573.17   **27.** $694.53   **29.** $860.20
31. a. $15,000   **b.** $60,000   **c.** $759   **d.** $213,240   **33. a.** $37,500   **b.** $112,500   **c.** $1,423.13   **d.** $399,825

Problem Set 5.5, pages 279–280

1. Deductible refers to the amount that the insured must pay before the company makes any payment for collision coverage.
3. The cash value of a whole life policy is a forced savings that is paid along with the premium.
5. Comprehensive coverage on an automobile policy pays for loss or damage to the insured's car due to something other than collision.
7. Extended coverage applies to home insurance and is protection against certain specific damages in addition to fire protection.   **9.** Answers vary.
11. a. 40/80/10 means the company will pay a maximum of $40,000 for bodily injury or death to one person, $80,000 if more than one person is injured, and $10,000 for property damage.
b. 100/200/20 means the company will pay a maximum of $100,000 for bodily injury or death to one person, $200,000 if more than one person is injured, and $20,000 for property damage.
13. $69.75   **15.** $43.75   **17.** $888.20   **19.** $55.65   **21.** $555   **23.** $35,130   **25.** $7,724   **27.** $10,204   **29.** $5,726

Chapter 5 Test, pages 281–283

1. a. $1,600   **b.** $3,660.96   **c.** 8.24%
2. a. Closed-end is an installment loan, and open-end is a credit card or revolving credit purchase.
b. Annual percentage rate; answers vary.
3. a. About $16\frac{1}{2}$% ($16\frac{1}{4}$%)   **b.** $1,633.16   **c.** About 17%   **4. a.** $9,065.40   **b.** $9,518.67   **c.** $9,971.94   **d.** $10,370
5. a. $7.64   **b.** $7.50   **c.** $8.25   **6. a.** $8,500   **b.** $17,000   **c.** $21,250   **7. a.** $906.53   **b.** $752.76   **c.** $655.99
8. a. $97,758.50   **b.** $128,520   **c.** $160,888   **9.** Lender A: 12.29%; Lender B: 12.44%; Lender A is better by .15%.
10. $19,310
11. Using the 36% factor, you could afford $671.40 in monthly payments. Since the loan described costs $10.29 per month per $1,000, you could afford to finance $65,247.81. This means that on an $85,000 home the down payment would need to be about $20,000 (actually $19,752.19).

Chapter 6:   Problem Set 6.1, pages 290–293

See text for **1, 3,** and **5.**   **7.** 7   **9.** Commutative   **11.** Commutative   **13.** Inverse   **15.** Commutative   **17.** Inverse

19.

Operation	Natural Numbers, N			
	+	×	−	÷
Closure	Yes	Yes	No	No
Associative	Yes	Yes	No	No
Identity	No	Yes	No	No
Inverse	—	No	—	—
Commutative	Yes	Yes	No	No
Distributive	× over + Yes		× over − Yes	

21.

Operation	Integers, Z			
	+	×	−	÷
Closure	Yes	Yes	Yes	No
Associative	Yes	Yes	No	No
Identity	Yes	Yes	No	No
Inverse	Yes	No	—	—
Commutative	Yes	Yes	No	No
Distributive	× over + Yes		× over − Yes	

23.

	Reals, R			
Operation	+	×	−	÷
Closure	Yes	Yes	Yes	Yes
Associative	Yes	Yes	No	No
Identity	Yes	Yes	No	No
Inverse	Yes	Yes	—	—
Commutative	Yes	Yes	No	No
Distributive	× over + Yes		× over − Yes	

25. ★

★	A	B	C	D	E	F	G	H
A	B	C	D	A	H	G	E	F
B	C	D	A	B	F	E	H	G
C	D	A	B	C	G	H	F	E
D	A	B	C	D	E	F	G	H
E	G	F	H	E	D	B	A	C
F	H	E	G	F	B	D	C	A
G	F	H	E	G	C	A	D	B
H	E	G	F	H	A	C	B	D

27. a. Yes; it is D.
b. Inverse of A is C, of B is D, of C is A, of D is D, of E is E, of F is F, of G is G, and the inverse of H is H.

Problem Set 6.2, pages 297–299

1. 15   **3.** 8   **5.** 5   **7.** $^-4$ **9.** $^-8$ **11.** 36   **13.** $^-56$ **15.** 3   **17.** 12   **19.** $^-14$ **21.** 0   **23.** 2   **25.** 4   **27.** 1   **29.** 0
31. 15   **33.** $\frac{20}{3}$ **35.** $^-6$ **37.** $^-2$ **39.** 8   **41.** 3   **43.** $^-15$ **45.** 14

Problem Set 6.3, pages 305–307

1. A prime number is a counting number that has exactly two divisors.
3. a. Not prime   **b.** Prime   **c.** Prime   **d.** Prime   **e.** Not prime   **5. a.** T   **b.** T   **c.** F   **d.** T   **e.** T
7. a. No   **b.** Yes
9. 2, 3, 5, 7, 11, 13, 17, 19, 23, 29, 31, 37, 41, 43, 47, 53, 59, 61, 67, 71, 73, 79, 83, 89, 97, 101, 103, 107, 109, 113, 127, 131, 137, 139, 149, 151, 157, 163, 167, 173, 179, 181, 191, 193, 197, 199, 211, 223, 227, 229, 233, 239, 241, 251, 257, 263, 269, 271, 277, 281, 283, and 293
11. Answers vary.   **13.** 1, 3, 7, 9, 13, 15, 21, 25, 31, 33, 37, 43, 49, 51, 63, 67, 69, 73, 75, 79, 87, 93, and 99
15. First look at some patterns:

$11 \cdot 14^1 + 1 = 155$
$11 \cdot 14^2 + 1 = 2,157$
$11 \cdot 14^3 + 1 = 30,185$
$11 \cdot 14^4 + 1 = 422,577$
$11 \cdot 14^5 + 1 = 5,916,065$
$\cdot$
$\cdot$
$\cdot$

If n is even, the number is divisible by 3.
If n is odd, the number is divisible by 5.
Conjecture: None are prime.

17. Let $2n + 1$, $2n + 3$, and $2n + 5$ be any three consecutive odd numbers. If $n = 1$, we have the prime triplets 3, 5, and 7. If $n > 1$, then one of the three numbers must be divisible by 3. Suppose the first is divisible by 3, then they are not prime triplets. If the first is not divisible by 3, then dividing it by 3 leaves a remainder of 1 or 2. If it leaves a remainder of 1, then the last number is divisible by 3. If it leaves a remainder of 2, then the middle one is divisible by 3. In any case, the numbers will not be prime triplets.

19.

7	61	43
73	37	1
31	13	67

Problem Set 6.4, pages 311–313

1. a. $2^3 \cdot 3$ **b.** $2 \cdot 3 \cdot 5$ **3. a.** $2^2 \cdot 7$ **b.** $2^2 \cdot 19$ **5. a.** $2^2 \cdot 3^3$ **b.** $2^2 \cdot 5 \cdot 37$ **7. a.** $2^3 \cdot 3 \cdot 5$ **b.** $2 \cdot 3^2 \cdot 5$
9. a. $2 \cdot 5 \cdot 7^2$ **b.** $2^4 \cdot 3^3 \cdot 11$ **11.** Prime   **13.** Prime   **15.** $13 \cdot 29$ **17.** $3 \cdot 5 \cdot 7$ **19.** Prime   **21.** $3^2 \cdot 5 \cdot 7$

23. Answers vary. **25.** $2^5 \cdot 3^3 \cdot 5 \cdot 7 \cdot 17$ **27.** $3^4 \cdot 7$ **29.** $19 \cdot 151$
31. $2 \cdot 3 \cdot 5 \cdot 7 \cdot 11 \cdot 13 + 1 = 30{,}031$, which is not prime, since $30{,}031 = 59 \cdot 509$.
33. $333{,}333{,}331 = 17 \cdot 19{,}607{,}843$ (See *International Mathematics Magazine*, February 1962, for a method of solution that does not involve calculators.)

Problem Set 6.5, pages 318–321

1. The greatest common factor is 1. **3. a.** $2^2 \cdot 3$ **b.** $2 \cdot 3^3$ **c.** $3^2 \cdot 19$ **d.** Prime **5.** 95 **7.** 1 **9.** 2 **11.** 3
13. 360 **15.** 2,052 **17.** 180 **19.** 252 **21.** 1,800
23. The product of the g.c.f. and l.c.m. for any two numbers equals the product of those numbers. **25.** Yes **27.** No
29. a. $\frac{6}{35}$ **b.** $\frac{3}{20}$ **c.** $\frac{5}{14}$ **d.** $\frac{7}{15}$ **31.** $\frac{11}{15}$ **33.** $\frac{-3}{4}$ **35.** $\frac{-53}{48}$ **37.** $\frac{-15}{56}$ **39.** $\frac{131}{25}$ **41.** $\frac{-92}{29}$ **43.** $-5\frac{23}{40}$ **45.** $\frac{18}{35}$ **47.** $\frac{4}{5}$
49. $\frac{22}{105}$ **51.** $\frac{-21}{40}$ **53.** $\frac{8}{3}$

Problem Set 6.6, pages 327–329

1. $\frac{2}{1}$ **3.** $\frac{4}{7}$ **5.** $\frac{1}{15}$ **7.** Yes **9.** Yes **11.** No **13.** Yes **15.** No **17.** 36 **19.** 56 **21.** 25 **23.** 20 **25.** $\frac{15}{4}$ **27.** $\frac{5}{3}$
29. 2 **31.** 7 **33.** 1 **35.** 350 **37.** $\frac{14}{1}$ or 14 miles to 1 gallon **39.** $\frac{1}{3}$ or the pitch of 1 ft to 3 ft **41.** $1.75
43. $.67 **45.** 10 gallons **47.** 2 gallons

Problem Set 6.7, pages 338–344

1. a. 3 **b.** 10 **c.** 3 **d.** 2 **3. a.** 12 **b.** 10 **c.** 8 **d.** 3 **5. a.** No solution **b.** 12
7. a. T **b.** F **c.** T **d.** F **9. a.** T **b.** T **c.** F **d.** F
11. a. 3 (mod 4) **b.** 3 (mod 5) **c.** 4 (mod 8) **d.** 2 (mod 9)
13. a. 3 (mod 5) **b.** 9 (mod 11) **c.** 0 (mod 2) **d.** 6 (mod 13)
15. a. 6 (mod 9) **b.** No solution **c.** 1 and 4 (mod 5) **d.** 5 (mod 13)
17. a. 5 (mod 6) **b.** 3 (mod 7) **c.** 0 (mod 4) **d.** 3 (mod 4)
19. On a 24-hour clock, $8 + 8k \equiv 0, 8, 16$ (mod 24) for any integral value of k; thus you will never have to take your medication between midnight and 8 A.M. The medication would be taken at 8:00 A.M., 4:00 P.M., and 12 midnight.
21. You must buy $12x$ inches of rope, where x is the number of multiples you will buy. Also, you need $7k + 80$ inches, where k is the number of pieces you will obtain. This means that $12x \equiv 80$ (mod 7). Solving, $x \equiv 2$ (mod 7). Thus, $x = 2, 9, 16, 23, \ldots$. If $x = 16$, you will have one piece 80 inches long and 16 pieces 7 inches long with no waste. Thus, you must buy 192 inches of rope (or 16 ft of rope).
23. April and July **25. a.** Yes **b.** Yes, 12 **c.** Yes

27.

×	0	1	2	3	4	5
0	0	0	0	0	0	0
1	0	1	2	3	4	5
2	0	2	4	0	2	4
3	0	3	0	3	0	3
4	0	4	2	0	4	2
5	0	5	4	3	2	1

a. *Closure:* Yes, since the table has no new entries.
b. *Associative:* Yes, try several examples.
c. *Identity:* Yes, the identity is 1.
d. *Inverse:* No $0 \cdot ? = 1$; has no inverse.
$\qquad$ $1 \cdot 1 = 1$; inverse of 1 is 1.
$\qquad$ 2 has no inverse.
$\qquad$ 3 has no inverse.
$\qquad$ 4 has no inverse.
$\qquad$ $5 \cdot 5 = 1$; inverse of 5 is 5.
e. *Commutative:* Yes, since the table is symmetric with respect to the principal diagonal.

30. a. Yes, 0 for addition and 1 for multiplication.   **31. a.** 41   **b.** 129   **c.** 75   **d.** 2,860

b. and c. *Number*	*Additive Inverse*	*Multiplicative Inverse*
0	0	None
1	10	1
2	9	6
3	8	4
4	7	3
5	6	9
6	5	2
7	4	8
8	3	7
9	2	5
10	1	10

Chapter 6 Test, pages 346–347

1. a. $a|b$ means that there exists a natural number k so that $b = ak$.   **b.** True, since $714 = 6 \cdot 119$.
c. False, since $8 \nmid 812$.   **d.** False, since $11 \nmid 095$ (095 was found by the scratch method).
e. False, $9 \nmid 35$ (35 was found by adding the digits).

2. a.

+	0	1	2	3
0	0	1	2	3
1	1	2	3	0
2	2	3	0	1
3	3	0	1	2

×	0	1	2	3
0	0	0	0	0
1	0	1	2	3
2	0	2	0	2
3	0	3	2	1

b. It satisfies the closure, associative, identity, inverse, and commutative properties for addition. It satisfies the closure, associative, identity, and commutative properties for multiplication. It does not satisfy the inverse property for multiplication. Therefore, it is not a field.

3. a. Prime   **b.** Prime   **c.** Prime   **d.** $7 \cdot 11 \cdot 13$ **e.** $3 \cdot 5^2 \cdot 7 \cdot 13$
4. a. g.c.f. is 3; l.c.m. is $2 \cdot 3^2 \cdot 5 \cdot 7 \cdot 11 = 6{,}930$ **b.** g.c.f. is 7; l.c.m. is $7^4 \cdot 11 \cdot 13 = 343{,}343$ **5.** $\frac{17}{19}$
6. a. $\frac{7}{72}$ **b.** $\frac{-7}{9}$ **c.** $\frac{708}{7{,}007}$ **d.** $\frac{1{,}396}{3{,}465}$ **7. a.** 3   **b.** 4   **c.** 2   **d.** $\frac{1}{4}$ **e.** 9   **8. a.** $\frac{-7}{3}$ **b.** 3   **c.** $\frac{1}{2}$ **d.** $\frac{11}{2}$ **e.** $\frac{17}{3}$
9. a. T   **b.** F   **c.** F   **d.** T   **e.** F   **10. a.** 2 (mod 8)   **b.** 1 (mod 2)   **c.** 5 (mod 7)   **d.** 3 (mod 6)   **e.** 4 (mod 7)
11. No (in 1941 May is the only month on which the 13th falls on a Friday).

Chapter 7:   Problem Set 7.1, pages 352–356

1. An axiom is accepted without proof; a theorem is a proved result.   **3. a.** No   **b.** No   **c.** No   **d.** No

5.   **7.**   **9.**

11. Answers may vary.

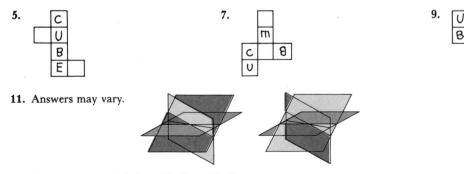

13. Answers vary.   **15.** B   **17.** D   **19.** D

Problem Set 7.2, pages 366–371

1. C **3.** B **5.** C **7.** B **9.** A **11.** ────────────

13. ──────────────────────────────

15. ────────────────── **17.** 14.9 cm

19. 948 mi **21.** 360 mi **23.** 10,200 mi **25.** 640 km **27.** 1,424 km **29.** 8,896 km **31.** $P = 110$ cm; $A = 450$ cm^2
33. $P = 32$ yd; $A = 48$ yd^2 **35.** $P = 20$ cm; $A = 25$ cm^2 **37.** $P = 440$ ft; $A = 7,560$ ft^2 **39.** $P = 92$ in.; $A = 500$ in.^{2}
41. $P = 22$ m; $A = 13.5$ m^2 **43.** $C = 62.8$ in.; $A = 314$ in.^{2}
45. Distance around the shaded portion is $\frac{1}{8}$ of the circumference + radius + radius or about 83.6 cm;
$A = \frac{1}{8}\pi(30)^2 \approx 353.43$ cm^2.
47. Distance around is $\frac{1}{2}$ of the circumference + 3 + 2 + 3 or about 11.14 cm; A is the area of the rectangle minus $\frac{1}{2}$ the
area of the circle or about 2.86 cm^2.

Problem Set 7.3, pages 374–375

1. $\angle A$, $\angle CAT$, $\angle TAC$, $\angle 5$ **3.** 55° **5.** 180° **7.** 125° **9.** Acute **11.** Acute **13.** Right **15.** Acute **17.** Right
19. Adjacent or supplementary angles **21.** Corresponding angles **23.** Alternate interior angles
25. Alternate exterior angles
27. $m\angle 1 = 110°$; $m\angle 2 = 70°$; $m\angle 3 = 110°$; $m\angle 4 = 70°$; $m\angle 5 = 110°$; $m\angle 6 = 70°$; $m\angle 8 = 70°$
29. $m\angle 1 = 151°$; $m\angle 2 = 19°$; $m\angle 3 = 151°$; $m\angle 4 = 19°$; $m\angle 5 = 151°$; $m\angle 7 = 151°$; $m\angle 8 = 19°$

Problem Set 7.4, pages 384–389

1. A **3.** B **5.** C **7.** B **9.** C **11.** C **13.** A **15.** A **17.** D **19.** A **21.** 512 in.^{3} **23.** 400 mm^3 **25.** 15.6 ℓ
27. 3.7 gal **29.** 24 ℓ **31.** **a.** 6,230 **b.** 450 **c.** 60 **d.** 4.8 **e.** 3.3
33. **a.** 62,500 **b.** 3,250 **c.** 4,280 **d.** 1.43 **e.** 25,000 **35.** 112

Problem Set 7.5, pages 395–397

1. Telegraph Hill **3.** Embarcadero **5.** Ferry Building **7.** Old Mint Building **9.** Coit Tower **11.** Niels Sovndal
13. Clint Stevenson **15.** Missy Smith **17.** **19.**

21.

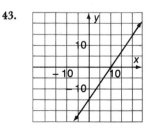

23. $(2, 9)$, $(3, 5)$, $(^-2, 5)$, $(1, 10)$, $(5, 11)$, $(5, 10)$, $(8, 10)$, $(9, 7)$, $(4, 6)$, $(5, 2)$, $(8, 0)$, $(5, ^-1)$, $(4, ^-3)$, $(4\frac{1}{2}, ^-5)$, $(2, ^-7)$, $(4, ^-7)$, $(5, ^-10)$, $(^-4, ^-7)$, $(^-3, ^-6)$, $(^-7, ^-7)$, $(^-9, ^-10)$, $(^-10, ^-6)$, $(^-8, ^-4)$, $(^-5, ^-5)$, $(^-5, ^-1)$, $(1, 5)$

25.

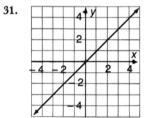

27.

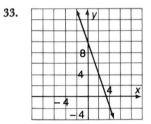

29.

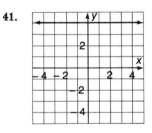

31.

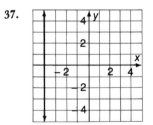

33.

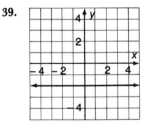

35.

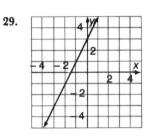

37.

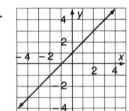

39.

41.

43.

45. a. *Option A: c = 30 + .4 m*
   Option B: c = 50
   c. They are the same for 50 miles.

b.

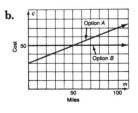

Problem Set 7.6, pages 400–404

1. Traversible   **3.** Not traversible   **5.** Traversible   **7.** Not traversible   **9.** Traversible
11. Not traversible, since there are more than two odd vertices
13. Transform the problem into the following network: **15.** Transform the problem into the following network:

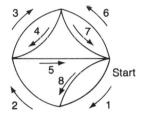

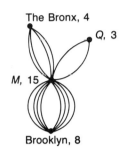

It is now traversible, as shown by the arrows.   There are two odd vertices, so it is traversible;
                                                            begin at either Queens or Manhattan.

17. Possible   **19.** Possible
21. If all rooms have an even number of doors, you can begin in any room. If there are two rooms with an odd number of
doors, you must begin in one of these rooms.
23. One edge   **25.** Two interlocking pieces result   **27.** Yes

Problem Set 7.7, pages 411–416

1. Saccheri quadrilateral   **3.** Not a Saccheri quadrilateral   **5.** One line
7. a. They never intersect.   **b.** Only great circles are considered to be lines.
9. The sum of the angles is greater than $180°$.
11. A and C are inside; B is outside.   **13.** A and C are inside; B is outside.   **15.** A and B are outside; C is inside.
17. It will cross an odd number of times for $\overline{BX}$ and $\overline{CX}$, and an even number of times for $\overline{AX}$.
19. If $\overline{AX}$ crosses the curve an even number of times, then A is outside; an odd number of times, A is inside (for some
point X, which is obviously outside).
21. Just up from the South Pole there is a parallel that has a circumference of exactly one mile.

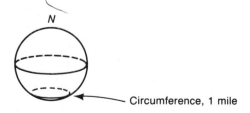

If you begin anywhere one mile above this circle, the conditions of the problems are satisfied.

Chapter 7 Test, pages 419–421

1. a. 0 **b.** 8 **c.** 12 **d.** 6 **e.** 1
2. a. Meter (length), liter (capacity), and gram (weight or mass)   **b.** Centi, $\frac{1}{100}$; deka, 10; kilo, 1,000; milli, $\frac{1}{1,000}$; deci, $\frac{1}{10}$

3. a. 83° **b.** 130° **c.** 97° **d.** 50° **e.** 180°

4. a. Corresponding angles **b.** Adjacent angles or supplementary angles **c.** Vertical angles **d.** Alternate interior angles **e.** Alternate exterior angles

5. D **6.** C **7.** E **8.** C **9.** E

10. a. Mass **b.** Length **c.** Capacity **d.** Capacity **e.** Length

11. a. Kilometer **b.** Centimeter **c.** Milliliter **d.** Cubic meters **e.** Gram

12. a. 4.3 **b.** $\frac{1}{1,000}$ **c.** 48 **d.** 2,880 **e.** .6

13. a. $P = 60$ in.; $A = 225$ in.^{2} **b.** $P = 28$ ft; $A = 33$ ft^2 **c.** $P = 26$ m; $A = 32$ m^2 **d.** $P = 12$ yd; $A = 6$ yd^2 **e.** $P = 18$ cm; $A = 6$ cm^2

14. a. 1,021 in.^{2} (1021.017613 on a calculator) **b.** 120 in.^{2} **15. a.** 896 dm^2 **b.** 1,848 in.^{2} **c.** 896 ℓ **d.** 8 gal

16. a.

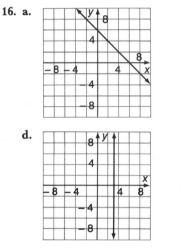

b.

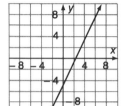

c.

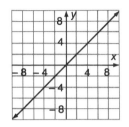

d.

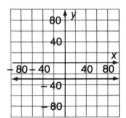

e.

17. a. Not traversible **b.** Traversible **18. a.** Possible **b.** Possible

19. A Saccheri quadrilateral is a quadrilateral $ABCD$ with base angles A and B right angles and with sides AC and BD with equal lengths. If the summit angles C and D are right angles, then the result is Euclidean geometry. If they are acute, then the result is hyperbolic geometry. If they are obtuse, the result is elliptic geometry.

20. Answers vary; see Table 7.5, page 409. **21. a.** Answers vary. **b.** Answers vary.

22. 2 edges, 2 sides; the result is two interlocking loops.

Chapter 8:   Problem Set 8.1, pages 426–430

1. If task A can be performed in m ways, and, after task A is performed, a second task, B, can be performed in n ways, then task A followed by task B can be performed in $m \cdot n$ ways.

3. 786,240 **5.** 2,730 **7.** 11 squares **9.** 60 **11. a.** 10 **b.** 8 **c.** 8 **13. a.** 9×10^8 **b.** 8.1×10^8

15. 47 **17.** Answers vary. **19.** No; all numbers but 13 are possible.

Problem Set 8.2, pages 435–437

1. 4,920 **3.** 90 **5.** 600 **7.** 5,040 **9.** 22,100 **11.** 720 **13.** 1,680 **15.** 40,296 **17.** 56 **19.** 84 **21.** 9

23. 504 **25.** 1 **27.** 132,600 **29.** 24 **31.** 95,040 **33.** 1,680 **35.** $\dfrac{g!}{(g-h)!}$ **37.** 52 **39.** 3,360

41. 12 **43.** Answers vary.

Problem Set 8.3, pages 440–443

1. 9 **3.** 84 **5.** 1 **7.** 22,100 **9.** 1 **11.** 35 **13.** 1,225 **15.** $\dfrac{g!}{h!(g-h)!}$ **17.** 1 **19.** 792 **21.** 792 **23.** 7! or 5,040
25. $\dfrac{6!}{2} = 360$ **27.** $2 \cdot 2 \cdot 2 = 8$ ways; use the Fundamental Counting Principle. **29.** $_6C_1 = 15$ **31.** $_{11}C_1 = 330$
33. $_5C_3 \cdot _6C_4 = 150$ **35.** 6 **37.** $_{13}C_5 = 1,287$ **39.** $_nC_r = \dfrac{_nP_r}{r!}$ or $\dfrac{n!}{r!(n-r)!}$

Problem Set 8.4, pages 448–451

1. In a permutation the order is important; in a combination the order of arrangement of the objects is not important.
3. a. Combination **b.** Neither **c.** Neither **d.** Permutation **e.** Permutation
5. a. Permutation **b.** Combination **c.** Combination **d.** Permutation **e.** Combination
7. a. 50 **b.** 33 **c.** 16 **d.** 67 **9.** 15 **11.** 63 **13.** $2^n - 1$
15. a. $_6P_4 = 6 \cdot 5 \cdot 4 \cdot 3 = 360$ **b.** $_{10}P_2 = 10 \cdot 9 = 90$ **c.** $_4P_3 = 4 \cdot 3 \cdot 2 = 24$ **d.** $_{52}C_2 = \dfrac{_{52}P_2}{2!} = \dfrac{52 \cdot 51}{2} = 1,326$
e. $_6P_5 = 6 \cdot 5 \cdot 4 \cdot 3 \cdot 2 = 720$
17. a. $_{15}P_4 = 15 \cdot 14 \cdot 13 \cdot 12 = 32,760$ **b.** $_{15}C_4 = \dfrac{15 \cdot 14 \cdot 13 \cdot 12}{4 \cdot 3 \cdot 2 \cdot 1} = 1,365$ **19. a.** 24,360 **b.** 4,060
21. 1,001,101

Problem Set 8.5, pages 455–457

1. Almost 4 days (nonstop)—actually, 3 days, 20 hours, 9 minutes, and 36 seconds.

3.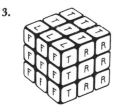

5.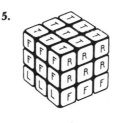

7.

9.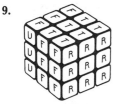

11.

13.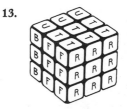

15.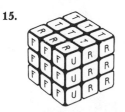

17.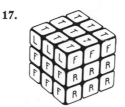

19.

21. 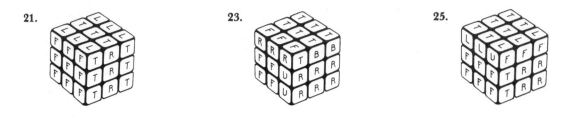 **23.** **25.**

Answers to Problems 27–33 are not unique.
27. F^{-1} **29.** F or $(F^{-1})^3$ **31.** BF^{-1} **33.** TF^{-1} **35.** No **37.** No; answers vary.

Chapter 8 Test, pages 458–459

1. a. 718 **b.** 4 **c.** 24 **d.** 6 **e.** 360 **2. a.** 10 **b.** 12 **c.** 1 **d.** 6 **e.** 495
3. Answers may vary. In a permutation the order of the arrangement is important, but in a combination the order of arrangement is not relevant. Permutations can be found in the following way:

$$\underbrace{{}_nP_r = n(n-1)(n-2) \cdots (n-r+1)}_{r \text{ factors}}$$

Combinations are related to permutations rather closely, as we see by writing the formula for finding combinations:

$$_nC_r = \frac{{}_nP_r}{r!}$$

4. a. 220 **b.** 120 **5.** *Happy:* 60; *College:* 1,260
6. The total number of ways possible is $_{10}P_{10} = 10! = 3,628,800 \approx 10,000$ years!

7. The number of ways three balls can be drawn is $_{10}C_3 = \dfrac{10 \cdot 9 \cdot 8}{3 \cdot 2 \cdot 1} = 120$

The number of ways of drawing no red balls (all white balls) is $_6C_3 = \dfrac{6 \cdot 5 \cdot 4}{3 \cdot 2 \cdot 1} = 20$

Thus the number of ways of drawing at least one red is $120 - 20 = 100$
8. a. 25 **b.** 15 **c.** 14 **d.** 26
9. Their claim is correct. There are eight ingredients, so the number of hamburgers is the number of subsets of a set with eight elements or $2^8 = 256$.
10. a.
b. $T^{-1}R^{-1}T^{-1}$
11. a. 24^5 or 7,962,624 arrangements
b. About 92 days (nonstop); actually 92 days, 3 hours, 50 minutes, and 24 seconds

Chapter 9:   Problem Set 9.1, pages 468–470

Since this section deals with experiments, the answers will, of course, vary.

Problem Set 9.2, pages 476–479

1. Empirical probability is the result of experimentation and could vary from experiment to experiment; it is dependent on the number of times the event is repeated. The theoretical probability is the numerical value obtained by using a mathematical model. The theoretical probability should be predictive of the empirical probability for a larger number of trials.
3. $\frac{3}{8}$ **5.** $\frac{3}{8}$ **7.** $\frac{4}{17}$ **9.** $\frac{1}{52}$ **11.** $\frac{1}{4}$ **13.** About .05 **15.** .19 **17. a.** $\frac{1}{2}$ **b.** $\frac{2}{3}$ **c.** $\frac{7}{12}$ **19. a.** $\frac{1}{12}$ **b.** $\frac{7}{12}$
21. $\frac{5}{36}$ **23.** $\frac{5}{36}$ **25.** $\frac{7}{36}$ **27.** $\frac{2}{9}$ **29.** $\frac{1}{9}$

31.

	1	2	3	4
1	(1, 1)	(1, 2)	(1, 3)	(1, 4)
2	(2, 1)	(2, 2)	(2, 3)	(2, 4)
3	(3, 1)	(3, 2)	(3, 3)	(3, 4)
4	(4, 1)	(4, 2)	(4, 3)	(4, 4)

33. a. $\frac{1}{4}$ **b.** $\frac{3}{16}$ **c.** $\frac{1}{8}$

Problem Set 9.3, pages 483–485

1. a. BBBB; BBBG; BBGB; BBGG; BGBB; BGBG; BGGB; BGGG; GBBB; GBBG; GBGB; GBGG; GGBB; GGBG; GGGB; GGGG **b.** $\frac{1}{16}$ **c.** $\frac{1}{4}$ **d.** $\frac{3}{8}$ **e.** 1
3. $\frac{3}{7}$ **5.** $\frac{1}{3}$ **7.** 0 **9.** $\frac{1}{3}$ **11.** $\frac{2}{11}$ **13.** 0 **15.** $\frac{1}{5}$ **17.** $\frac{1}{3}$ **19.** $\frac{1}{3}$ **21.** $\frac{1}{13}$ **23.** $\frac{1}{17}$ **25.** $\frac{13}{51}$ **27.** $\frac{25}{51}$ **29.** $\frac{167}{1,000} \approx .17$
31. $\frac{240}{643} \approx .37$ **33.** $\frac{403}{643} \approx .63$ **35.** $\frac{167}{407} \approx .41$

Problem Set 9.4, pages 492–496

1. \$.83 **3.** The barber's expectation is about \$.09, but he is most likely to receive a nickel. **5.** \$.02 **7.** \$.50 **9.** 1 to 3
11. $\frac{9}{10}$ **13.** \$.95 **15.** \$.95 **17.** \$.95 **19.** \$.92 **21.** \$.95 **23.** \$2,515
25. They should sink the test well since the expectation is positive. **27.** Yes, since the expectation is about \$.12.
29. Horse number 1, probability of winning: $\frac{1}{3}$; number 2: $\frac{1}{16}$; number 3: $\frac{2}{5}$; number 4: $\frac{5}{12}$; and number 5: $\frac{1}{2}$.
Note: At a horse race the odds are determined by the track betting and the sum of these probabilities is not necessarily 1.
31. \$.02 **33.** \$1.05 **35.** \$1.03

Problem Set 9.5, pages 500–505

1. .998 **3.** .952 **5.** .998
7. a. $\frac{_{13}C_5}{_{52}C_5} \approx .0004951981$ **b.** .0019807923 **c.** No, not quite because in the table of probabilities of poker hands, the straight flushes and the royal flushes are subtracted; that is, in part a we found 5,148 favorable events (flushes), but did not subtract the 36 straight flushes and 4 royal flushes to obtain 5,108 as shown in the table.
9. a. $\frac{_4C_3 \cdot {_4C_2}}{_{52}C_5} \approx .0000092345$ **b.** No, because in part a we are finding *a particular* full house, and the entry in the table of probabilities of poker hands is finding *all* full houses.

11. $\frac{1}{32}$ **13.** P(Win) $\approx$ .58; P(Lose) $\approx$ .42; $E \approx$ .16, so you should play.

15. Answers vary; (1) you do not have unlimited funds and this system assumes unlimited resources, (2) you run the risk of losing large sums of money to win $1, and (3) the game has a limit that prevents you from doubling at some point in time.

17. a. .002 **b.** .1 **c.** .025 **d.** .0025 **e.** .01 **f.** .0045 **g.** .0045 **h.** .00075 **19.** $\frac{25}{64}$ **21.** $\frac{15}{32}$ **23.** $\frac{5}{14}$ **25.** $\frac{15}{28}$

27. $\frac{1}{4} =$ .25 **29.** $\frac{1}{16} =$ .0625 **31.** .004

Chapter 9 Test, pages 506–507

1. If an event E can occur in any one of n mutually exclusive and equally likely ways, and if s of those ways are considered successful, then the probability of the event is given by $P(E) = s/n$.

2. a. $\frac{1}{4}$ **b.** $\frac{1}{52}$ **c.** $\frac{4}{13}$ **3.** $\frac{10}{11}$ **4. a.** $\frac{1}{16}$ **b.** $\frac{1}{3}$ **c.** 0 **5.** $\frac{29}{36}$ **6.** $\frac{11}{36}$; $\frac{1}{36}$

7. $E =$ $.62; no; he should be willing to pay about 62¢ each for the tickets. **8. a.** $\frac{1}{2}$ **b.** 1 to 1 **9.** .50 **10. a.** $\frac{3}{5}$ **b.** $\frac{1}{2}$

11. Die C; answers vary. **12.** P $(A$ beats $B) = \frac{3}{4}$; P$(B$ beats $C) = \frac{3}{4}$; P$(C$ beats $D) = \frac{5}{8}$; P$(D$ beats $A) = \frac{5}{8}$

Chapter 10: Problem Set 10.1, pages 513–519

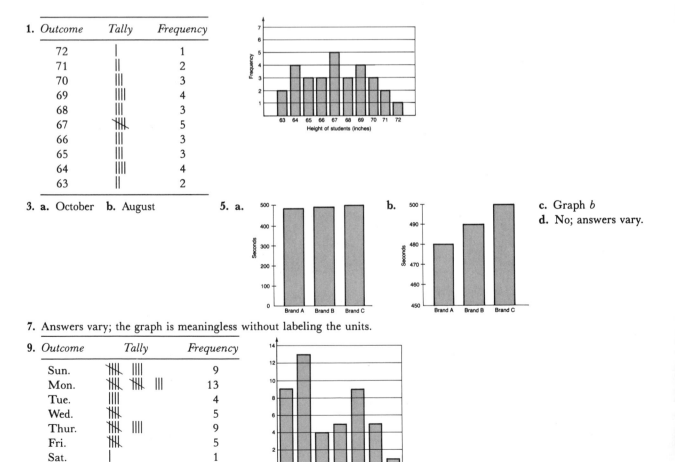

1.

Outcome	Tally	Frequency
72			1
71				2
70					3
69						4
68					3
67	⊮	5
66					3
65					3
64						4
63				2

3. a. October **b.** August

5. a. **b.** **c.** Graph b **d.** No; answers vary.

7. Answers vary; the graph is meaningless without labeling the units.

9.

Outcome	Tally	Frequency
Sun.	⊮ ||||	9
Mon.	⊮ ⊮ |||	13
Tue.	||||	4
Wed.	⊮	5
Thur.	⊮ ||||	9
Fri.	⊮	5
Sat.			1

11. a. 1980   **b.** 1982   **c.** 1982   **13.**

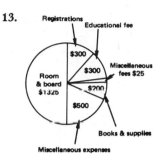

Problem Set 10.2, pages 527–530

1. Mean = 3; median = 3; no mode; range = 4; var = 2.5; $\sigma \approx 1.58$
3. Mean = 105; median = 105; no mode; range = 4; var = 2.5; $\sigma \approx 1.58$
5. Mean = 6; median = 7; mode = 7; range = 8; var ≈ 6.67; $\sigma \approx 2.58$
7. Mean = 10; median = 8; no mode; range = 18; var = 52; $\sigma \approx 7.21$
9. Mean = 91; median = 95; mode = 95; range = 17; var = 50.5; $\sigma \approx 7.11$
11. Mean = 3; median = 3; mode = 3; range = 4; var ≈ 1.67; $\sigma \approx 1.29$
13. Same range, variance, and standard deviation
15. Mean = $7,900; median = $8,000; mode = $5,000   **17.** Mean ≈ 68.1; median = 70; mode = 70
19. Mean = 56; median = 65; mode = 70; range = 90   **21.** Examples vary.
23. Mean = $34,000; median = $27,000; bimodal at $20,000 and $30,000; the median is most descriptive.
25. Mean = 5.05; median = 4; mode = 4; the mean is most descriptive. Range = 13; $\sigma = 3.32$
27. Answers vary.   **29.** Var = 96.2; $\sigma \approx 9.8$ **31.** Var = 714; $\sigma \approx 26.7$ **33.** Var = 21.6; $\sigma \approx 4.6$
35. Mean = 67.1333333; Var = 6.395402; St. dev. ≈ 2.53

Problem Set 10.3, pages 533–535

1. 34   **3.** 8   **5.** 25   **7.** 1   **9.** .022   **11.** .978   **13.** It will exceed 40.5 inches about 2.2% of the time.   **15.** .022
17. a. 1   **b.** 22   **c.** 136   **d.** 682   **e.** 136   **f.** 22   **g.** 1   **19.** 12.9 oz mean weight

Problem Set 10.4, pages 537–538

This problem set deals with sampling, and answers will vary.

Chapter 10 Test, pages 539–541

1. Answers vary.   **2. a.** 25   **b.** 25   **c.** No mode   **d.** 4   **e.** 2.5   **f.** 1.581139
3. a. 7.8   **b.** 8   **c.** 8   **d.** 4   **e.**

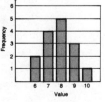

4. a. Mean = 13.5; median = 14; mode = 16; mode is the most useful.   **b.** Mean = 79; median = 74; no mode; mean is the most representative.   **c.** Mean = $14,666; median = $15,000; mode = $18,000; median is the most representative.
5. a. Not possible, unless every person had *exactly* the same sales.
b. Answers vary; we don't know how many were surveyed; we don't know how the sample was chosen; we don't know what percentage of dentists didn't respond; we don't know how the question was asked.
6. Range = 19; variance = 72.5; standard deviation about 85.15   **7. a.** 68.2%   **b.** About 220 women
8. a. The class with the smaller standard deviation would be more homogeneous and consequently easier to teach.
b. She did better on her math test because she was two standard deviations above the mean, whereas in history she was only one standard deviation above the mean.
9. a. Lower score is 65; upper score is 81.   **b.** About 22 children   **10.** Answers vary.
11. Answers vary; player A has a better lifetime batting average, but player B has a better batting average *each and every year.*

Chapter 11: Problem Set 11.1, pages 550–551

1. A is 3×3; B is 3×1; C is 3×2; D is 4×1; E is 2×3; F is 1×4; G is 1×2; H is 2×2
3. a. No, since January has 31 days and 7 columns   **b.** Friday, since the first must be a Sunday
5.

	Insects	Nuts	Roots	Honey	Scorpions	Mammals
Chimpanzee	1	1	0	1	0	1
Langur	1	0	0	0	0	0
Macaque	1	1	1	0	0	0
Baboon	1	1	1	0	1	1

7. $\begin{bmatrix} 3 & 3 & 7 \\ 5 & 2 & 1 \\ -4 & 5 & 1 \end{bmatrix}$ **9.** $\begin{bmatrix} 1 & -3 & 7 \\ 3 & -4 & 5 \\ 6 & 1 & -1 \end{bmatrix}$ **11.** $\begin{bmatrix} 2 \\ 2 \\ 9 \end{bmatrix}$ **13.** $[8 \quad 3 \quad 7]$ **15.** $\begin{bmatrix} 5 & 3 & 14 \\ 9 & 1 & 4 \\ -3 & 8 & 1 \end{bmatrix}$ **17.** $\begin{bmatrix} 2 \\ 5 \\ 21 \end{bmatrix}$

19. Not conformable   **21.** $[20 \quad 1 \quad 7]$ **23.** $[8 \quad 8 \quad 4]$

Problem Set 11.2, pages 556–558

1. $\begin{bmatrix} 2 & 4 & 2 \\ 6 & -2 & 4 \\ 2 & 2 & 5 \end{bmatrix}$ **3.** $\begin{bmatrix} -3 & 6 & 10 \\ -4 & 15 & 8 \\ 7 & 15 & 2 \end{bmatrix}$ **5.** $\begin{bmatrix} -6 & 12 & 20 \\ -8 & 30 & 16 \\ 14 & 30 & 4 \end{bmatrix}$ **7.** $\begin{bmatrix} 35 & 5 & 18 \\ 46 & 2 & 16 \\ 39 & -7 & 26 \end{bmatrix}$ **9.** $\begin{bmatrix} 13 & 25 & 20 \\ 25 & 31 & 23 \\ 42 & 24 & 33 \end{bmatrix}$

11. $\begin{bmatrix} -1 & 37 & 32 \\ 4 & 14 & 45 \\ 27 & 9 & 28 \end{bmatrix}$ **13.** $\begin{bmatrix} 34 & 117 & 44 \\ 45 & 143 & 97 \\ 109 & 100 & 151 \end{bmatrix}$ **15.** $\begin{bmatrix} 132 & 83 & 70 \\ 89 & 59 & 77 \\ 172 & 23 & 150 \end{bmatrix}$ **17.** $\begin{bmatrix} 2 & 8 \\ 16 & 8 \\ -7 & 7 \\ 7 & 7 \end{bmatrix}$ **19.** $\begin{bmatrix} 14 & 14 \\ -7 & 7 \end{bmatrix}$

21. $\begin{bmatrix} 14 & 4 & 8 & 30 \\ -7 & -1 & -9 & 3 \end{bmatrix}$ **23.** Not conformable for addition   **25.** Not conformable for multiplication

27. $\begin{bmatrix} 1 & 0 & 0 & 0 \\ 0 & 1 & 0 & 0 \\ 0 & 0 & 1 & 0 \\ 0 & 0 & 0 & 1 \end{bmatrix}$ **29.** When the dimension of the matrices to be added are the same

To

	A	B	C	D	E
A	2	0	1	0	0
B	1	0	0	0	0
C	0	1	0	1	1
D	0	1	0	1	1
E	0	1	0	1	1

31. From

33. a. $[8,050]$

b. Total cost of raw materials for the total production for a day's demand

35. $\begin{bmatrix} 230 \\ 520 \\ 280 \end{bmatrix}$ Outside bottling, \$230; produced and bottled at the winery, \$520; estate bottled, \$280

Problem Set 11.3, pages 562–563

1. $\begin{bmatrix} 1 & 5 & 3 & | & -2 \\ -1 & 3 & -2 & | & 3 \\ -2 & 2 & 4 & | & 8 \end{bmatrix}$ **3.** $\begin{bmatrix} -1 & 1 & 2 & | & 4 \\ -1 & 3 & -2 & | & 3 \\ 1 & 5 & 3 & | & -2 \end{bmatrix}$ **5.** $\begin{bmatrix} -2 & 2 & 4 & | & 8 \\ -2 & -2 & -5 & | & 5 \\ 1 & 5 & 3 & | & -2 \end{bmatrix}$ **7.** $\begin{bmatrix} 6 & -2 & 4 & | & 6 \\ 1 & -2 & 9 & | & -7 \\ 9 & -12 & 0 & | & -9 \end{bmatrix}$

9. $\begin{bmatrix} 6 & -2 & 4 & | & 6 \\ 1 & -2 & 9 & | & -7 \\ 0 & -2 & 27 & | & -18 \end{bmatrix}$ **11.** $(1, 2)$ **13.** $(5, 3)$ **15.** $(3, -2)$ **17.** $(-1, 3)$ **19.** $(6, 3)$ **21.** $(0, -1, 2)$

23. $(3, 2, 5)$ **25.** $(2, -1, 0)$ **27.** $(-6, 3, 7)$ **29.** Mix 3 containers of spray I and 4 containers of spray II.

Problem Set 11.4, pages 568–569

1. Inverses   **3.** Inverses   **5.** Inverses   **7.** Not inverses   **9.** Inverses   **11.** $\begin{bmatrix} 2 & 7 \\ 1 & 4 \end{bmatrix}$ **13.** $\begin{bmatrix} 2 & -5 \\ -1 & 3 \end{bmatrix}$

15. $\frac{1}{6}\begin{bmatrix} 0 & 3 \\ 2 & -1 \end{bmatrix}$ **17.** $\frac{1}{22}\begin{bmatrix} 2 & -3 \\ 1 & 4 \end{bmatrix}$ **19.** $\begin{bmatrix} 4 & 3 \\ 2 & 2 \end{bmatrix}$ **21.** $\begin{bmatrix} 3 & -17 & -20 \\ 3 & -18 & -20 \\ -1 & 6 & 7 \end{bmatrix}$ **23.** $\begin{bmatrix} 3 & 3 & -1 \\ -2 & -2 & 1 \\ -4 & -5 & 2 \end{bmatrix}$

25. $\frac{1}{12}\begin{bmatrix} -2 & 2 & 2 \\ 8 & -8 & 4 \\ 7 & -1 & -1 \end{bmatrix}$

Problem Set 11.5, pages 577–579

1. Strictly determined game; $[1 \quad 0]$ and $\begin{bmatrix} 0 \\ 1 \end{bmatrix}$;   the value is 0.   **3.** $[\frac{1}{2} \quad \frac{1}{2}]$ and $\begin{bmatrix} \frac{1}{2} \\ \frac{1}{2} \end{bmatrix}$;   the value is 0.

5. Strictly determined game; $[1 \quad 0]$ and $\begin{bmatrix} 0 \\ 1 \end{bmatrix}$;   the value is -2.

7. Strictly determined game; $[1 \quad 0]$ and $\begin{bmatrix} 0 \\ 1 \end{bmatrix}$;   the value is -1.   **9.** $[\frac{1}{3} \quad \frac{2}{3}]$ and $\begin{bmatrix} \frac{1}{6} \\ \frac{5}{6} \end{bmatrix}$;   the value is $-\frac{1}{6}$.

11. Strictly determined game; $[1 \quad 0 \quad 0]$ and $\begin{bmatrix} 1 \\ 0 \\ 0 \end{bmatrix}$;   the value is 2.   **13.** $[\frac{1}{6} \quad 0 \quad \frac{5}{6}]$ and $\begin{bmatrix} 0 \\ \frac{2}{3} \\ \frac{1}{3} \end{bmatrix}$;   the value is $-\frac{1}{3}$.

15. Strictly determined game; $[0 \quad 1 \quad 0]$ and $\begin{bmatrix} 0 \\ 1 \\ 0 \end{bmatrix}$; the value is $^-1$. **17.** $[\frac{1}{4} \quad 0 \quad \frac{3}{4}]$ and $\begin{bmatrix} \frac{1}{4} \\ \frac{3}{4} \\ 0 \\ 0 \end{bmatrix}$; the value is $-\frac{1}{4}$.

19. Me

	Opponent 10¢	25¢	50¢	
10¢	10	$^-15$	$^-40$	
25¢	$^-15$	25	$^-25$	Not a strictly determined game
50¢	$^-40$	$^-25$	50	

21. Attacking forces

	Defending forces Two in the north	One in each direction	Two in the south	
Two regiments from north	0	1	1	
One force in each direction	$^-1$	0	$^-1$	Not a strictly determined game
Two regiments from south	1	1	0	

23. $[\frac{1}{5} \quad \frac{4}{5}]$ and $\begin{bmatrix} \frac{1}{10} \\ \frac{9}{10} \end{bmatrix}$; the value is $\frac{11}{5}$.

Chapter 11 Test, pages 581–582

1. A is 2×2; B is 2×2; C is 3×3; D is 3×3; E is 1×4

2. a. $\begin{bmatrix} ^-1 & ^-1 & 5 \\ 4 & 9 & ^-14 \\ 4 & ^-5 & 7 \end{bmatrix}$ **b.** $\begin{bmatrix} 14 & ^-12 \\ 13 & ^-2 \end{bmatrix}$ **c.** 4 **3. a.** $\begin{bmatrix} 0 & 10 \\ ^-3 & 9 \end{bmatrix}$ **b.** $\begin{bmatrix} 5 & 2 \\ ^-5 & 4 \end{bmatrix}$ **c.** Not conformable

4. a. $\begin{bmatrix} 10 & ^-8 \\ 12 & ^-6 \end{bmatrix}$ **b.** $\begin{bmatrix} ^-5 & 0 \\ 0 & ^-5 \end{bmatrix}$ **c.** $\begin{bmatrix} ^-8 & ^-9 & 5 \\ 1 & 20 & ^-26 \\ ^-21 & ^-22 & 14 \end{bmatrix}$ **5. a.** $\begin{bmatrix} 0 & \frac{1}{3} \\ ^-\frac{1}{2} & \frac{2}{3} \end{bmatrix}$ **b.** $\frac{1}{11}\begin{bmatrix} ^-55 & 0 & 22 \\ ^-45 & ^-1 & 15 \\ ^-12 & ^-1 & 4 \end{bmatrix}$

6. $(5, 4)$ **7.** $(^-2, 5, 3)$ **8.** They need to mix 4 units of batch A and 1 unit of batch B.
9. Not a strictly determined game

10. Us

	Competitor Uptown	Downtown	Suburbs
Uptown	2	4	$^-2^*$
Downtown	1	$^-3^*$	$^-3^*$
Suburbs	3	1	$^-4^*$

Strictly determined and the value is $^-2$; that is, the *best* we can do is trail by \$2,000.

11. $[\frac{13}{20} \quad \frac{7}{20}]$ and $\begin{bmatrix} \frac{13}{20} \\ \frac{7}{20} \end{bmatrix}$; the value is $\frac{9}{10}$.

Credits

Page 5. Photo of Polya courtesy of New York Public Library.

Page 8. Title page of an arithmetic book of Petrus Apianus, Ingolstadt. It shows Pascal's triangle as it was first printed in 1527.

Page 10. Photo by Richard Hagberg.

Page 14. Melancolia by Albrecht Dürer. Courtesy of the Bettmann Archive, Inc.

Page 20. Designs reprinted from *The Mathematics Teacher,* January 1982 (vol. 75, p. 87), copyright 1982 by the National Council of Teachers of Mathematics. Reprinted by permission.

Page 20. Wizard of Id reprinted by permission of Johnny Hart and Field Enterprises, Inc.

Page 22. B.C. cartoon reprinted by permission of Johnny Hart and Field Enterprises, Inc.

Page 27. From MENSA, the high IQ Society, 1701 W. 3rd Street, Brooklyn, NY 11223. Reprinted by permission. MENSA is an organization whose members have scored at or above the 98th percentile of the general population on any standardized IQ test.

Page 28. Peanuts cartoon © 1967 United Feature Syndicate, Inc. Reprinted by permission.

Page 35. Problem 40 adapted from "Kitsch and Tell" in *Games Magazine,* September/October, 1981, p. 52. Reprinted by permission.

Page 37. Quote from *In Mathematical Circles,* by H. Eves. Copyright 1969 by Prindle, Weber, & Schmidt, Inc. Reprinted by permission.

Page 38. Bathers by Georges Seurat. Courtesy of the Trustees, The National Gallery, London.

Page 40. David by Michelangelo. Firenze, Galleria dell'Accademia. Courtesy of the Gallerie Fiorentine.

Page 40. Dynamic Symmetry of a Human Face by Leonardo da Vinci, and *Proportions of the Human Body* by Albrecht Dürer. © VEB Bibliographisches Institut Leipzig. Reprinted by permission.

Page 48. Photo courtesy of Photo Researchers.

Page 53. Quiz on "The Empty Set" courtesy of Robert L. Walter, The University of South Dakota.

Page 64. Photo of Boole courtesy of The New York Public Library.

Page 66. Cartoon by Robert Mankoff from *Saturday Review,* June 11, 1977. Copyright © 1977 by Robert Mankoff. Reprinted by permission of Robert Mankoff.

Page 70. B.C. cartoon reprinted by permission of Johnny Hart and Field Enterprises, Inc.

Page 71. Photo of Einstein courtesy of The New World Library of World Literature, Inc., The University and Dr. Einstein.

Page 74. Pixie cartoon © 1972 United Feature Syndicate, Inc. Reprinted by permission.

Page 80. Photo courtesy of Vermont Agency of Transportation.

Page 86. Cartoon by Sidney Harris, from *American Scientist.* Copyright © 1977 by Sidney Harris. Reprinted by permission.

Page 89. Beetle Bailey cartoon © 1974 by King Features Syndicate, Inc. Reprinted by permission.

Page 92. Photo of Dodgson courtesy of David Eugene Smith Collection, Rare Book and Manuscript Library, Columbia University.

Page 93. Buz Sawyer cartoon © 1969 by King Features Syndicate, Inc. Reprinted by permission.

Page 94. "Whodunit?" from *Brain Puzzlers Delight,* by Eric Emmet. © 1967, 1970 by E. R. Emmet. Reprinted by permission of the publisher, Emerson Books, Inc., and Macmillan, London and Basingstoke.

Page 99. Street sign photo by Ken Glass.

Page 104. Cartoon by Joe Short. Reprinted by permission of DATAMATION® magazine, © copyright by Technical Publishing Company, A Division of DUN-DONNELLEY PUBLISHING CORPORATION, A DUN & BRADSTREET COMPANY, 1969—all rights reserved.

Page 105. Photo of Napier's rods courtesy of IBM.

Page 106. Pascal quotation from *The Computer Prophets,* by J. M. Rosenberg, Ph.D., p. 4. Copyright © 1969 by Jerry M. Rosenberg. Reprinted by permission of the publisher, Macmillan Publishing Company.

Page 106. Photo of Pascal's calculator courtesy of IBM.

Page 107. Leibniz quotation from *The Computer Prophets,* by J. M. Rosenberg, Ph.D., p. 48. Copyright © 1969 by Jerry M. Rosenberg. Reprinted by permission of the publisher, Macmillan Publishing Company.

Page 107. Photo of Leibniz' reckoning machine courtesy of IBM.

Page 107. Photos of Jacquard's loom and punched card courtesy of IBM.

Page 108. Photo of Babbage's Difference Engine courtesy of IBM. Photo of Analytical Engine reprinted with permission of British Crown Copyright. Science Museum, London.

Page 108. Slide rule photo courtesy of Pickett Industries.

Page 109. Photo of Wiener courtesy of The New York Public Library.

Page 109. Photo from *A Computer Glossary.* © 1968 International Business Machines. Reprinted by permission.

Page 110. Photos (a) and (b) courtesy of Juliet Lee.

Page 110. Photos (c)–(f) by Ben Rose, from "Evolution of Microelectronics," *Scientific American,* February 1970. Reprinted by permission.

Page 110. Photo of integrated circuit courtesy of IBM.

Page 111. Photo copyright 1982, Apple Computer, Inc. Used by permission of Apple Computer, Inc., 20525 Mariani Ave., Cupertino, CA. 95014.

Page 111. Cartoon © 1978 by Sidney Harris-Datamation Magazine. Reprinted by permission.

Page 113. Quote from *In Mathematical Circles,* by H. Eves. Copyright 1969 by Prindle, Weber, & Schmidt, Inc. Reprinted by permission.

Page 115. Photo (a) courtesy of Sharp Electronics Corporation. Photo (b) courtesy of Texas Instruments, Inc. Photo (c) courtesy of Hewlett-Packard.

Page 117. Cartoon © 1976 by Howie Schneider. Reprinted by permission.

Page 118. Problem 14 from "Mathematical Games Department," by M. Gardner, *Scientific American,* July 1976, p. 131. Reprinted by permission of Scientific American, Inc., and W. H. Freeman and Company, Publishers.

Page 121. Photos from *A Computer Glossary.* © 1968 International Business Machines. Reprinted by permission.

Page 121. Photos of computer and input/output device courtesy of Digital Equipment Corporation.

Page 122. Photo of TRS-80 Model II Computer, courtesy of Tandy Corporation.

Page 123. News article © 1972 United Press International.

Page 123. © 1973 Newhouse News Service. Reprinted by permission.

Page 123. Advertisement courtesy of General Motors.

Page 125. Cartoon by Bob Schochet. Reprinted by permission.

Page 127. © Cartoons by Johns. Reprinted by permission.

Page 127. Lion courtesy of Honeywell Information Systems.

Page 128. "A Programmer's Lament" from *The Compleat Computer,* by Dennie Van Tassel. Copyright © 1983, 1976, Science Research Associates, Inc. Reprinted by permission of the publisher.

Page 131. "Computer Career Opportunities" courtesy of Honeywell, Inc.

Page 131. Margin note reprinted with the permission of *The Wall Street Journal,* © Dow Jones & Company, Inc. (1973).

Page 133. Flowchart from General Library Information Leaflet No. 1. University of California, Davis. Reprinted by permission.

Page 136. "Glossary of Computer Terms" adapted from "Educator's Lexicon of Computerese," by

Jerry Johnson, *Arithmetic Teacher,* February 1983.

Page 140. Limerick from *Cybernetic Serendipity,* edited by J. Reichardt. © 1968 by W. J. Mackay & Co., Ltd.; London: *Studio International;* New York: Frederick A. Prager. Reprinted by permission.

Page 141. © Michael L. Kim 1977. First published in *Creative Computing.* Reprinted by permission.

Page 141. Computer composition from *Cybernetic Serendipity,* edited by J. Reichardt. © 1968 by W. J. Mackay & Co., Ltd.; London: *Studio International;* New York: Frederick A. Prager. Reprinted by permission.

Page 142. Cartoon by Mal, © 1971 Washington Star Syndicate, Inc. Reprinted by permission.

Pages 143–144. News article © 1976 United Press International.

Page 149. Figure courtesy of Dr. R. E. Herron. University of Illinois, Champaign. Reprinted by permission.

Page 150. "Machine Translation" from *The Compleat Computer,* by Dennie Van Tassel. Copyright © 1983, 1976, Science Research Associates, Inc. Reprinted by permission of the publisher.

Page 152. Cartoon © 1978 by Sidney Harris-*Datamation Magazine.* Reprinted by permission.

Page 152. Computer poems from *Cybernetic Serendipity,* edited by J. Reichardt. © 1968 by W. J. Mackay & Co., Ltd.; London: *Studio International;* New York: Frederick A. Prager. Reprinted by permission.

Page 155. Flowers (left) courtesy of Computra, Inc., Computer Art for People, 2513 Moore Road, Muncie, IN 47302. Graphic displays courtesy of IMLAC Corporation, Needham, MA. Bottom (center) courtesy of IEEE Computer Society.

Pages 156–157. News article © 1974 United Press International. Reprinted by permission.

Page 159. Floorplan developed by Saphier, Lerner, Schindler, Environetics, Inc., 600 Madison Avenue, New York, NY 10017. Reprinted by permission.

Page 160. Cartoon by Stew Burgess. Reprinted by permission of DATAMATION® magazine, © copyright by TECHNICAL PUBLISHING COMPANY, A Division of DUN-DONNELLEY PUBLISHING CORPORATION, A DUN & BRADSTREET COMPANY, 1971—all rights reserved.

Page 160. Problem 22 from "How to Solve It with the Computer," by Donald T. Piele in *Creative Computing,* March 1981. Reprinted by permission.

Page 166. Peanuts cartoon © 1963 United Feature Syndicate, Inc. Reprinted by permission.

Page 169. Photo from *Science Awakening,* by B. L. Van Der Vaerden, 1961, Oxford University Press. Courtesy of the Berlin Museum and

Wolters-Noordhoff Publishing Company.

Page 170. News article © 1972 United Press International. Reprinted by permission.

Page 174. Mayan numerals from *An Introduction to the Study of the Mayan Hieroglyphics,* by S. G. Morley. Reprinted by permission of the Scholarly Press.

Page 175. From *Margarita Philosophica Nova,* 1512. Courtesy of Museum of the History of Science, University of Oxford.

Page 177. News article used by permission of the Associated Press, © 1980.

Page 178. Quote from *In Mathematical Circles,* by H. Eves. Copyright 1969 by Prindle, Weber, & Schmidt, Inc. Reprinted by permission.

Page 182. Dennis the Menace cartoon © 1972. Reprinted by permission of Hank Ketcham.

Page 183. Graffiti by Leary. Reprinted with permission of The McNaught Syndicate, Inc.

Page 184. Photo © 1973 by United Press International. Reprinted by permission.

Page 184. Peanuts cartoon © 1963 United Feature Syndicate, Inc.

Page 186. News article © 1971 United Press International. Reprinted by permission.

Page 191. Euler quotation from *In Mathematical Circles,* by H. Eves. Copyright 1969 by Prindle, Weber, & Schmidt, Inc. Reprinted by permission.

Page 193. Blondie cartoon © 1974 by King Features Syndicate, Inc. Reprinted by permission.

Page 193. Cartoon reprinted by permission of the National Enquirer.

Page 196. Ascending/Descending by M. C. Escher. Reproduction © BEELDRECHT, Amsterdam/V.A.G.A., New York. Collection Haags Gemeentemuseum—The Hague. Reprinted by permission.

Page 198. Problem 38 adapted from "Beer Glass Puzzle" by Mel Stover in *Games Magazine,* November/December, 1980, p. 15. Reprinted by permission.

Page 209. B.C. cartoon reprinted by permission of Johnny Hart and Field Enterprises, Inc.

Page 211. B.C. cartoon reprinted by permission of Johnny Hart and Field Enterprises, Inc.

Page 213. Peanuts cartoon © 1966 United Feature Syndicate, Inc. Reprinted by permission.

Page 221. B.C. cartoon reprinted by permission of Johnny Hart and Field Enterprises, Inc.

Pages 222–223. Illustrations courtesy of The British Library, London.

Page 230. Table reprinted with permission of Macmillan Publishing Company from *Number: The Language of Science,* 4th Edition, by Tobias Dantzig. Copyright 1930, 1933, 1939, 1954 by Macmillan Publishing Co., Inc.; copyright renewed 1958, 1961, 1967 by Anna G. Dantzig; renewed 1982 by Mildred B. Dantzig.

Page 231. β cartoon courtesy of Patrick J. Boyle.

Page 243. Pascal's triangle from *Science and Civilization in China* (Vol. 3), by J. Needham. Reprinted by permission of the publisher, Cambridge University Press.

Page 243. Poem from a 19th century autograph book as published in Gaye LeBaron's column in the *Press Democrat*, Santa Rosa, CA. Reprinted by permission.

Page 244. Blondie cartoon © 1972 by King Features Syndicate, Inc. Reprinted by permission.

Page 256. MasterCard logo used courtesy of MasterCard International Inc.

Page 259. Finance statement courtesy of Sears, Roebuck and Co.

Page 261. Purchasing Guide from *The Car Book* published by U.S. Department of Transportation, National Highway Traffic Safety Administration, Washington, D.C. 20590.

Page 276. Born Loser cartoon reprinted by permission. © 1974 NEA, Inc.

Page 294. "Mystery Thriller" from *4 The Mathematical Wizard*, by Louis Grant Brandes, © 1962 by J. Weston Walch, Publisher.

Page 301. Photo courtesy of AMR International, Inc.

Page 303. Postmark courtesy of Donald B. Gillies, University of Illinois.

Pages 304–305. Copyright, 1982, *Los Angeles Times.* Reprinted by permission.

Page 307. Photos courtesy of Los Alamos Scientific Laboratory.

Page 316. Cartoon by Bill Knowlton. First published in *Herald Tribune* syndicate 12–16 circa 1960. Copyright Bill Knowlton, reprinted by permission of Bill Knowlton/Toni Mendez Inc.

Page 318. Photo of Noether courtesy of the Archive of Georg-August, Universität Göttingen.

Pages 320–321. Problem 56 courtesy of Mark Wassmansdorf of the Philadelphia School District.

Page 329. Peanuts cartoon © 1979 United Feature Syndicate, Inc. Reprinted by permission.

Page 335. Table from "An Aid to the Superstitious," by G. L. Ritter, S. R. Lowry, H. B. Woodruff, and T. L. Isenhour, *The Mathematics Teacher*, May 1977 (vol. 70, p. 456), copyright 1977 by the National Council of Teachers of Mathematics. Used by permission.

Pages 343–344. Modular art designs from "Using Mathematical Structures to Generate Artistic Designs," by S. Forseth and A. P. Troutman, *The Mathematics Teacher*, May 1974 (vol. 67, pp. 393–398), copyright 1974 by the National Council of Teachers of Mathematics. Used by permission.

Page 349. Poem by William Whewell from *Fantasia Mathematica*, by C. Fadiman. Copyright © 1958 by Clifton Fadiman. Reprinted by permission of Simon & Schuster, Inc.

Page 350. Grid Transformation from *On Growth and Form* (Cambridge, London: Cambridge University Press, 1917) by D'Arcy Wentworth Thompson.

Page 351. Cartoon © 1971. Reprinted by permis-

sion of *Saturday Review* and Henry Martin.

Page 351. Apollo photo courtesy of Rockwell International Corporation. Snowflake photo courtesy of the National Oceanic and Atmospheric Administration.

Page 352. Photo of *Allegorical Garden of Geometry* courtesy of The Metropolitan Museum of Art, The Elisha Whittelsey Collection, The Elisha Whittelsey Fund, 1951.

Page 352. Winged Lion photo by William Abbenseth. Asian Art Museum of San Francisco, The Avery Brundage Collection.

Page 353. Photo of Book of Kells courtesy of Trinity College, Dublin, Ireland.

Pages 353, 355. Problems 4–9 and 17–21 adapted from "Faces of a Cube," by Ruth Butler and Robert W. Clark, *The Mathematics Teacher*, March 1979, copyright 1979 by the National Council of Teachers of Mathematics. Used by permission.

Page 355. Photo of "Mirror Image" by Diane Dawson, reprinted from *Games Magazine,* May/June, 1979. Reprinted by permission.

Page 356. Mad Magazine cover copyright © 1964 by E. C. Publications, Inc. Reprinted by permission. All rights reserved.

Page 357. Photo courtesy of Department of Transportation, Ashland, Ohio.

Page 366. Graffiti by Leary. Reprinted with permission of The McNaught Syndicate, Inc.

Page 367. Newsclipping reprinted with permission from the February 1977 *Reader's Digest.*

Page 369. Wizard of Id cartoon reprinted by permission of Johnny Hart and Field Enterprises, Inc.

Page 369. Photos courtesy of Sears, Roebuck and Co.

Page 385. Limerick and cartoon from *The Metrics Are Coming! The Metrics Are Coming!,* by R. Cardnell. Copyright © 1975 by Dorrance & Company. Reprinted by permission.

Page 388. Radiolaria forms from *Report on the Scientific Results of the Voyage of H.M.S. Challenge* (vol. 18), 1887.

Page 399. Historical note adapted from *Mathematical Circles Adieu*, by Howard W. Eves. Copyright © 1977 by Prindle, Weber, & Schmidt. Reprinted by permission.

Page 402. Möbius Strip II, by M. C. Escher. Reproduction © BEELDRECHT, Amsterdam/V.A.G.A., New York. Collection Haags Gemeentemuseum—The Hague. Reprinted by permission.

Pages 402–403. Paper-boot puzzle from *Topology,* by D. Johnson and W. Glen. Copyright 1960. Reprinted by permission.

Page 403. Problem 30 adapted from "The Amazing Mirror Maze" by Walter Wick in *Games Magazine,* September/October, 1981, p. 25. Reprinted by permission.

Page 405. Duccio's *Last Supper* courtesy of Opera del Duomo, Siena.

Page 405. Designer of the Lying Woman, by

Albrecht Dürer. Reproduction courtesy of Trustees of The British Museum.

Page 408. Photo courtesy of Donald M. Welch.

Page 410. Photo courtesy of Donald M. Welch.

Page 412. Poem from *Collected Poems,* Harper & Row. Copyright 1923, 1951 by Edna St. Vincent Millay and Norma Millay Ellis. Reprinted by permission.

Page 414. "Four Colors Suffice" postmark courtesy of the University of Illinois.

Page 415. Advertisement by Daniel Wade Arthur, taken from the *Los Angeles Times.*

Page 423. "Threes" by John Atherton. Reprinted by permission; © 1957 The New Yorker Magazine, Inc.

Page 424. Cartoon by Doug. Reprinted by permission from The Saturday Evening Post Company, © 1976.

Page 425. Photos courtesy of Donald M. Welch.

Page 436. Peanuts cartoon © 1973 United Feature Syndicate, Inc. Reprinted by permission.

Page 441. Photos by Richard Hagberg.

Page 444. Photo courtesy of the Master Lock Company.

Page 450. "Match the Mayors" from the *Press Democrat,* Santa Rosa, CA, January 24, 1971. Reprinted by permission.

Page 450. Advertisement courtesy of US JVC CORP. Reprinted by permission.

Page 450. Peanuts cartoon © 1974 United Feature Syndicate, Inc. Reprinted by permission.

Page 451. Problem 25 from *The American Mathematical Monthly,* January 1970. Reprinted by permission.

Page 455. Photo courtesy of Juliet Lee.

Page 455. Cartoon reprinted by permission of Michael Dater.

Page 459. Advertisement courtesy of Wendy's International, Inc.

Page 461. Limerick by Sir Arthur Eddington from *Fantasia Mathematica,* by C. Fadiman. Copyright © 1958 by Clifton Fadiman. Reprinted by permission of Simon & Schuster, Inc.

Page 465. Ancient Greek and Roman Dice reproduced by courtesy of the Trustees of the British Museum.

Page 470. Photo courtesy of Juliet Lee.

Page 471. Le Trente-et-un, ou la maison de pret sur nantissement, by Darcis. Courtesy of Bibliotheque National, Paris.

Page 476. Photo by Ron Pietersma, *The Dairyman,* January 1974. Reprinted by permission.

Page 478. B.C. cartoon reprinted by permission of Johnny Hart and Field Enterprises, Inc.

Page 481. Photo courtesy of WIDE WORLD PHOTOS.

Page 484. B.C. cartoon reprinted by permission of Johnny Hart and Field Enterprises, Inc.

Page 493. Lottery tickets courtesy of Hospital's Trust, Dublin, Ireland; Ohio Lottery Commission;

Subject Index

Index of Special Problems

Ideas for special projects and reports are listed throughout this book. References, sources, and specific ideas are provided. This list summarizes those ideas and gives a few additional suggestions. You should also use a library to locate additional references.